Fundamentals of Electrical Circuit Analysis

Md. Abdus Salam · Quazi Mehbubar Rahman

Fundamentals of Electrical Circuit Analysis

Springer

Md. Abdus Salam
Electrical and Electronic Engineering Programme Area, Faculty of Engineering
Universiti Teknologi Brunei
Bandar Seri Begawan
Brunei Darussalam

Quazi Mehbubar Rahman
Department of Electrical and Computer Engineering, Faculty of Engineering
The University of Western Ontario
London, ON
Canada

ISBN 978-981-10-8623-6 ISBN 978-981-10-8624-3 (eBook)
https://doi.org/10.1007/978-981-10-8624-3

Library of Congress Control Number: 2018934454

Printed on acid-free paper

This Springer imprint is published by the registered company Springer Nature Singapore Pte Ltd. part of Springer Nature.
The registered company address is: 152 Beach Road, #21-01/04 Gateway East, Singapore 189721, Singapore

To my parents, wife Asma Ara Bagum, children Syeed Hasan, Yusra Salam, and Sundus Salam for their constant support, and all of my teachers' and well-wishers who have helped me grow professionally over the years.

Md. Abdus Salam

To my wife Munadil Rahman Tina, and our children Mehrin Rahman and Mayaaz Rahman for their selfless love, kindness and endless support.

Quazi Mehbubar Rahman

Preface

Four-year electrical and electronic engineering curriculums generally incorporate two courses on electrical circuit analysis to provide students with a strong foundation of electrical engineering practices. The contents of the *Fundamentals of Electrical Circuit Analysis* book have been arranged in a logical way to enable students to gradually make transition from simple topics to more complex ones. Therefore, this book is an essential resource for the undergraduate electrical engineering students especially in the first year of their studies. Electrical engineers need to be critical thinkers, and be competent with the design, construction and production of electrical equipment and systems for the technological advancement of society. In order to achieve such goals, proper fundamental knowledge in electrical circuit analysis is crucial. This knowledge is also important for other branches of electrical engineering such as power and energy, communication, mechatronics, robotics and control system engineering. In addition to electrical engineering, this book also covers areas such as mechanical, mechatronic and civil engineering. This book addresses all the necessary materials for accreditation given by Accreditation Board for Engineering and Technology (ABET), USA, Institution of Engineering and Technology (IET), UK and Canadian Engineering Accreditation Board (CEAB).

The main features of this book are as follows:

- Easy and simple presentation of each topic.
- Supporting mathematical equations for the presented circuit parameters.
- Step-by-step approach in worked out examples, preceded by the corresponding theoretical background.
- Practice problems after each topic.
- Inclusion of fundamentals of PSpice simulation.
- Some worked out examples in PSpice simulation.
- Significant number of exercise problems at the end of each chapter.
- Answers to all the exercises problems.

Organization of the Book

Chapter 1 focusses on the fundamentals of electrical parameters, measuring equipment and circuit parameters drawing and labelling in PSpice.

Chapter 2 deals with basic circuit laws, series–parallel circuit, delta–wye conversion, short and open circuit and source conversion technique.

Chapter 3 includes mesh analysis, nodal analysis, supermesh and supernode.

Chapter 4 discusses network theorems such as linearity property, superposition theorem, Thevenin's theorem, Norton's theorem and maximum power transfer theorem.

Chapter 5 looks after capacitor, inductor, charging and discharging activities, RL and RC circuit with step response.

Chapter 6 focusses on alternating current parameters, rms value, average values, form factor, peak factor, phasor algebra, AC circuits with resistance, inductance and capacitance and AC circuit with Kirchhoff's laws.

Chapter 7 deals with AC circuits with mesh analysis, nodal analysis, superposition, Thevenin's and Norton's theorem.

Chapter 8 discusses instantaneous power, average power, apparent power, reactive power, complex power, power factor and maximum power transfer theorem.

Chapter 9 deals with three-phase voltage generation, balanced and unbalanced wye and delta connections, wye and delta connections with balanced and unbalanced loads, three-phase power and power measurement.

Chapter 10 talks about frequency response with resistance, inductance and capacitance, transfer function, different types of filters and Bode plots.

The authors would like to thank all the faculty members who have contributed to the preparation of this book and the staff of Springer Publishers who have tirelessly worked to publish this book successfully.

Bandar Seri Begawan, Brunei Darussalam — Md. Abdus Salam
London, Canada — Quazi Mehbubar Rahman

Contents

Chapter 1
Fundamental Electrical Terms

1.1 Introduction

It is essential to understand the fundamental of electrical terms to comprehend any electrical phenomena. This chapter presents these necessary concepts and components as a first stepping stone in understanding any underlying electrical principles. In addition, it introduces the core measuring equipment in the electrical domain along with charge, current, voltage, power, energy, resistance, semiconductor and insulator.

1.2 Charge

Charge is an electrical property of matter, measured in coulombs (C). The charge in motion represents the current and the stationary charge represents static electricity. There are two types of charge namely the positive charge and the negative charge. Like charges repel each other and opposite charges attract each other. The charge is represented by the letter Q or q and its SI derived unit is coulomb (C). Coulomb is a large unit of the charge and, therefore, generally the smaller units of charge (pC, nC and μC) are used in practice. Note: charge on an electron, which is negative, has the magnitude of 1.602×10^{-19} C. In one Coulomb of charge, there are $1/(1.602 \times 10^{-19}) = 6.24 \times 10^{18}$ electrons [1–3].

1.3 Current

The flow of electrons results in current. The direction of the current is opposite to the flow of electrons (negative charges) as shown in Fig. 1.1. The current is represented by the letter I and its SI unit is ampere (A) in honour of French Mathematician and Physicist Andre-Marie Ampere (1775–1836).

Md. A. Salam and Q. Mehbubar Rahman, *Fundamentals of Electrical Circuit Analysis*,
https://doi.org/10.1007/978-981-10-8624-3_1

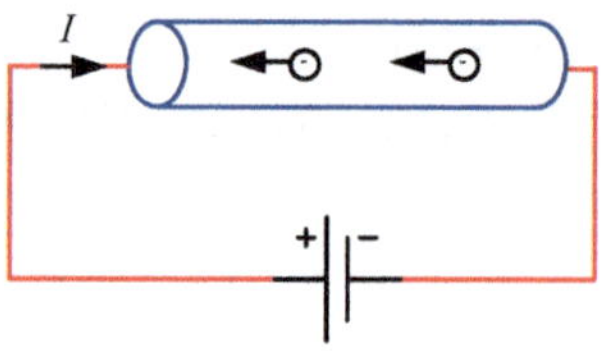

Fig. 1.1 Direction of current

In general, the current is defined as the total charge Q transferred in time t and it is expressed as,

$$I = \frac{Q}{t} \tag{1.1}$$

The rate of change of charge transferred at any particular time is known as instantaneous current, i, and it is expressed as,

$$i = \frac{dq}{dt} \tag{1.2}$$

$$dq = idt \tag{1.3}$$

$$q = \int_{t_1}^{t_2} idt \tag{1.4}$$

Equation (1.4) provides the expression for the total charge transferred through a conductor between time t_1 and t_2.

Example 1.1 The expression of charge at a terminal is given by $q = 2t^3 + 5t^2 + t + 1$ C. Find the current at $t = 0.5$ s.

Solution:

$$i = \frac{dq}{dt} = \frac{d}{dt}(2t^3 + 5t^2 + t + 1) = 6t^2 + 10t + 1 \text{ A} \tag{1.5}$$

At $t = 0.1$ s, the value of the current is,

$$i = 6(0.5)^2 + 10(0.5) + 1 = 7.5 \text{ A} \tag{1.6}$$

Practice Problem 1.1 The total charge entering a terminal is given by $q = 2e^{-2} + 3t^2 + t$ C. Find the current at $t = 0.1$ s.

Example 1.2 A current as shown in Fig. 1.2 passes through a wire. Determine the value of the charge.

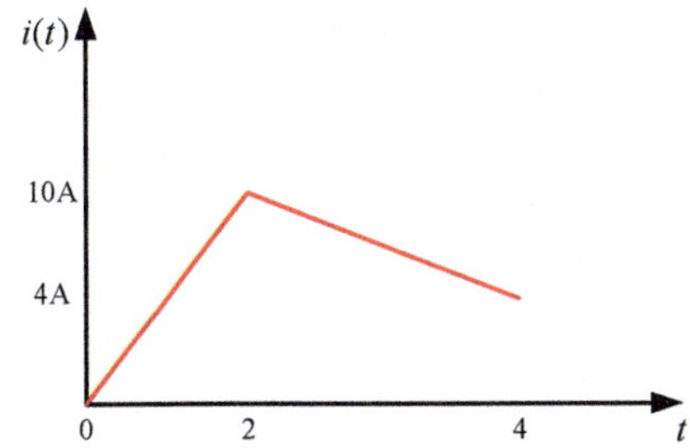

Fig. 1.2 Variation of current with time

Solution:
The expression of current from 0 to 2 s is,

$$i(t) = \frac{10}{2}t = 5t \text{ A} \tag{1.7}$$

The expression of current from 2 to 4 s can be written as,

$$\frac{i(t) - 4}{4 - 10} = \frac{t - 4}{4 - 0} \tag{1.8}$$

$$i(t) - 4 = -\frac{6}{4}(t - 4) \tag{1.9}$$

$$i(t) = 4 - \frac{6}{4}(t - 4) = 10 - 1.5t \tag{1.10}$$

The value of the charge from 0 to 4 s can be calculated as,

$$q = \int_0^2 5t\text{d}t + \int_2^4 (10 - 1.5t)\text{d}t \tag{1.11}$$

$$q = 2.5(2^2 - 0) + 10(4 - 2) - 0.75(4^2 - 2^2) = 21 \text{ C} \tag{1.12}$$

Practice Problem 1.2 The value of an instantaneous current varying with time is shown in Fig. 1.3. Calculate the value of the charge.

Fig. 1.3 Triangular shape of current

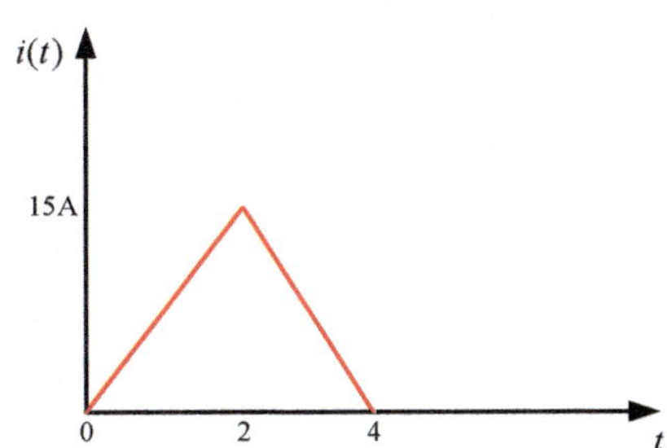

1.4 Direct and Alternating Currents

There are two types of current: direct current, which is abbreviated as DC and alternating current, which is abbreviated as AC [4, 5]. When the magnitude of the current does not change with time, the current is known as direct current, while, in alternating current, the magnitude of the current changes with respect to the time. Both of these currents are shown in Fig. 1.4.

1.5 Conductor, Insulator and Semiconductor

The concept of conductor, insulator and semiconductor can be presented in terms of energy band models from the solid-state physics [6]. The band model consists of two energy bands, valance band and conduction band, and the energy gap between these two bands. The lower energy valance band contains the charges at their lowest energy levels while the highest energy conduction band, being generally empty, contains the energy-excited charges to make conduction. In case of a conductor, a large number of charges populate the conduction band even at room temperature, and as a result, a conductor provides least resistance to the flow of charges. In this case, there is no energy gap in between the valence band and the conduction band, and that means these two energy bands overlap with each other as shown in Fig. 1.5a. For a conductor, the charges in the conduction band are loosely bound with the parent atoms and hence, they can move easily under the influence of external energy source such as electric field.

In the insulator, the energy gap in between the valence and conduction bands is very high as a result, no charge can gain enough energy to jump to the conduction band as shown in Fig. 1.5b. As a result, an insulator provides most resistance to the flow of charges.

A semiconductor falls in between the conductor and insulator and it offers a moderate resistance to flow the charge. A semiconductor has smaller energy gap in between the conduction band and the valence band, when compared to the insulators, as shown in Fig. 1.5c.

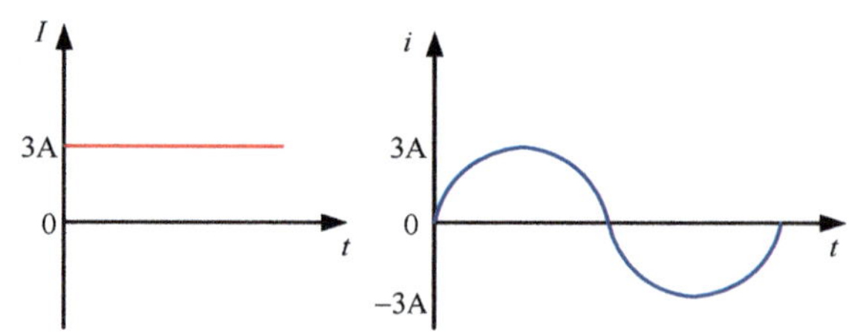

Fig. 1.4 Direct and alternating currents

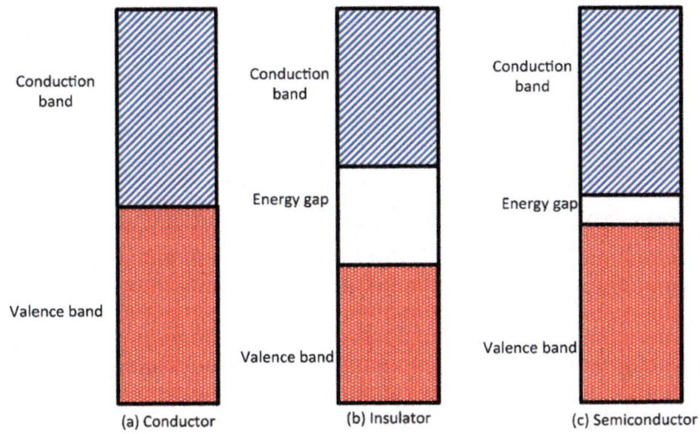

Fig. 1.5 Energy bands conductor, insulator and semiconductor

1.6 Resistance and Conductance

The property of a material, which opposes the flow of electric current through it, is known as resistance. The resistance is represented by the letter R and its SI unit is ohm (Ω). The symbol of a resistor is shown in Fig. 1.6. The resistance of any material is directly proportional to the length (l), and inversely proportional to the cross-sectional area (A) of the material as shown in Fig. 1.7, and can be written as,

$$R \propto l \tag{1.13}$$

$$R \propto \frac{1}{A} \tag{1.14}$$

Combining Eqs. (1.13) and (1.14) yields,

$$R \propto \frac{l}{A} \tag{1.15}$$

$$R = \rho \frac{l}{A} \tag{1.16}$$

where ρ is the proportionality constant which is known as the resistivity of a material and its unit is Ω-m. A good conductor has a very low resistivity, while an insulator has a very high resistivity.

Fig. 1.6 Symbol of the resistor

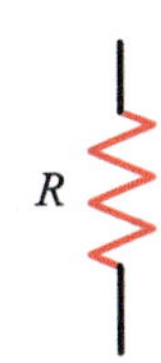

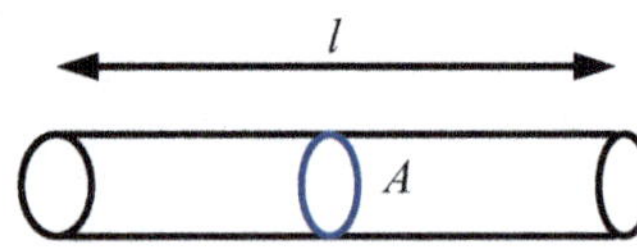

Fig. 1.7 A material with cross-sectional area A

In practice, the common commercial resistors are manufactured in a way so that their resistance values can be determined by the colour bands printed on their surfaces. Different coded values for different colours are shown in Table 1.1. The manufacturing tolerance for each resistor is also colour coded on its surface; Table 1.2 shows these coded values. From the colour bands, the value of a particular resistance can be determined as,

$$R = xy \times 10^z\ \Omega \tag{1.17}$$

where x is the first colour band, closest to the edge of the resistor, y is the second colour band next to x and z is the third colour band. The fourth colour band (if any) represents the tolerance in percent.

Tolerance is the percentage of error in the resistance value that says how much more or less one can expect a resistor's actual measured resistance to be from its stated resistance.

As an example, the resistance value of a resistor shown in Fig. 1.8 can be determined from its colour code, with the help of Tables 1.1 and 1.2 as,

$$R = 52 \times 10^6 = 52\ \text{M}\Omega \tag{1.18}$$

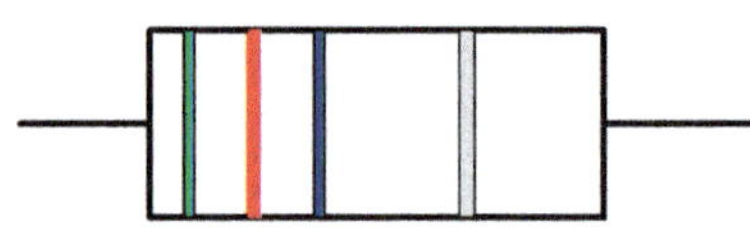

Fig. 1.8 Resistance with different colours

Table 1.1 Resistance colour codes with values

Colours name	Colours	Value
Black		0
Brown		1
Red		2
Orange		3
Yellow		4
Green		5
Blue		6
Violet		7
Grey		8
White		9

Table 1.2 Tolerance colour codes with values

Colours name	Colours	Value (%)
Orange		0.01
Yellow		0.001
Brown		1
Red		2
Gold		5
Silver		10
No colour		20

The tolerance is calculated as,

$$\text{Tolerance} = 52 \times 10^6 \times 10\% = 5.2\ \text{M}\Omega \tag{1.19}$$

The measured resistance may vary between (52 + 5.2) and (52 − 5.2) MΩ.

Conductance, which is the reciprocal of the resistance, is a characteristic of materials that promotes the flow of electric charge. It is represented by the letter G and its SI unit is Siemens (S). Mathematically, it is expressed as,

$$G = \frac{1}{R} \tag{1.20}$$

Dependence of Resistance on Temperature The resistance of a good conductor such as, copper aluminium, etc., increases with an increase in temperature. Whereas the resistance of electrolyte (*An electrolyte is a substance that produces an electrically conducting solution when dissolved in a polar solvent, such as water*), alloy (*An alloy is a metal, made of the combination of two or more metallic elements*) and insulating material decreases with an increase in temperature. Let us assume that the resistance of a conductor at temperature T and T_0 degree centigrade are R and R_0, respectively. In this case, the change in resistance is,

$$\Delta R = R - R_0 \tag{1.21}$$

While the change in temperature is,

$$\Delta T = T - T_0 \tag{1.22}$$

The ratio of change in resistance to the resistance at T_0 degree centigrade is directly proportional to the change in temperature and it can be expressed as,

$$\frac{R - R_0}{R_0} \infty (T - T_0) \tag{1.23}$$

$$\frac{R - R_0}{R_0} = \alpha(T - T_0) \tag{1.24}$$

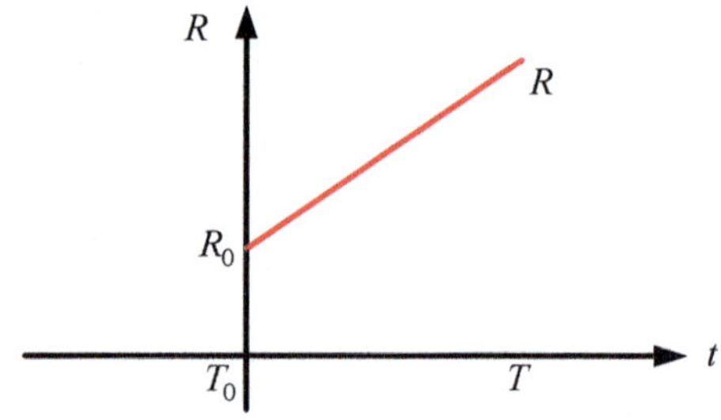

Fig. 1.9 Resistance with different temperatures

$$R - R_0 = \alpha R_0 (T - T_0) \tag{1.25}$$

$$R = [R_0 + \alpha R_0 (T - T_0)] \tag{1.26}$$

$$R = R_0 \, [1 + \alpha (T - T_0)] \tag{1.27}$$

where α is the proportionality constant which is known as the temperature coefficient of the resistance. Equation (1.27) provides the relationship between the resistance and the temperature, which has been presented in Fig. 1.9. In this figure, the temperature coefficient α represents the slope of the line.

Example 1.3 The length and diameter of a copper wire are found to be 100 and 0.002 m, respectively. Consider that the resistivity of the copper wire is 1.72×10^{-8} Ω-m. Calculate the value of the resistance.

Solution:
The area of the wire is,

$$A = \pi d^2 = \pi (0.002)^2 = 1.26 \times 10^{-5} \text{ m}^2 \tag{1.28}$$

The value of the resistance can be calculated as,

$$R = \rho \frac{l}{A} = 1.72 \times 10^{-8} \times \frac{100}{1.26 \times 10^{-5}} = 0.14 \, \Omega \tag{1.29}$$

Practice Problem 1.3 The resistance of a 120 m length aluminium wire is found to be 10 Ω. The resistivity of the aluminium wire is 2.8×10^{-8} Ω-m. Determine the diameter of the wire.

1.7 Voltage

The voltage is one kind of force required to move a charge in the conductor. Voltage is defined as the work done per unit charge when a charge to move it from one point to another point in a conductor. It is represented by the letter V and its unit is volts (V) in honour of Italian Scientist Alessandro Giuseppe Antonio Anastasio

Volta (February 1745–March 1827) who invented electric battery. Mathematically, the voltage is expressed as,

$$v = \frac{dw}{dq} \tag{1.30}$$

In general, the expression of voltage is,

$$V = \frac{W}{Q} \tag{1.31}$$

As common analogy, a water tank can be used to describe the charge, voltage and current. In this analogy, the amount of water in the tank represents the charge, the water pressure represents the voltage and the flow of water through the hose-pipe, which is connected at the bottom of the tank, represents the current as shown in Fig. 1.10.

The voltage difference is the difference in potential between the two points in a conductor, which is also known as potential difference.

Example 1.4 A charge is required to move between the two points in a conductor. In this case, the expression of work done is given by $w = 3q^2 + q + 1$ J. Find the voltage to move the charge $q = 0.5$ C.

Solution:
The expression of voltage is,

$$v = \frac{dw}{dq} = \frac{d}{dq}(3q^2 + q + 1) = 6q + 1 \text{ V} \tag{1.32}$$

The value of the voltage for $q = 0.5$ C is,

$$v = 6q + 1 = 6 \times 0.5 + 1 = 4 \text{ V} \tag{1.33}$$

Practice Problem 1.4 The work done is given by $w = 5q^2 + 2q$ J when a charge is moved between the two points in a conductor. Determine the value of the charge if the voltage is found to be 10 V.

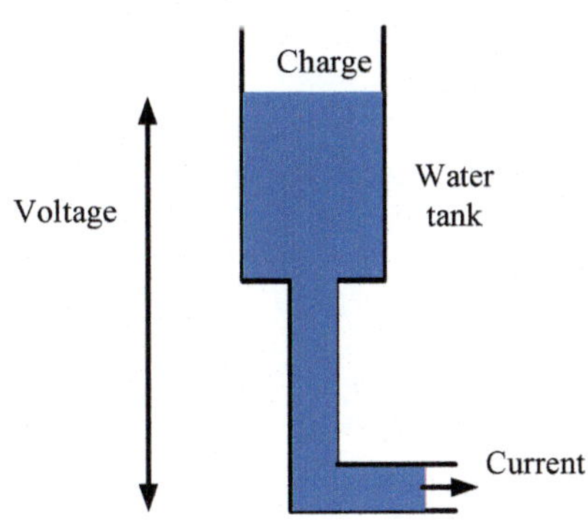

Fig. 1.10 A tank with water

1.8 Voltage and Current Sources

Any electrical source (voltage or current) provides energy to the elements connected to this source. An ideal voltage source is a two-terminal circuit element that provides a specific magnitude of voltage across its terminals regardless of the current flowing through it. An ideal current source is a two-terminal circuit element that maintains constant current through its terminals regardless of the voltage across those terminals. Ideal voltage and current sources are shown in Fig. 1.11.

An active circuit element usually supplies voltage or current to the passive elements like resistance, inductance and capacitance. Active element is classified as either independent or dependent (controlled) source of energy. An independent voltage or current source is an active element that can generate specified voltage or current without depending on any other circuit elements. The symbols for independent voltage and current sources are shown in Fig. 1.12. A dependent source is an active element whose magnitude is fully dependent on other voltage or current value in the circuit. The symbols for dependent voltage and current sources are shown in Fig. 1.13. The dependent sources are classified into four types, and these are,

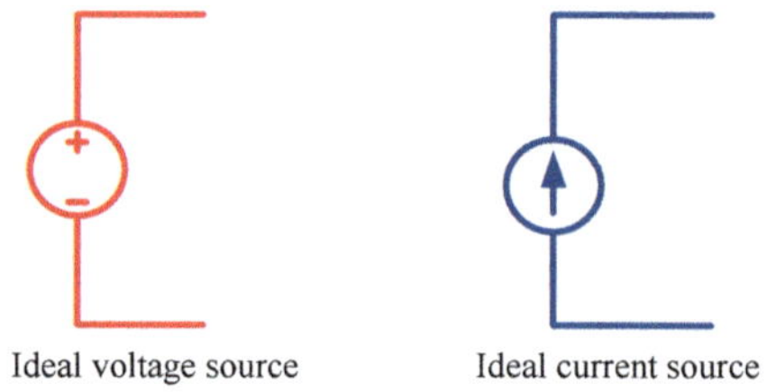

Fig. 1.11 Ideal voltage and current sources

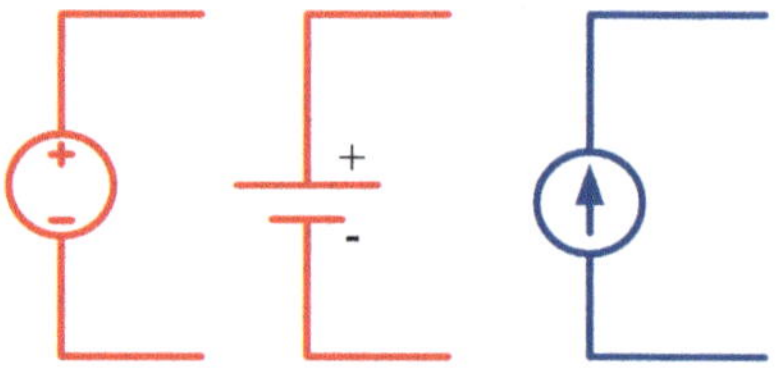

Fig. 1.12 Independent sources

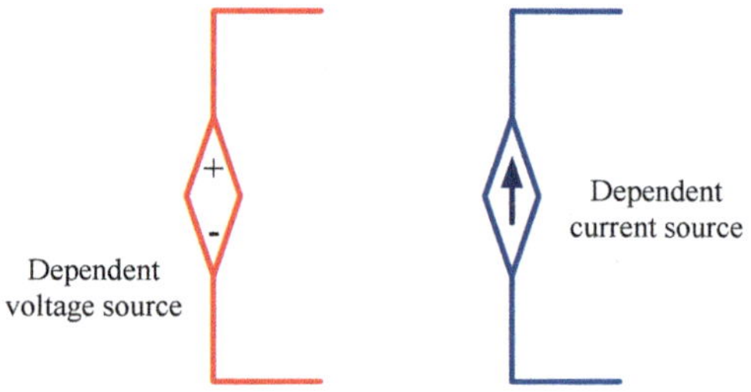

Fig. 1.13 Dependent sources

(1) Voltage-controlled voltage source (VCVS)
(2) Voltage-controlled current source (VCCS)
(3) Current-controlled voltage source (CCVS)
(4) Current-controlled current source (CCCS)

Different controlled (dependent) sources along with the corresponding application circuits are shown in Figs. 1.14, 1.15, 1.16 and 1.17. In Fig. 1.14, the magnitude of the voltage in the VCVS is α times the voltage across any element, and in the corresponding application circuit, this magnitude is equal to 0.2 (the value of α) times the voltage across the resistance R_1.

In the VCCS application circuit in Fig. 1.15, the magnitude of the voltage source depends on the current through resistance R_1. Similar explanation can be offered for CCVS and CCCS and their application circuits are shown in Figs. 1.16 and 1.17, respectively.

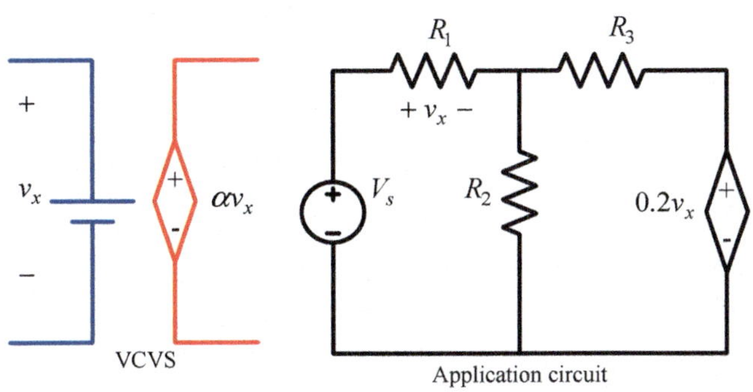

Fig. 1.14 Voltage-controlled voltage source and its application

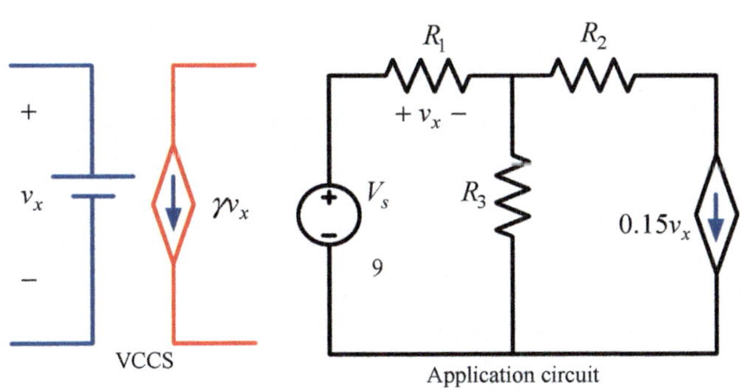

Fig. 1.15 Voltage-controlled current source and its application

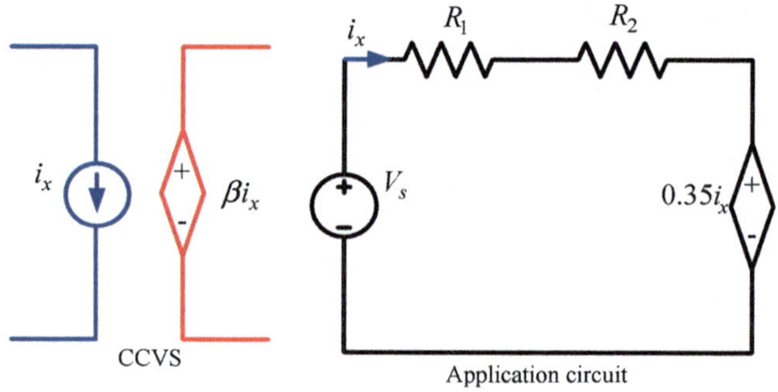

Fig. 1.16 Current-controlled voltage source and its application

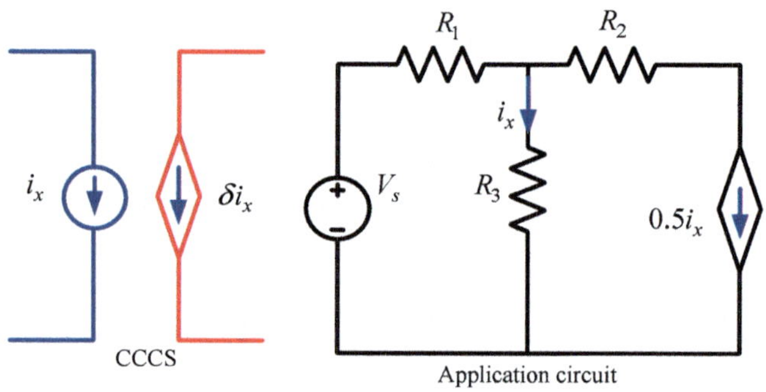

Fig. 1.17 Current-controlled current source and its application

1.9 Electric Power and Energy

The electric power is defined as the rate of receiving or delivering energy from one circuit to another circuit. The power is represented by the letter p and its unit is joules per sec (J/s) or watts (W) in honour to British Scientist James Watt (1736–1819). Mathematically, the expression of power can be written as,

$$p = \frac{\mathrm{d}w}{\mathrm{d}t} \tag{1.34}$$

where p is the power in watts (W), w is the energy in joules (J) and t is the times in seconds (s). In general, the total electric power can be written as,

$$P = \frac{W}{t} \tag{1.35}$$

Equation (1.34) can be re-arranged as,

$$p = \frac{\mathrm{d}w}{\mathrm{d}q}\frac{\mathrm{d}q}{\mathrm{d}t} \tag{1.36}$$

Substituting Eqs. (1.2) and (1.30) into Eq. (1.36) yields,

$$p = vi \tag{1.37}$$

Equations (1.1) and (1.31) can be re-arranged as,

$$t = \frac{Q}{I} \tag{1.38}$$

$$W = VQ \tag{1.39}$$

Substituting Eqs. (1.38) and (1.39) into Eq. (1.35) yields,

$$P = \frac{W}{t} = \frac{VQ}{\frac{Q}{I}} \tag{1.40}$$

$$P = VI \tag{1.41}$$

Both Eqs. (1.37) and (1.41) show that the power associated to any basic circuit element is the product of the voltage across that element and the current through that element. If the power, associated to a circuit element, has positive sign, then the power is absorbed by that element. Whereas, in a circuit element, the power with negative sign represents that the element delivers power to the other elements. The phenomena of power delivery and power absorption by the any element can be visualized with the aid of Fig. 1.18.

According to the circuit in Fig. 1.18, the expressions of power can be expressed as,

$$P_d = V(-I) = -VI \tag{1.42}$$

$$P_r = VI \tag{1.43}$$

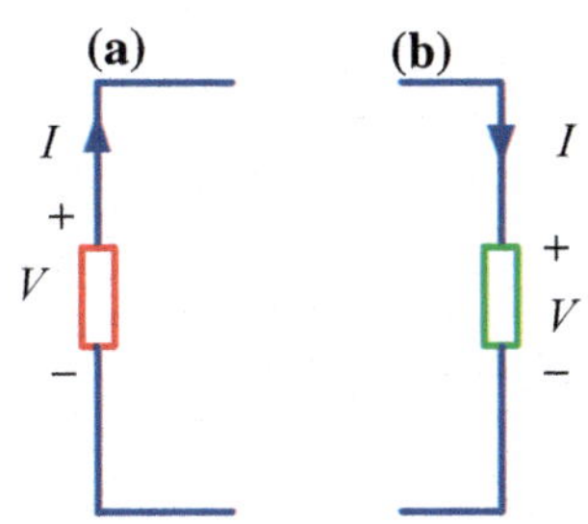

Fig. 1.18 Power delivered and absorbed circuits

The negative sign in Eq. (1.42) indicates that the power is being delivered by the element, while the positive sign in Eq. (1.43) indicates the power is being absorbed by the element.

Energy is the capacity to do work. Since, the SI unit of work is joule (J), and for the same reason, it is represented by the letter w. From Eq. (1.34), the expression of energy can be written as,

$$\mathrm{d}w = p\mathrm{d}t \tag{1.44}$$

For a time window between t_0 and t, the total energy can be calculated as,

$$w = \int_{t_0}^{t} p\mathrm{d}t \tag{1.45}$$

In practice, the energy supplied to the consumers is rated the term of kilowatt hour (kWh). One kWh is defined as the electrical power of 1 kW consumed in an hour, and it is expressed as,

$$E = P \times t \tag{1.46}$$

where E is the electrical energy in kWh, P is the power in kW and t is the time in hour.

Example 1.5 The expression of work done is given by $w = 5t^2 + 4t + 1$ J. Calculate the power at $t = 0.1$ s.

Solution:
The expression of power is,

$$p = \frac{\mathrm{d}w}{\mathrm{d}t} = \frac{\mathrm{d}}{\mathrm{d}t}(5t^2 + 4t + 1) = 10t + 4 \text{ W} \tag{1.47}$$

Power at $t = 0.1$ s is,

$$p = 10(0.1) + 4 = 5 \text{ W} \tag{1.48}$$

Example 1.6 A power with an expression of $p = 5 + 5t^2$ W is used from time 0.1 to 0.4 s to complete a specific task. Calculate the corresponding energy.

Solution:

$$\begin{aligned} w &= \int_{0.1}^{t=0.4} (5 + 5t^2)\mathrm{d}t = 5[t]_{0.1}^{0.4} + \frac{5}{3}[t^3]_{0.1}^{0.4} \\ &= 5(0.4 - 0.1) + 1.67(0.4^3 - 0.1^3) = 1.61 \text{ J} \end{aligned} \tag{1.49}$$

Example 1.7 A 500 W electric toaster operates for 30 min, a 7 kW DVD player operates for 1 h and a 700 W electric iron operates for 45 min. Find the total energy used, and the associated cost if the energy price is 5 cents per kWh.

Solution:
The total energy used is,

$$\begin{aligned} E &= \frac{500}{1000} \times \frac{30}{60} + 7 \times 1 + \frac{700}{1000} \times \frac{45}{60} \\ &= 0.500 \times 0.5 + 7 + 0.700 \times 0.75 = 7.78\ \text{kWh} \end{aligned} \tag{1.50}$$

The cost of the energy used is,

$$C = 7.78 \times \frac{5}{100} = 0.39\,\$ \tag{1.51}$$

Practice Problem 1.5 The power consumed at time t, for the work $w = 5t^2 + 4t + 1$ J, is found to be 20 W. Calculate time t.

Practice Problem 1.6 A power of $p = 10e^{-t} + 2$ W is used to complete a task. Calculate the energy for the time interval 0.01 to 0.2 s.

Practice Problem 1.7 Five 10 W energy-saving light bulbs operate for 8 h, a 60 W smart TV operates for 5 h and an 800 W hair dryer operates for 30 min. Calculate the total energy used, and the associated cost if the energy price is 20 cents per kWh.

1.10 Measuring Equipment

Measuring equipment are used to measure the electrical parameters accurately, both analogue and digital metres are available in practice. In an analogue metre, a pointer moves on a scale to read out the reading. Whereas the digital metre can display the reading directly in the form of numerical digits. Some of these widely used metres have been discussed below.

1.10.1 Ohmmeter

An ohmmeter, shown in Fig. 1.19, can measure the resistance of an electrical component. To measure the resistance of any component, it needs to be disconnected from the circuit. Ohmmeter is also used to identify the continuity of a circuit and the correct terminal ends in the application field where long cables are used. In the measurement process, the metre supplies current to the resistance, then measures the voltage drop across that resistance. Finally, the metre gives the value of the resistance.

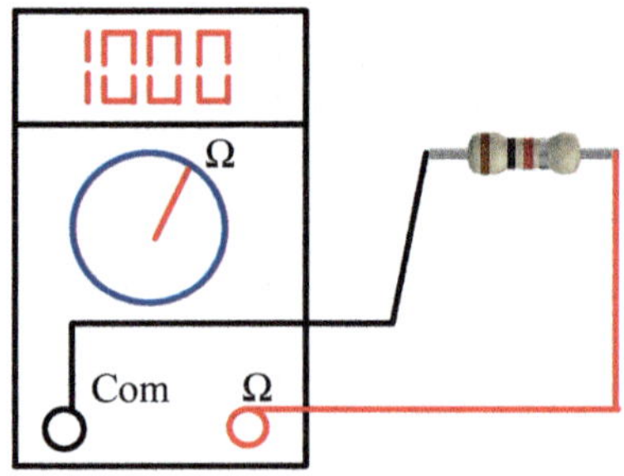

Fig. 1.19 Schematic of an ohmmeter

1.10.2 Ammeter

An ammeter is used to measure the current in the circuit. Ammeter is always connected in series with the circuit under test. This metre offers low internal resistance, which does not affect the actual measurement. The metre symbol and a circuit under test are shown in Fig. 1.20.

1.10.3 Voltmeter

A voltmeter is used to measure the voltage across any circuit element. A voltmeter is always connected in parallel to the circuit element under test. Voltmeters can be designed to measure AC or DC. The internal resistance of a voltmeter is usually kept very high to reduce the current flowing through it to a negligible small amount. Most of the voltmeters have several scales including 0–250, 0–500 V. The metre symbol and a circuit under test are shown in Fig. 1.21.

Fig. 1.20 Symbol and connection diagram of an ammeter

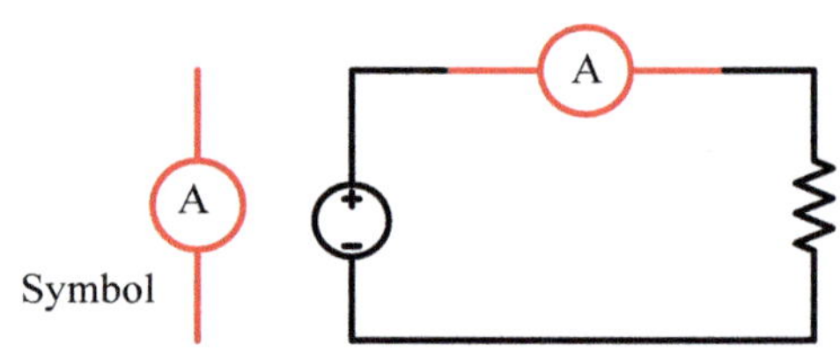

Fig. 1.21 Symbol and connection diagram of a voltmeter

1.10.4 Wattmeter

A wattmeter is used to measure either the power delivered by an electrical source or the power absorbed by a load. As shown in Fig. 1.22, a wattmeter has two coils, the current coil (CC) and the voltage coil (VC).

The current coil with very low resistance is connected in series with the load and responds to the load current, while the voltage coil is connected in parallel to the load circuit and responds to the load voltage. Depending on the test section (source or load), one terminal of each coil is shorted and connected either to the source or to the load as shown in Fig. 1.23. In an AC circuit, wattmeter is widely used to measure the single-phase and three-phase real power as shown in Fig. 1.24.

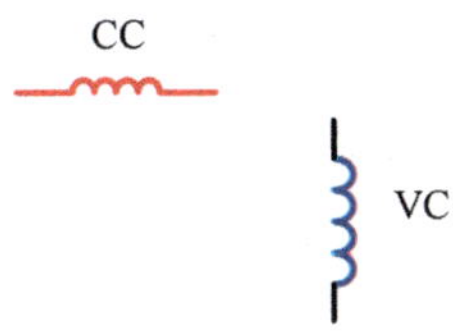

Fig. 1.22 Current coil and voltage coil of a wattmeter

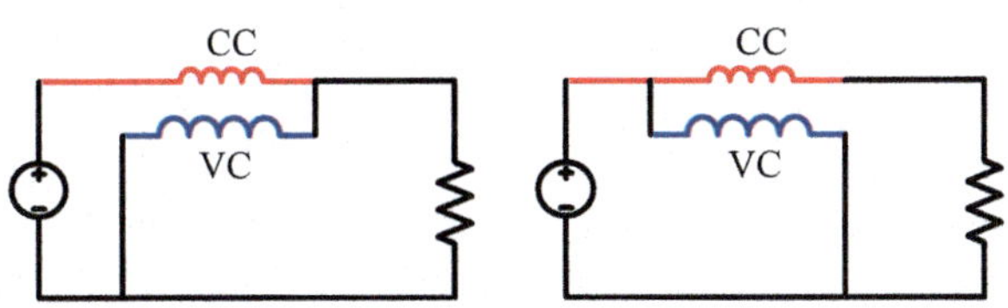

Fig. 1.23 Schematics of a wattmeter connection

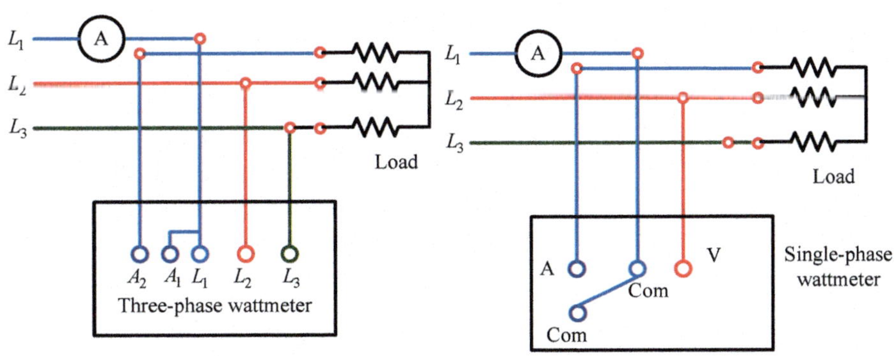

Fig. 1.24 Connection diagrams of three-phase and single-phase wattmeter

1.11 Electricity Bill

The generated power is usually transmitted and distributed by the power utility company to different types of consumers namely, residential, commercial and industrial for a set price. This price is set by the utility company, by taking into account the energy will charge based on production cost, transmission and distribution cost and the profit margin. The energy consumed by the customers is represented by kWh, based on which the consumers incur the associated costs. The rate at which electrical power is sold to the consumer is generally known as tariff. The main objective of the tariff is to recover the associated costs outlined above. There are different types of tariff namely, two-part tariff, three-part tariff, block rate tariff, maximum demand tariff and power factor tariff. The discussion on these different types of tariff is out of the scope of this text. Generally, the tariff is expressed as,

$$\begin{aligned}\text{Tariff} = a + b \times \text{energy use for first block} \\ + c \times \text{energy use for 2nd block} \\ + d \times \text{energy use for next block}\end{aligned} \tag{1.52}$$

where

a is the fixed charge in \$,
b is the charge (in \$) per kWh for the first block,
c is the charge (in \$) per kWh for the second block,
d is the char kWh for the next block

Example 1.8 The energy of a commercial building is supplied by a local power utility company, and this building consumes 1000 kWh of energy per month. The fixed charge of this demand is \$10. The cost for the first block of 100 kWh is \$ 0.10/kWh, the second block of 300 kWh is \$ 0.05/kWh and the next block is \$ 0.02/kWh. Determine the tariff of this demand.

Solution:
The tariff of this demand can be calculated as,

$$\text{Tariff} = 10 + 100 \times 0.1 + 300 \times 0.05 + 600 \times 0.02 = \$47 \tag{1.53}$$

Practice Problem 1.8 A local power utility company supplies power to a spinning mill that consumes 1500 kWh per month. The fixed charge of this demand is \$100. The cost for the first block of 500 kWh is \$ 0.15/kWh, and the second block of 1000 kWh is \$ 0.02/kWh. Calculate the tariff of this demand.

1.12 Efficiency of a System

The efficiency of a system is defined as the ratio of output power to the input power. The input power is equal to the output power plus the loss as shown in Fig. 1.25. Mathematically, the efficiency η is expressed as,

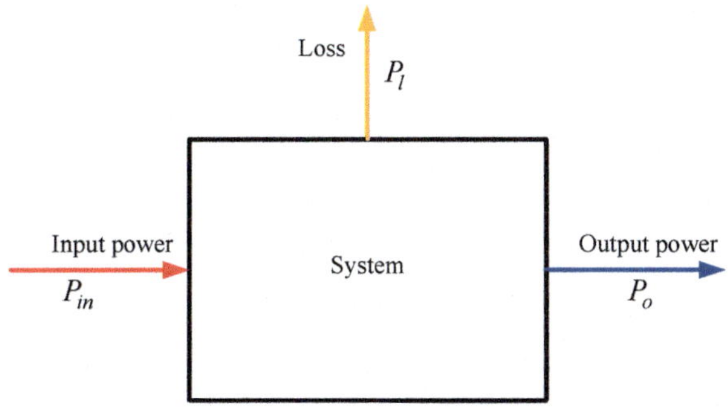

Fig. 1.25 System with different powers

$$\eta = \frac{P_o}{P_{in}} \times 100 \tag{1.54}$$

$$P_{in} = P_o + P_l \tag{1.55}$$

where

P_{in} is the input power in the system in W,
P_o is the output power in the system in W,
P_l is the loss in the system in W.

Substituting Eq. (1.55) into Eq. (1.54) yields,

$$\eta = \frac{P_{in} - P_l}{P_{in}} \times 100 \tag{1.56}$$

$$\eta = \left(1 - \frac{P_l}{P_{in}}\right) \times 100 \tag{1.57}$$

Again, Eq. (1.54) can be written as,

$$\eta = \frac{P_o}{P_o + P_l} \times 100 \tag{1.58}$$

Example 1.9 The input power and the loss of a system are found to be 150 and 30 W, respectively. Determine the output power and the efficiency of the system.

Solution:
The output of the system can be determined as,

$$150 = P_o + 30 \tag{1.59}$$

$$P_o = 150 - 30 = 120 \text{ W} \tag{1.60}$$

The efficiency is calculated as,

$$\eta = \frac{P_{\mathrm{o}}}{P_{\mathrm{in}}} \times 100 = \frac{120}{150} \times 100 = 80\% \quad (1.61)$$

Practice Problem 1.9 The output power and the loss of a system are found to be 100 and 10 W, respectively. Calculate the input power and the efficiency of the system.

1.13 PSpice Fundamentals

In PSpice simulation, user needs to open the schematic of PSpice and select circuit elements like resistance, voltage source, current source and analogue ground as shown in Fig. 1.26. The circuit elements can be rotated by holding the control (ctrl) key, and clicking on the specific element that needs to be rotated. According to circuit configuration, all elements need to be connected using the pencil tool. The value of any circuit element can be changed by double-clicking on it.

Sometimes, the software library does not contain the necessary circuit parts. In this case, the circuit part can be added by clicking on the *Libraries* button and clicking on the *Part Browser Advanced Window* as shown in Fig. 1.27.

For example, to find a DC voltage source, write the part name underneath the *Part Name* option as shown in Fig. 1.28, and browse the *Part Browser Advanced* window. Once the part appears, click on the place and close the button to get this part in the PSpice schematic window. In the PSpice simulation, the ground (*agnd*) of the circuit is very important. The simulation cannot be carried out without the ground connection. All parts of the circuit need to be placed along with the *agnd* and then are connected using the pencil tool as shown in Fig. 1.29. Then, the final circuit under simulation needs to be saved and then is run by clicking on the simulated button to calculate the voltage and current.

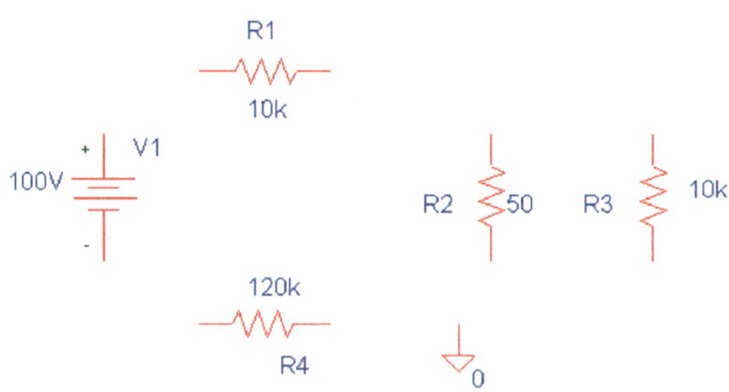

Fig. 1.26 Different circuit elements

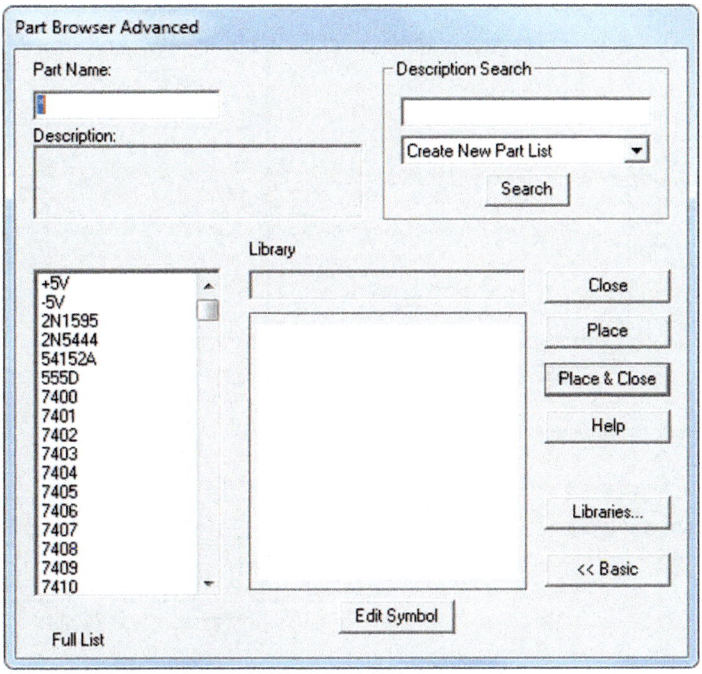

Fig. 1.27 Circuit part browser window

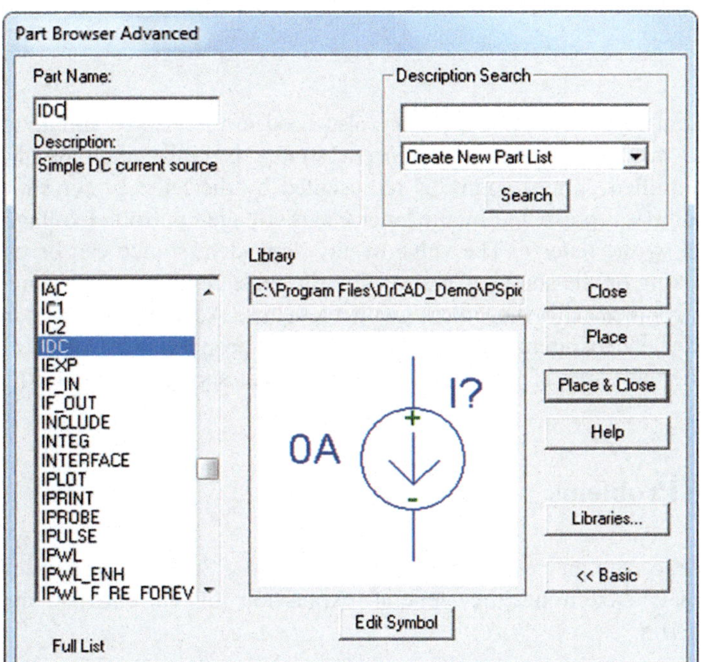

Fig. 1.28 Missing part in part browser window

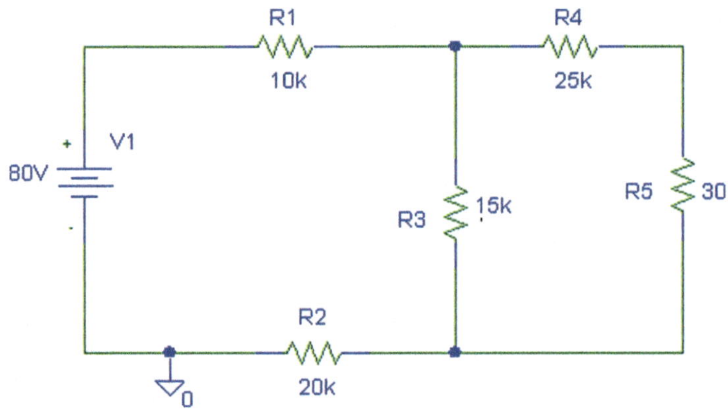

Fig. 1.29 Final for PSpice simulation

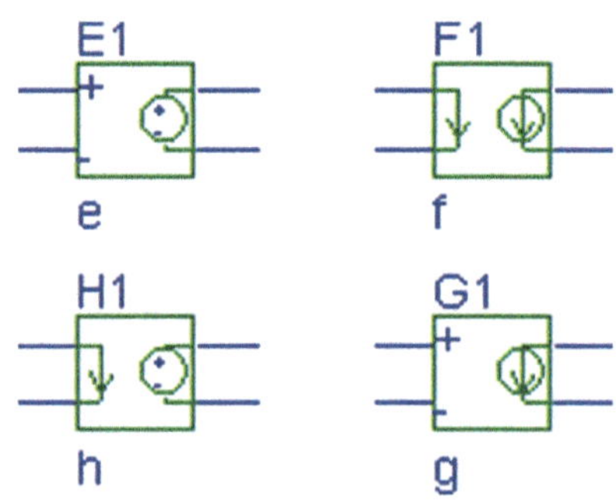

Fig. 1.30 Different types of dependent sources

Dependent or controlled sources are also used in the PSpice simulation. In this simulation, the current-controlled current source is represented by the letter *f*, voltage-controlled current source is represented by the letter *g*, current-controlled voltage source is represented by the letter *h* and voltage-controlled voltage source is represented by the letter *e*. The value of any dependent source can be changed by double-clicking on the source. Careful attention must be given while connecting the dependent sources with the other circuit elements. Accurate results can only be obtained if the dependent sources are connected properly with other circuit elements. Different types of the dependent sources are shown in Fig. 1.30.

Exercise Problems

1.1 A charge with an expression of $q = e^{-2t} + 6t^2 + 3t + 2$ C is found in a circuit terminal. Determine the general expression of current and its vale at $t = 0.01$ s.

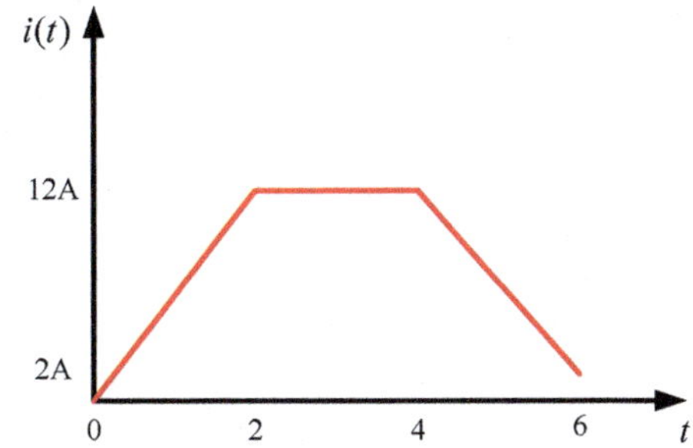

Fig. 1.31 Current waveform with times

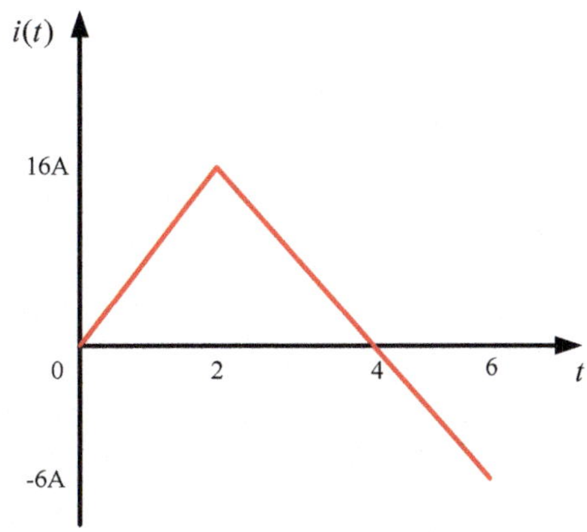

Fig. 1.32 Current waveform with times

1.2 A current of 6 A passes through a wire, where the expression of charge is found to be $q = e^{2t} + 3t$ C. Calculate the value of the time.

1.3 A charge of 8 C moves through a wire for 2 s. Determine the value of the current.

1.4 A current as shown in Fig. 1.31 passes through a wire. Find the expressions of current for different times and the value of the charge.

1.5 Fig. 1.32 shows a current waveform that passes through a wire. Calculate the expression of current and the value of the associated charge.

1.6 The resistivity of a 200 m long and 1.04×10^{-6} m^2 copper wire is 1.72×10^{-8} Ω-m. Calculate the value of the resistance.

1.7 The resistance of a copper wire is found to be 10 Ω. If the area and the resistivity of the wire are 0.95×10^{-7} m^2 and 1.72×10^{-8} Ω-m, respectively, then find the length of the wire.

1.8 The resistivity of a 200 m long aluminium wire is 2.8×10^{-8} Ω-m. Determine the diameter of the wire, if the resistance of the wire is 6 Ω.

1.9 A work with an expression of $w = 5q^3 - 3q^2 + q$ J is required to move a charge from one point to another point of a conductor. Calculate the voltage when $q = 0.01$ C.

1.10 The work done for a time t s is given by $w = 2t^2 + 10t$ J. Find the value of the time for the corresponding power of 26 W.

1.11 A power $p = 2e^{-2t} + 5$ W is used for a time period 0.01 to 0.03 to complete a task. Calculate the corresponding energy.

1.12 A house uses twelve 40 W bulbs for 3 h, a 730 W iron for 1.5 h and a 40 W computer for 5 h. Calculate the total energy used in the house and the associated cost if the energy price is 10 cents per kWh.

References

1. C.K. Alexander, M.N.O. Sadiku, in *Fundamentals of Electric Circuits*, 6th edn. (McGraw-Hill Higher Education, Jan 2016)
2. J. David Irwin, R. Mark Nelms, in *Basic Engineering Circuit Analysis*, 11th edn. (Wiley, USA, 2015)
3. J.W. Nilsson, S.A. Riedel, in *Electric Circuits*, 10th edn. (Prentice Hall International Edition, 2015)
4. R.L. Boylestad, in *Introductory Circuit Analysis*, 13th edn. (Pearson, 2016)
5. M. Abdus Salam, in *Basic Electrical Circuits*, 2nd edn. (Shroff Publishers & Distributors Pvt. Ltd, India, 2007)
6. B. Anderson, R. Anderson, in *Fundamentals of Semiconductor Devices*, 2nd edn. (McGraw-Hill Higher Education, Feb 2017)

Chapter 2
Electrical Laws

2.1 Introduction

Electrical laws are necessary to analyse any electrical circuit effectively and efficiently by determining different circuit parameters such as current, voltage power and resistance. These laws include Ohms law, Kirchhoff's current and voltage laws, and voltage and current division rules. The knowledge of series and parallel circuit orientations, delta–wye and wye–delta–wye transformations are also required to analyse electrical circuits. In this chapter, different electrical laws, delta–wye and wye–delta–wye transformations, source conversion technique and Wheatstone bridge circuit have been discussed.

2.2 Ohm's Law

Ohm's law is the most important law in the electric circuit analysis that can be applied to any electrical network in any time frame. Ohm's law states that the current flow in a conductor is directly proportional to the voltage across the conductor. A German Physicist, George Simon Ohm (1787–1854), established this relationship between the current and the voltage for a given resistor. According to his name, it is known as Ohm's law. Let a current I flows in a resistor R as shown in Fig. 2.1 resulting in a voltage drop V across the resistor. According to Ohm's law, the relationship between the current and the resultant voltage can be written as,

$$I \infty V \tag{2.1}$$

$$V = IR \tag{2.2}$$

where R is the proportionality constant, which is known as the resistance of the circuit element.

Md. A. Salam and Q. Mehbubar Rahman, *Fundamentals of Electrical Circuit Analysis*,
https://doi.org/10.1007/978-981-10-8624-3_2

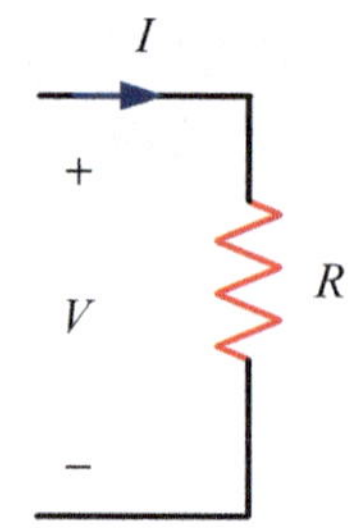

Fig. 2.1 Current in a resistor

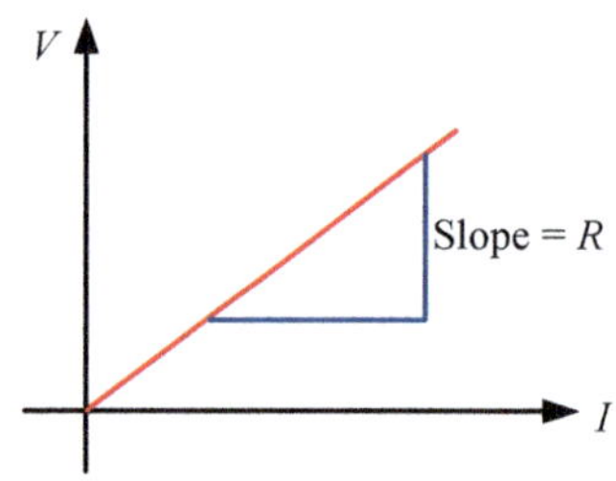

Fig. 2.2 Variation of voltage with current

Equation (2.2) can be expressed as,

$$I = \frac{V}{R} \tag{2.3}$$

From Eq. (2.3), Ohm's law can be restated as follows: the current flowing in a resistor is directly proportional to the voltage across the resistor and inversely proportional to the resistance. According to Ohm's law, if the current increases, the voltage will increase as can be seen in Fig. 2.2, where we assume that the resistance is constant.

Example 2.1 A hairdryer is connected to a 220 V source and draws 3 A current. Determine the value of the resistance.

Solution:
The value of the resistance can be determined as,

$$R = \frac{V}{I} = \frac{220}{3} = 73.33\,\Omega \tag{2.4}$$

Example 2.2 A 10 Ω resistance is connected to a voltage source as shown in Fig. 2.3. Calculate the current drawn and power absorbed by the resistance.

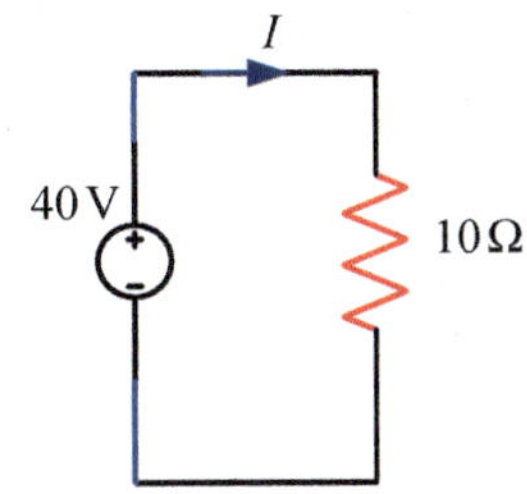

Fig. 2.3 Circuit for Example 2.2

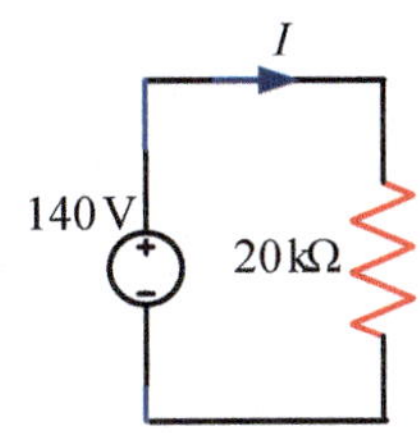

Fig. 2.4 Circuit for Practice Problem 2.2

Solution:
The current drawn by the resistance can be determined as,

$$I = \frac{V}{R} = \frac{40}{10} = 4\,\text{A} \tag{2.5}$$

The power absorbed by the resistance is,

$$P = I^2 R = 4^2 \times 10 = 160\,\text{W} \tag{2.6}$$

Practice Problem 2.1
A 600 W electric blender is connected to a 220 V source. Calculate the value of the current and resistance.

Practice Problem 2.2
A 20 kΩ resistance is connected to a voltage source as shown in Fig. 2.4. Find the current.

2.3 Kirchhoff's Current Law

In 1845, German physicist, Gustav Robert Kirchhoff, developed the relationship between different types of currents and voltages in an electrical circuit. These laws, known as Kirchhoff's laws, are used to calculate the current and voltage in an electrical circuit. Kirchhoff's current law states that the algebraic sum of the currents around a node (node refers to any point on a circuit where two or more circuit elements meet) equals to zero. The circuit in Fig. 2.5 illustrates Kirchhoff's current

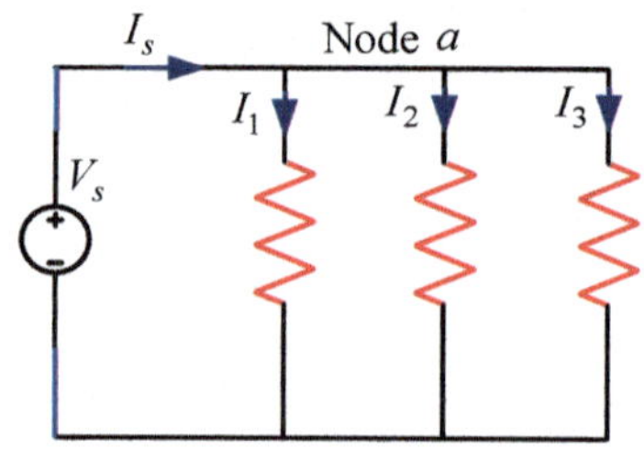

Fig. 2.5 Circuit for KCL

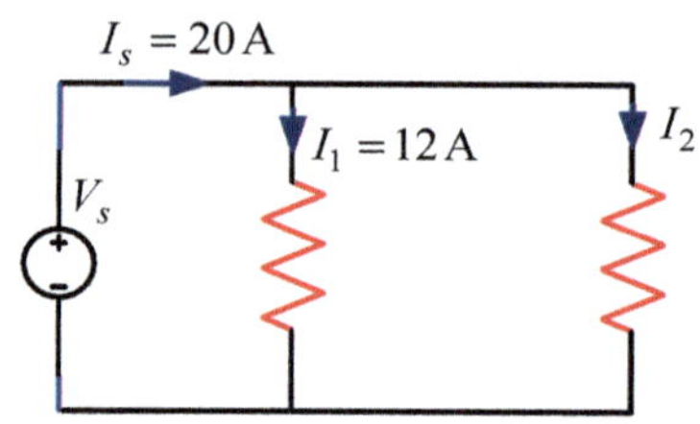

Fig. 2.6 Circuit for Example 2.3

law (KCL). Here, the currents entering node a are considered positive while the currents coming out of the node are considered negative.

From Fig. 2.5, for node a, the following equation can be written:

$$I_s - I_1 - I_2 - I_3 = 0 \tag{2.7}$$

$$I_s = I_1 + I_2 + I_3 \tag{2.8}$$

$$\sum I_{\text{entering}} = \sum I_{\text{leaving}} \tag{2.9}$$

From Eq. (2.9), the KCL can be stated as the sum of the entering currents at a node is equal to the current leaving from the same node.

Example 2.3 The current distribution in a circuit is shown in Fig. 2.6. Determine the value of the unknown current.

Solution:
Applying KCL at the node of the circuit in Fig. 2.6 yields,

$$20 = 12 + I_2 \tag{2.10}$$

$$I_2 = 20 - 12 = 8\,\text{A} \tag{2.11}$$

Practice Problem 2.3
Figure 2.7 shows a circuit with the current distribution. Calculate the values of the unknown currents.

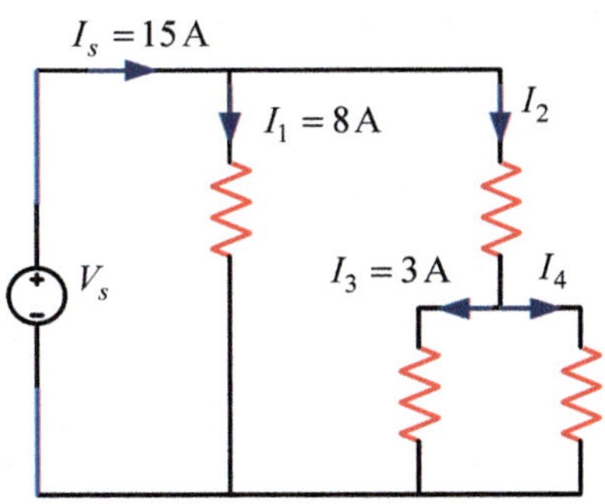

Fig. 2.7 Circuit for Practice Problem 2.3

2.4 Kirchhoff's Voltage Law

Kirchhoff's voltage law (KVL) states that the algebraic sum of the voltages around any loop (loop refers to a closed path in a circuit) of a circuit is equal to zero. The circuit in Fig. 2.8 is considered to explain the KVL.

Applying KVL to the only loop of the circuit in Fig. 2.8 yields,

$$-V_{s1} - Vs_2 + V_1 + V_2 + V_3 = 0 \tag{2.12}$$

$$V_{s1} + V_{s2} = V_1 + V_2 + V_3 \tag{2.13}$$

$$\sum V_{\text{rises}} = \sum V_{\text{drops}} \tag{2.14}$$

From Eq. (2.14), the KVL can be restated as the sum of the voltage rises is equal to the sum of the voltage drops.

Example 2.4 The voltage distribution of a circuit is shown in Fig. 2.9. Determine the value of V_2, source current I_s and resistance R_3.

Solution:
Applying KVL to the circuit in Fig. 2.9 yields,

$$-40 - 10 + 4 + 6 + V_2 = 0 \tag{2.15}$$

$$V_2 = 40\,\text{V} \tag{2.16}$$

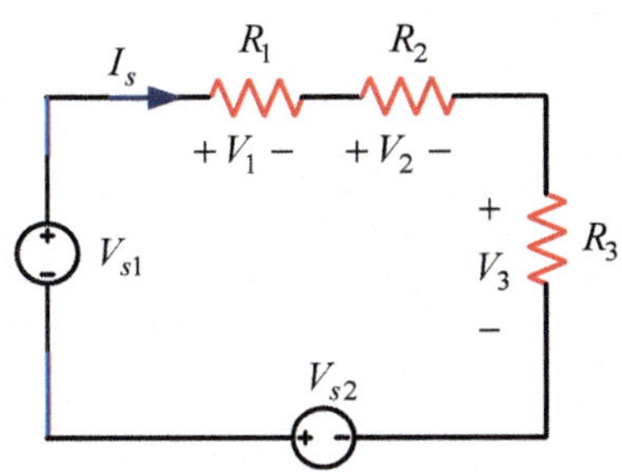

Fig. 2.8 Circuit for KVL

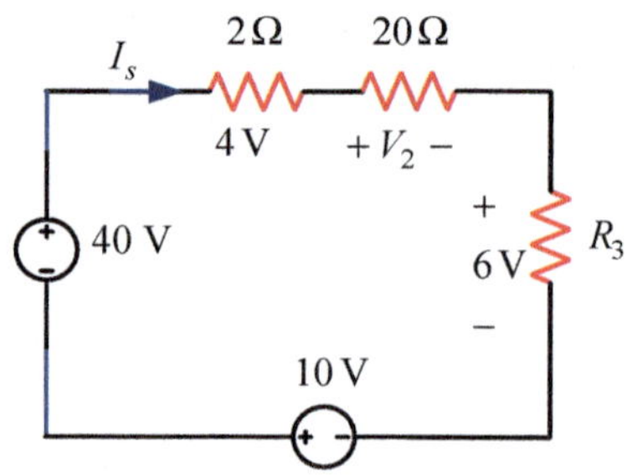

Fig. 2.9 Circuit for Example 2.4

The value of the current in the circuit is determined as,

$$V_2 = I_s \times 4 = 40 \tag{2.17}$$

$$I_s = 10\,\text{A} \tag{2.18}$$

The value of the resistance R_3 is determined as,

$$R_3 = \frac{V_3}{I_s} = \frac{6}{10} = 0.6\,\Omega \tag{2.19}$$

Practice Problem 2.4

Figure 2.10 shows an electrical circuit with different parameters. Calculate the value of the source current, voltage across 6 Ω and 12 Ω resistors.

Example 2.5 An electrical circuit with a voltage-controlled voltage source is shown in Fig. 2.11. Find the power absorbed by the 6 Ω resistor.

Solution:
Applying KVL to the circuit in Fig. 2.11 yields,

$$-40 + 2I_s + 6I_s + 8I_s + 2V_1 = 0 \tag{2.20}$$

$$2I_s - V_1 = 0 \tag{2.21}$$

$$V_1 = 2I_s \tag{2.22}$$

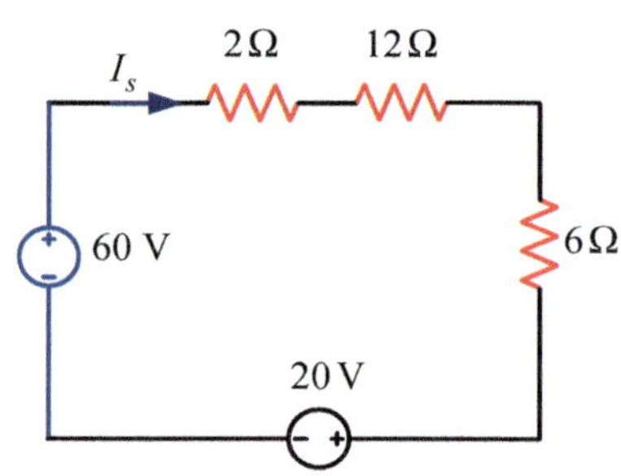

Fig. 2.10 Circuit for Practice Problem 2.4

Fig. 2.11 Circuit for Example 2.5

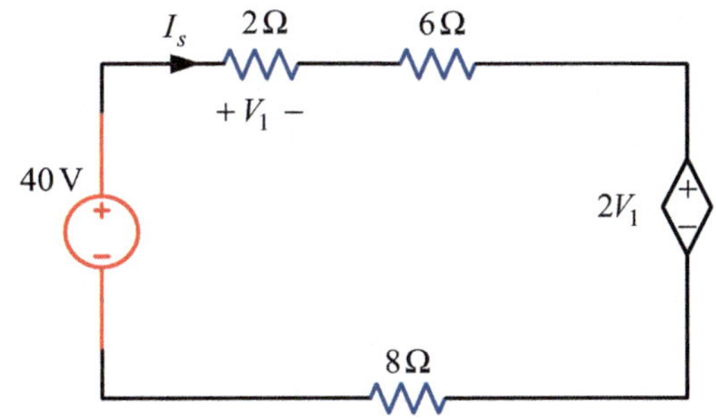

Fig. 2.12 Circuit for Practice Problem 2.5

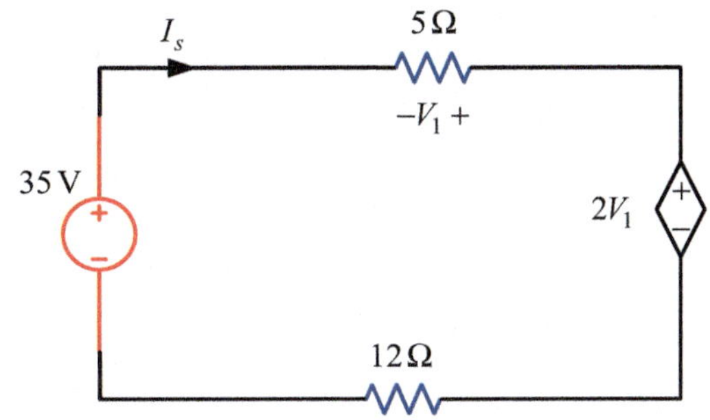

Substituting Eq. (2.22) into Eq. (2.20) yields,

$$-40 + 2I_s + 6I_s + 8I_s + 2 \times 2I_s = 0 \tag{2.23}$$

$$I_s = \frac{40}{20} = 2\,\text{A} \tag{2.24}$$

The power absorbed by the 6 Ω resistor is,

$$P_{6\Omega} = I_s^2 \times 6 = 2^2 \times 6 = 24\,\text{W} \tag{2.25}$$

Practice Problem 2.5

An electrical circuit with a voltage-controlled voltage source is shown in Fig. 2.12. Calculate the voltage across the 12 Ω resistor.

2.5 Series Resistors and Voltage Division Rule

In a series circuit, the circuit elements such as voltage source, resistors, etc. are connected in an end-to-end connection, where the same current flows through each element. Let us consider that three resistors R_1, R_2, and R_3 are connected in series with a voltage source V_s as shown in the circuit in Fig. 2.13.

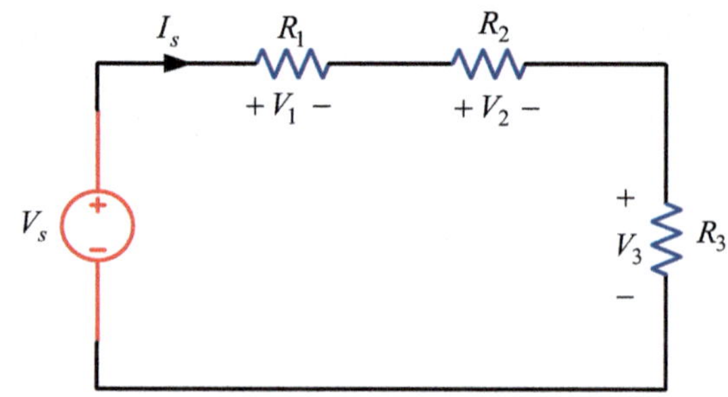

Fig. 2.13 A series circuit with three resistors

If the resulting current I_s flows through this series circuit, according to Ohm's law, as the corresponding voltage drops V_1, V_2, and V_3 across three resistors R_1, R_2, and R_3, respectively, can be written as,

$$V_1 = I_s R_1 \tag{2.26}$$

$$V_2 = I_s R_2 \tag{2.27}$$

$$V_3 = I_s R_3 \tag{2.28}$$

Applying KVL to the circuit in Fig. 2.13 yields,

$$-V_s + V_1 + V_2 + V_3 = 0 \tag{2.29}$$

$$V_s = V_1 + V_2 + V_3 \tag{2.30}$$

Substituting Eqs. (2.26), (2.27) and (2.28) into Eq. (2.30) yields,

$$V_s = I_s R_1 + I_s R_2 + I_s R_3 \tag{2.31}$$

$$V_s = I_s(R_1 + R_2 + R_3) \tag{2.32}$$

$$\frac{V_s}{I_s} = (R_1 + R_2 + R_3) \tag{2.33}$$

$$R_s = (R_1 + R_2 + R_3) \tag{2.34}$$

In Eq. (2.34), R_s represents the series equivalent circuit. Following the similar trend, it can be concluded that if a circuit contains N number of series resistors, then the equivalent resistance can be expressed as,

$$R_s = R_1 + R_2 + R_3 + \cdots + R_N = \sum_{n=1}^{N} R_n \tag{2.35}$$

The voltage divider rule or voltage division rule is closely related to the series circuit [1, 2]. Substituting Eq. (2.34) to Eq. (2.33) yields the source current in terms of equivalent resistance as,

$$I_s = \frac{V_s}{R_s} \tag{2.36}$$

Equation (2.33) can be rearranged as,

$$I_s = \frac{V_s}{R_1 + R_2 + R_3} \tag{2.37}$$

Substituting Eq. (2.37) into Eqs. (2.26), (2.27) and (2.28) yields,

$$V_1 = \frac{R_1}{R_1 + R_2 + R_3} V_s \tag{2.38}$$

$$V_2 = \frac{R_2}{R_1 + R_2 + R_3} V_s \tag{2.39}$$

$$V_3 = \frac{R_3}{R_1 + R_2 + R_3} V_s \tag{2.40}$$

By observing Eqs. (2.38) to (2.40), it can be concluded that in a series circuit with resistors and voltage source, the voltage drop across any resistor is equal to the source voltage times the ratio of the resistance of that resistor to the total resistance; this is the voltage divider rule. According to voltage divider rule, the voltage drop V_n across the nth resistor in a circuit with N number of series resistors and voltage source V_s is given by,

$$V_n = \frac{R_n}{R_1 + R_2 + R_3 + \cdots + R_N} V_s \tag{2.41}$$

Example 2.6 A series circuit is shown in Fig. 2.14. Calculate the voltage across each resistor using voltage divider rule.

Solution:
The voltage across 2 Ω resistor is,

$$V_{2\Omega} = \frac{R_1}{R_1 + R_2 + R_3} V_s = \frac{2}{16} \times 80 = 10\,\text{V} \tag{2.42}$$

$$V_{6\Omega} = \frac{R_2}{R_1 + R_2 + R_3} V_s = \frac{6}{16} \times 80 = 30\,\text{V} \tag{2.43}$$

$$V_{8\Omega} = \frac{R_3}{R_1 + R_2 + R_3} V_s = \frac{8}{16} \times 80 = 40\,\text{V} \tag{2.44}$$

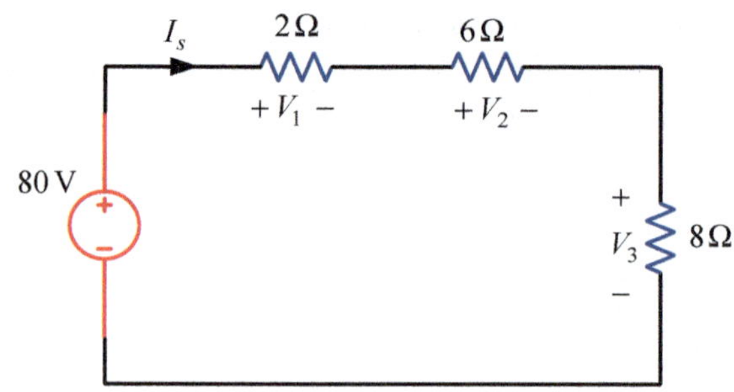

Fig. 2.14 Circuit for Example 2.6

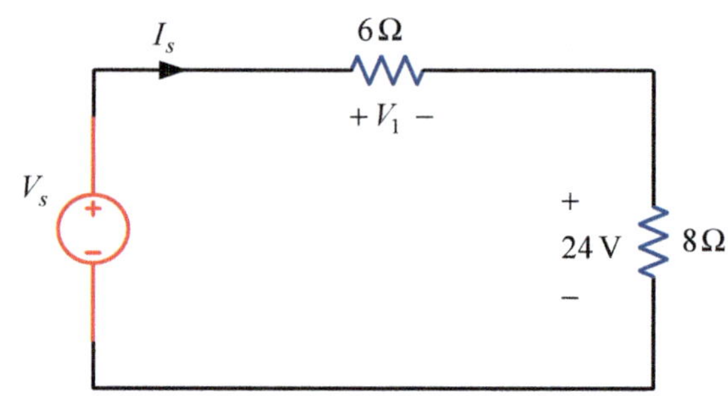

Fig. 2.15 Circuit for Practice Problem 2.6

Practice Problem 2.6
Figure 2.15 shows a series circuit. Use voltage divider rule to find the source voltage and the voltage across the 6 Ω resistor.

2.6 Parallel Resistors and Current Division Rule

Resistors are sometimes connected in parallel in a circuit. Two or more circuit elements are said to be in parallel, when they share same pair of nodes. In a parallel circuit, the voltage across each of the parallel circuit elements is same, but the current through each of the elements might be different. Let us consider three resistors are connected in a parallel to a source voltage V_p as shown in the circuit in Fig. 2.16. In this case, the voltage across each of these resistors will be V_p too.

According to Ohm's law, the currents I_1, I_2, and I_2 through the corresponding resistors R_1, R_2, and R_3 can be written as,

$$I_1 = \frac{V_p}{R_1} \tag{2.45}$$

$$I_2 = \frac{V_p}{R_2} \tag{2.46}$$

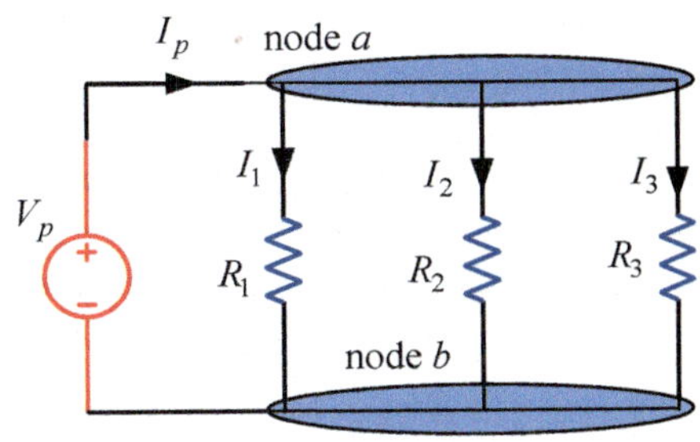

Fig. 2.16 Parallel circuit with three resistances

$$I_3 = \frac{V_p}{R_3} \tag{2.47}$$

Applying KCL to the circuit in Fig. 2.16 yields,

$$I_p = I_1 + I_2 + I_3 \tag{2.48}$$

Substituting Eqs. (2.45), (2.46), (2.47) into Eq. (2.48) yields,

$$I_p = \frac{V_p}{R_1} + \frac{V_p}{R_2} + \frac{V_p}{R_3} \tag{2.49}$$

$$I_p = V_p\left(\frac{1}{R_1} + \frac{1}{R_2} + \frac{1}{R_3}\right) \tag{2.50}$$

$$\frac{1}{\left(\frac{1}{R_1} + \frac{1}{R_2} + \frac{1}{R_3}\right)} = \frac{V_p}{I_p} \tag{2.51}$$

$$R_p = \frac{1}{\left(\frac{1}{R_1} + \frac{1}{R_2} + \frac{1}{R_3}\right)} \tag{2.52}$$

Equation (2.52) provides the equivalent resistance for the resistors connected in parallel. For n number of resistors, the above equation can be generalized as:

$$R_p = \frac{1}{\frac{1}{R_1} + \frac{1}{R_2} + \frac{1}{R_3} + \cdots + \frac{1}{R_N}} \tag{2.53}$$

Let us consider a special case, where two resistors R_1 and R_2 are connected in parallel. In this case, the equivalent resistance R_p will be:

$$R_p = \frac{1}{\frac{1}{R_1} + \frac{1}{R_2}} \tag{2.54}$$

$$R_p = \frac{1}{\frac{R_1 + R_2}{R_1 R_2}} \tag{2.55}$$

$$R_p = \frac{R_1 R_2}{R_1 + R_2} \tag{2.56}$$

From Eq. (2.56), it can be concluded that the equivalent resistance for two resistors in parallel is equal to the ratio of the product of two resistances to their sum.

If $R_1 = R_2 = R$, then Eq. (2.56) becomes,

$$R_p = \frac{R \times R}{2R} = \frac{R}{2} \tag{2.57}$$

From Eq. (2.57), it is observed that for two resistors, with equal resistance values connected in parallel, will result in the equivalent resistance of half of the value of the single resistor.

Now with the aid of the concept on parallel resistance and the circuit in Fig. 2.17, the current divider rule can be explained. From the circuit in Fig. 2.17, the equivalent resistance R_t can be written as,

$$R_t = \frac{R_1 R_2}{R_1 + R_2} \tag{2.58}$$

The total voltage drop across the two resistors will be,

$$V_t = I_t R_t \tag{2.59}$$

Substituting Eq. (2.58) into Eq. (2.59) yields,

$$V_t = \frac{R_1 R_2}{R_1 + R_2} I_t \tag{2.60}$$

Now that both R_1 and R_2 are in parallel, both the resistors will exhibit the same voltage drop across each of those. In this case, the current in the resistance R_1 will be,

Fig. 2.17 A circuit with two parallel resistors

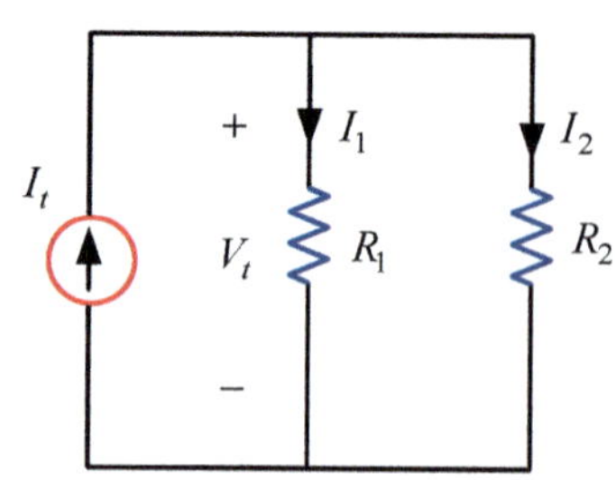

$$I_1 = \frac{V_t}{R_1} \tag{2.61}$$

Substituting Eq. (2.60) into Eq. (2.61) yields,

$$I_1 = \frac{1}{R_1}\frac{R_1 R_2}{R_1 + R_2} I_t \tag{2.62}$$

$$I_1 = \frac{R_2}{R_1 + R_2} I_t \tag{2.63}$$

The current in the resistance R_2 will be,

$$I_2 = \frac{V_t}{R_2} \tag{2.64}$$

Substituting Eq. (2.60) into Eq. (2.64) yields,

$$I_2 = \frac{1}{R_2}\frac{R_1 R_2}{R_1 + R_2} I_t \tag{2.65}$$

$$I_2 = \frac{R_1}{R_1 + R_2} I_t \tag{2.66}$$

From Eqs. (2.63) and (2.66), it is observed that the current through any resistor in a circuit with two parallel resistors is equal to the total current times the ratio of the opposite resistance to the sum of the two resistances. This is known as current divider rule.

Example 2.7 A series–parallel circuit is shown in Fig. 2.18. Calculate the source current and the branch currents.

Solution:
The total resistance of the circuit is,

$$R_t = 2 + \frac{3 \times (1+5)}{3+6} = 4\,\Omega \tag{2.67}$$

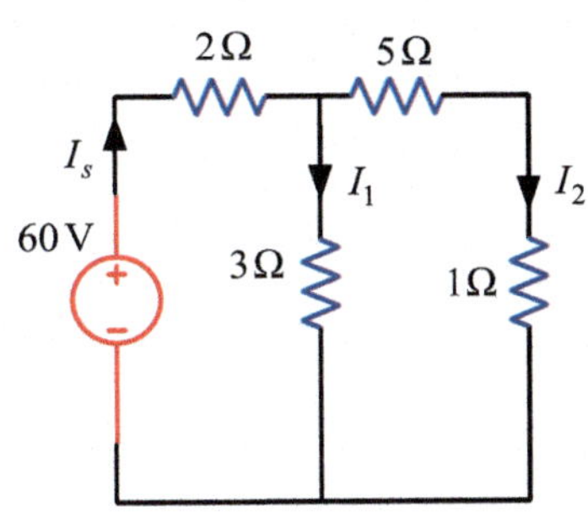

Fig. 2.18 Circuit for Example 2.7

The value of the source current is,

$$I_s = \frac{60}{4} = 15\,\mathrm{A} \tag{2.68}$$

The branch currents I_1 and I_2 can be calculated as,

$$I_1 = \frac{(5+1)}{6+3} \times 15 = 10\,\mathrm{A} \tag{2.69}$$

$$I_2 = \frac{3}{6+3} \times 15 = 5\,\mathrm{A} \tag{2.70}$$

Example 2.8 Figure 2.19 shows a series–parallel circuit with a jumper wire. Find the total resistance and the source current.

Solution:
In this circuit, 3 Ω and 6 Ω resistors are connected in parallel, and that results in the equivalent resistance as,

$$R_1 = \frac{3 \times 6}{3+6} = 2\,\Omega \tag{2.71}$$

Again, 4 Ω and R_1 (2 Ω) resistors are in series, and this equivalent resistance is parallel to 12 Ω resistor. The overall equivalent resistance is,

$$R_2 = \frac{(4+2) \times 12}{4+2+12} = 4\,\Omega \tag{2.72}$$

Finally, R_2 (4 Ω) and 2 Ω resistors are in series and this equivalent resistance is in parallel to 12 Ω resistor. Finally, this equivalent resistance is in series with 8Ω resistor. The final equivalent circuit resistance is,

$$R_t = 8 + \frac{(4+2) \times 12}{4+2+12} = 12\,\Omega \tag{2.73}$$

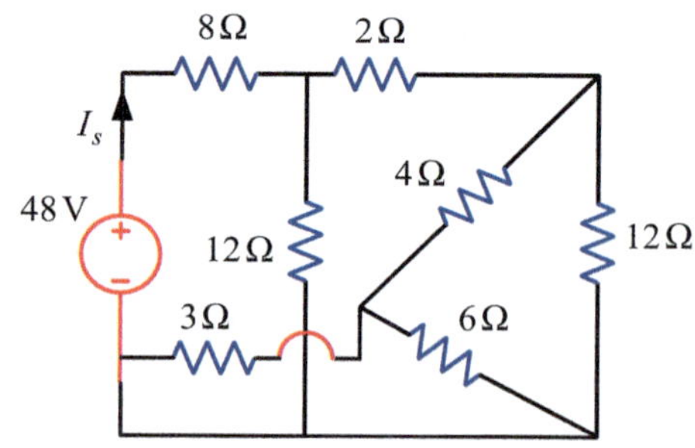

Fig. 2.19 Circuit for Example 2.8

The value of the source current is,

$$I_s = \frac{48}{12} = 4\,\text{A} \tag{2.74}$$

Example 2.9 A circuit with a current-controlled current source is shown in Fig. 2.20. Determine the branch currents and power absorbed by the $4\,\Omega$ resistor.

Solution:
Applying KCL to the circuit in Fig. 2.20 yields,

$$I_x + I_1 + 2I_x = 14 \tag{2.75}$$

Applying Ohm's law to the circuit in Fig. 2.20 yields,

$$I_x = \frac{V_s}{2} \tag{2.76}$$

$$I_1 = \frac{V_s}{4} \tag{2.77}$$

Substituting Eqs. (2.76) and (2.77) into Eq. (2.75) yields,

$$14 = 3\frac{V_s}{2} + \frac{V_s}{4} \tag{2.78}$$

$$14 = \frac{7V_s}{4} \tag{2.79}$$

$$V_s = \frac{14 \times 4}{7} = 8\,\text{V} \tag{2.80}$$

Substituting Eq. (2.80) into Eqs. (2.76) and (2.77) yields,

$$I_x = \frac{8}{2} = 4\,\text{A} \tag{2.81}$$

$$I_1 - \frac{8}{4} = 2\,\text{A} \tag{2.82}$$

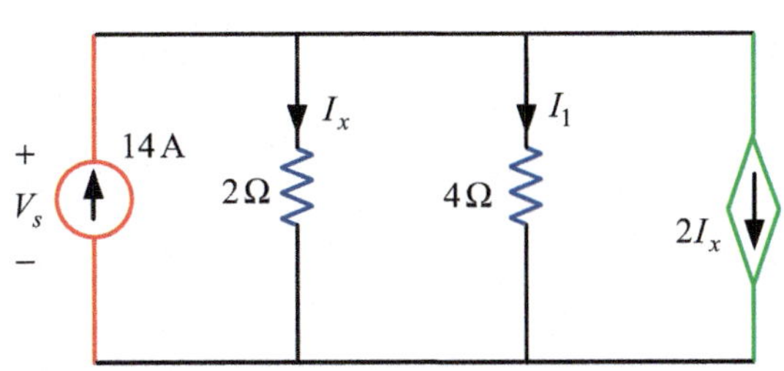

Fig. 2.20 Circuit for Example 2.9

The power absorbed by the 4 Ω resistance is,

$$P_{4\Omega} = 2^2 \times 4 = 16\,\text{W} \tag{2.83}$$

Practice Problem 2.7

A series–parallel circuit is shown in Fig. 2.21. Calculate the equivalent resistance and the current through 8 Ω resistor.

Practice Problem 2.8

A series–parallel circuit with two jumper wires is shown in Fig. 2.22. Determine the equivalent circuit resistance and the source current.

Practice Problem 2.9

A circuit with a voltage-controlled current source is shown in Fig. 2.23. Calculate the current I_1 and the power absorbed by the 3 Ω resistor.

Fig. 2.21 Circuit for Practice Problem 2.7

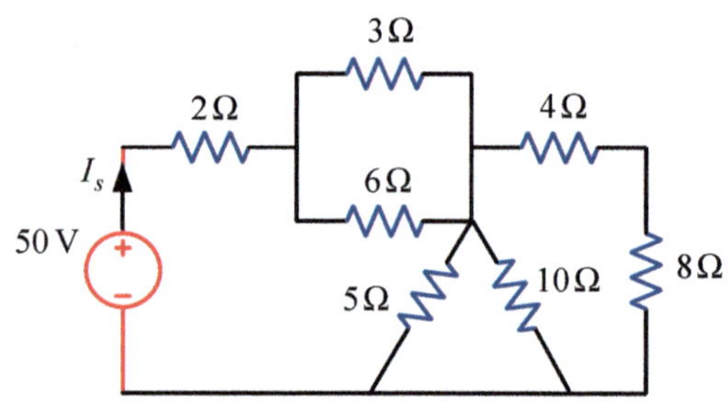

Fig. 2.22 Circuit for Practice Problem 2.8

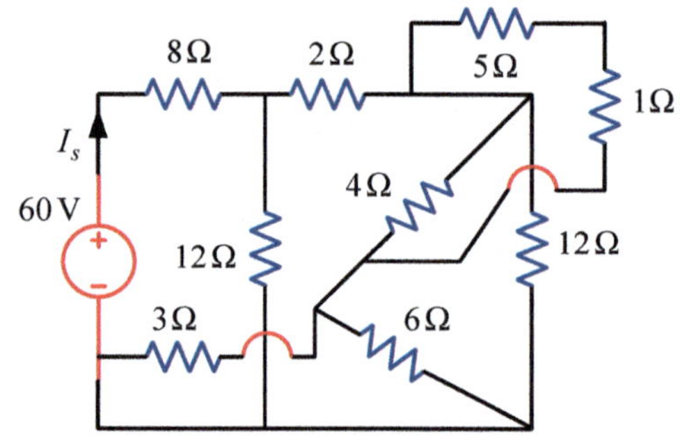

Fig. 2.23 Circuit for Practice Problem 2.9

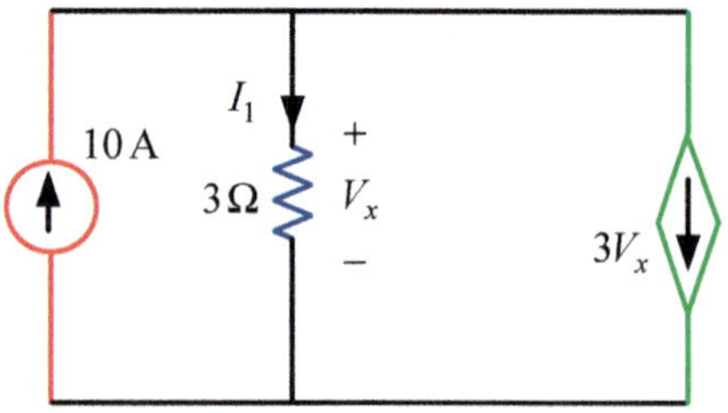

2.7 Delta–Wye Transformation

Resistors are sometimes either not connected in series or in parallel. In this case, a special technique needs to be used for simplifying an electrical circuit. This technique is known as delta–wye or wye–delta transformation. Consider three resistors which are connected in delta (Δ) and wye (Y) forms as shown in Fig. 2.24.

For delta connection, the total resistances between terminals 1 and 2, terminals 2 and 3, and terminals 3 and 1 are,

$$R_{12(\Delta)} = \frac{R_b(R_a + R_c)}{R_a + R_b + R_c} \tag{2.84}$$

$$R_{23(\Delta)} = \frac{R_c(R_a + R_b)}{R_a + R_b + R_c} \tag{2.85}$$

$$R_{31(\Delta)} = \frac{R_a(R_b + R_c)}{R_a + R_b + R_c} \tag{2.86}$$

For wye connection, the total resistances between terminals 1 and 2, terminals 2 and 3, and terminals 3 and 1 are,

$$R_{12(Y)} = R_1 + R_2 \tag{2.87}$$

$$R_{23(Y)} = R_2 + R_3 \tag{2.88}$$

$$R_{31(Y)} = R_3 + R_1 \tag{2.89}$$

Electrically, the total resistances between terminals 1 and 2, terminals 2 and 3, and terminals 3 and 1 are the same for delta and wye connections [3–5]. This results in the following relationships:

$$R_{12(Y)} = R_{12(\Delta)} \tag{2.90}$$

$$R_{23(Y)} = R_{23(\Delta)} \tag{2.91}$$

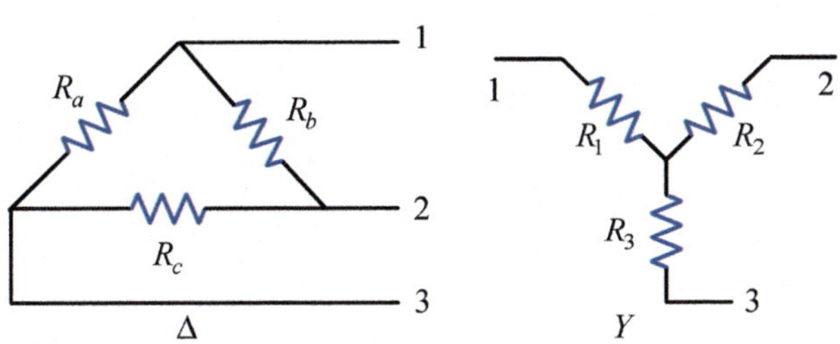

Fig. 2.24 Circuit for delta–wye transformation

$$R_{31(Y)} = R_{31(\Delta)} \tag{2.92}$$

Substituting Eqs. (2.84) and (2.87) into Eq. (2.90) yields,

$$R_1 + R_2 = \frac{R_b(R_a + R_c)}{R_a + R_b + R_c} \tag{2.93}$$

Substituting Eqs. (2.85) and (2.88) into Eq. (2.91) yields,

$$R_2 + R_3 = \frac{R_c(R_a + R_b)}{R_a + R_b + R_c} \tag{2.94}$$

And, substituting Eqs. (2.86) and (2.89) into Eq. (2.92) yields,

$$R_3 + R_1 = \frac{R_a(R_b + R_c)}{R_a + R_b + R_c} \tag{2.95}$$

Adding Eqs. (2.93), (2.94) and (2.95) yields,

$$2(R_1 + R_2 + R_3) = \frac{2(R_aR_b + R_bR_c + R_cR_a)}{R_a + R_b + R_c} \tag{2.96}$$

$$R_1 + R_2 + R_3 = \frac{R_aR_b + R_bR_c + R_cR_a}{R_a + R_b + R_c} \tag{2.97}$$

Subtracting Eqs. (2.94), (2.95) and (2.93) from Eq. (2.97) individually yields,

$$R_1 = \frac{R_aR_b}{R_a + R_b + R_c} \tag{2.98}$$

$$R_2 = \frac{R_bR_c}{R_a + R_b + R_c} \tag{2.99}$$

$$R_3 = \frac{R_cR_a}{R_a + R_b + R_c} \tag{2.100}$$

2.8 Wye–Delta Transformation

To transform wye-connected resistors into delta-connected ones; the formulae can be derived from Eqs. (2.98), (2.99) and (2.100). Multiplying Eq. (2.98) by (2.99), Eq. (2.99) by (2.100) and Eq. (2.100) by (2.98) yields,

$$R_1R_2 = \frac{R_aR_b^2R_c}{(R_a+R_b+R_c)^2} \tag{2.101}$$

$$R_2R_3 = \frac{R_aR_bR_c^2}{(R_a+R_b+R_c)^2} \tag{2.102}$$

$$R_3R_1 = \frac{R_a^2R_bR_c}{(R_a+R_b+R_c)^2} \tag{2.103}$$

Adding Eqs. (2.101), (2.102) and (2.103) yields,

$$R_1R_2+R_2R_3+R_3R_1 = \frac{R_aR_b^2R_c+R_aR_bR_c^2+R_a^2R_bR_c}{(R_a+R_b+R_c)^2} \tag{2.104}$$

$$R_1R_2+R_2R_3+R_3R_1 = \frac{R_aR_bR_c(R_a+R_b+R_c)}{(R_a+R_b+R_c)^2} \tag{2.105}$$

$$R_1R_2+R_2R_3+R_3R_1 = \frac{R_aR_bR_c}{R_a+R_b+R_c} \tag{2.106}$$

Dividing Eq. (2.106) by Eqs. (2.99), (2.100) and (2.98) individually yields,

$$R_a = \frac{R_1R_2+R_2R_3+R_3R_1}{R_2} \tag{2.107}$$

$$R_b = \frac{R_1R_2+R_2R_3+R_3R_1}{R_3} \tag{2.108}$$

$$R_c = \frac{R_1R_2+R_2R_3+R_3R_1}{R_1} \tag{2.109}$$

Example 2.10 A delta–wye circuit is shown in Fig. 2.25. Calculate the total circuit resistance, the source current and the power absorbed by the 4 Ω resistor.

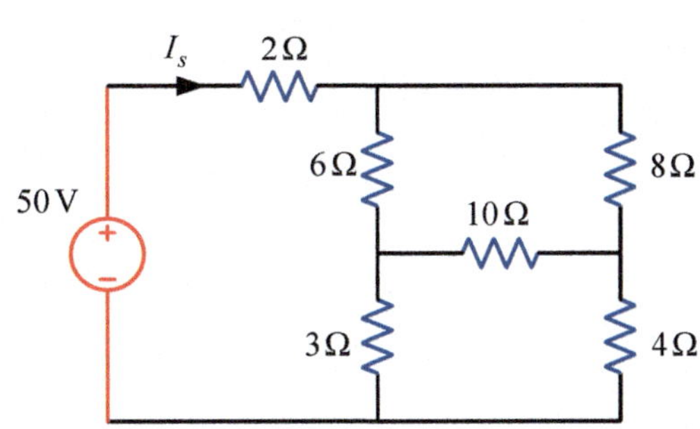

Fig. 2.25 Circuit for Example 2.10

Solution:
Converting the upper part delta circuit to wye circuit as shown in Fig. 2.26. The values of the resistors are,

$$R_1 = \frac{6 \times 8}{6 + 10 + 8} = 2\,\Omega \tag{2.110}$$

$$R_2 = \frac{10 \times 8}{6 + 10 + 8} = 3.33\,\Omega \tag{2.111}$$

$$R_3 = \frac{10 \times 6}{6 + 10 + 8} = 2.5\,\Omega \tag{2.112}$$

Now, the circuit in Fig. 2.26 can be redrawn as shown in Fig. 2.27. The R_3 and $3\,\Omega$ are in series, and R_2 and $4\,\Omega$ are in series, then in parallel to each other. In this case, the equivalent resistance is,

$$R_4 = \frac{(R_2 + 4)(R_3 + 3)}{R_2 + 4 + R_3 + 3} = \frac{(3.33 + 4)(2.5 + 3)}{3.33 + 4 + 2.5 + 3} = 3.14\,\Omega \tag{2.113}$$

From Fig. 2.27, the total circuit resistance is determined as,

$$R_t = 2 + R_1 + R_4 = 2 + 2 + 3.14 = 7.14\,\Omega \tag{2.114}$$

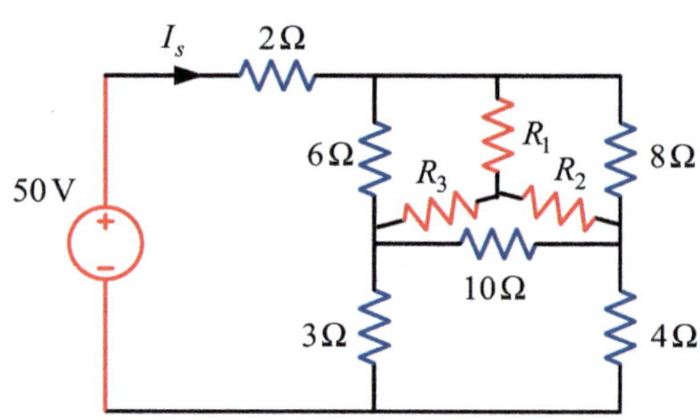

Fig. 2.26 Circuit for conversion delta to wye

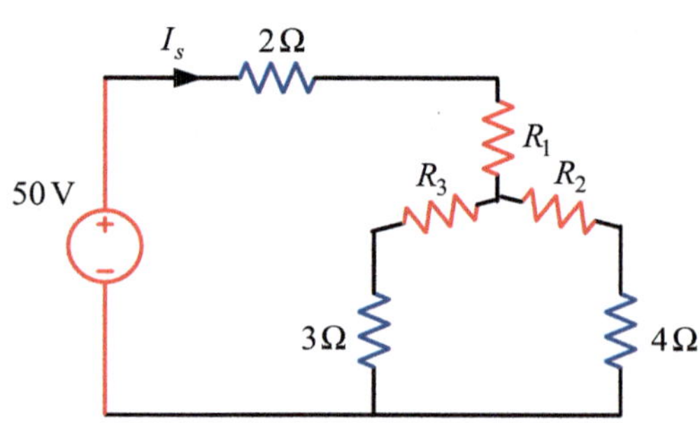

Fig. 2.27 A series–parallel circuit for Example 2.10

The value of the source current is,

$$I_s = \frac{50}{7.14} = 7\,\text{A} \tag{2.115}$$

The current through 4 Ω resistor is,

$$I_{4\Omega} = 7 \times \frac{(3+2.5)}{5.5+3.33+4} = 3\,\text{A} \tag{2.116}$$

The power absorbed by 4 Ω resistor is,

$$P_{4\Omega} = I_{4\Omega}^2 \times 4 = 3^2 \times 4 = 36\,\text{W} \tag{2.117}$$

Example 2.11 Figure 2.28 shows a delta–wye circuit. Determine the total circuit resistance and the source current.

Solution:
Convert the wye connection to a delta connection as shown in Fig. 2.29. The values of the resistances are,

$$R_1 = \frac{7 \times 10 + 10 \times 3 + 3 \times 7}{3} = 40.33\,\Omega \tag{2.118}$$

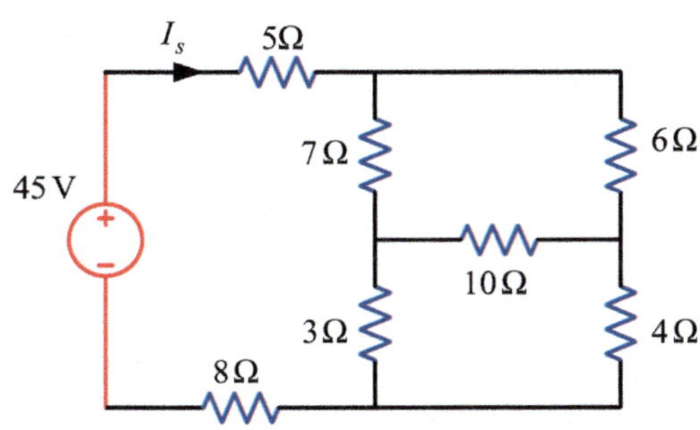

Fig. 2.28 Circuit for Example 2.11

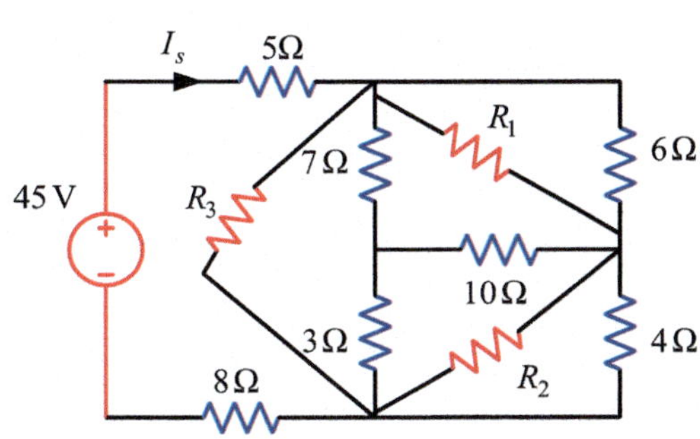

Fig. 2.29 Circuit conversion from wye to delta

$$R_2 = \frac{7 \times 10 + 10 \times 3 + 3 \times 7}{7} = 17.29\,\Omega \tag{2.119}$$

$$R_3 = \frac{7 \times 10 + 10 \times 3 + 3 \times 7}{10} = 12.1\,\Omega \tag{2.120}$$

From Fig. 2.30, it is seen that R_1 and $6\,\Omega$ are in parallel, and R_2 and $4\,\Omega$ are in parallel, and then they are in series. The value of the equivalent resistance R_4 is,

$$R_4 = \frac{R_1 \times 6}{R_1 + 6} + \frac{R_2 \times 4}{R_2 + 4} = \frac{40.33 \times 6}{40.33 + 6} + \frac{17.29 \times 4}{17.29 + 4} = 8.47\,\Omega \tag{2.121}$$

The R_3 and R_4 ($8.47\,\Omega$) resistors are in parallel, then in series with $5\,\Omega$ and $8\,\Omega$ resistors as shown in Fig. 2.31. Finally, the equivalent circuit resistance is,

$$R_t = \frac{12.1 \times 8.47}{12.1 + 8.47} + 5 + 8 = 17.98\,\Omega \tag{2.122}$$

The value of the source current can be determined as,

$$I_s = \frac{45}{17.98} = 2.5\,\text{A} \tag{2.123}$$

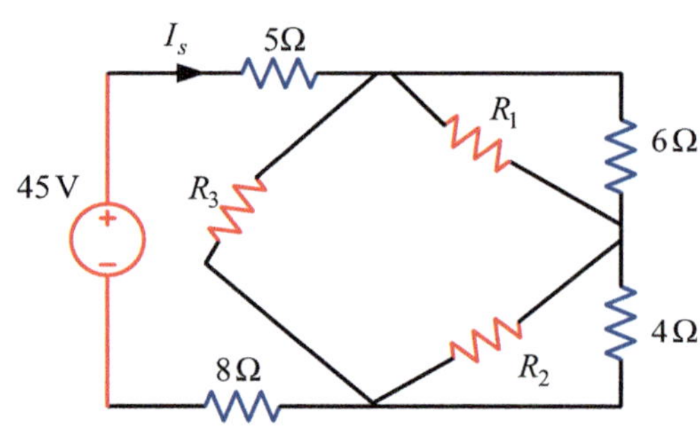

Fig. 2.30 First series–parallel circuit for Example 2.11

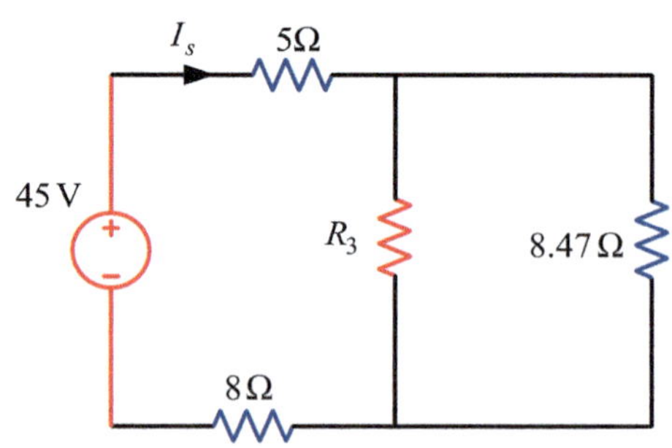

Fig. 2.31 Second series–parallel circuit for Example 2.11

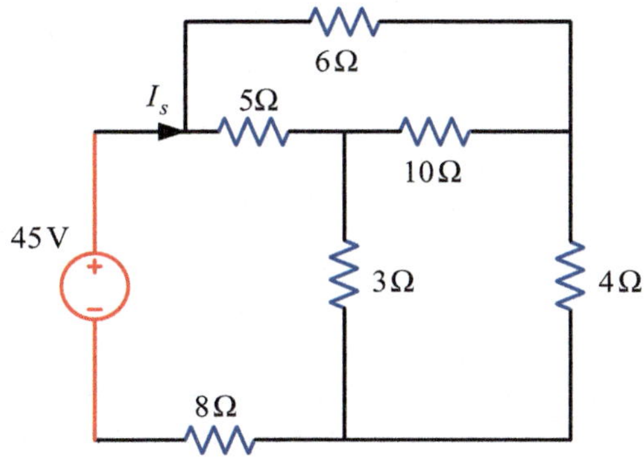

Fig. 2.32 Circuit for Practice Problem 2.10

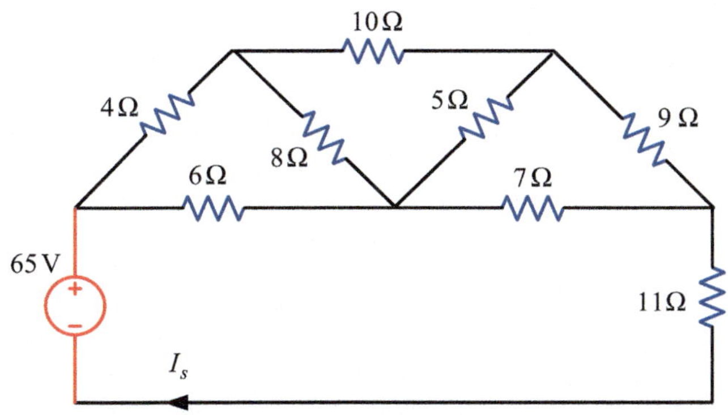

Fig. 2.33 Circuit for Example 2.11

Practice Problem 2.10
Figure 2.32 shows a delta–wye circuit. Calculate the equivalent circuit resistance and the source current.

Practice Problem 2.11
Figure 2.33 shows a delta–wye circuit. Determine the equivalent circuit resistance and the source current.

2.9 Short Circuit and Open Circuit

A circuit is said to be a short circuit when this allows current to flow through an inadvertent path between two nodes with a negligible low (zero) resistance. The voltage across a short circuit is zero. The short-circuit condition can happen both in DC and AC circuits. When two terminals of a dry cell or an automobile battery are

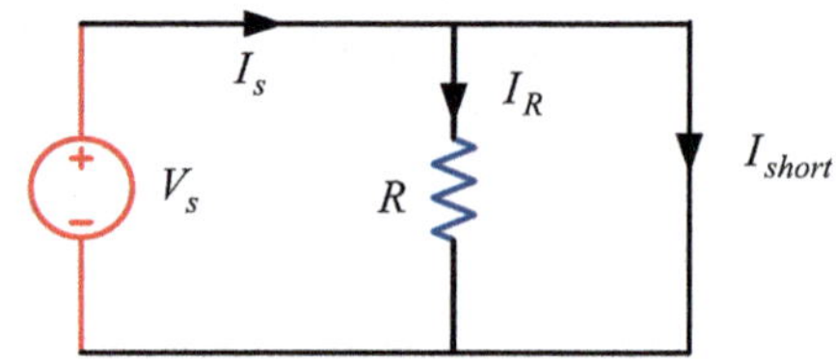

Fig. 2.34 A resistance with a short circuit

shorted by a piece of conducting wire, then the battery will discharge through the wire, which in turn heats up the wire. The principle of an arc welding is based on the application of heating due to a short circuit. Figure 2.34 shows a circuit to explain the short-circuit phenomena. Applying current divider rule the currents I_R and I_{short} are found to be,

$$I_R = I_s \times \frac{0}{R+0} = 0 \tag{2.124}$$

$$I_{\text{short}} = I_s \times \frac{R}{R+0} = I_s \tag{2.125}$$

The voltage in the short circuit is,

$$V_{\text{short}} = I_{\text{short}} \times R_{\text{short}} = I_s \times 0 = 0 \tag{2.126}$$

where R_{short} represents the resistance of the short-circuit branch.

A circuit is said to be an open circuit when it contains an infinite resistance between two nodes. Figure 2.35 illustrates the concept of an open circuit. From Fig. 2.34, the expression of the current can be written as,

$$I_s = \frac{V_s}{R} = \frac{V_s}{\infty} = 0 \tag{2.127}$$

Example 2.12 A short and open circuit is shown in Fig. 2.36. Determine the equivalent circuit resistance, and the source current when the terminals a and b are short and open circuited.

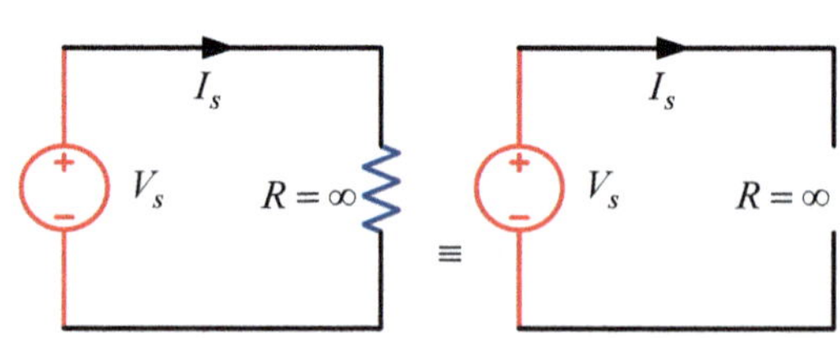

Fig. 2.35 An open

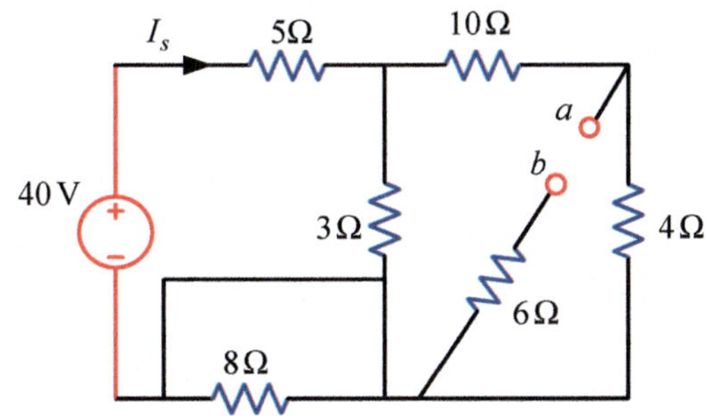

Fig. 2.36 Circuit for Example 2.12

Solution:
Consider that the terminals *a* and *b* are short circuited. Then, the resistances 6 Ω and 4 Ω are in parallel, and then in series with 10 Ω resistance. The value of the resistance is,

$$R_1 = \frac{4 \times 6}{4 + 6} + 10 = 12.4\,\Omega \tag{2.128}$$

Again, 12.4 Ω and 3 Ω are in parallel. In this case, the value of the resistance is,

$$R_2 = \frac{12.4 \times 3}{12.4 + 3} = 2.42\,\Omega \tag{2.129}$$

The total circuit resistance is,

$$R_{ts} = 5 + 2.42 = 7.42\,\Omega \tag{2.130}$$

The value of the source current is,

$$I_{ss} = \frac{40}{7.42} = 5.39\,\text{A} \tag{2.131}$$

Again, consider the terminals *a* and *b* are open circuited. The resistances 4 Ω and 10 Ω are in series, then in parallel with 3 Ω resistance. The values of the resistance are,

$$R_3 = 4 + 10 = 14\,\Omega \tag{2.132}$$

$$R_4 = \frac{14 \times 3}{14 + 3} = 2.47\,\Omega \tag{2.133}$$

The total circuit resistance is,

$$R_{to} = 5 + 2.47 = 7.47\,\Omega \tag{2.134}$$

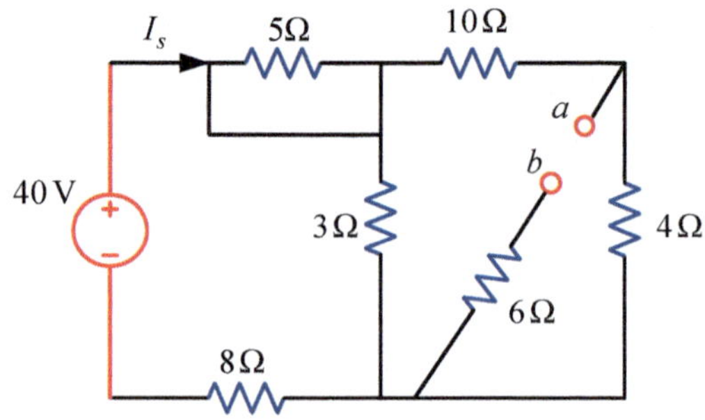

Fig. 2.37 Circuit for Practice Problem 2.12

The value of the source current is,

$$I_{so} = \frac{40}{7.47} = 5.35\,\text{A} \tag{2.135}$$

Practice Problem 2.12
Figure 2.37 shows a short and open circuit. Find the equivalent circuit resistance and the source current.

2.10 Source Configuration

The voltage and current sources are usually connected either in series or in parallel configurations. There are six cells in an automobile battery where each cell provides a voltage of two volts.

These cells are connected in series to generate a voltage of twelve volts. Two voltage sources are connected in series as shown in Fig. 2.38. Consider the voltage source V_{s1} is greater than the voltage source V_{s2}. Then, the following equations from Fig. 2.38 can be written:

$$V_a = V_{s1} + V_{s2} \tag{2.136}$$

$$V_b = V_{s1} - V_{s2} \tag{2.137}$$

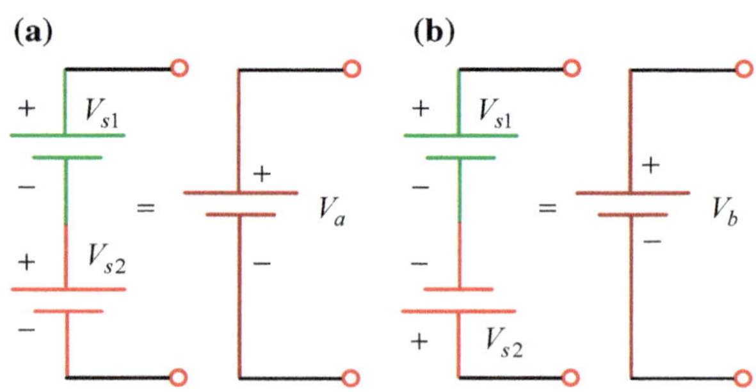

Fig. 2.38 Voltage sources are in series

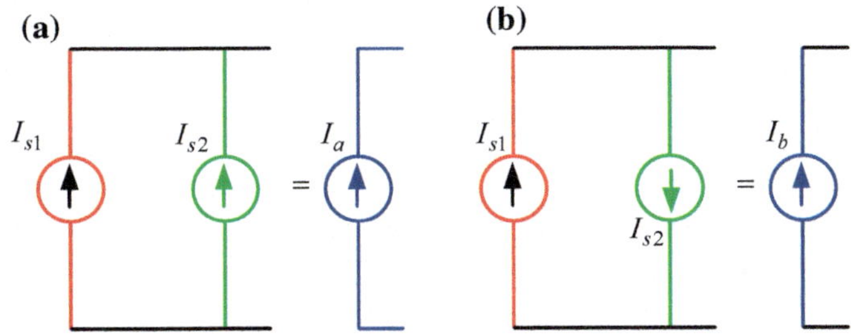

Fig. 2.39 Current sources are in parallel

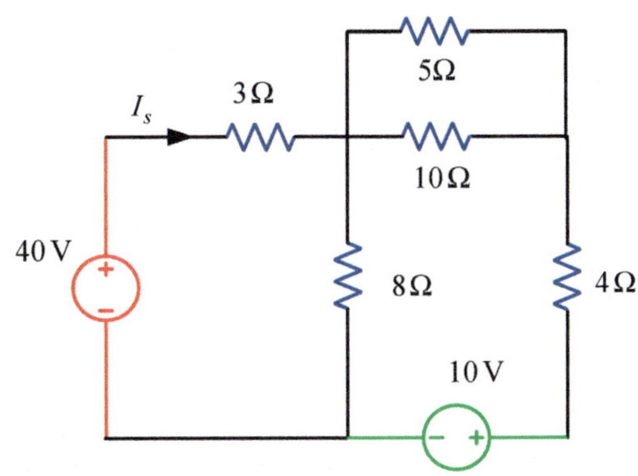

Fig. 2.40 Circuit for Example 2.13

The current sources are not recommended for series connection, but they are usually connected in a parallel in the circuit as shown in Fig. 2.39.

Consider the current source I_{s1} is greater than the current source I_{s2}. From Fig. 2.39, the following equations can be written:

$$I_a = I_{s1} + I_{s2} \tag{2.138}$$

$$I_b = I_{s1} - I_{s2} \tag{2.139}$$

Example 2.13 A series–parallel circuit with two voltage sources is shown in Fig. 2.40. Calculate the equivalent circuit resistance, source current and the power absorbed by 10 Ω resistor.

Solution:
Two voltage sources are connected in series but in an opposite direction. Therefore, the value of the resultant voltage is,

$$V_s = 40 - 10 = 30\,\text{V} \tag{2.140}$$

The 5 Ω and 10 Ω resistors are in parallel, and then in series with 4 Ω. The value of the resistance is,

$$R_1 = \frac{5 \times 10}{5 + 10} + 4 = 7.33\,\Omega \tag{2.141}$$

R_1 (7.33 Ω) from Eq. (2.141) is in parallel with 8 Ω resistor, and then in series with 3 Ω resistance. The equivalent circuit resistance is,

$$R_t = \frac{7.33 \times 8}{7.33 + 8} + 3 = 6.83\,\Omega \tag{2.142}$$

The value of the source current is,

$$I_s = \frac{30}{6.83} = 4.39\,\Omega \tag{2.143}$$

The current in the 4 Ω resistor is,

$$I_{4\Omega} = 4.39 \times \frac{8}{8 + 4 + 3.33} = 2.29\,\text{A} \tag{2.144}$$

The current through 10 Ω resistor is,

$$I_{10\Omega} = 2.29 \times \frac{5}{5 + 10} = 0.76\,\text{A} \tag{2.145}$$

Power absorbed by 10 Ω resistor is,

$$P_{10\Omega} = 0.76^2 \times 10 = 5.78\,\text{W} \tag{2.146}$$

Example 2.14 A parallel circuit with two current sources is shown in Fig. 2.41. Find the current in the 8 Ω resistor.

Solution:
The value of the equivalent current source is,

$$I_s = 20 - 8 = 12\,\text{A} \tag{2.147}$$

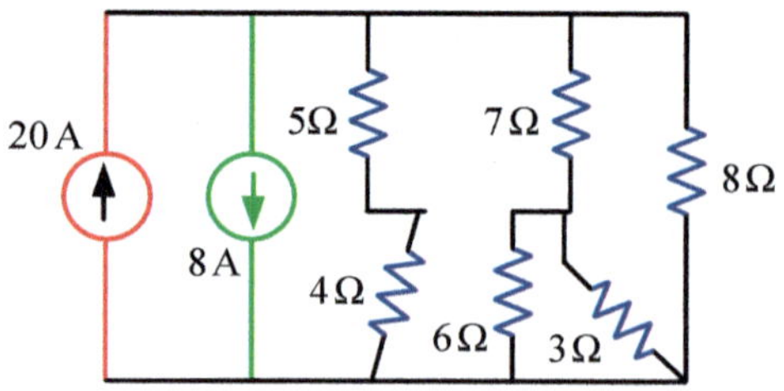

Fig. 2.41 Circuit for Example 2.14

The $6\,\Omega$ and $3\,\Omega$ resistors are in parallel, then series with $7\,\Omega$ resistor. Here, the equivalent resistance is,

$$R_1 = \frac{3 \times 6}{3+6} + 7 = 9\,\Omega \tag{2.148}$$

Then, R_1 ($9\,\Omega$) is in parallel with $5+4=9\,\Omega$ resistor, and the value of the equivalent resistance is,

$$R_2 = \frac{9 \times 9}{9+9} = 4.5\,\Omega \tag{2.149}$$

The current through $8\,\Omega$ resistor is,

$$I_{8\Omega} = 12 \times \frac{4.5}{4.5+8} = 4.32\,\text{A} \tag{2.150}$$

Practice Problem 2.13
Figure 2.42 shows a series–parallel circuit with two voltage sources. Determine the equivalent resistance, source current and the current in the $6\,\Omega$ resistor.

Practice Problem 2.14
Figure 2.43 shows a parallel circuit with two current sources. Determine the current in the $8\,\Omega$ resistance.

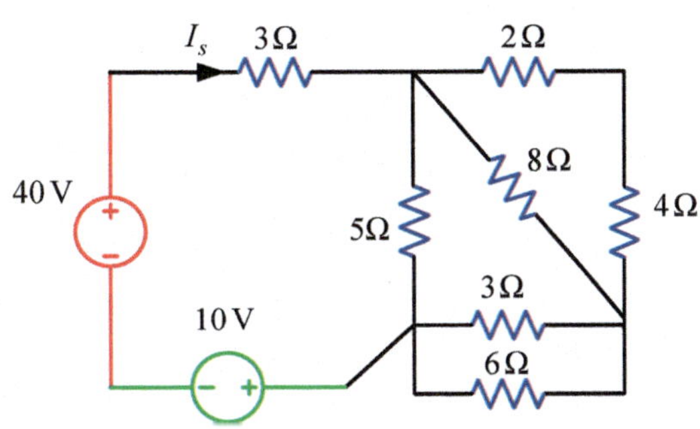

Fig. 2.42 Circuit for Practice Problem 2.13

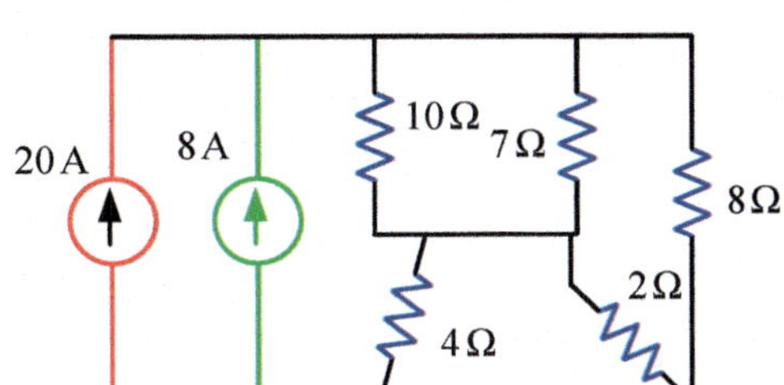

Fig. 2.43 Circuit for Practice Problem 2.14

2.11 Source Conversion Technique

Voltage and current sources are connected in an electrical circuit based on the application requirement. For circuit analysis, sometimes these sources need to be converted from one form to another to simplify the circuit configuration. This task can be performed using a source conversion technique. In this case, a resistance needs to be connected in series with a voltage source while a resistance needs to be connected in parallel with a current source. Figure 2.44 is considered to explain the source conversion technique.

From Fig. 2.44, the following expressions can be written:

$$I_s = \frac{V_s}{R} \tag{2.151}$$

$$V_s = I_s R \tag{2.152}$$

Example 2.15 A series–parallel circuit with voltage and current sources is shown in Fig. 2.45. Use source conversion technique to find the current in the 8 Ω resistor.

Solution:
The 30 V and 10 V voltage sources are converted to a current source as,

$$I_{s1} = \frac{30}{3} = 10\,\text{A} \tag{2.153}$$

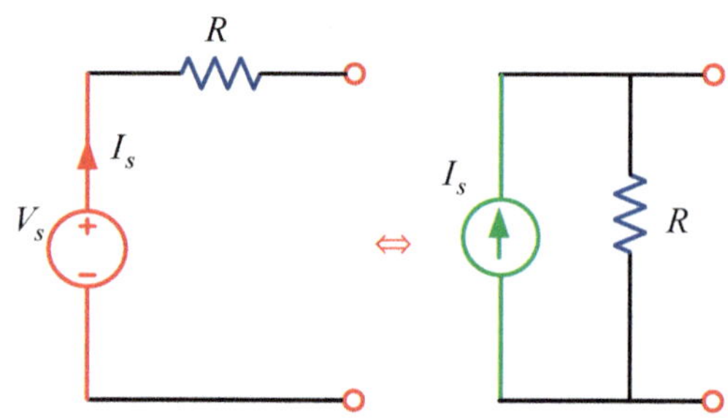

Fig. 2.44 Voltage and current sources with resistance

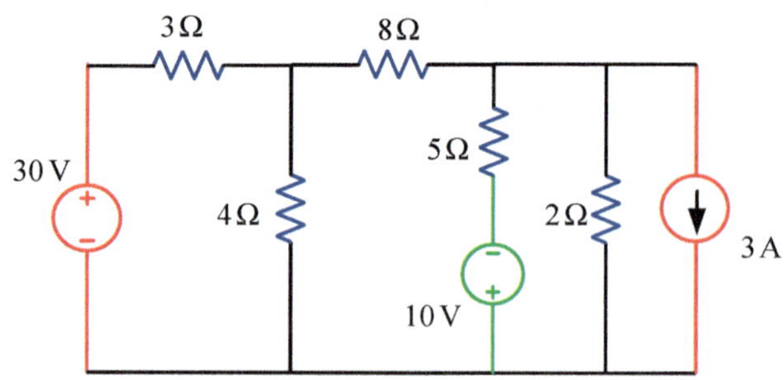

Fig. 2.45 Circuit for Example 2.15

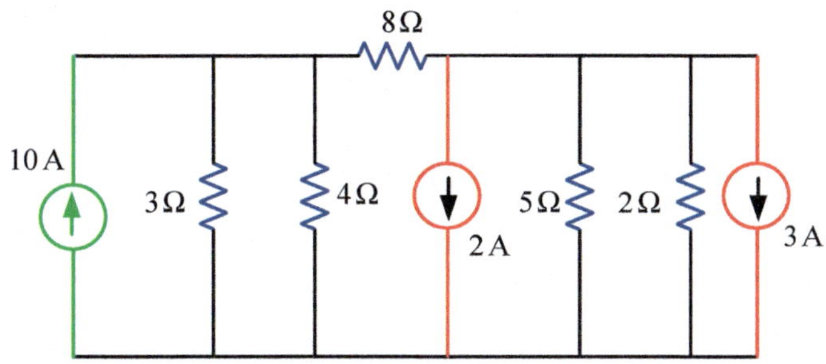

Fig. 2.46 Current sources with resistors

$$I_{s1} = \frac{10}{5} = 2\,\text{A} \tag{2.154}$$

The equivalent circuit is shown in Fig. 2.46. The resistors 3 Ω and 4 Ω, 5 Ω and 2 Ω are in parallel. Their equivalent resistances are,

$$R_1 = \frac{3 \times 4}{3 + 4} = 1.71\,\Omega \tag{2.155}$$

$$R_2 = \frac{5 \times 2}{5 + 2} = 1.43\,\Omega \tag{2.156}$$

The two current sources are added together and the value of the equivalent current source is,

$$I_s = 2 + 3 = 5\,\text{A} \tag{2.157}$$

In this case, the circuit is converted to an equivalent circuit as shown in Fig. 2.47.

Fig. 2.47 Two current sources with resistors

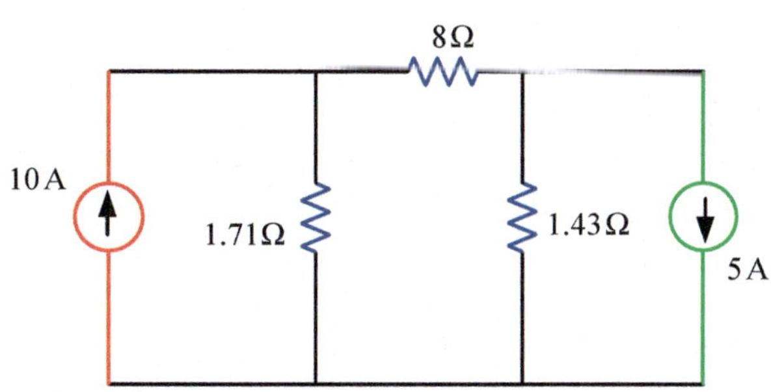

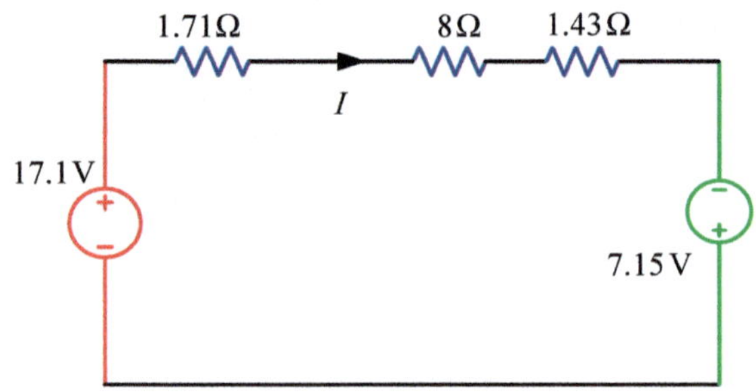

Fig. 2.48 Voltage sources with resistors

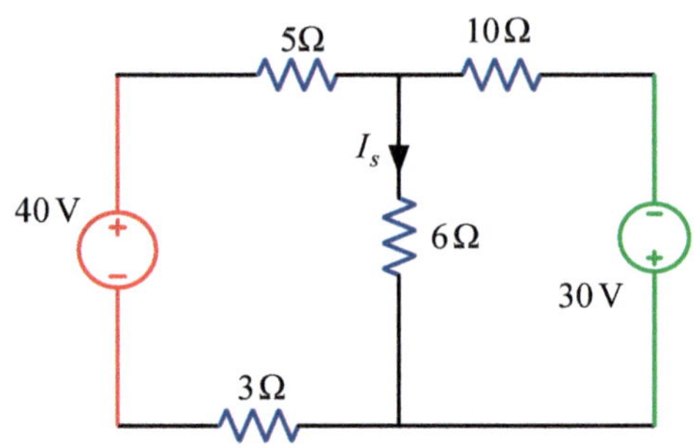

Fig. 2.49 Circuit for Practice Problem 2.15

Again, the current sources are converted to voltage sources as shown in Fig. 2.48, where the values of voltage sources are found to be,

$$V_{s1} = 10 \times 1.71 = 17.1\,\text{V} \tag{2.158}$$

$$V_{s2} = 5 \times 1.43 = 7.15\,\text{V} \tag{2.159}$$

Applying KVL to the circuit in Fig. 2.48 yields,

$$-17.1 + I(1.71 + 8 + 1.43) - 7.15 = 0 \qquad (2.160)$$

$$I = \frac{24.25}{1.71 + 8 + 1.43} = 2.18\,\text{A} \qquad (2.160)$$

Practice Problem 2.15

Figure 2.49 shows a series–parallel circuit with two voltage sources. Use source conversion technique to find the current in 6 Ω resistor.

Exercise Problems

2.1 A 750 W electric iron is connected to a 220 V voltage source. Determine the current drawn by the iron.

2.2 A 15 Ω resistor is connected to a 240 V voltage source. Calculate the power absorbed by the resistor.

2.3 A circuit with different branches along with currents is shown in Fig. 2.50. Determine the values of the unknown currents.

2.4 Figure 2.51 shows a circuit with some unknown currents. Determine the unknown currents.

2.5 Some known and unknown voltages of a circuit are shown in Fig. 2.52. Use KVL to determine the unknown voltages.

2.6 Figure 2.53 shows a circuit with known and unknown voltages. Use KVL to find the voltages V_1 and V_2.

2.7 A voltage-controlled voltage source is shown in Fig. 2.54. Use KVL to determine the voltage V_x and the voltage between the points (nodes) a and b.

Fig. 2.50 Circuit for Problem 2.3

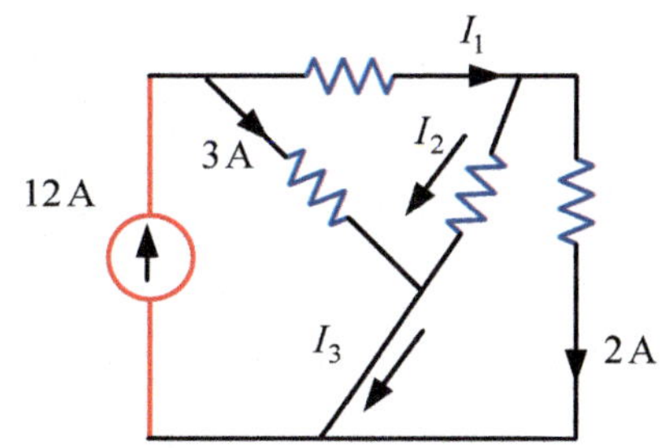

Fig. 2.51 Circuit for Problem 2.4

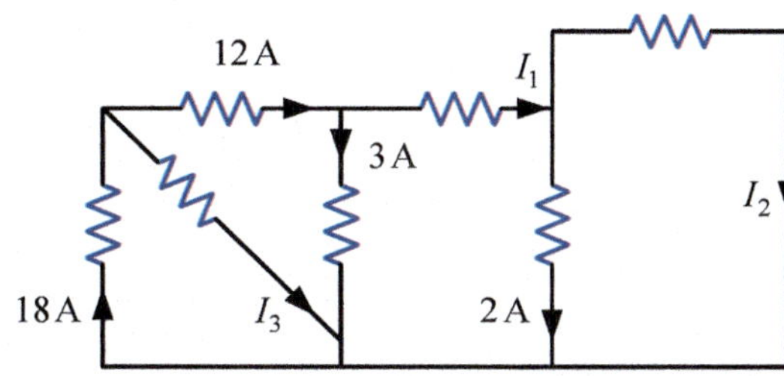

Fig. 2.52 Circuit for Problem 2.5

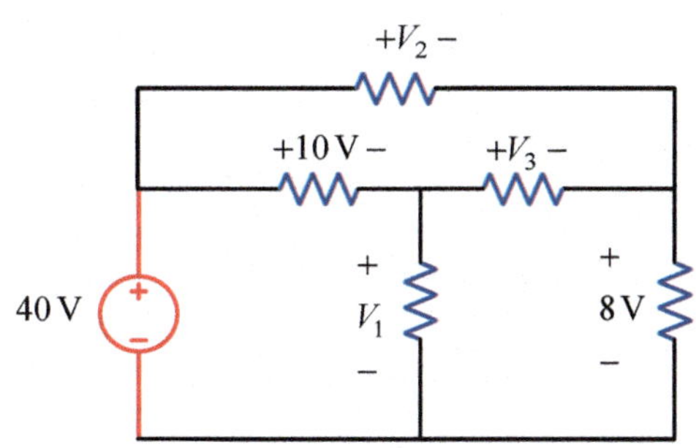

Fig. 2.53 Circuit for Problem 2.6

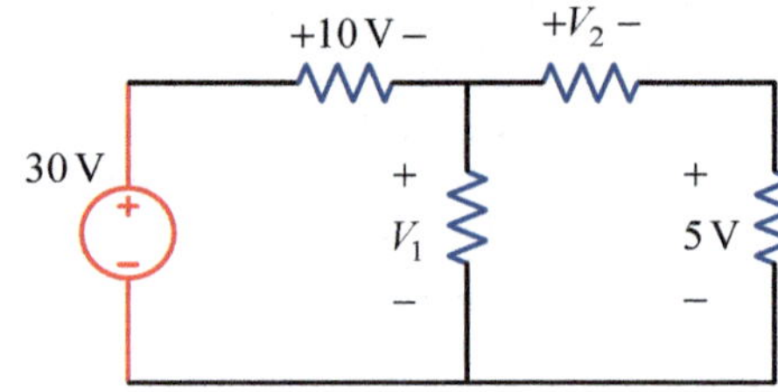

Fig. 2.54 Circuit for Problem 2.7

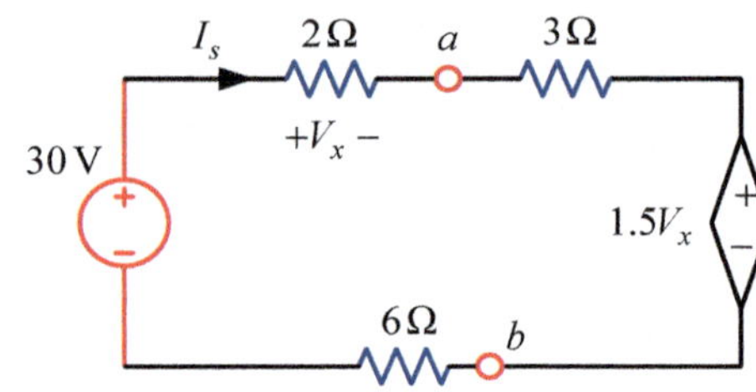

Fig. 2.55 Circuit for Problem 2.8

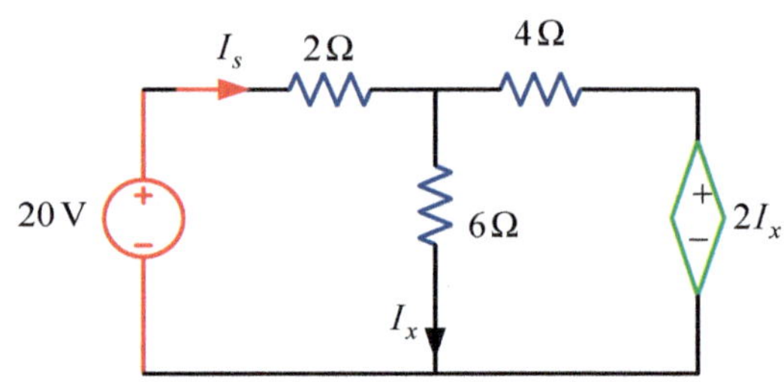

Fig. 2.56 Circuit for Problem 2.9

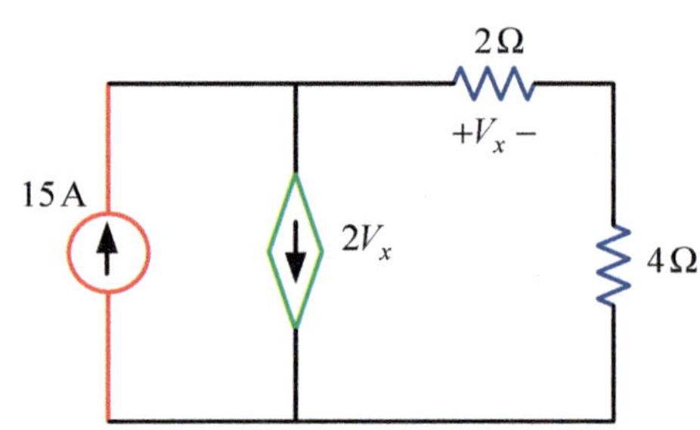

2.8 A current-controlled voltage source is shown in Fig. 2.55. Determine the currents I_s and I_x using KVL.

2.9 A voltage-controlled current source is shown in Fig. 2.56. Calculate the voltage drop across the 4 Ω resistor.

2.10 Figure 2.57 shows a current-controlled current source. Use KCL to determine the power absorbed by the 8 Ω resistor.

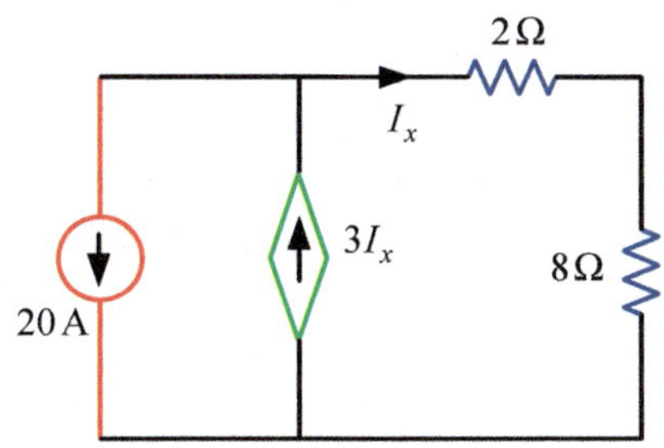

Fig. 2.57 Circuit for Problem 2.10

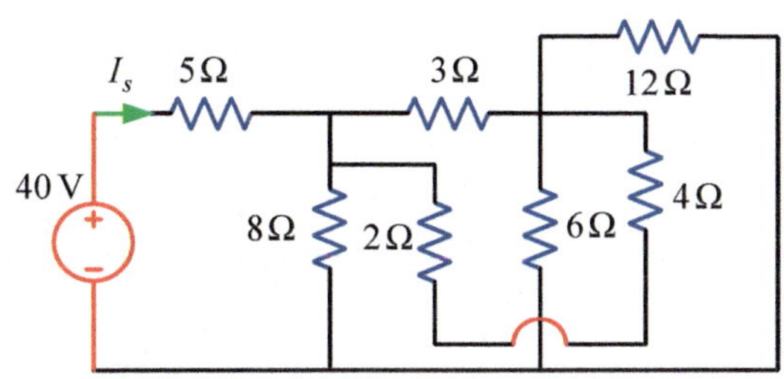

Fig. 2.58 Circuit for Problem 2.11

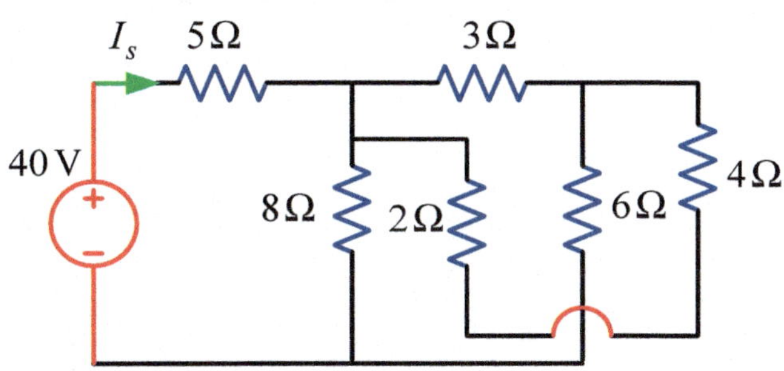

Fig. 2.59 Circuit for Problem 2.12

2.11 A series–parallel circuit with a voltage source is shown in Fig. 2.58. Calculate the total circuit resistance and the source current.

2.12 Figure 2.59 shows a series–parallel circuit with a voltage source. Calculate the total circuit resistance and the source current.

2.13 A series–parallel circuit with a voltage source is shown in Fig. 2.60. Determine the total circuit resistance and the source current.

2.14 A series–parallel circuit is shown in Fig. 2.61. Calculate the total circuit resistance and the power absorbed by the 4 Ω resistor.

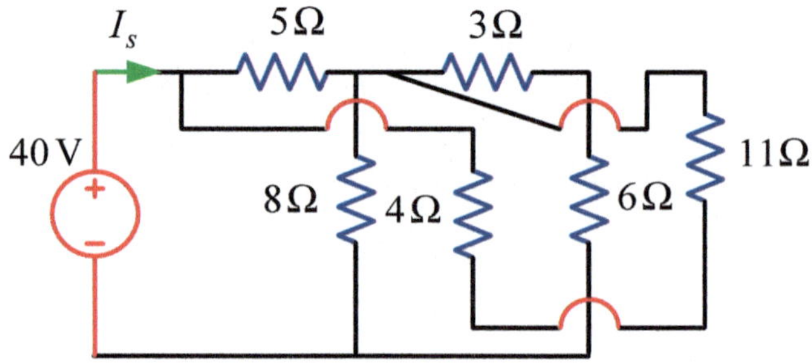

Fig. 2.60 Circuit for Problem 2.13

Fig. 2.61 Circuit for Problem 2.14

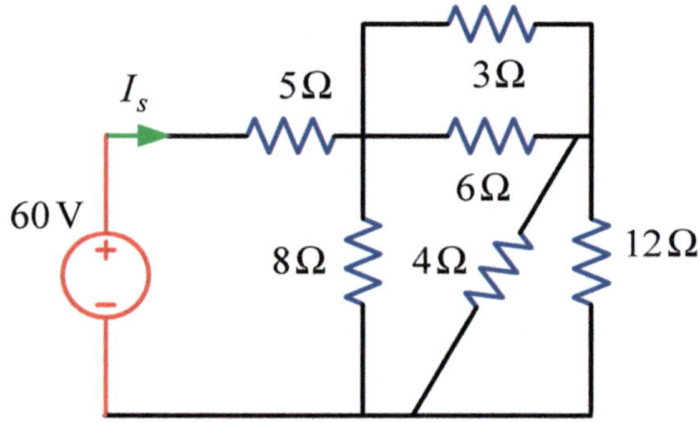

Fig. 2.62 Circuit for Problem 2.15

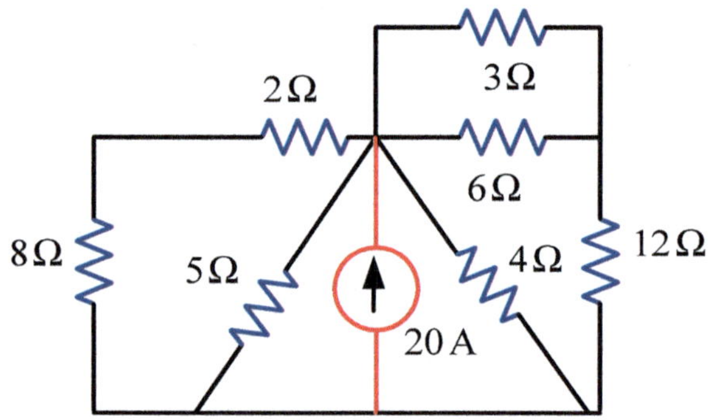

2.15 Find the current through the 4 Ω resistor of the circuit shown in Fig. 2.62.

2.16 A series–parallel circuit is shown in Fig. 2.63. Calculate the source current and the voltage drop across the 4 Ω resistor.

2.17 Figure 2.64 shows a series–parallel circuit. Determine the source current and the voltage drop across the 8 Ω resistor.

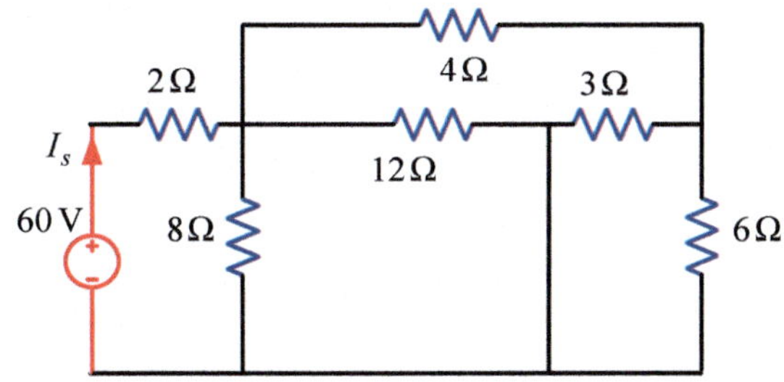

Fig. 2.63 Circuit for Problem 2.16

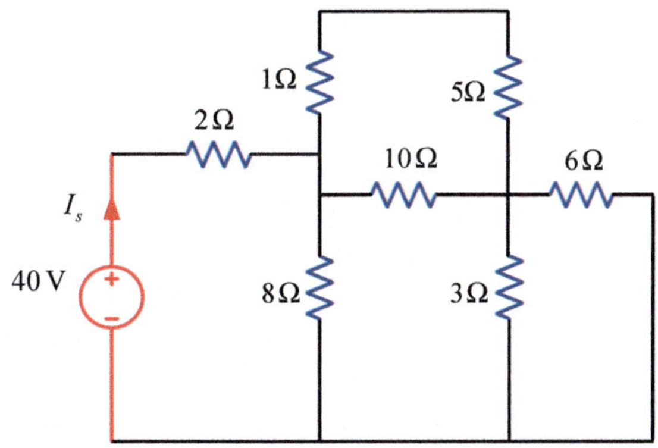

Fig. 2.64 Circuit for Problem 2.17

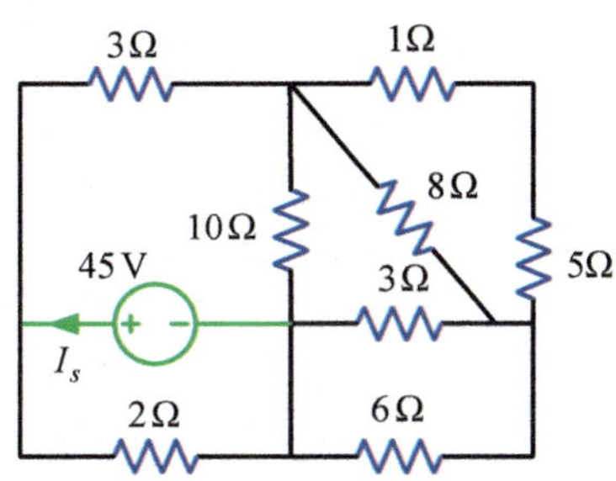

Fig. 2.65 Circuit for Problem 2.18

2.18 Figure 2.65 shows a series–parallel circuit. Find the total circuit resistance and the source current.

2.19 A series–parallel circuit is shown in Fig. 2.66. Calculate the total circuit resistance and the source current.

2.20 Figure 2.67 shows a series–parallel circuit. Calculate the total circuit resistance and the source current.

2.21 A series–parallel electrical circuit is shown in Fig. 2.68. Find the total circuit resistance and the source current.

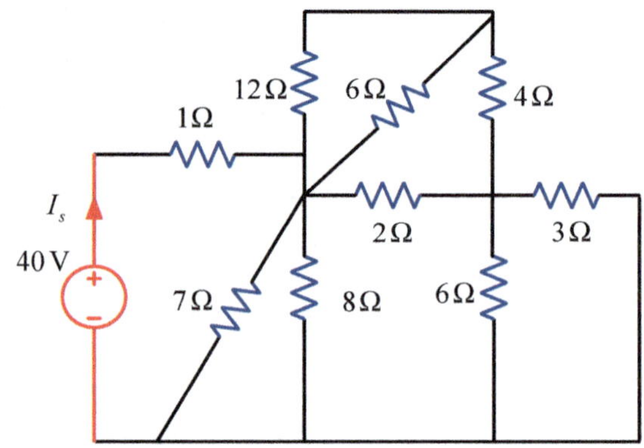

Fig. 2.66 Circuit for Problem 2.19

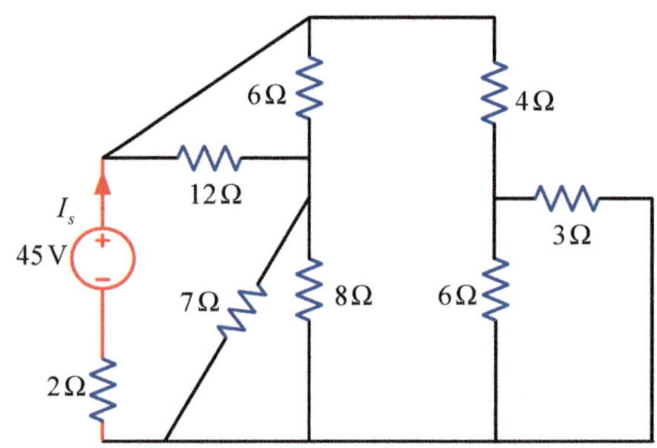

Fig. 2.67 Circuit for Problem 2.20

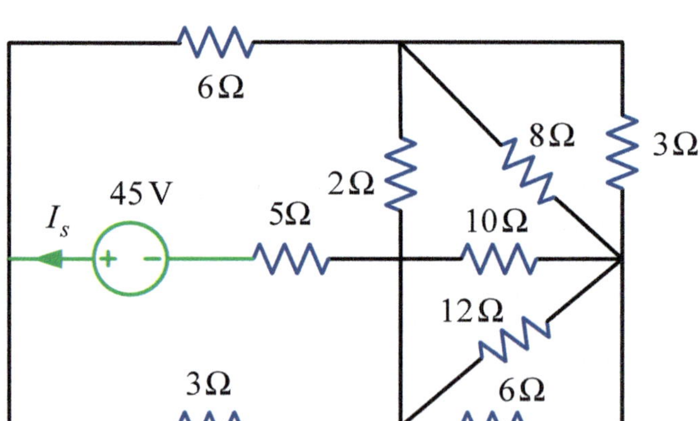

Fig. 2.68 Circuit for Problem 2.21

2.22 Figure 2.69 shows a series–parallel electrical circuit. Calculate the total circuit resistance and the source current.

2.23 A series–parallel circuit is shown in Fig. 2.70. Determine the total circuit resistance and the source current.

2.24 A series–parallel circuit is shown in Fig. 2.71. Calculate the total circuit resistance and the source current.

2.25 An electrical circuit is shown in Fig. 2.72. Determine the total circuit resistance and the power absorbed by the 2 Ω resistor.

2.26 Figure 2.73 shows an electrical circuit. Calculate the total circuit resistance and the current through the 5 Ω resistor.

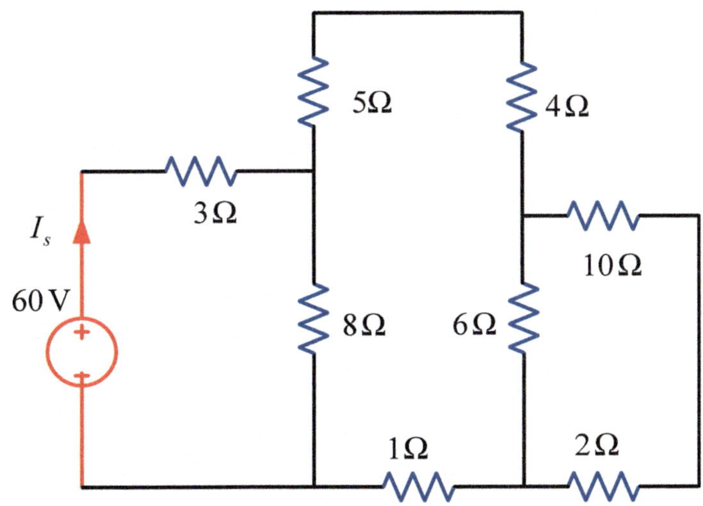

Fig. 2.69 Circuit for Problem 2.22

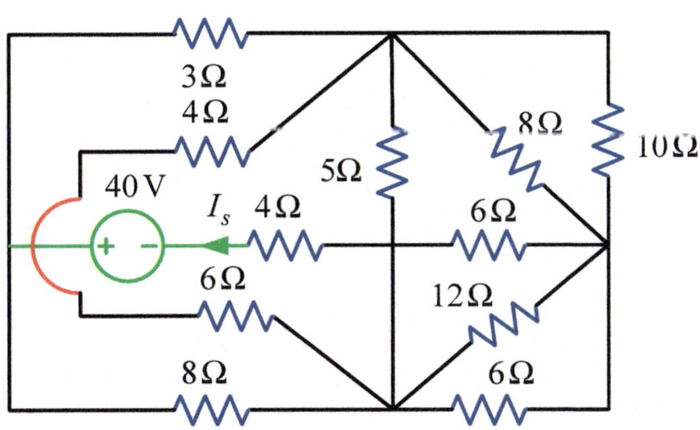

Fig. 2.70 Circuit for Problem 2.23

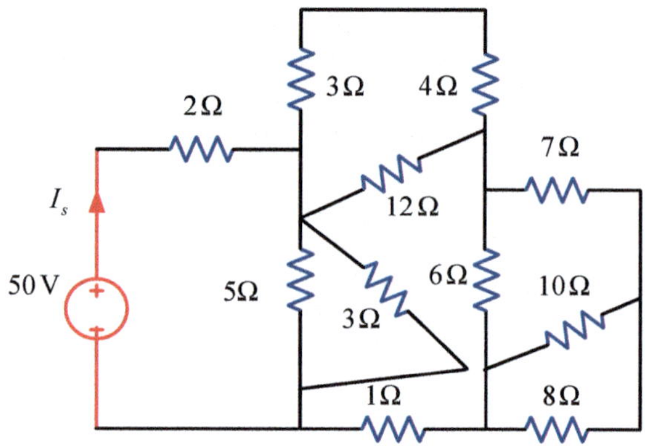

Fig. 2.71 Circuit for Problem 2.24

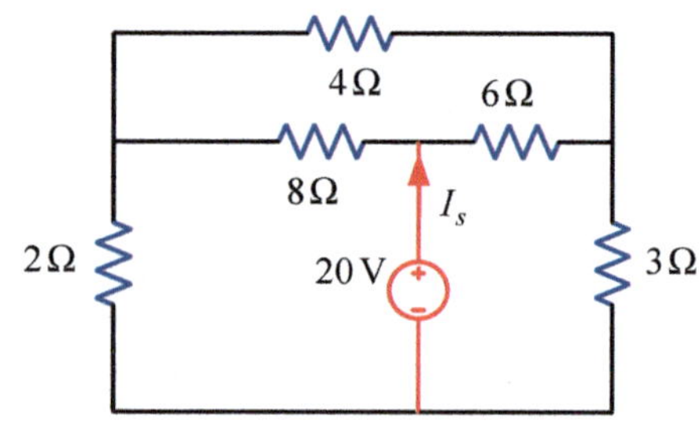

Fig. 2.72 Circuit for Problem 2.25

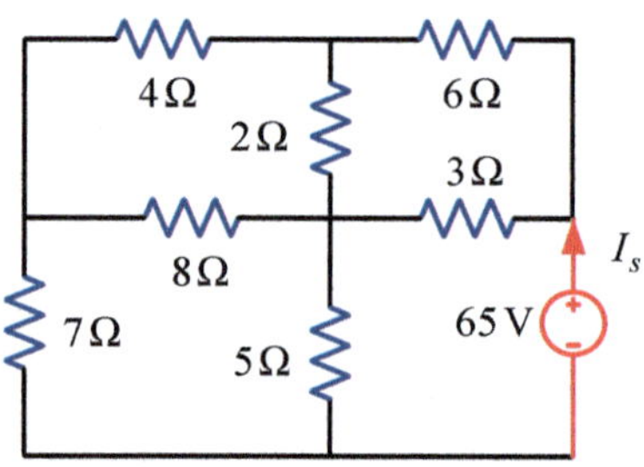

Fig. 2.73 Circuit for Problem 2.26

2.27 An electrical circuit is shown in Fig. 2.74. Calculate the total circuit resistance and the source current in the circuit.

2.28 A delta–wye-connected electrical circuit is shown in Fig. 2.75. Determine the total circuit resistance, source current and the voltage drop across the 3 Ω resistor.

2.29 An electrical circuit is shown in Fig. 2.76. Use delta–wye conversion to determine the total circuit resistance and the source current in the circuit.

2.30 A series–parallel electrical circuit is shown in Fig. 2.77. Calculate the power absorbed by the 4 Ω resistor.

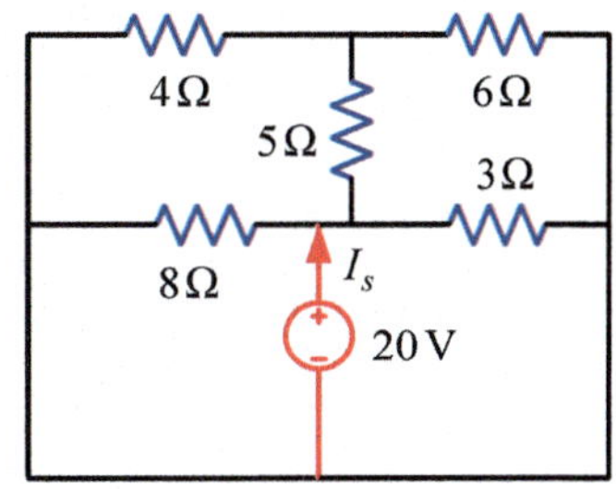

Fig. 2.74 Circuit for Problem 2.27

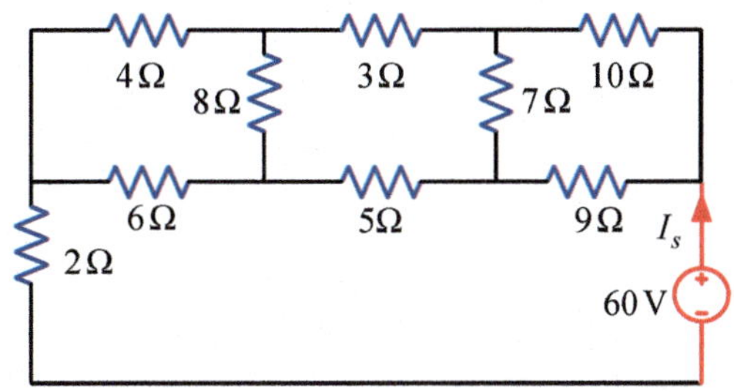

Fig. 2.75 Circuit for Problem 2.28

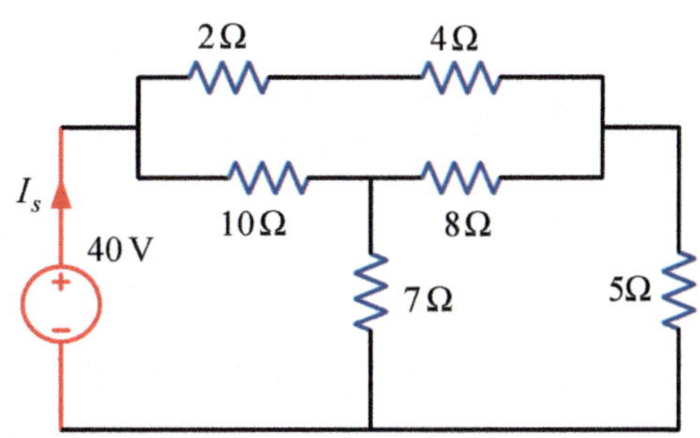

Fig. 2.76 Circuit for Problem 2.29

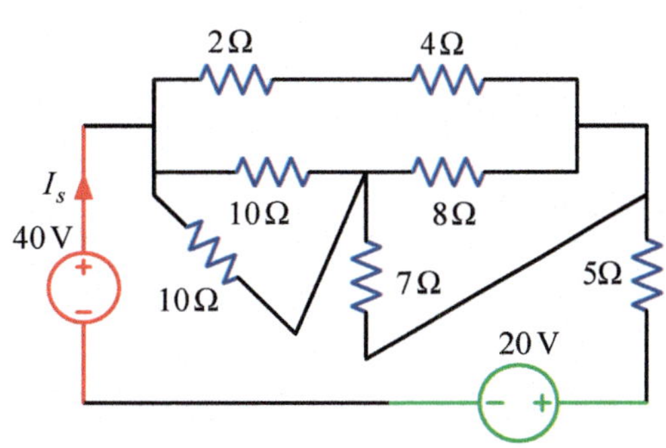

Fig. 2.77 Circuit for Problem 2.30

2.31 Figure 2.78 shows a series–parallel electrical circuit. Calculate the current in the 4 Ω resistor.

2.32 Figure 2.79 shows a series–parallel electrical circuit. Calculate the value of the voltage V_x.

2.33 A series–parallel electrical circuit with two current sources is shown in Fig. 2.80. Determine the current in the 5 Ω resistor.

2.34 A series–parallel electrical circuit is shown in Fig. 2.81. Calculate the voltage drop across the 4 Ω resistor.

2.35 Figure 2.82 shows a series–parallel electrical circuit. Determine the power absorbed by the 10 Ω resistor.

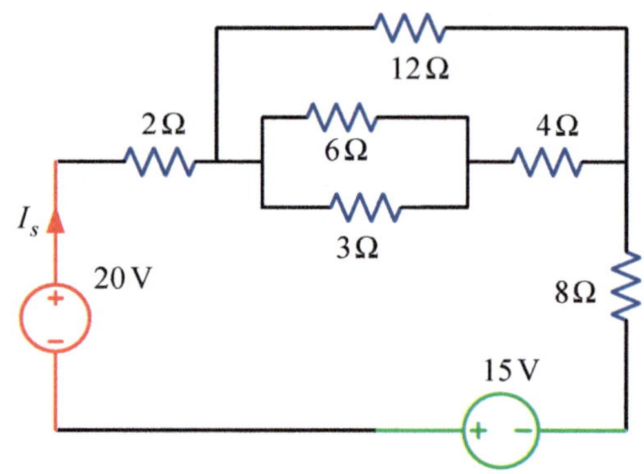

Fig. 2.78 Circuit for Problem 2.31

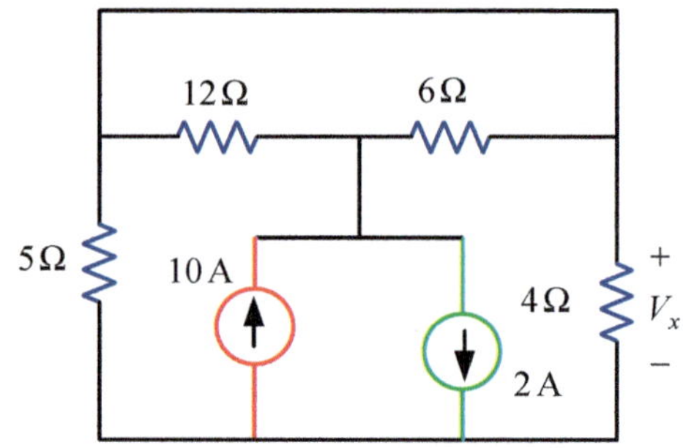

Fig. 2.79 Circuit for Problem 2.32

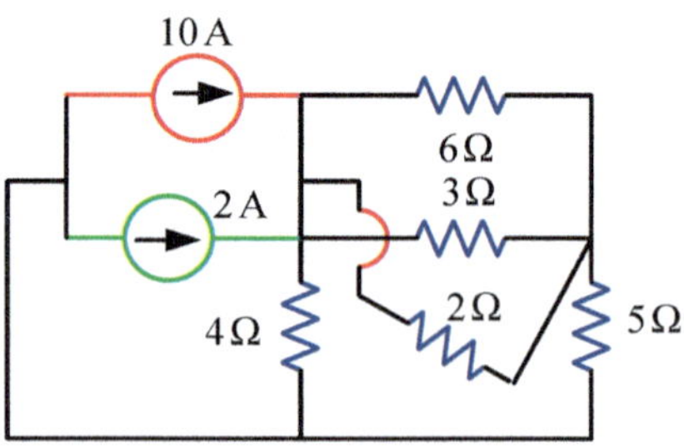

Fig. 2.80 Circuit for Problem 2.33

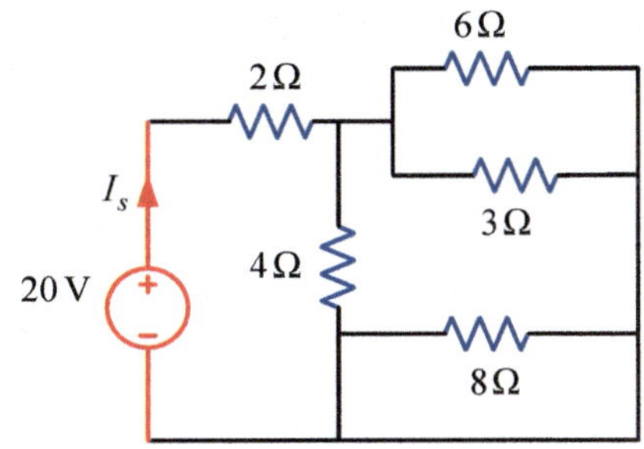

Fig. 2.81 Circuit for Problem 2.34

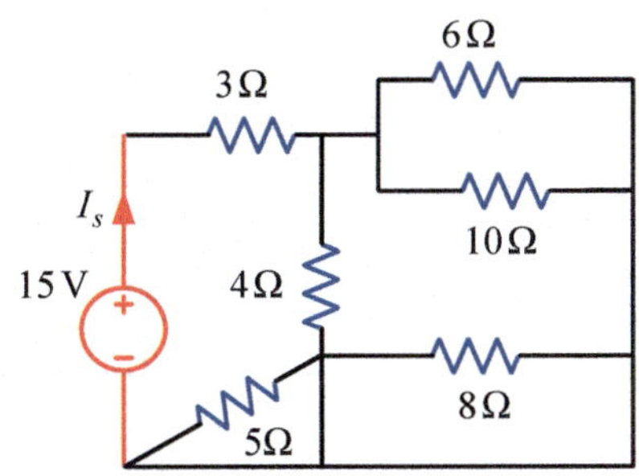

Fig. 2.82 Circuit for Problem 2.35

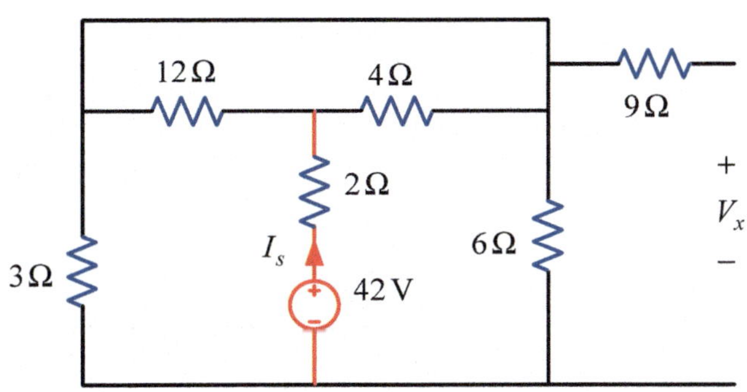

Fig. 2.83 Circuit for Problem 2.36

2.36 Figure 2.83 shows a series–parallel electrical circuit. Determine the value of the voltage, V_x.

2.37 Calculate the value of the voltage, V_0 of the circuit as shown in Fig. 2.84.

2.38 An electrical circuit is shown in Fig. 2.85. Calculate the voltage, V_0, of the circuit.

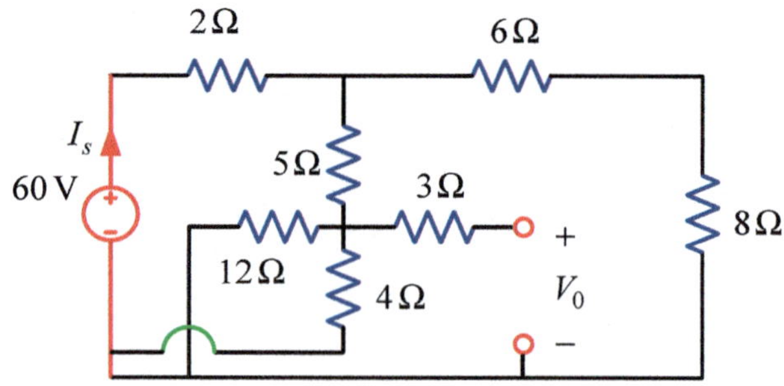

Fig. 2.84 Circuit for Problem 2.37

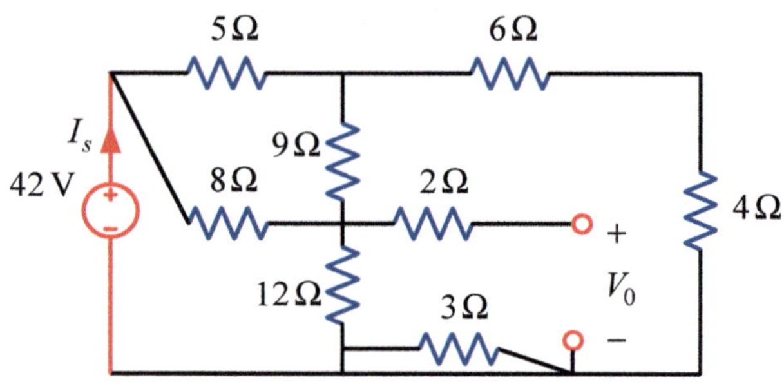

Fig. 2.85 Circuit for Problem 2.38

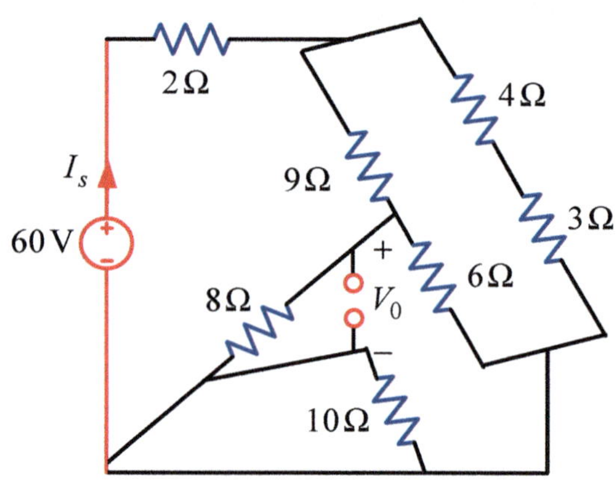

Fig. 2.86 Circuit for Problem 2.39

2.39 An electrical circuit is shown in Fig. 2.86. Determine the voltage, V_0, of the circuit.

2.40 Figure 2.87 shows an electrical circuit. Calculate the voltage, V_0, of the circuit.

2.41 An electrical circuit is shown in Fig. 2.88. Find the voltage, V_0, of the circuit.

2.42 An electrical circuit is shown in Fig. 2.89. Determine the voltage, V_0, of the circuit.

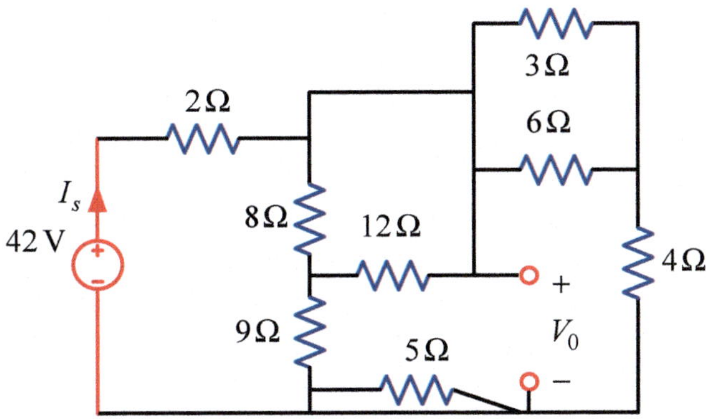

Fig. 2.87 Circuit for Problem 2.40

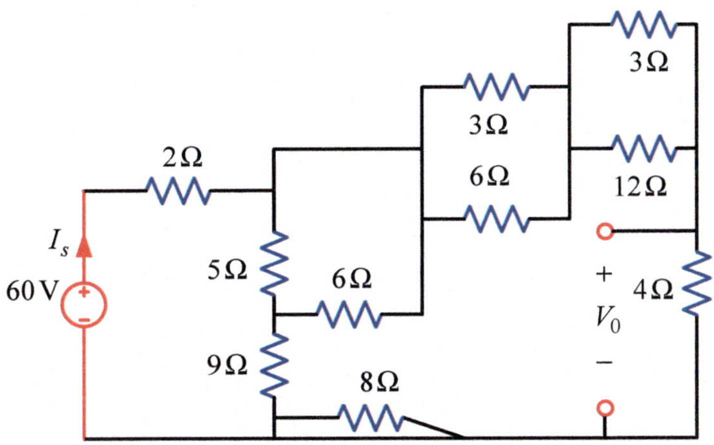

Fig. 2.88 Circuit for Problem 2.41

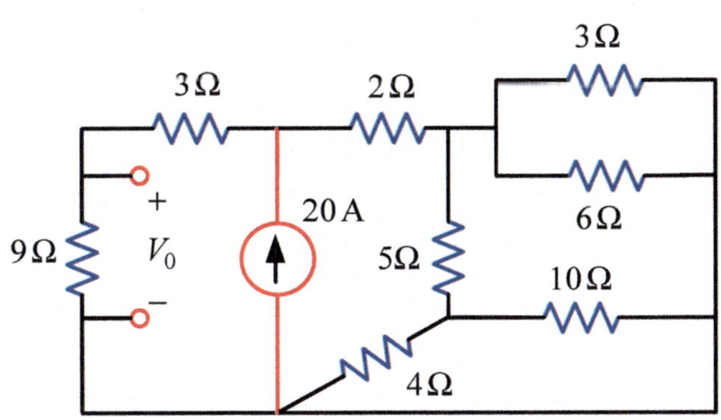

Fig. 2.89 Circuit for Problem 2.42

2.43 An electrical circuit is shown in Fig. 2.90. Calculate the voltage, V_0, of the circuit.

2.44 Figure 2.91 shows an electrical circuit. Determine the voltage, V_x, of the circuit.

2.45 An electrical circuit is shown in Fig. 2.92. Calculate the voltage, V_0, of the circuit.

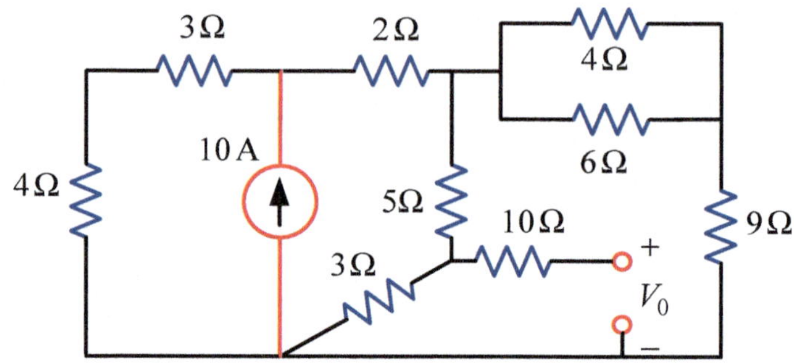

Fig. 2.90 Circuit for Problem 2.43

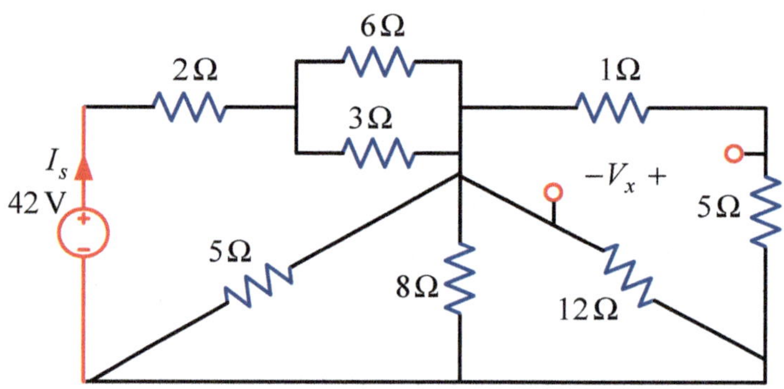

Fig. 2.91 Circuit for Problem 2.44

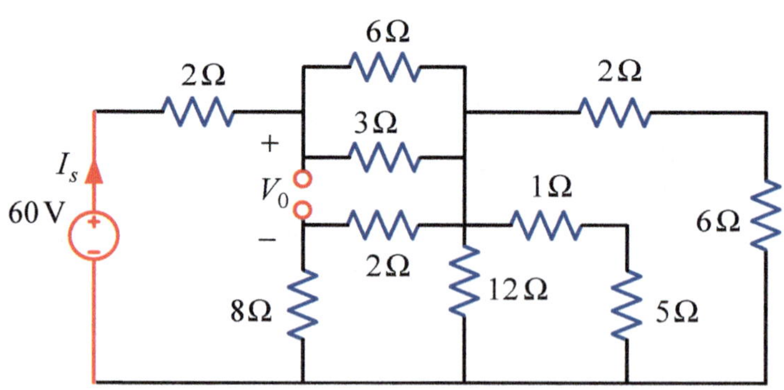

Fig. 2.92 Circuit for Problem 2.45

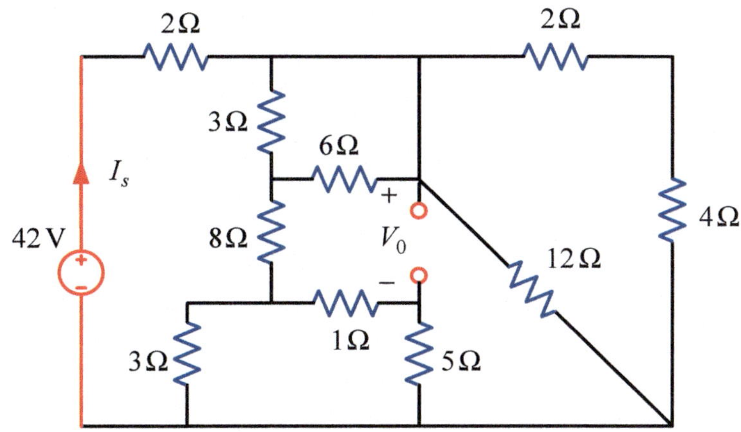

Fig. 2.93 Circuit for Problem 2.46

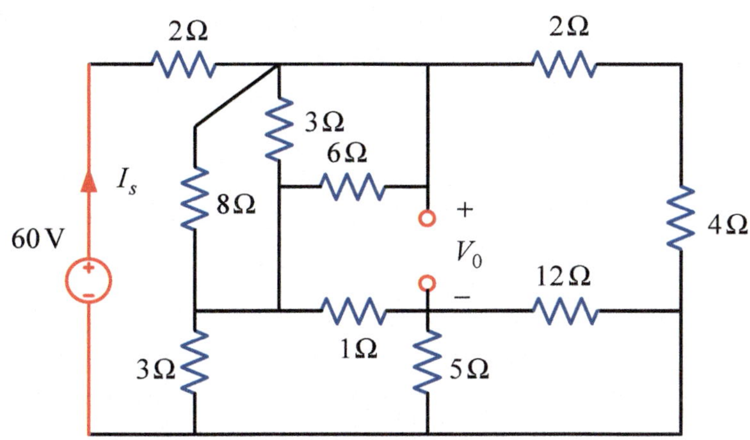

Fig. 2.94 Circuit for Problem 2.47

2.46 Figure 2.93 shows an electrical circuit. Find the voltage, V_0, of the circuit.

2.47 Figure 2.94 shows an electrical circuit. Calculate the voltage, V_0, of the circuit.

2.48 An electrical circuit is shown in Fig. 2.95. Use source conversion technique to calculate the voltage drop across the 5 Ω resistor.

2.49 Figure 2.96 shows an electrical circuit. Determine the voltage drop across the 2 Ω resistor by using source conversion technique.

2.50 Use source conversion technique to find the voltage drop across the 3 Ω resistor of the circuit in Fig. 2.97.

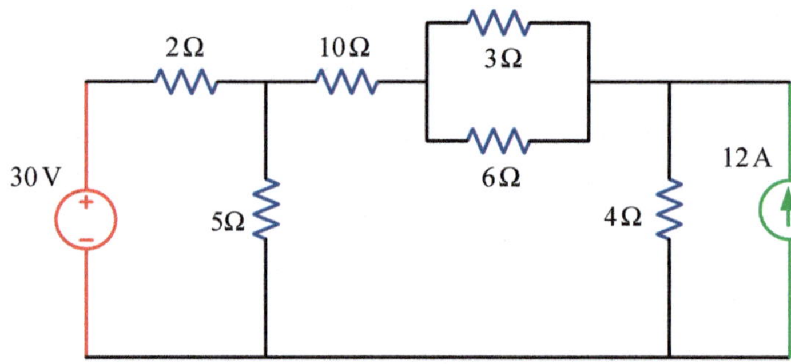

Fig. 2.95 Circuit for Problem 2.48

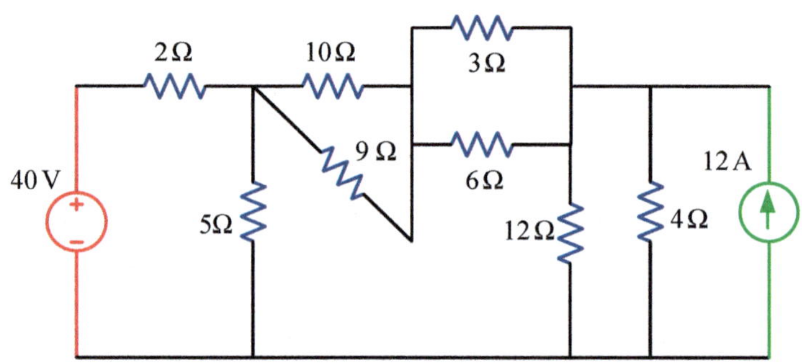

Fig. 2.96 Circuit for Problem 2.49

Fig. 2.97 Circuit for Problem 2.50

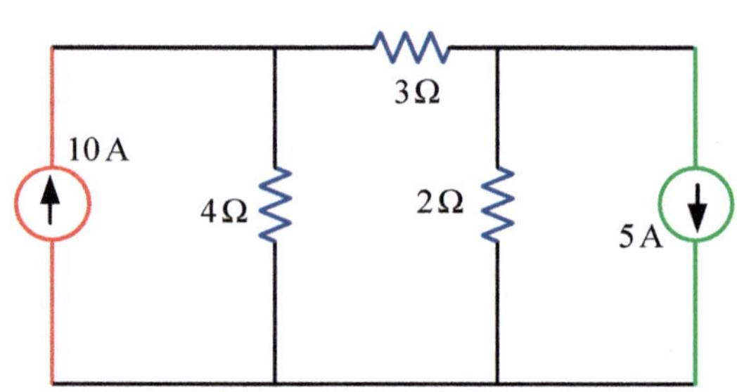

References

1. C.K. Alexander, M.N.O. Sadiku, *Fundamentals of Electric Circuits*, 6th Edn, (McGraw-Hill Higher Education, New York, January 2016)
2. R.L. Boylestad, *Introductory Circuit Analysis*, vol. 13 (Pearson, London, 2016)
3. J. David Irwin, R. Mark Nelms, *Basic Engineering Circuit Analysis*, 11th edn. (Wiley, USA, 2015)
4. J.W. Nilsson, S.A. Riedel, *Electric Circuits*, vol. 10 (Prentice Hall International Edition, New Jersey, 2015)
5. Md. Abdus Salam, *Basic Electrical Circuits*, 2nd edn. (Shroff Publishers & Distributors Pvt. Ltd, India, 2007)

Chapter 3
Different Methods for Circuit Analysis

3.1 Introduction

Fundamental electrical laws and the associated parameters have already been discussed in the previous chapters. Additional analytical methods have been developed to analyse practical electrical circuits with more than one source. These methods include mesh analysis and node voltage analysis, also known as nodal analysis. Kirchhoff's voltage law (KVL) is used in the mesh analysis while Kirchhoff's current law (KCL) is used in the nodal analysis. In these analytical methods, Cramer's rule appears to be a handy tool that helps to solve the system of linear equations after the application of Kirchhoff's laws in an electrical circuit. Upon introducing Cramer's rule, this chapter presents the theoretical foundations on supermesh, supernode, mesh and nodal analysis with dependent and independent sources.

3.2 Cramer's Rule

In 1750, Gabriel Cramer developed an algebraic rule for solving unknown parameters. According to his name, it is known as Cramer's rule. It is an efficient rule for a system of two or more equations. Consider the three linear equations as follows:

$$a_1x + b_1y + c_1z = d_1 \tag{3.1}$$

$$a_2x + b_2y + c_2z = d_2 \tag{3.2}$$

$$a_3x + b_3y + c_3z = d_3 \tag{3.3}$$

Md. A. Salam and Q. Mehbubar Rahman, *Fundamentals of Electrical Circuit Analysis*,
https://doi.org/10.1007/978-981-10-8624-3_3

The determinant of those equations is,

$$D = \begin{vmatrix} a_1 & b_1 & c_1 \\ a_2 & b_2 & c_2 \\ a_3 & b_3 & c_3 \end{vmatrix} \tag{3.4}$$

The values of the three unknown parameters x, y and z can be determined as,

$$x = \frac{D_x}{D} \tag{3.5}$$

$$y = \frac{D_y}{D} \tag{3.6}$$

$$z = \frac{D_z}{D} \tag{3.7}$$

The values of D_x, D_y and D_z are determined as,

$$D_x = \begin{vmatrix} d_1 & b_1 & c_1 \\ d_2 & b_2 & c_2 \\ d_3 & b_3 & c_3 \end{vmatrix} \tag{3.8}$$

$$D_y = \begin{vmatrix} a_1 & d_1 & c_1 \\ a_2 & d_2 & c_2 \\ a_3 & d_3 & c_3 \end{vmatrix} \tag{3.9}$$

$$D_z = \begin{vmatrix} a_1 & b_1 & d_1 \\ a_2 & b_2 & d_2 \\ a_3 & b_3 & d_3 \end{vmatrix} \tag{3.10}$$

Example 3.1 Three simultaneous equations of currents of a multisource electrical circuit are given by $2I_1 + 3I_2 + 6I_3 = 5$, $3I_1 + 5I_2 - 9I_3 = 8$, $4I_1 + 6I_2 + 8I_3 = 10$. Determine the unknown currents.

Solution:

The determinant of three simultaneous equations is,

$$\begin{aligned} D = \begin{vmatrix} 2 & 3 & 6 \\ 3 & 5 & -9 \\ 4 & 6 & 8 \end{vmatrix} &= 2(40 + 54) - 3(24 + 36) + 6(18 - 20) \\ &= -4 \end{aligned} \tag{3.11}$$

Other related parameters can be determined as,

$$D_1 = \begin{vmatrix} 5 & 3 & 6 \\ 8 & 5 & -9 \\ 10 & 6 & 8 \end{vmatrix} = 5(40+54) - 3(64+90) + 6(48-50) \\ = -4 \tag{3.12}$$

$$D_2 = \begin{vmatrix} 2 & 5 & 6 \\ 3 & 8 & -9 \\ 4 & 10 & 8 \end{vmatrix} = 2(64+90) - 5(24+36) + 6(30-32) \\ = -4 \tag{3.13}$$

$$D_3 = \begin{vmatrix} 2 & 3 & 5 \\ 3 & 5 & 8 \\ 4 & 6 & 10 \end{vmatrix} = 2(50-48) - 3(30-32) + 5(18-20) \\ = 0 \tag{3.14}$$

The values of the currents can be determined as,

$$I_1 = \frac{D_1}{D} = \frac{-4}{-4} = 1\,\mathrm{A} \tag{3.15}$$

$$I_2 = \frac{D_2}{D} = \frac{-4}{-4} = 1\,\mathrm{A} \tag{3.16}$$

$$I_3 = \frac{D_3}{D} = \frac{0}{-4} = 0\,\mathrm{A} \tag{3.17}$$

Practice Problem 3.1 Three simultaneous equations of currents of a multisource electrical circuit are given by $I_1 + 3I_2 - 4I_3 = 5$, $2I_1 + 3I_2 - 4I_3 = 5$, $3I_1 + 2I_2 + 5I_3 = 7$. Calculate the unknown currents.

3.3 Mesh Analysis with Independent Source

An electrical circuit with an independent voltage sources is shown in Fig. 3.1. In mesh analysis, the current sources are required to be converted to voltage sources with appropriate polarity as identified by the current directions of the current sources. Applying KVL to the circuit in Fig. 3.1 yields,

$$-V_1 + I_1R_1 + R_3(I_1 - I_2) + I_1R_2 = 0 \tag{3.18}$$

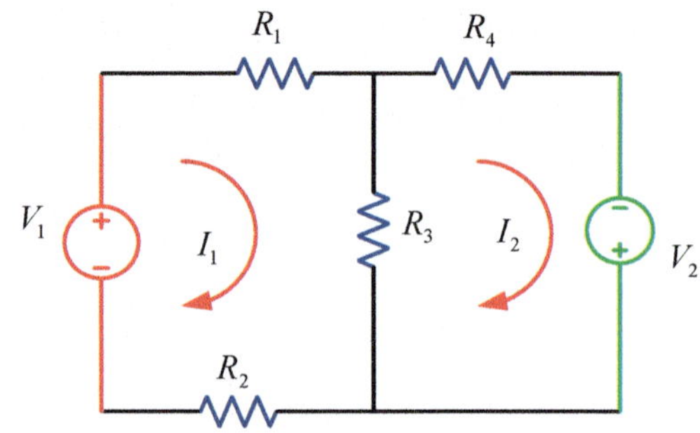

Fig. 3.1 Circuit with two voltage sources

$$I_1(R_1+R_3+R_2)-I_2R_3=V_1 \tag{3.19}$$

$$-V_2+R_3(I_2-I_1)+I_2R_4=0 \tag{3.20}$$

$$-I_1R_3+(R_3+R_4)I_2=V_2 \tag{3.21}$$

Equation (3.21) can be written as,

$$I_1=\frac{R_3+R_4}{R_3}I_2-\frac{V_2}{R_3} \tag{3.22}$$

Substituting Eq. (3.22) into Eq. (3.19) yields,

$$\left(\frac{R_3+R_4}{R_3}I_2-\frac{V_2}{R_3}\right)(R_1+R_3+R_2)-I_2R_3=V_1 \tag{3.23}$$

Equations (3.22) and (3.23) can be used to determine the mesh currents.

Example 3.2 Determine the power absorbed by the 6 Ω resistor using mesh analysis for the circuit in Fig. 3.2.

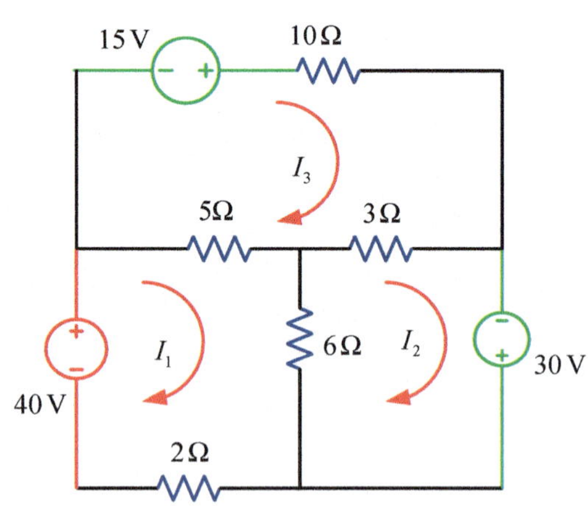

Fig. 3.2 Circuit for Example 3.2

Solution:
Applying KVL to the meshes 1, 2 and 3 yields,

$$13I_1 - 6I_2 - 5I_3 = 40 \tag{3.24}$$

$$-6I_1 + 9I_2 - 3I_3 = 30 \tag{3.25}$$

$$-2I_1 + 3I_2 - I_3 = 10 \tag{3.26}$$

$$-5I_1 - 3I_2 + 18I_3 = 15 \tag{3.27}$$

Determinant of the three simultaneous equations is calculated as,

$$D = \begin{vmatrix} 13 & -6 & -5 \\ -2 & 3 & -1 \\ -5 & -3 & 18 \end{vmatrix} = 13(54-3) + 6(-36-5) - 5(6+15) \\ = 312 \tag{3.28}$$

Other related parameters can be determined as,

$$D_1 = \begin{vmatrix} 40 & -6 & -5 \\ 10 & 3 & -1 \\ 15 & -3 & 18 \end{vmatrix} = 40(54-3) + 6(180+15) - 5(-30-45) \\ = 3585 \tag{3.29}$$

$$D_2 = \begin{vmatrix} 13 & 40 & -5 \\ -2 & 10 & -1 \\ -5 & 15 & 18 \end{vmatrix} = 13(180+15) - 40(-36-5) - 5(-30+50) \\ = 4075 \tag{3.30}$$

The values of the currents are calculated as,

$$I_1 = \frac{D_1}{D} = \frac{3585}{312} = 11.49 \text{ A} \tag{3.31}$$

$$I_2 = \frac{D_2}{D} = \frac{4075}{312} = 13.06 \text{ A} \tag{3.32}$$

The power absorbed by the 6 Ω resistor is,

$$P_{6\Omega} = (I_1 - I_2)^2 \times 6 = (11.49 - 13.06)^2 \times 6 = 14.79 \text{ W} \tag{3.33}$$

Practice Problem 3.2 Calculate the voltage drop across the 4 Ω resistor using mesh analysis for the circuit in Fig. 3.3.

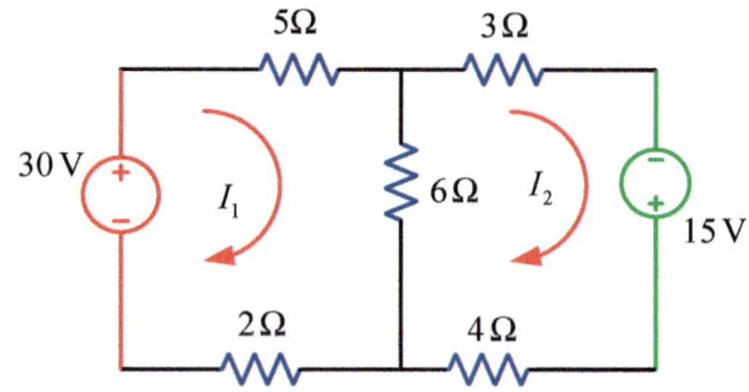

Fig. 3.3 Circuit for Practice Problem 3.2

3.4 Mesh Analysis with Dependent Source

An electrical circuit with a dependent source is shown in Fig. 3.4. Here, voltage-controlled voltage source is considered for mesh analysis. Based on the assigned directions of mesh currents as shown in Fig. 3.4, the application of KVL yields,

$$(R_1 + R_2 + R_4)I_1 - R_2I_2 = V_1 \tag{3.34}$$

$$-R_2I_1 + (R_2 + R_3)I_2 = mV_x \tag{3.35}$$

According to the circuit configuration,

$$V_x - I_1R_1 = 0 \tag{3.36}$$

$$V_x = I_1R_1 \tag{3.37}$$

Substituting Eq. (3.37) into Eq. (3.35) yields,

$$-R_2I_1 + (R_2 + R_3)I_2 = I_1R_1 \tag{3.38}$$

$$-(R_1 + R_2)I_1 + (R_2 + R_3)I_2 = 0 \tag{3.39}$$

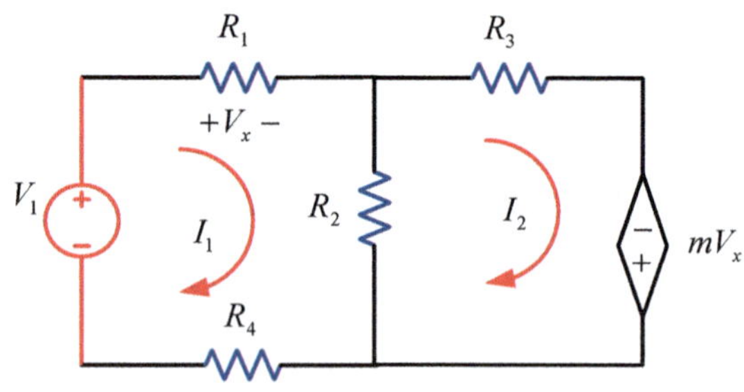

Fig. 3.4 Circuit mesh analysis with a dependent source

The determinant can be determined as,

$$D = \begin{vmatrix} R_1 + R_2 + R_4 & -R_2 \\ -(R_1 + R_2) & R_2 + R_3 \end{vmatrix} = (R_1 + R_2 + R_4)(R_2 + R_3) - (R_1 + R_2)R_2 \quad (3.40)$$

According to Cramer's rule, the other parameters can be written as,

$$D_1 = \begin{vmatrix} V_1 & -R_2 \\ 0 & R_2 + R_3 \end{vmatrix} = V_1(R_2 + R_3) \quad (3.41)$$

$$D_2 = \begin{vmatrix} R_1 + R_2 + R_4 & V_1 \\ -(R_1 + R_2) & 0 \end{vmatrix} = (R_1 + R_2)V_1 \quad (3.42)$$

Then mesh currents can be determined as,

$$I_1 = \frac{D_1}{D} = \frac{V_1(R_2 + R_3)}{(R_1 + R_2 + R_4)(R_2 + R_3) - (R_1 + R_2)R_2} \quad (3.43)$$

$$I_2 = \frac{D_2}{D} = \frac{V_1(R_1 + R_2)}{(R_1 + R_2 + R_4)(R_2 + R_3) - (R_1 + R_2)R_2} \quad (3.44)$$

Example 3.3 Determine the power absorbed by the 6 Ω resistor and voltage drop across the 4 Ω resistor using mesh analysis for the circuit in Fig. 3.5. Apply PSpice simulation to find the mesh currents.

Solution:
Applying KVL to the circuit in Fig. 3.5 yields,

$$11I_1 - 4I_2 = 15 \quad (3.45)$$

$$-4I_1 + 13I_2 = 2V_x \quad (3.46)$$

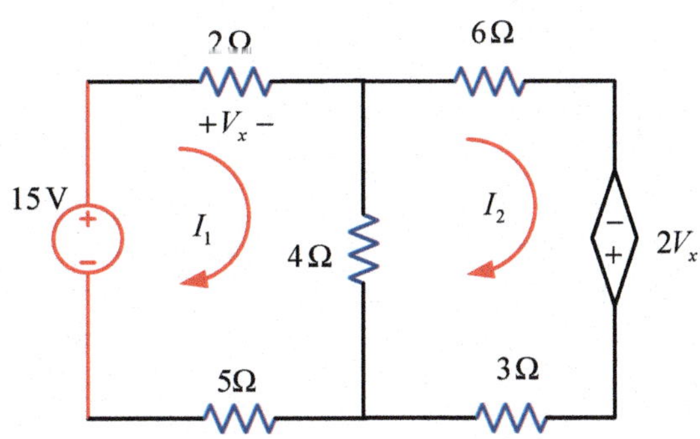

Fig. 3.5 Circuit for Example 3.3

According to the given circuit,

$$V_x = 2I_1 \tag{3.47}$$

Substituting Eq. (3.47) into Eq. (3.47) yields,

$$-4I_1 + 13I_2 = 4I_1 \tag{3.48}$$

$$I_1 = \frac{13I_2}{8} = 1.63I_2 \tag{3.49}$$

Substituting Eq. (3.49) into Eq. (3.45) yields,

$$11(1.63I_2) - 4I_2 = 15 \tag{3.50}$$

$$I_2 = \frac{15}{13.93} = 1.08 \text{ A} \tag{3.51}$$

Substituting Eq. (3.51) into Eq. (3.49) yields,

$$I_1 = 1.63 \times 1.08 = 1.76 \text{ A} \tag{3.52}$$

The power absorbed by the 6 Ω resistor is,

$$P_{6\,\Omega} = I_2^2 \times 6 = 1.08^2 \times 6 = 7 \text{ W} \tag{3.53}$$

The voltage drop across the 4 Ω resistor is,

$$V_{4\,\Omega} = 4 \times (I_1 - I_2) = 4 \times (1.76 - 1.08) = 2.72 \text{ V} \tag{3.54}$$

The PSpice simulation circuit is shown in Fig. 3.6. The simulation mesh currents are the same as the calculated currents.

Practice Problem 3.3 Calculate the mesh currents in the circuit shown in Fig. 3.7. Verify the results by the PSpice simulation.

3.5 Supermesh Circuit

In an electric circuit, when a current source is contained in a branch, which is common between two meshes, these two meshes are jointly called a supermesh. The supermesh is treated in a way as if there is no current source. In a supermesh circuit, both KCL and KVL are applied to get the simultaneous equations. Figure 3.8 shows an electrical circuit, which contains a supermesh.

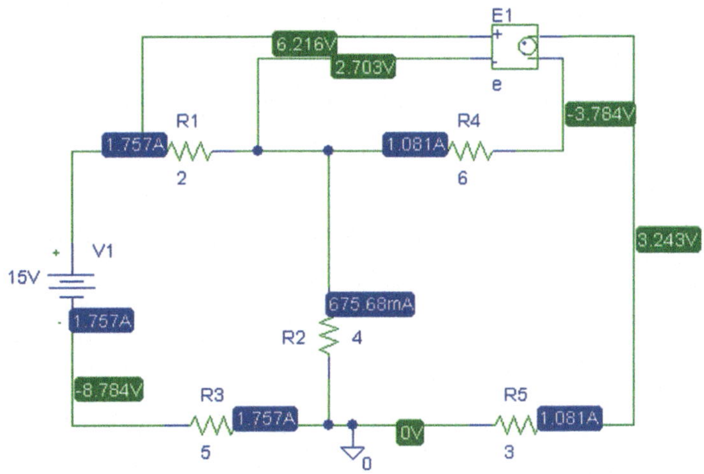

Fig. 3.6 PSpice simulation circuit for Example 3.3

Fig. 3.7 Circuit for Practice Problem 3.3

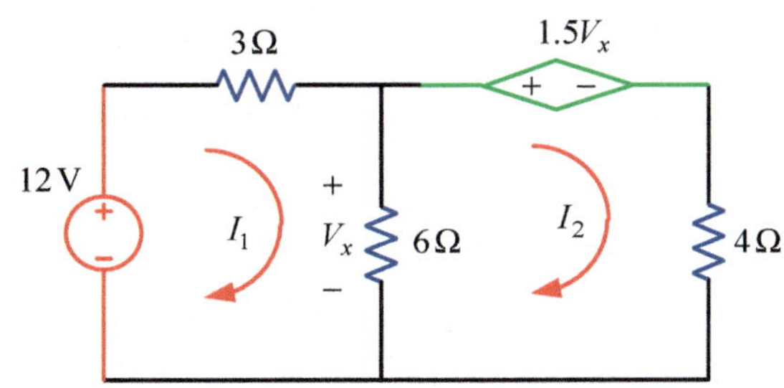

Fig. 3.8 Circuit with a common current source

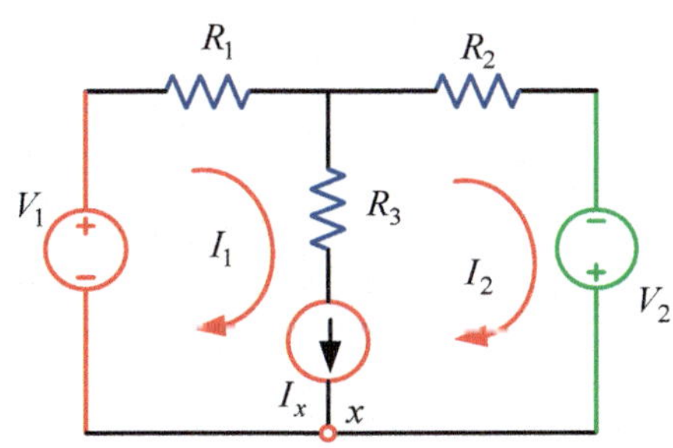

Applying KCL at the node x yields,

$$I_1 = I_x + I_2 \tag{3.55}$$

$$I_x = I_1 - I_2 \tag{3.56}$$

Now in the supermesh, after removing the current source as shown in Fig. 3.9, the applying of KVL yields,

$$-V_1 + I_1R_1 + I_2R_2 - V_2 = 0 \tag{3.57}$$

$$I_1R_1 + I_2R_2 = V_1 + V_2 \tag{3.58}$$

The mesh currents can be calculated from Eqs. (3.56) and (3.58), if the other electrical quantities are known.

Example 3.4 Find the power absorbed by the 6 Ω resistor and voltage drop across the 1 Ω resistor using mesh analysis for the circuit in Fig. 3.10. Apply PSpice simulation to find the mesh currents.

Solution:
Applying KCL at the node x yields,

$$I_1 = 4 + I_2 \tag{3.59}$$

The supermesh circuit is shown in Fig. 3.11. Applying KVL to the circuit in Fig. 3.11 yields,

$$-12 + 3I_1 + 10I_2 - 16 = 0 \tag{3.60}$$

$$3I_1 + 10I_2 = 28 \tag{3.61}$$

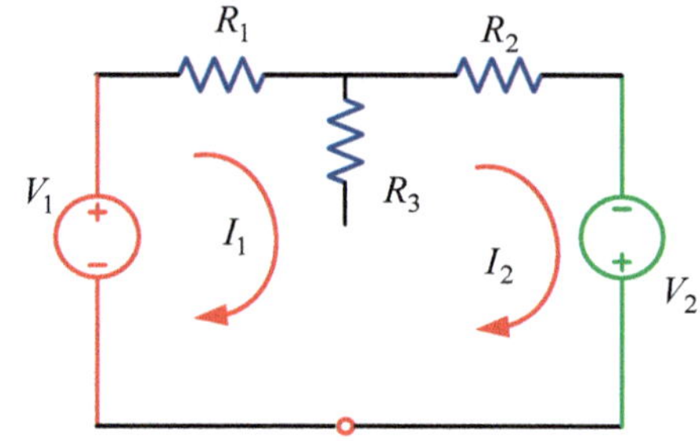

Fig. 3.9 A supermesh circuit

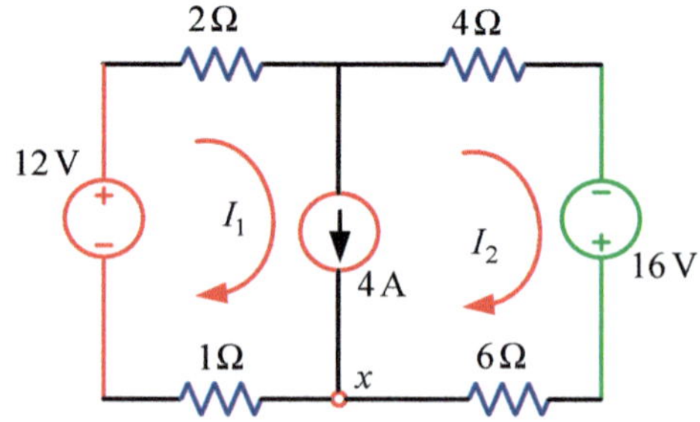

Fig. 3.10 A circuit for Example 3.4

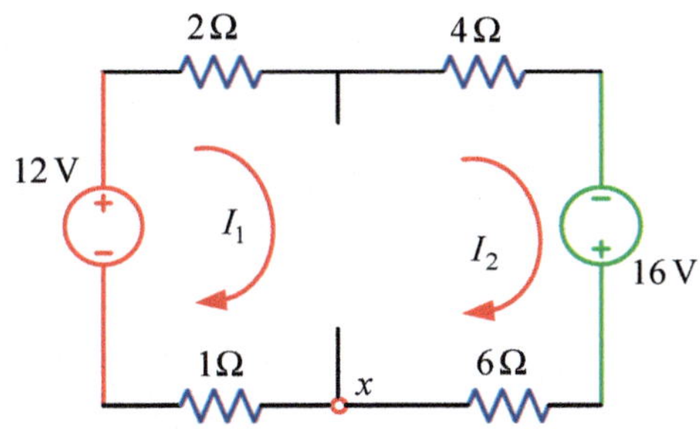

Fig. 3.11 Supermesh circuit for Example 3.4

Substituting Eq. (3.59) into Eq. (3.61) yields,

$$3(4+I_2)+10I_2=28 \tag{3.61}$$

$$13I_2=16 \tag{3.62}$$

$$I_2=\frac{16}{13}=1.23\text{ A} \tag{3.63}$$

Substituting Eq. (3.63) into Eq. (3.59) yields,

$$I_1=4+1.23=5.23\text{ A} \tag{3.64}$$

The power absorbed by the 6 Ω resistor is,

$$P_{6\,\Omega}=1.23^2\times 6=9.08\text{ W} \tag{3.65}$$

The voltage drop across the 1 Ω resistor is,

$$V_{1\,\Omega}=5.23\times 1=5.23\text{ V} \tag{3.66}$$

The PSpice simulation circuit is shown in Fig. 3.12. The values of the calculated mesh currents are the same as the simulation mesh currents.

Example 3.5 Determine the voltage drop across the 2 Ω resistor using mesh analysis for the circuit in Fig. 3.13. Apply PSpice simulation to find the mesh currents.

Solution:
Applying KCL at the node x yields,

$$I_1=4+I_2 \tag{3.67}$$

The circuit for the supermesh is shown in Fig. 3.14. Applying KVL to the circuit in Fig. 3.14 yields,

$$-16 + 6I_1 + 10I_2 + 1.5V_x = 0 \tag{3.68}$$

According to the given circuit,

$$V_x = 4I_1 \tag{3.69}$$

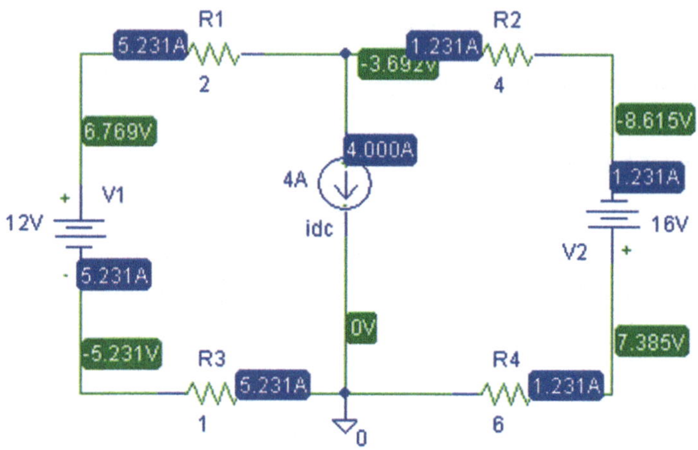

Fig. 3.12 Simulation circuit for Example 3.4

Fig. 3.13 A circuit for Example 3.5

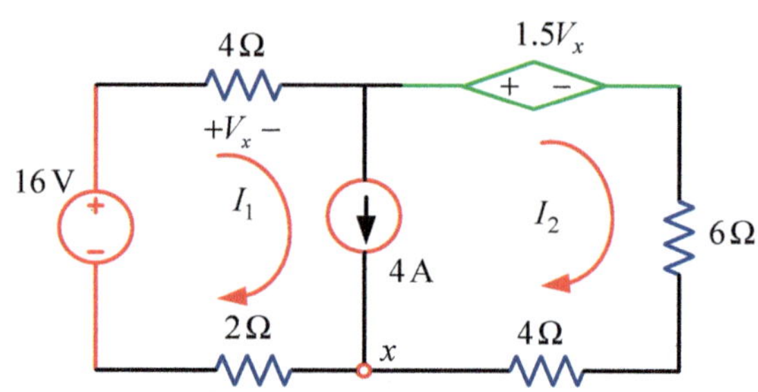

Fig. 3.14 A supermesh circuit for Example 3.5

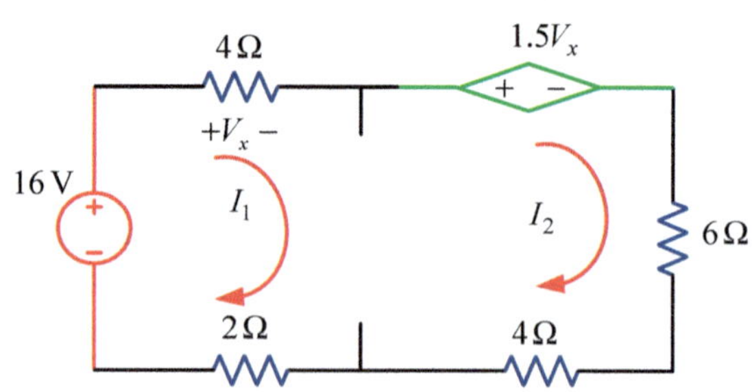

Substituting Eqs. (3.67) and (3.69) into Eq. (3.68) yields,

$$-16 + 6I_1 + 10I_2 + 1.5 \times 4I_1 = 0 \tag{3.70}$$

$$12(4 + I_2) + 10I_2 = 16 \tag{3.71}$$

$$I_2 = \frac{-32}{22} = -1.45 \text{ A} \tag{3.72}$$

Substituting Eq. (3.72) into Eq. (3.67) yields,

$$I_1 = 4 - 1.45 = 2.55 \text{ A} \tag{3.73}$$

The voltage drop across the 2 Ω resistor is,

$$V_{2\Omega} = 2 \times 2.3 = 4.6 \text{ V} \tag{3.74}$$

The simulation circuit is shown in Fig. 3.15. The values of the simulation mesh currents are the same as the calculated mesh currents.

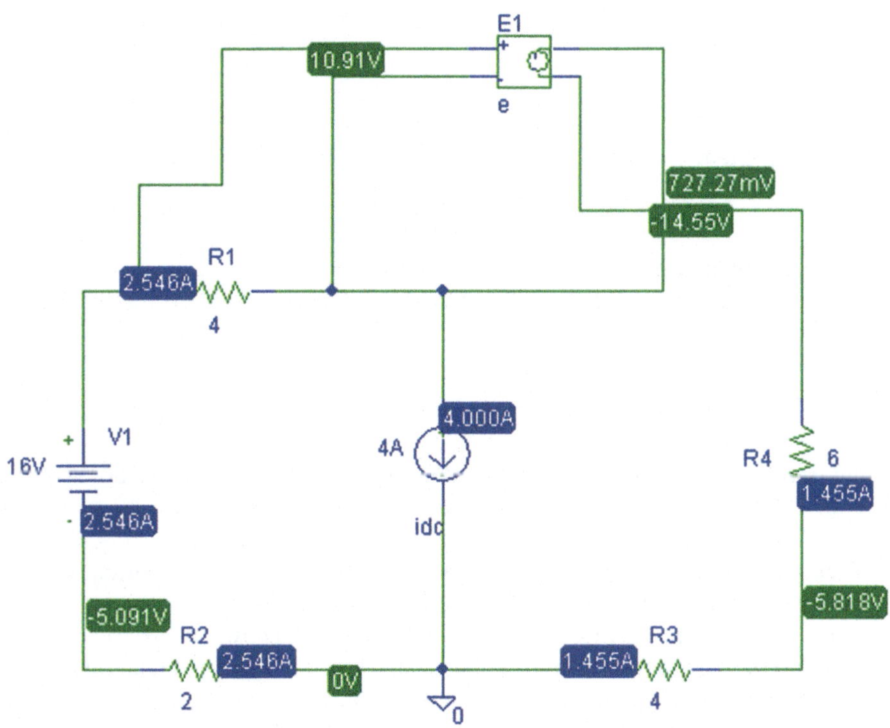

Fig. 3.15 Simulation circuit for Example 3.5

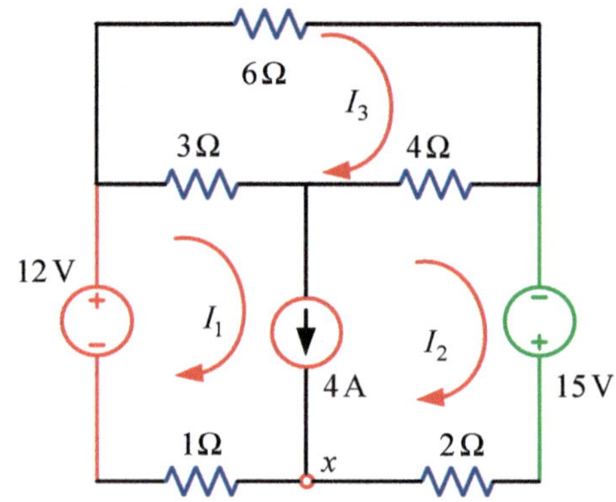

Fig. 3.16 A circuit for Practice Problem 3.4

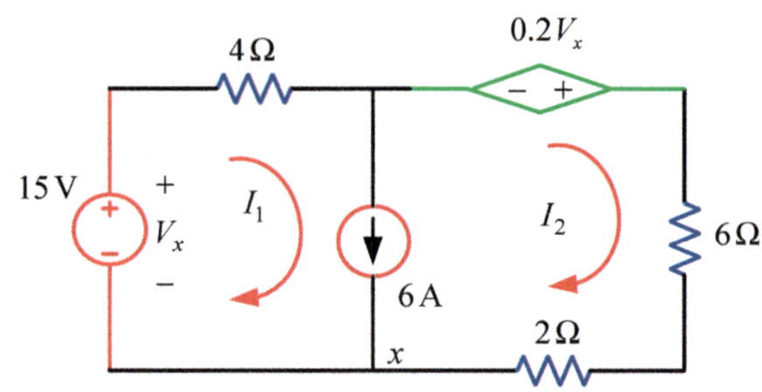

Fig. 3.17 A circuit for Practice Problem 3.5

Practice Problem 3.4 Calculate the power absorbed by the 3 Ω resistor using mesh analysis of the circuit in Fig. 3.16. Use PSpice simulation to find the mesh currents.

Practice Problem 3.5 Find the power absorbed by the 6 Ω resistor using mesh analysis for the circuit in Fig. 3.17. Use PSpice simulation to find the mesh currents.

3.6 Nodal Analysis

Nodal analysis is another method to calculate the electrical parameters such as current, voltage and power of a circuit that contains more than one source. In this method, reference and non-reference nodes need to be identified.

Then, if possible, any voltage source needs to be converted to a current source. Finally, KCL is applied in each node to get the simultaneous equations, which are solved to calculate the electrical parameters. Figure 3.18 shows an electrical circuit with voltage and current sources. A voltage source is converted to a current source. A circuit with reference ($V = 0$) and non-reference nodes (V_1, V_2) is shown in Fig. 3.19. Then, applying KCL at nodes 1 (V_1) and 2 (V_2) yields,

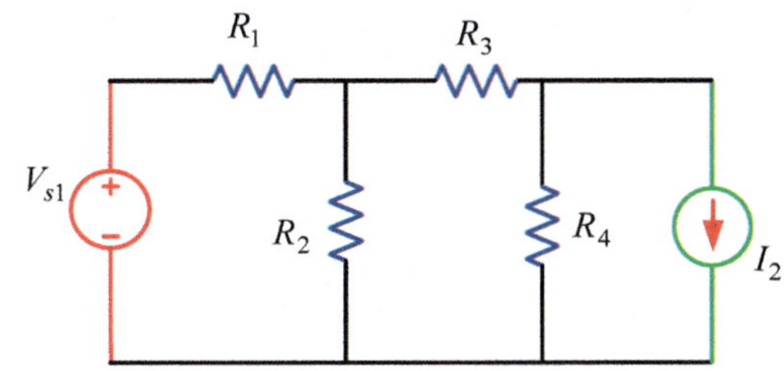

Fig. 3.18 First circuit for nodal analysis

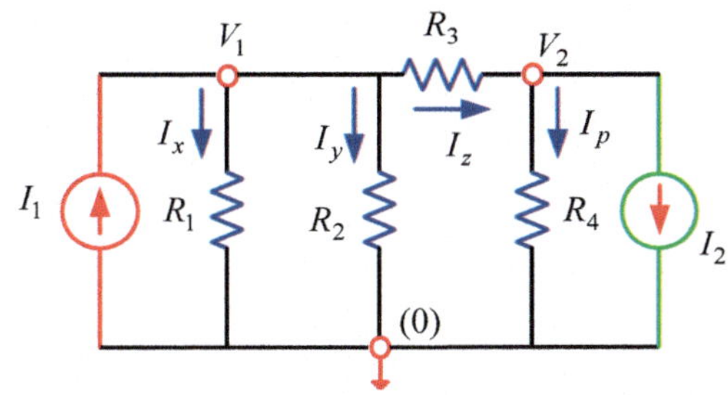

Fig. 3.19 Second circuit for nodal analysis

$$I_1 = I_x + I_y + I_z \tag{3.75}$$

$$I_z = I_p + I_2 \tag{3.76}$$

Equations (3.75) and (3.76) can be expanded in terms of node voltages as,

$$I_1 = \frac{V_1}{R_1} + \frac{V_1}{R_2} + \frac{V_1 - V_2}{R_3} \tag{3.77}$$

$$\left(\frac{1}{R_1} + \frac{1}{R_2} + \frac{1}{R_3}\right)V_1 - \frac{1}{R_3}V_2 = I_1 \tag{3.78}$$

$$\frac{V_1 - V_2}{R_3} = \frac{V_2}{R_4} + I_2 \tag{3.79}$$

$$\frac{1}{R_3}V_1 - \left(\frac{1}{R_3} + \frac{1}{R_4}\right)V_2 = I_2 \tag{3.80}$$

The values of node voltages can be calculated from Eqs. (3.78) and (3.80) by any methods if the other electrical quantities are known.

Example 3.6 Find the power absorbed by the 6 Ω resistor using nodal analysis of the circuit in Fig. 3.20. Use PSpice simulation to find the node voltages.

Solution:
Applying KCL at nodes 1, 2 and 3 yields,

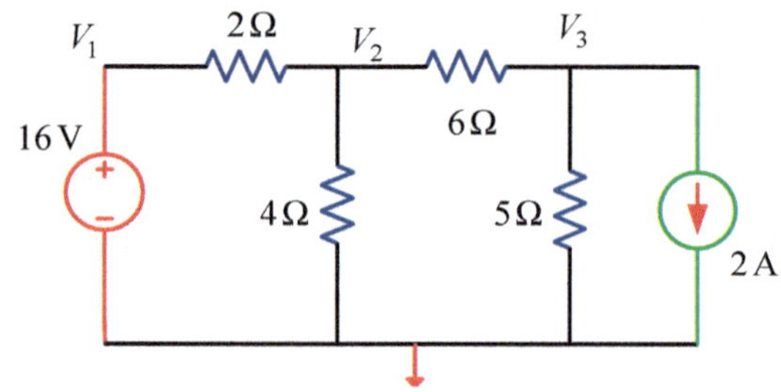

Fig. 3.20 An electrical circuit for Example 3.6

$$V_1 = 16\text{ V} \tag{3.81}$$

$$\frac{V_1 - V_2}{2} = \frac{V_2}{4} + \frac{V_2 - V_3}{6} \tag{3.82}$$

$$6V_1 - 6V_2 = 3V_2 + 2V_2 - 2V_3 \tag{3.83}$$

$$11V_2 - 2V_3 = 96 \tag{3.84}$$

$$\frac{V_2 - V_3}{6} = \frac{V_3}{5} + 2 \tag{3.85}$$

$$5V_2 - 5V_3 = 6V_3 + 60 \tag{3.86}$$

$$5V_2 - 11V_3 = 60 \tag{3.87}$$

From Eqs. (3.84) and (3.87), the node voltage, V_3 can be determined as,

$$V_3 = \frac{-180}{111} = -1.62\text{ V} \tag{3.88}$$

The value of the node voltage, V_2 can be determined as,

$$V_2 = \frac{60 + 11 \times (1.62)}{5} = 8.44\text{ V} \tag{3.89}$$

The current through the 6 Ω resistor is calculated as,

$$I_{6\,\Omega} = \frac{8.44 - (-1.62)}{6} = 1.68\text{ A} \tag{3.90}$$

The power absorbed by the 6 Ω resistor is calculated as,

$$P_{6\,\Omega} = 1.68^2 \times 6 = 16.93\text{ W} \tag{3.91}$$

The PSpice simulation circuit is shown in Fig. 3.21. The simulation node voltages are the same as the calculated node voltages.

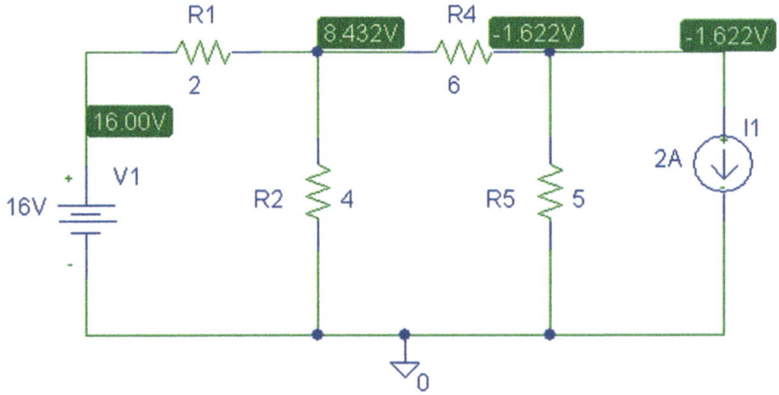

Fig. 3.21 PSpice simulation circuit for Example 3.6

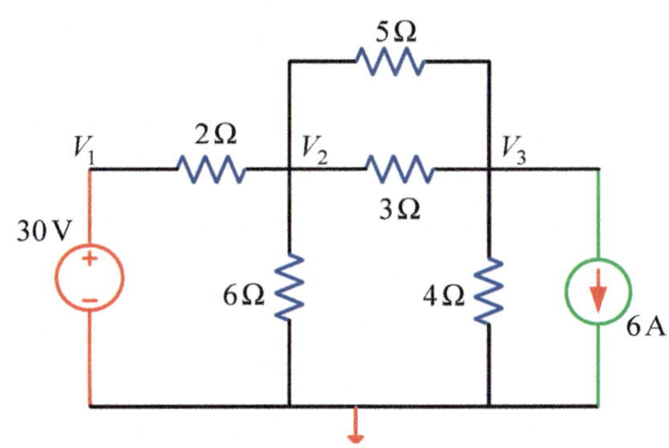

Fig. 3.22 Circuit for Practice Problem 3.6

Practice Problem 3.6 Determine the power absorbed by the 6 Ω resistor using nodal analysis for the circuit in Fig. 3.22. Use PSpice simulation to find the node voltages.

3.7 Nodal Analysis with Dependent Source

Sometimes, an electrical circuit may contain dependent source such as the one shown in Fig. 3.23 with current-controlled current source. In this circuit, applying KCL at node 1 yields,

$$I_{s1} = I_x + I_y \tag{3.92}$$

$$I_{s1} = \frac{V_1}{R_1} + \frac{V_1 - V_2}{R_2} \tag{3.93}$$

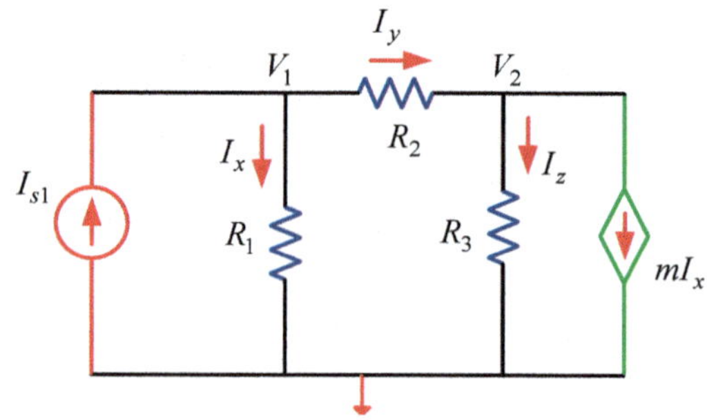

Fig. 3.23 Circuit with dependent current source for nodal analysis

$$I_{s1} = \left(\frac{1}{R_1} + \frac{1}{R_2}\right)V_1 - \frac{1}{R_2}V_2 \tag{3.94}$$

$$I_{s1} = (G_1 + G_2)V_1 - G_2V_2 \tag{3.95}$$

Applying KCL at node 2 yields,

$$I_y = I_z + mI_x \tag{3.96}$$

$$\frac{V_1 - V_2}{R_2} = \frac{V_2}{R_3} + m\frac{V_1}{R_1} \tag{3.97}$$

$$0 = \left(\frac{1}{R_2} - m\frac{1}{R_1}\right)V_1 - \left(\frac{1}{R_2} + \frac{1}{R_3}\right)V_2 \tag{3.98}$$

$$0 = (G_2 - mG_1)V_1 - (G_2 + G_3)V_2 \tag{3.99}$$

From Eqs. (3.95) and (3.99), the values of node voltages can be determined if other electrical quantities are given.

Example 3.7 Calculate the current in the 6 Ω resistor using nodal analysis for the circuit in Fig. 3.24. Use PSpice simulation to find the node voltages.

Solution:
Applying KCL at node 1 yields,

$$10 = I_x + I_y + I_n \tag{3.100}$$

$$10 = \frac{V_1}{2} + \frac{V_1 - V_2}{3} + \frac{V_1 - V_2}{4} \tag{3.101}$$

$$120 = 6V_1 + 4V_1 - 4V_2 + 3V_1 - 3V_2 \tag{3.102}$$

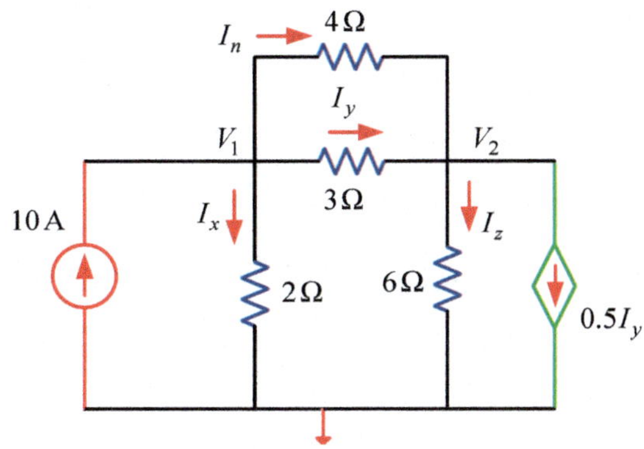

Fig. 3.24 Circuit for Example 3.7

$$13V_1 - 7V_2 = 120 \tag{3.103}$$

Applying KCL at node 2 yields,

$$I_y + I_n = I_z + 0.5I_y \tag{3.104}$$

$$0.5I_y + I_n = I_z \tag{3.105}$$

$$0.5\frac{V_1 - V_2}{3} + \frac{V_1 - V_2}{4} = \frac{V_2}{6} \tag{3.106}$$

$$2V_1 - 2V_2 + 3V_1 - 3V_2 = 2V_2 \tag{3.107}$$

$$V_1 = \frac{7}{5}V_2 \tag{3.108}$$

Substituting Eq. (3.108) into Eq. (3.103) yields,

$$13\frac{7}{5}V_2 - 7V_2 = 120 \tag{3.109}$$

$$V_2 = \frac{600}{56} = 10.71 \text{ V} \tag{3.110}$$

The current in the 6 Ω resistor is,

$$I_z = \frac{V_2}{6} = \frac{10.71}{6} = 1.79 \text{ A} \tag{3.111}$$

The PSpice simulation circuit is shown in Fig. 3.25. The simulation current in the 6 Ω resistor is same as the calculated current.

Practice Problem 3.7 Use nodal analysis to calculate the node voltages for the circuit in Fig. 3.26. Use PSpice simulation to verify the results.

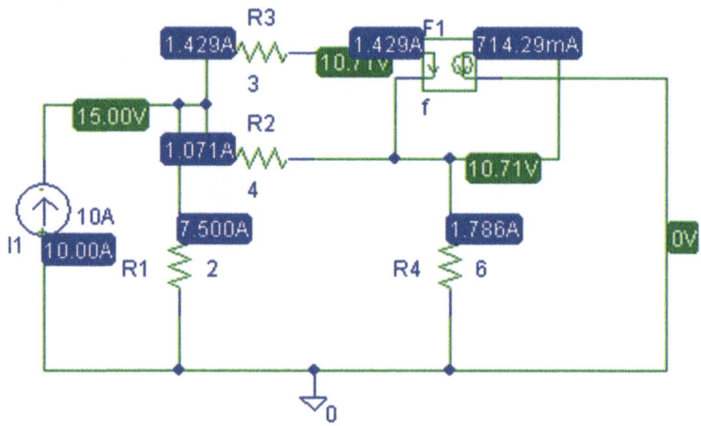

Fig. 3.25 PSpice simulation circuit for Example 3.7

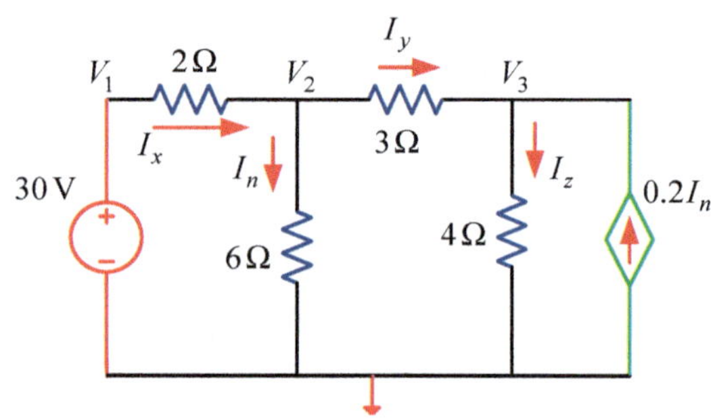

Fig. 3.26 Circuit for Practice Problem 3.7

3.8 Supernode

Sometimes, in practical electrical circuits, a voltage source with or without series resistance is connected between the two non-reference (or one reference and one non-reference) nodes. When a voltage source is connected between two nodes, then it is known as a supernode. Both KVL and KCL are applied to the supernode to get simultaneous equations [1–5]. An electrical circuit with a supernode is shown in Fig. 3.27.

Applying KCL at node 1 (V_1) yields,

$$I_{s1} = I_x + I_y \tag{3.112}$$

$$I_{s1} = \frac{V_1 - V_2}{R_1} + \frac{V_1 - V_3}{R_4} \tag{3.113}$$

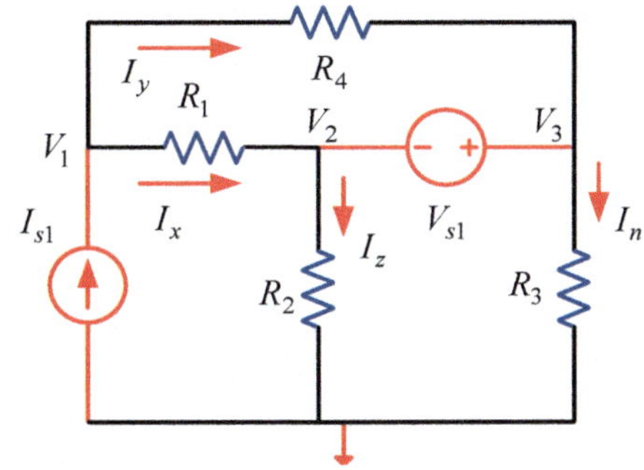

Fig. 3.27 An electrical circuit with supernode

$$(G_1 + G_4)V_1 - G_1V_2 - G_4V_3 = I_{s1} \tag{3.114}$$

Applying KCL at the supernode (the node with V_2, V_3 and V_{s1}) yields,

$$I_x = I_z + I_n \tag{3.115}$$

$$\frac{V_1 - V_2}{R_1} = \frac{V_2}{R_2} + \frac{V_3}{R_3} \tag{3.116}$$

$$G_1V_1 - (G_1 + G_2)V_2 - G_3V_3 = 0 \tag{3.117}$$

Applying KVL at the supernode yields,

$$-V_2 - V_{s1} + V_3 = 0 \tag{3.118}$$

$$0V_1 - V_2 + V_3 = V_{s1} \tag{3.119}$$

From Eqs. (3.114), (3.117) and (3.119), the values of node voltages can be calculated if the other electrical quantities are given.

Example 3.8 An electrical circuit is shown in Fig. 3.28. Use nodal analysis to calculate the node voltages and the current in the 3 Ω resistor. Use PSpice simulation to determine the node voltages.

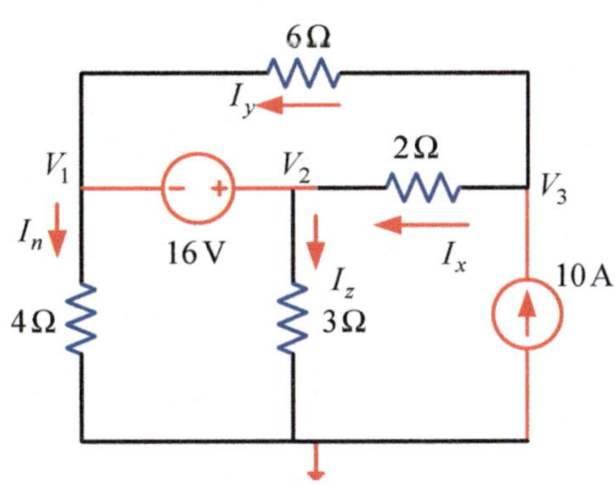

Fig. 3.28 Circuit for Example 3.8

Solution:
Applying KVL to the supernode yields,

$$-V_1 - 16 + V_2 = 0 \quad (3.120)$$

$$-V_1 + V_2 + 0V_3 = 16 \quad (3.121)$$

Applying KCL to the supernode yields,

$$I_x + I_y = I_n + I_z \quad (3.122)$$

$$\frac{V_3 - V_2}{2} + \frac{V_3 - V_1}{6} = \frac{V_1}{4} + \frac{V_2}{3} \quad (3.123)$$

$$6V_3 - 6V_2 + 2V_3 - 2V_1 - 3V_1 - 4V_2 = 0 \quad (3.124)$$

$$-5V_1 - 10V_2 + 8V_3 = 0 \quad (3.125)$$

Applying KCL to the node 3 yields,

$$10 = I_x + I_y \quad (3.126)$$

$$10 = \frac{V_3 - V_2}{2} + \frac{V_3 - V_1}{6} \quad (3.127)$$

$$60 = 3V_3 - 3V_2 + V_3 - V_1 \quad (3.128)$$

$$-V_1 - 3V_2 + 4V_3 = 60 \quad (3.129)$$

The value of the determinant of three simultaneous equations is,

$$D = \begin{vmatrix} -1 & 1 & 0 \\ -5 & -10 & 8 \\ -1 & -3 & 4 \end{vmatrix} = -1(-40 + 24) - 1(-20 + 8) = 28 \quad (3.130)$$

To evaluate V_2, D_2 parameter is calculated as,

$$D_2 = \begin{vmatrix} -1 & 16 & 0 \\ -5 & 0 & 8 \\ -1 & 60 & 4 \end{vmatrix} = -1(0 - 480) - 16(-20 + 8) = 672 \quad (3.131)$$

The value of the voltage at node 2 is determined as,

$$V_2 = \frac{D_2}{D} = \frac{672}{28} = 24 \text{ V} \quad (3.132)$$

From Eqs. (3.121) and (3.125), the values of other node voltages can be determined as,

$$V_1 = 24 - 16 = 8\ \text{V} \tag{3.133}$$

$$V_3 = \frac{5 \times 8 + 10 \times 24}{8} = 35\ \text{V} \tag{3.134}$$

The value of the current in the 3 Ω resistor is,

$$I_{3\,\Omega} = \frac{V_2}{3} = \frac{24}{3} = 8\ \text{A} \tag{3.135}$$

The PSpice simulation circuit is shown in Fig. 3.29. The simulation current is found to be the same as the calculated current.

Practice Problem 3.8 Figure 3.30 shows an electrical circuit. Calculate the node voltages and the power absorbed by the 5 Ω resistor. Use PSpice simulation to determine the node voltages.

Example 3.9 Figure 3.31 shows an electrical circuit. Use nodal analysis to calculate the node voltages and the current in the 6 Ω resistor. Use PSpice simulation to determine the node voltages.

Solution:
The value of the voltage at node 1 is,

$$V_1 = 20\ \text{V} \tag{3.136}$$

Applying KVL at the supernode of the circuit in Fig. 3.32 yields,

$$-V_2 + 2I_x + V_3 = 0 \tag{3.137}$$

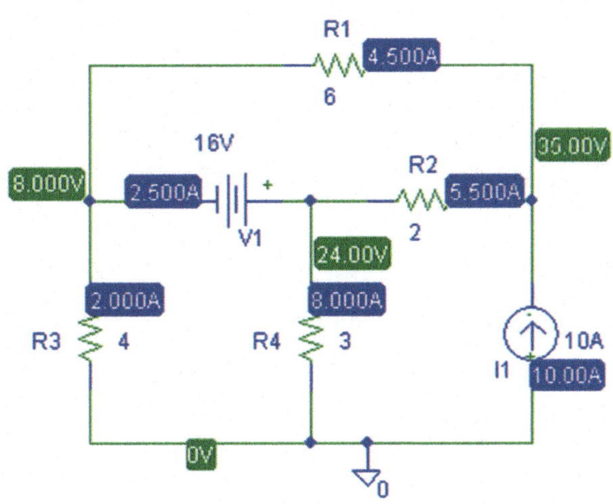

Fig. 3.29 PSpice simulation circuit for Example 3.8

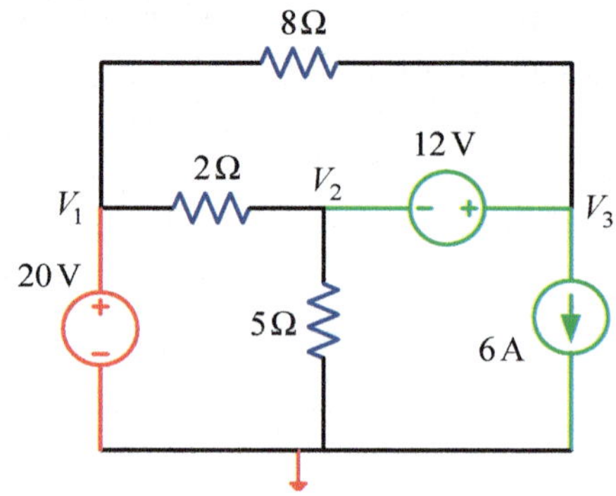

Fig. 3.30 Circuit for Practice Problem 3.8

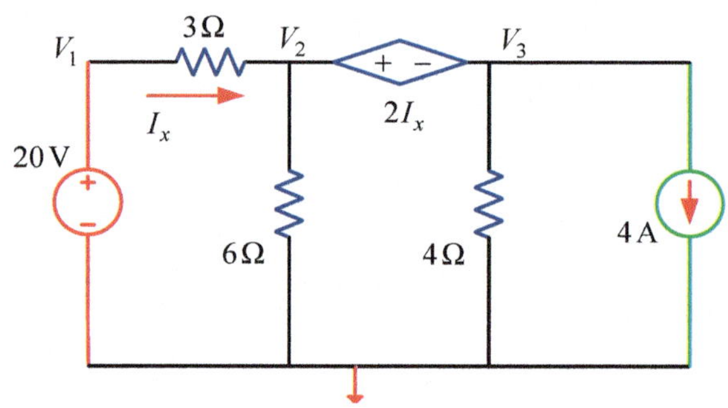

Fig. 3.31 An electrical circuit for Example 3.9

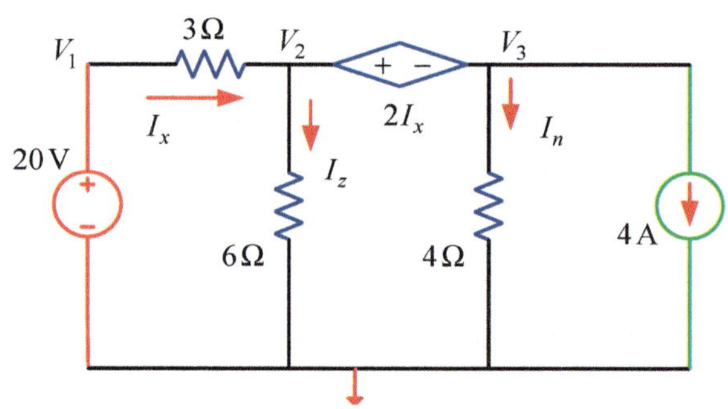

Fig. 3.32 An electrical circuit with directions of currents for Example 3.9

The current I_x through 3 Ω resistor can be written as,

$$I_x = \frac{20 - V_2}{3} \tag{3.138}$$

Substituting Eq. (3.138) into an Eq. (3.137) yields,

$$-V_2 + 2\frac{20 - V_2}{3} + V_3 = 0 \tag{3.139}$$

$$-5V_2 + 3V_3 = -40 \tag{3.140}$$

Applying KCL at the supernode of Fig. 3.32 yields,

$$I_x = I_n + I_z + 4 \tag{3.141}$$

$$\frac{20 - V_2}{3} = \frac{V_2}{6} + \frac{V_3}{4} + 4 \tag{3.142}$$

$$80 - 4V_2 = 2V_2 + 3V_3 + 48 \tag{3.143}$$

$$6V_2 + 3V_3 = 32 \tag{3.144}$$

From Eqs. (3.140) and (3.144), the value of the node voltage can be determined as,

$$V_2 = \frac{72}{11} = 6.54 \text{ V} \tag{3.145}$$

Substituting Eq. (3.145) into Eq. (3.144) yields,

$$V_3 = \frac{32 - 6 \times 6.54}{3} = -2.41 \text{ V} \tag{3.146}$$

The current in the 6 Ω resistor is calculated as,

$$I_{6\,\Omega} = \frac{V_2}{6} = \frac{6.54}{6} = 1.09 \text{ A} \tag{3.147}$$

The PSpice simulation circuit is shown in Fig. 3.33. The simulation node voltages and current are found to be the same as the calculated results.

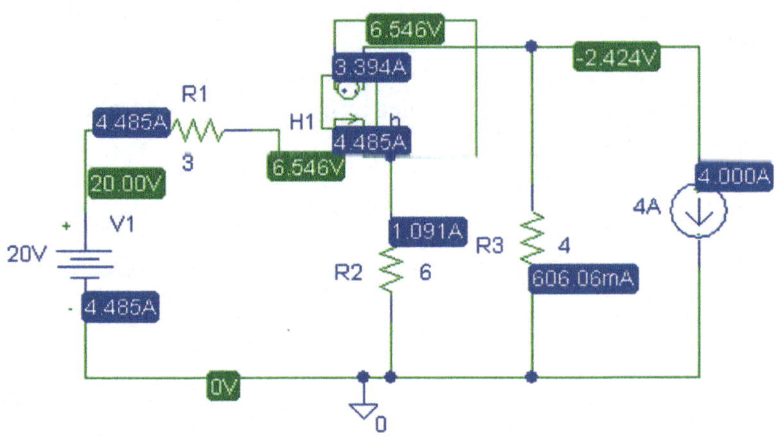

Fig. 3.33 PSpice simulation circuit for Example 3.9

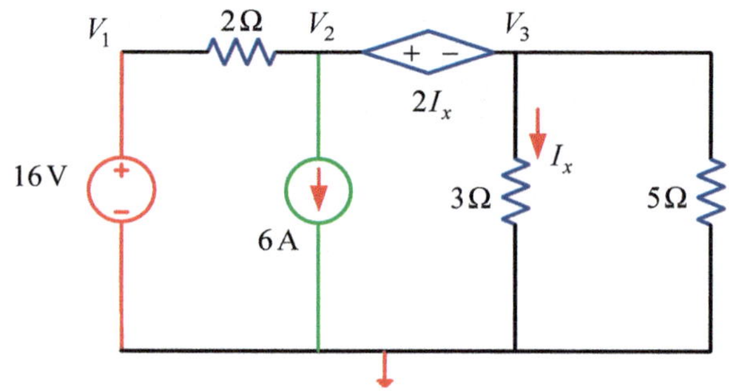

Fig. 3.34 Circuit for Practice Problem 3.9

Practice Problem 3.9 An electrical circuit is shown Fig. 3.34. Determine the node voltages and the current in the 5 Ω resistor using nodal analysis. Use PSpice simulation to determine the node voltages.

Exercise Problems

3.1 Three simultaneous equations of voltages of a multisource electrical circuit are given by $2V_1+4V_2-5V_3=8$, $V_1-2V_2+3V_3=6$, $2V_1+3V_2+4V_3=8$. Determine the unknown voltages.

3.2 Calculate the current in the 3 Ω resistor of the circuit in Fig. 3.35 using mesh analysis. Use PSpice simulation to verify the result.

3.3 Use mesh analysis to determine the current in the 4 Ω resistor of the circuit in Fig. 3.36 and verify the result by PSpice simulation.

3.4 A multisource electrical circuit is shown in Fig. 3.37. Determine the current in the 1 Ω resistor using mesh analysis. Verify the result by PSpice simulation.

3.5 An electrical circuit with multisource is shown in Fig. 3.38. Use mesh analysis to calculate the current in the 4 Ω resistor and verify the result by PSpice simulation.

3.6 Fig. 3.39 shows a multisource electrical circuit. Calculate the current in the 5 Ω resistor using mesh analysis and verify the result by PSpice simulation.

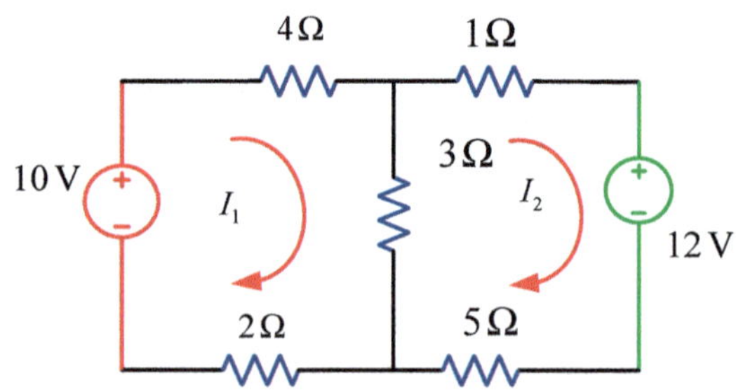

Fig. 3.35 Circuit for exercise problem 3.2

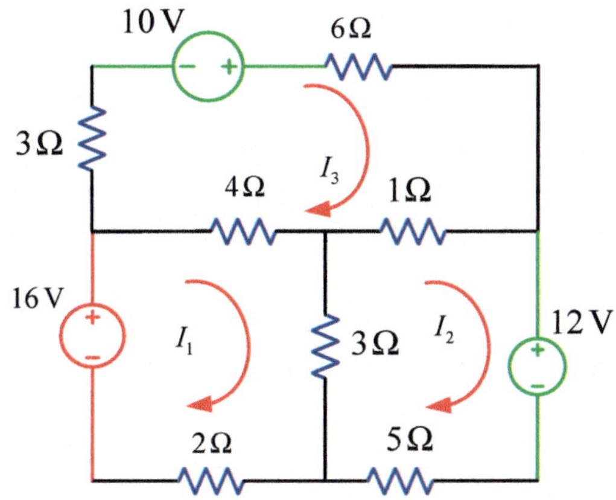

Fig. 3.36 Circuit for exercise problem 3.3

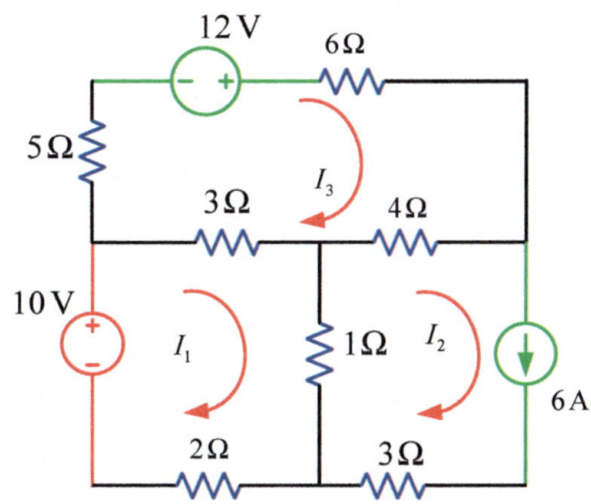

Fig. 3.37 Circuit for exercise problem 3.4

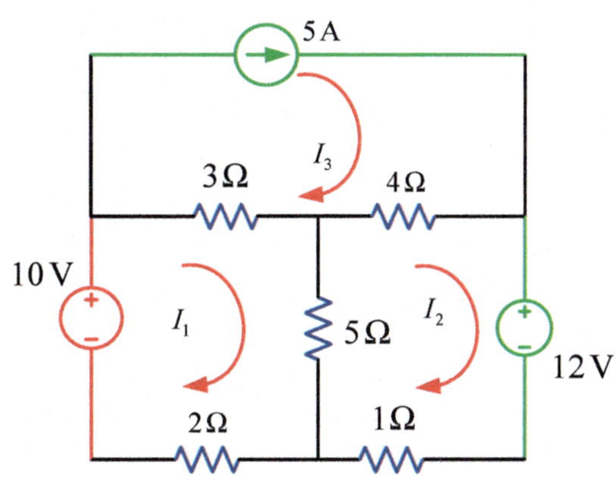

Fig. 3.38 Circuit for exercise problem 3.5

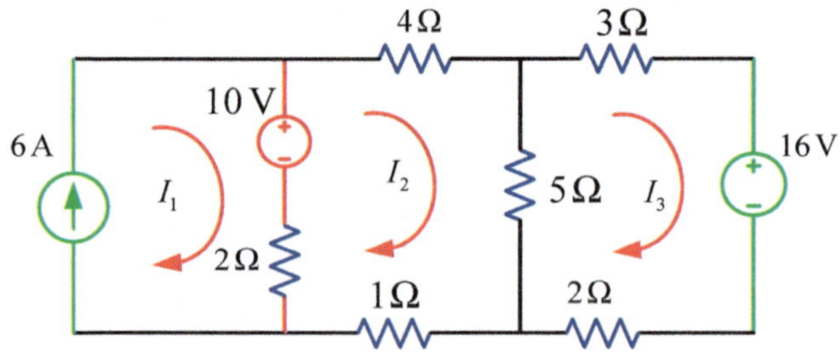

Fig. 3.39 Circuit for exercise problem 3.6

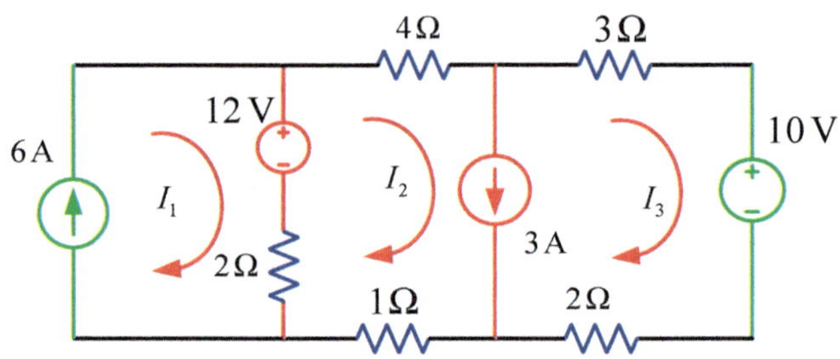

Fig. 3.40 Circuit for exercise problem 3.7

Fig. 3.41 Circuit for exercise problem 3.8

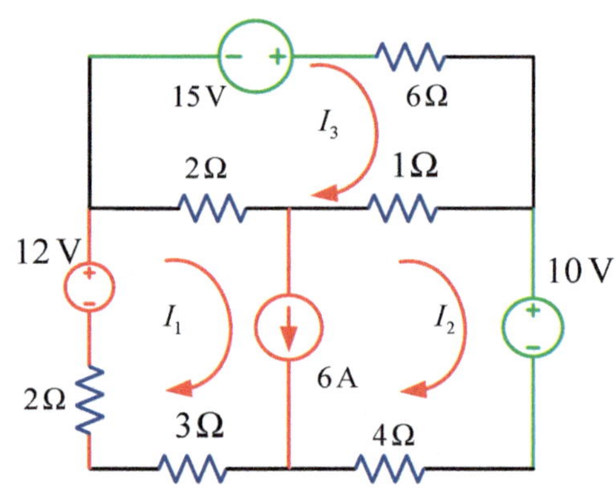

3.7 Use mesh analysis to calculate the power absorbed by the 3 Ω resistor of the circuit in Fig. 3.40 and calculate the mesh currents by PSpice simulation.

3.8 Fig. 3.41 shows an electrical circuit. Use mesh analysis to calculate the current in the 1 Ω resistor and calculate the mesh currents by PSpice simulation.

3.9 Use mesh analysis to calculate the current in the 2 Ω resistor in the circuit in Fig. 3.42. Calculate the mesh currents by PSpice simulation.

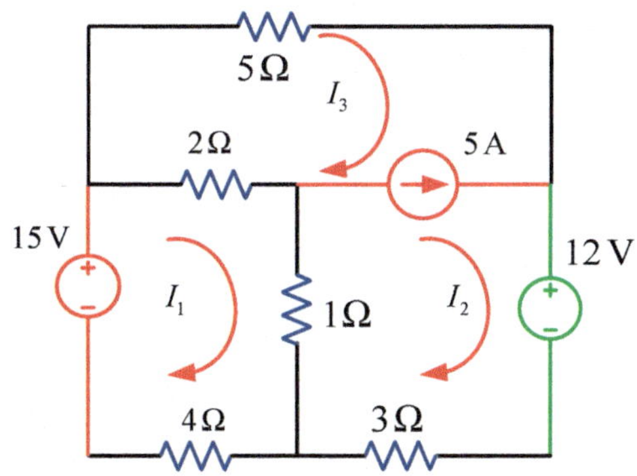

Fig. 3.42 Circuit for exercise problem 3.9

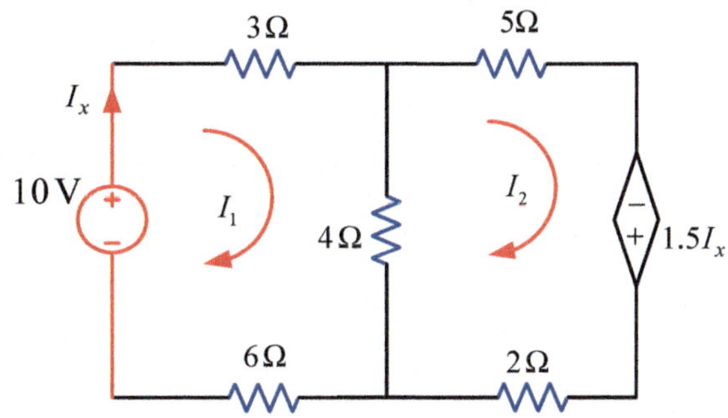

Fig. 3.43 Circuit for exercise problem 3.10

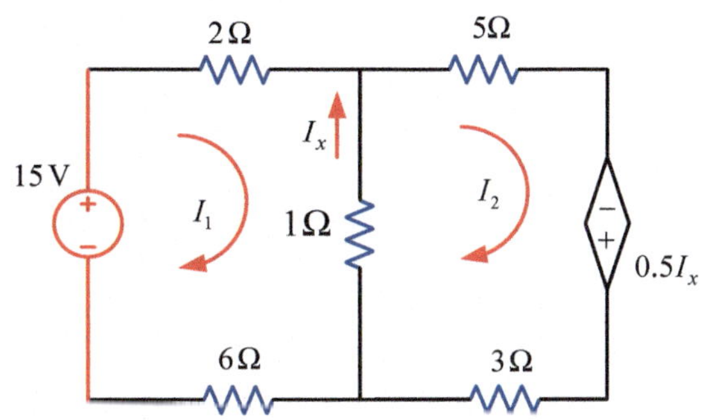

Fig. 3.44 Circuit for exercise problem 3.11

3.10 An electrical circuit is shown in Fig. 3.43. Use mesh analysis to calculate the current in the 4 Ω resistor. Find the mesh currents by PSpice simulation.

3.11 Fig. 3.44 shows an electrical circuit. Determine the current in the 1 Ω resistor using mesh analysis. Find the mesh currents by PSpice simulation.

3.12 Use mesh analysis to determine the current in the 3 Ω resistor of the circuit in Fig. 3.45. Find the mesh currents by PSpice simulation.

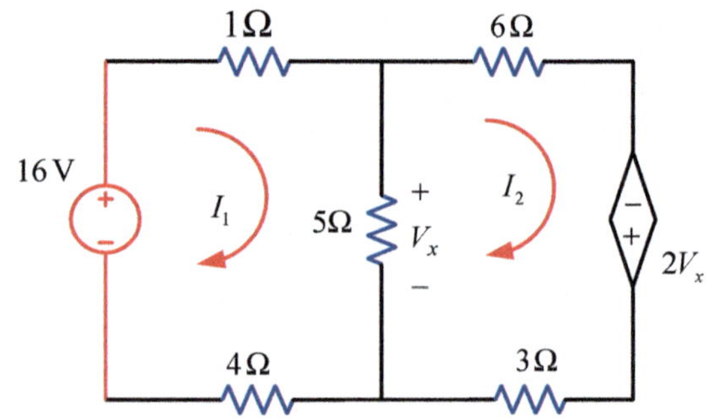

Fig. 3.45 Circuit for exercise problem 3.12

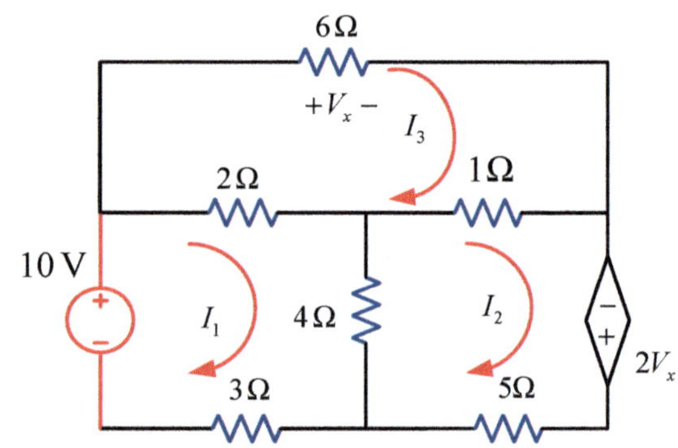

Fig. 3.46 Circuit for exercise problem 3.13

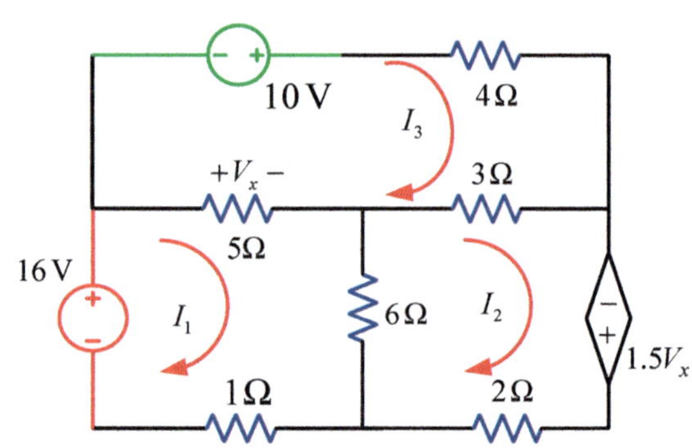

Fig. 3.47 Circuit for exercise problem 3.14

3.13 Fig. 3.46 shows a circuit. Use mesh analysis to determine the current in the 2 Ω resistor and find the mesh currents by PSpice simulation.

3.14 Use mesh analysis to determine the current in the 2 Ω resistor of the circuit in Fig. 3.47.

3.15 Fig. 3.48 shows an electrical circuit. Determine the mesh currents and verify the results by PSpice simulation.

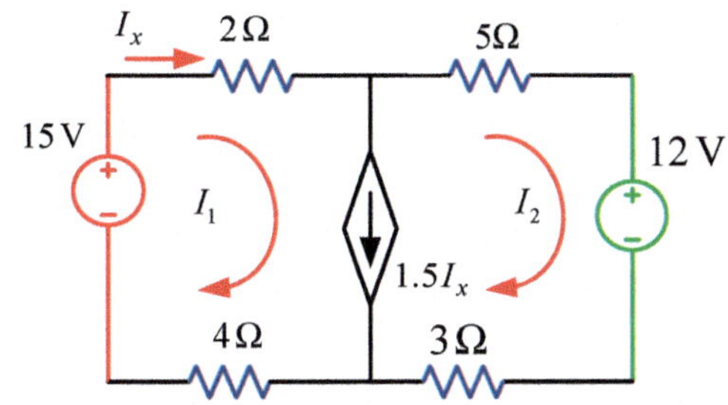

Fig. 3.48 Circuit for exercise problem 3.15

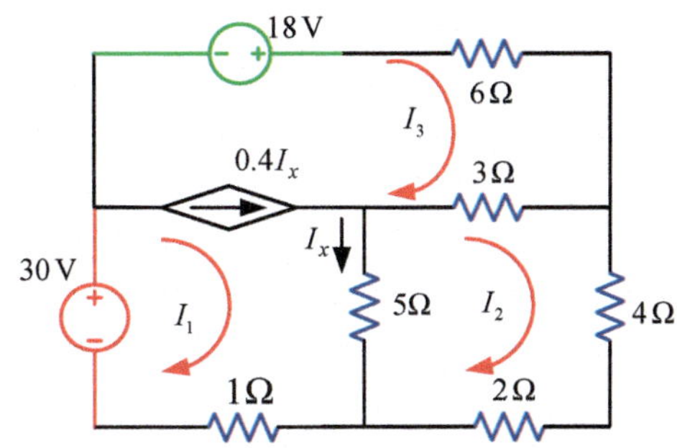

Fig. 3.49 Circuit for exercise problem 3.16

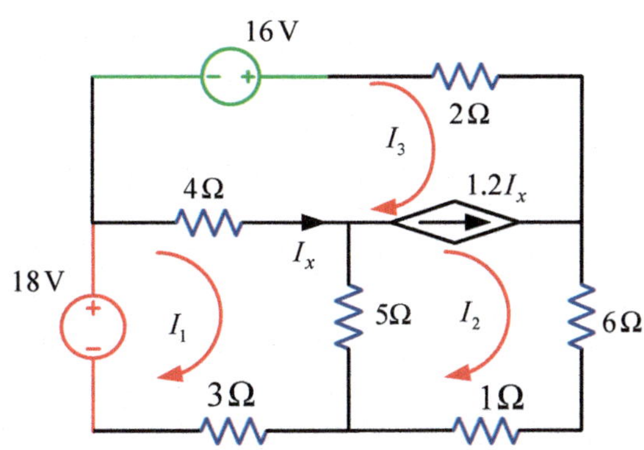

Fig. 3.50 Circuit for exercise problem 3.17

3.16 An electrical circuit is shown in Fig. 3.49. Use mesh analysis to find the current in the 3 Ω resistor. Find the mesh currents by PSpice simulation.
3.17 Use mesh analysis to find the current in the 5 Ω resistor of the circuit in Fig. 3.50. Find the mesh currents by PSpice simulation.
3.18 Use nodal analysis to find the current in the 6 Ω resistor of the circuit in Fig. 3.51. Determine the node voltages by PSpice simulation.
3.19 An electrical circuit is shown in Fig. 3.52. Use nodal analysis to find the current in the 6 Ω resistor and find the node voltages by PSpice simulation.

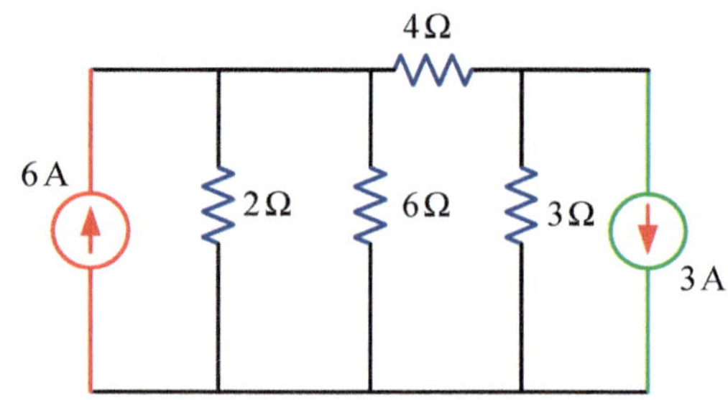

Fig. 3.51 Circuit for exercise problem 3.18

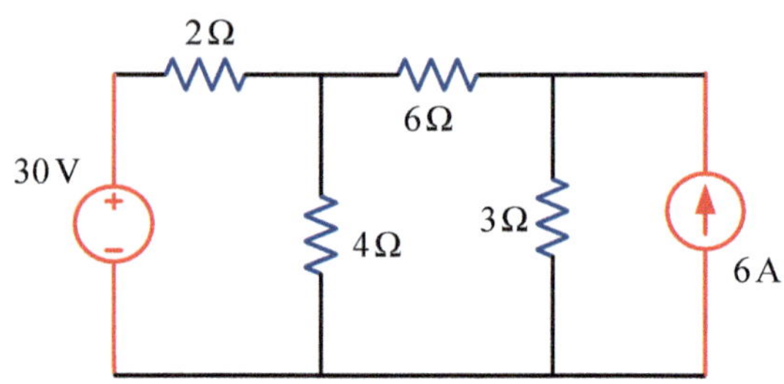

Fig. 3.52 Circuit for exercise problem 3.19

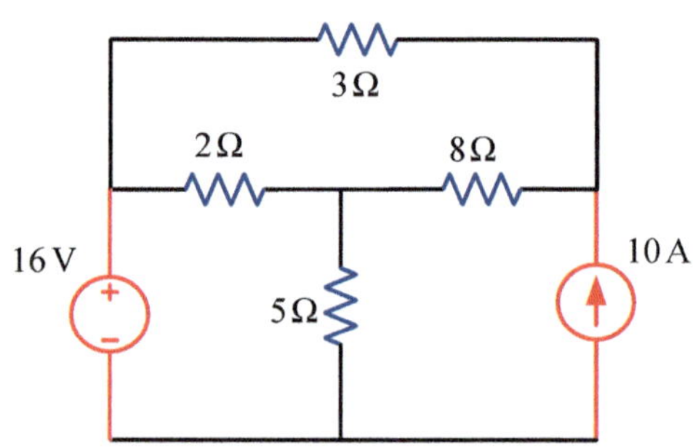

Fig. 3.53 Circuit for exercise problem 3.20

3.20 Fig. 3.53 shows an electrical circuit. Find the current in the 5 Ω resistor using nodal analysis and verify the result by PSpice simulation.

3.21 An electrical circuit is shown in Fig. 3.54. Calculate the current in the 2 Ω resistor using nodal analysis and verify the result by PSpice simulation.

3.22 Use nodal analysis to calculate the current in the 3 Ω resistor of the circuit in Fig. 3.55. Verify the result by PSpice simulation.

3.23 An electrical circuit is shown in Fig. 3.56. Use nodal analysis to calculate the current in the 5 Ω resistor. Verify the result by PSpice simulation.

3.24 Fig. 3.57 shows an electrical circuit. Use nodal analysis to calculate the current in the 3 Ω resistor. Verify the result by PSpice simulation.

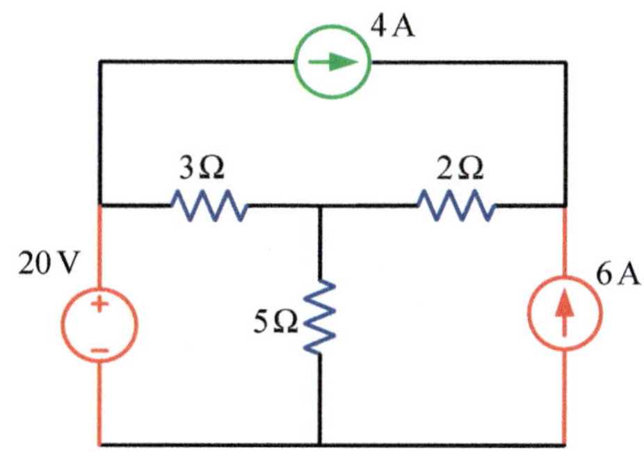

Fig. 3.54 Circuit for exercise problem 3.21

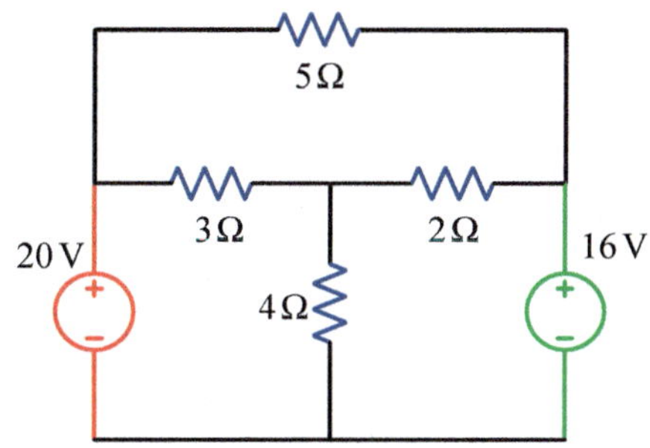

Fig. 3.55 Circuit for exercise problem 3.22

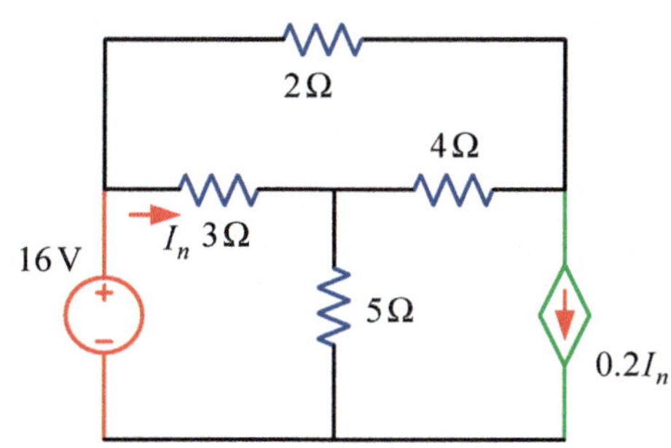

Fig. 3.56 Circuit for exercise problem 3.23

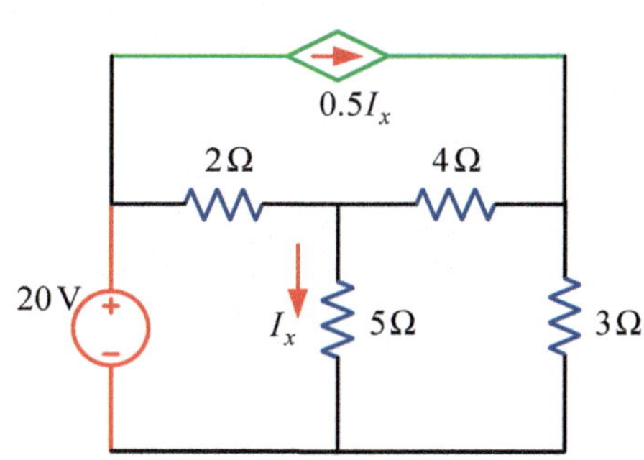

Fig. 3.57 Circuit for exercise problem 3.24

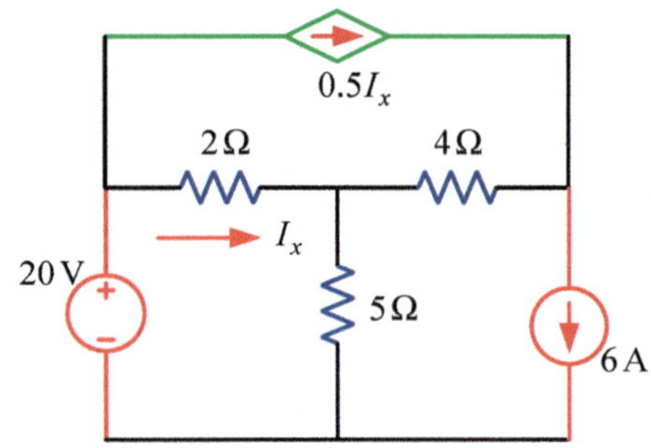

Fig. 3.58 Circuit for exercise problem 3.25

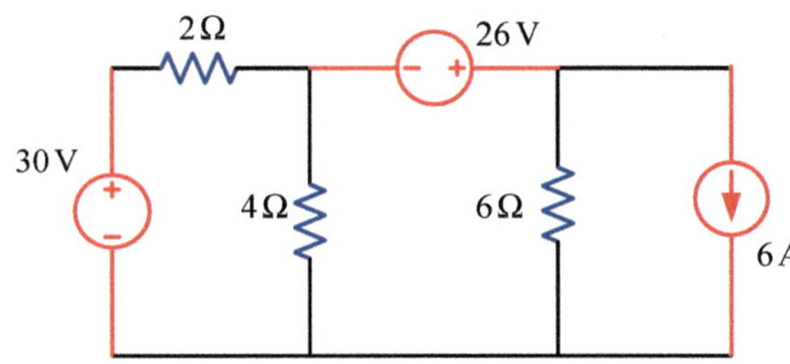

Fig. 3.59 Circuit for exercise problem 3.26

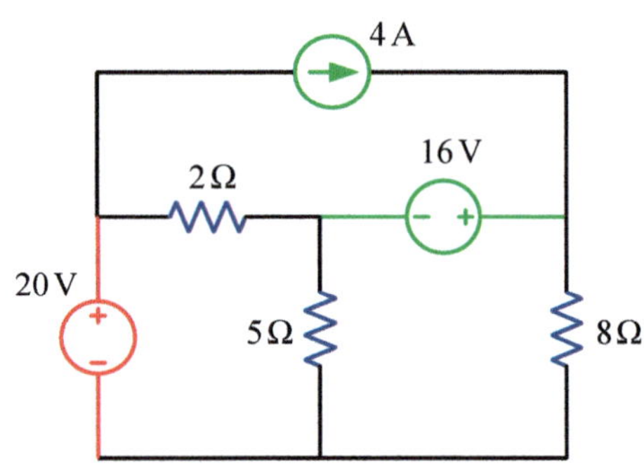

Fig. 3.60 Circuit for exercise problem 3.27

3.25 Fig. 3.58 shows an electrical circuit. Use nodal analysis to calculate the current in the 4 Ω resistor. Verify the result by PSpice simulation.

3.26 Use nodal analysis to calculate the current in the 4 Ω resistor of the circuit in Fig. 3.59. Verify the result by PSpice simulation.

3.27 Fig. 3.60 shows an electrical circuit. Determine the current in the 5 Ω resistor using nodal analysis and verify the result by PSpice simulation.

3.28 An electrical circuit is shown in Fig. 3.61. Calculate the current in the 2 Ω resistor using nodal analysis and verify the result by PSpice simulation.

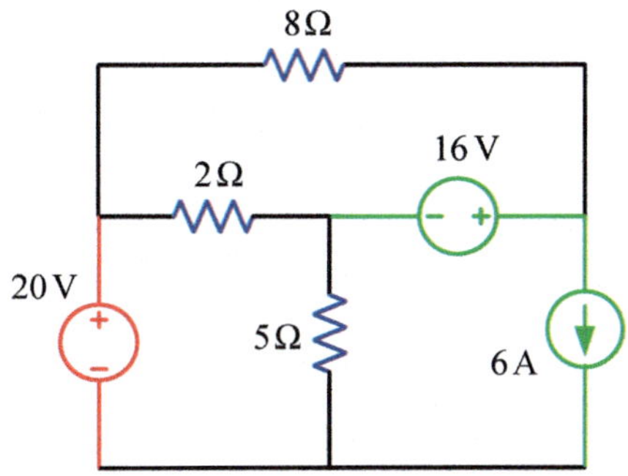

Fig. 3.61 Circuit for exercise problem 3.28

References

1. R.L. Boylestad, *Introductory Circuit Analysis*, 13th edn. (Pearson, 2016)
2. J. David Irwin, R. Mark Nelms, *Basic Engineering Circuit Analysis*, 11th edn. (Wiley, USA, 2015)
3. C.K. Alexander, M.N.O. Sadiku, *Fundamentals of Electric Circuits*, 6th edn. (McGraw-Hill Higher Education, Jan 2016)
4. J.W. Nilsson, S.A. Riedel, *Electric Circuits*, 10th edn. (Prentice Hall International Edition, 2015)
5. M. Abdus Salam, *Basic Electrical Circuits*, 2nd edn. (Shroff Publishers & Distributors Pvt. Ltd, India, 2007)

Chapter 4
Network Theorems

4.1 Introduction

Fundamental electrical laws and the associated methods of analysis, which have been discussed in the previous chapters, need tedious mathematical manipulation. These cumbersome mathematical analyses can be simplified by using advanced techniques known as network or circuit theorems. These include linearity property, superposition theorem, Thevenin's theorem, Norton's theorem and maximum power transfer theorem. Here, most of these theorems will be discussed with independent and dependent sources. In addition, PSpice simulation will also be used in some cases to verify the analytical results.

4.2 Linearity Property

Linearity is the property of a system or an element that contains homogeneity (scaling) and additive properties. The homogeneity property states that if the input of a system is multiplied by a constant term, then the output would be obtained by multiplying the same constant. To explain this (scaling/homogeneity) property, the following equation is considered, where the number 4 represents constant term.

$$v = 4i \tag{4.1}$$

If $i = 1\,\text{A}$, the voltage is,

$$v = 4\,\text{V} \tag{4.2}$$

If $i = 2\,\text{A}$, the voltage is,

$$v = 8\,\text{V} \tag{4.3}$$

Md. A. Salam and Q. Mehbubar Rahman, *Fundamentals of Electrical Circuit Analysis*,
https://doi.org/10.1007/978-981-10-8624-3_4

From the above equations, it can be inferred that according to the homogeneity/scaling property presented in Eq. (4.1), if the current increases by a factor of n, then the associated voltage increases by the same factor. This results in a linear relationship between the current and voltage as shown in Fig. 4.1.

Again, consider the following equation:

$$v = 4i + 2 \tag{4.4}$$

If $i = 1\,\text{A}$, then the voltage is,

$$v = 6\,\text{V} \tag{4.5}$$

If $i = 2\,\text{A}$, then the voltage is,

$$v = 10\,\text{V} \tag{4.6}$$

It is seen that for a 1 A current, the voltage is found to be 6 V. If the current increases by a factor of 2, then the voltage should be $v = 2 \times 6 = 12\,\text{V}$, which is not the case according to Eq. (4.6). Hence, it is concluded that Eq. (4.4) does not hold the homogeneity property and therefore, it is not a linear equation, as can be shown in Fig. 4.2.

The additive property states that the response of a system to a sum of the inputs is same as the sum of the responses of the system when each of the inputs acts separately to the system.

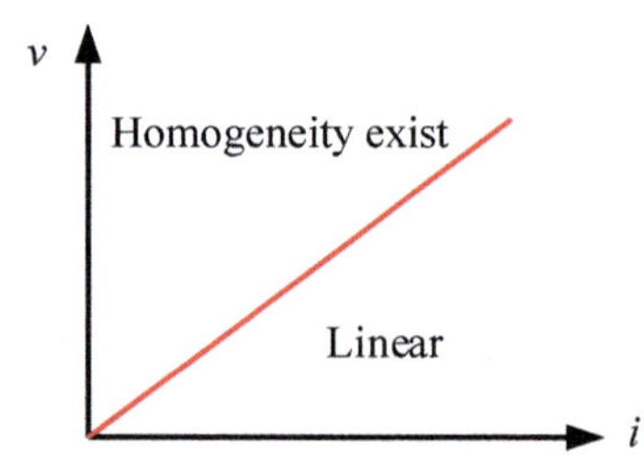

Fig. 4.1 Linear system

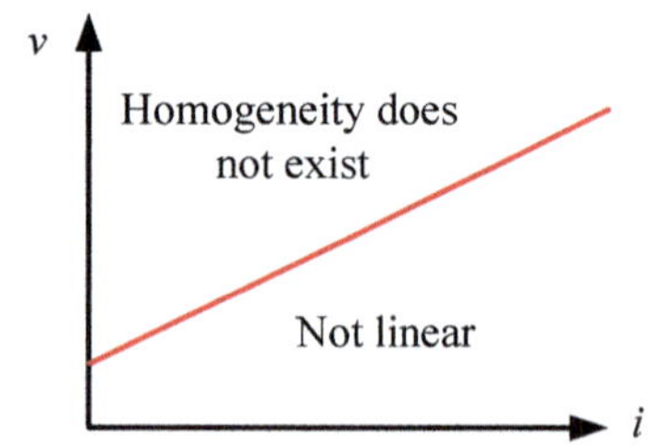

Fig. 4.2 Nonlinear system

In general, a function f for two inputs x_1 and x_2 will be linear if the following additive property exists:

$$f(x_1 + x_2) = f(x_1) + f(x_2) \tag{4.7}$$

This property can be explained by considering the circuit shown in Fig. 4.3 where two voltage sources V_{s1} and V_{s2} are acting on the resistive circuit with resistance R.

Applying KVL to the circuit in Fig. 4.3 yields,

$$-V_{s1} - V_{s2} + IR = 0 \tag{4.8}$$

$$I = \frac{V_{s1} + V_{s2}}{R} \tag{4.9}$$

where I is the resultant current. Now to find the individual responses of this resistive circuit by considering input voltage sources separately, let us refer to Figs. 4.4 and 4.5. From Fig. 4.4, the resultant current with the application of the voltage source V_{s1} is given by,

$$I_1 = \frac{V_{s1}}{R} \tag{4.10}$$

While, from Fig. 4.5, the resultant current is given by,

$$I_2 = \frac{V_{s2}}{R} \tag{4.11}$$

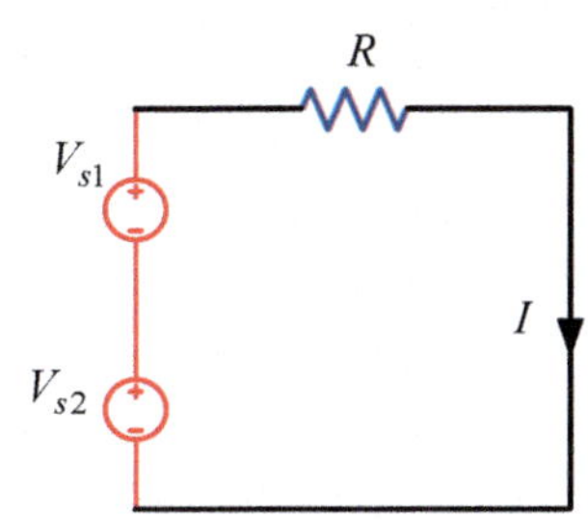

Fig. 4.3 A circuit with two voltage sources

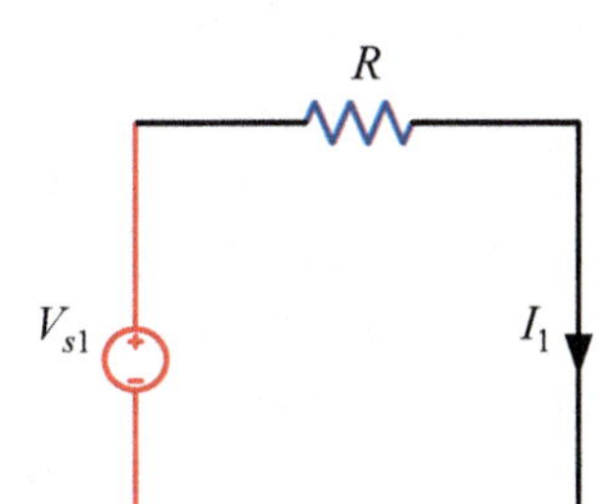

Fig. 4.4 A circuit with first voltage source

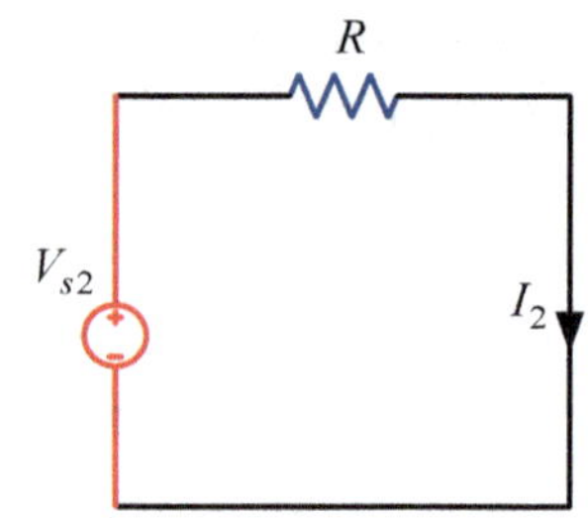

Fig. 4.5 A circuit with second voltage source

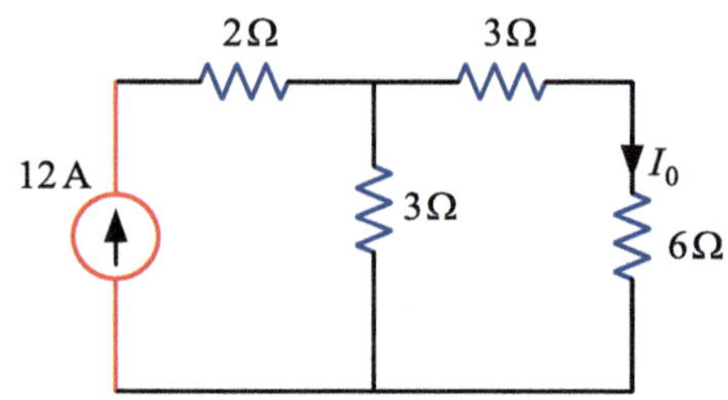

Fig. 4.6 A circuit for Example 4.1

Adding Eqs. (4.10) and (4.11) yields,

$$I = I_1 + I_2 = \frac{V_{s1}}{R} + \frac{V_{s2}}{R} = \frac{V_{s1} + V_{s2}}{R} \tag{4.12}$$

Equation (4.12) is same as Eq. (4.9) which was obtained using the original circuit. Therefore, it holds the additive property, and hence, the circuit is a linear circuit. In a linear circuit, the output of the circuit is directly proportional to its input.

Example 4.1 An electrical circuit is shown in Fig. 4.6. Assume $I_0 = 1\,\mathrm{A}$. Use linearity property to find the actual value of I_0.

Solution:
The voltage V_1 as can be seen in Fig. 4.7 can be determined as,

$$V_1 = (3 + 6)1 = 9\,\mathrm{V} \tag{4.13}$$

The current I_1 is calculated as,

$$I_1 = \frac{V_1}{3} = \frac{9}{3} = 3\,\mathrm{A} \tag{4.14}$$

The source current is,

$$I_s = 3 + 1 = 4\,\mathrm{A} \tag{4.15}$$

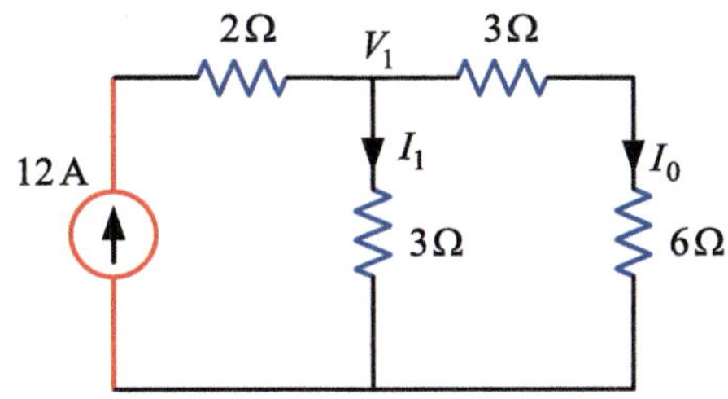

Fig. 4.7 Circuit for Example 4.1 with a node voltage

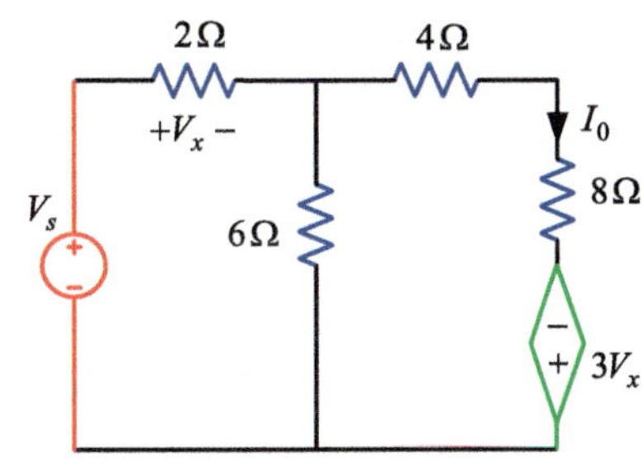

Fig. 4.8 Circuit for Example 4.2

Here, the assumption that $I_0 = 2$ A resulted in a source current of 4 A. However, since the actual source current is 12 A, the actual value of the current I_0 is,

$$I_0 = \frac{12}{4} = 3\,\text{A} \tag{4.16}$$

Example 4.2 Figure 4.8 shows an electrical circuit with a dependent source. The value of the source voltage is assigned to 18 and 36 V, respectively, for two different cases. Use linearity property to find the actual value of I_0.

Solution:
The mesh currents are assigned as shown in Fig. 4.9. The mesh equations can be written as

$$-V_s + 2I_1 + 6(I_1 - I_2) = 0 \tag{4.17}$$

$$V_s = 8I_1 - 6I_2 \tag{4.18}$$

$$18I_2 - 3V_x = 0 \tag{4.19}$$

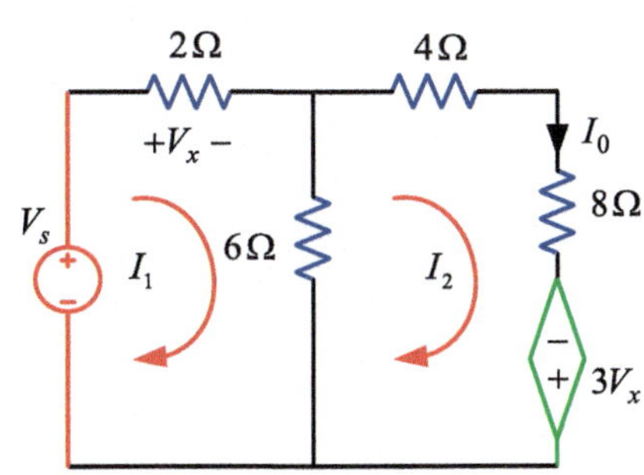

Fig. 4.9 Circuit for Example 4.2 with mesh currents

According to the given circuit,

$$V_x = 2I_1 \tag{4.20}$$

Substituting Eq. (4.20) into Eq. (4.19) yields,

$$18I_2 - 6I_1 = 0 \tag{4.21}$$

$$I_1 = 3I_2 \tag{4.22}$$

Substituting Eq. (4.22) into Eq. (4.18) yields,

$$V_s = 24I_1 - 6I_2 \tag{4.23}$$

$$I_2 = \frac{V_s}{18} \tag{4.24}$$

When $V_s = 18\,\text{V}$, the current is,

$$I_0 = I_2 = \frac{18}{18} = 1\,\text{A} \tag{4.25}$$

When $V_s = 36\,\text{V}$, the current is,

$$I_0 = I_2 = \frac{36}{18} = 2\,\text{A} \tag{4.26}$$

Practice Problem 4.1
Figure 4.10 shows an electrical circuit with the source voltage $V_s = 36\,\text{V}$. Determine the actual voltage V_0 when $V_0 = 1\,\text{V}$.

Practice Problem 4.2
An electrical circuit with a dependent source is shown in Fig. 4.11. Calculate the current I_0 when $I_s = 5\,\text{A}$ and $I_s = 10\,\text{A}$.

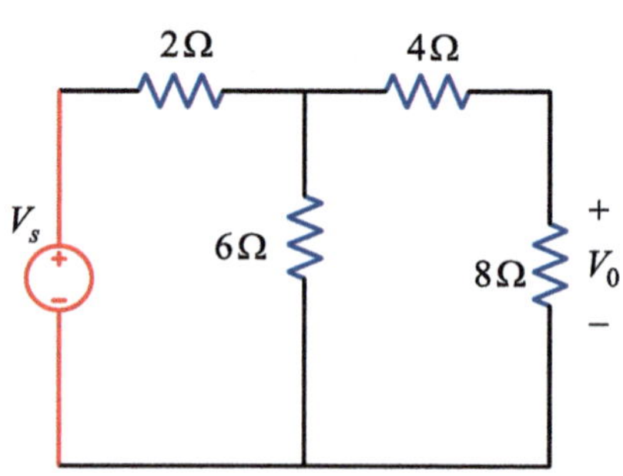

Fig. 4.10 Circuit for Practice Problem 4.1

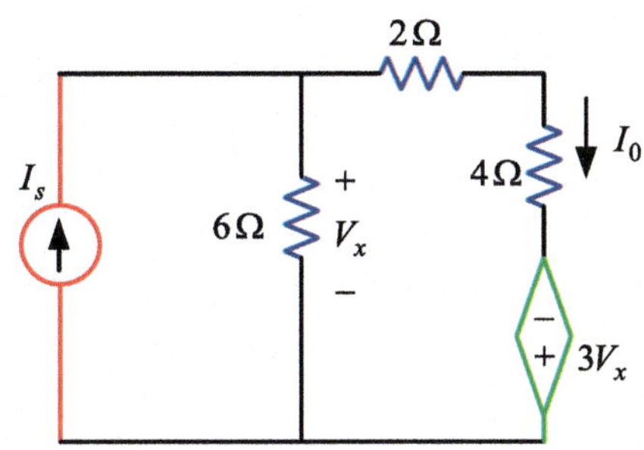

Fig. 4.11 Circuit for Practice Problem 4.2

4.3 Superposition Theorem

The principle of superposition is usually applied to a linear circuit network, which contains more than one source. The superposition theorem states as 'in any linear network, the current through or voltage across any element is the algebraic sum of the currents through or voltages across that element due each independent source acting alone'. To apply this theorem to any linear network, two important points need to be considered:

- Only one independent source at any given time will remain active, while all other independent sources are turned off. To turn off a voltage source, it is short circuited, while to turn off a current source, it is open circuited.
- Since the dependent sources are controlled by other circuit variables they are left as is.

The following steps are carried out in applying superposition theorem to any linear circuit:

- Except for one independent source, turn off all the other independent sources.
- Calculate the output (current or voltage) due to active source.
- Repeat above steps for each independent source.
- Calculate the total contribution by algebraically adding all the contributions resulting from the independent sources.

Let us apply the superposition theorem in the circuit shown in Fig. 4.12 to find current I_3. Consider that the current source is open circuited (turned off), while the voltage source is acting alone in the circuit as shown in Fig. 4.13. Now to evaluate the current I_s due to the independent voltage source, we need to find dent the total resistance of this circuit, which is,

$$R_t = R_1 + \frac{R_2(R_3 + R_4)}{R_2 + R_3 + R_4} \tag{4.27}$$

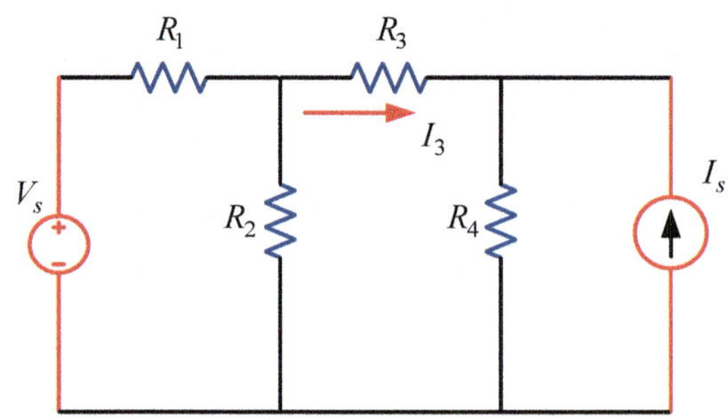

Fig. 4.12 Circuit with independent sources for Superposition theorem

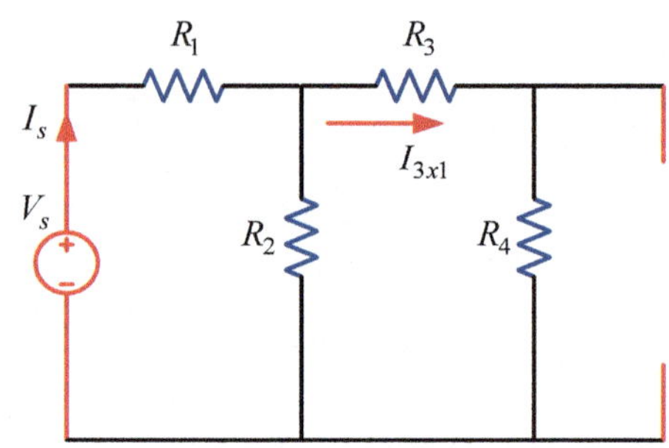

Fig. 4.13 A circuit with active voltage source

In this case, the current I_s will be,

$$I_s = \frac{V_s}{R_t} \tag{4.28}$$

Substituting Eq. (4.31) into Eq. (4.32) yields,

$$I_s = \frac{Vs(R_2 + R_3 + R_4)}{R_1(R_2 + R_3 + R_4) + R_2(R_3 + R_4)} \tag{4.29}$$

The current through the resistance R_3 for the active voltage source V_s is,

$$I_{3\times 1} = I_s \frac{R_2}{R_2 + R_3 + R_4} \tag{4.30}$$

Again, consider that the current source is active and the voltage source is short circuited (turned off) as shown in Fig. 4.14. In this case, the resistors R_1 and R_2 are in parallel, which results in the equivalent resistance R_p, given by,

$$R_p = \frac{R_1 R_2}{R_1 + R_2} \tag{4.31}$$

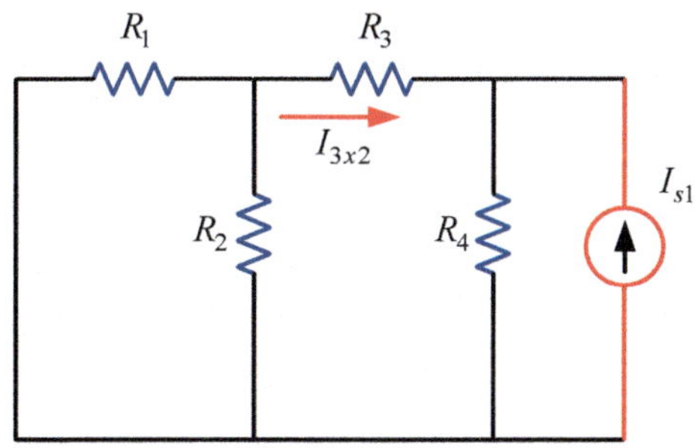

Fig. 4.14 Circuit with active current source

The value of the current through R_3 for the active current source is found using current divider rule as,

$$-I_{3\times 2} = I_{s1} \times \frac{R_4}{R_p + R_3 + R_4} \tag{4.32}$$

$$I_{3\times 2} = -I_{s1} \times \frac{R_4}{R_p + R_3 + R_4} \tag{4.33}$$

So, the actual value of the current through the resistor R_3 is,

$$I_3 = I_{3\times 1} + I_{3\times 2} \tag{4.34}$$

Example 4.3 Calculate the current through $4\,\Omega$ resistor in the circuit shown in Fig. 4.15 using Superposition theorem.

Solution:
Consider that the voltage source is active and the current source is open circuited as shown in Fig. 4.16. The total resistance of this circuit is,

$$R_t = 2 + \frac{4 \times 6}{4 + 6} = 4.4\,\Omega \tag{4.35}$$

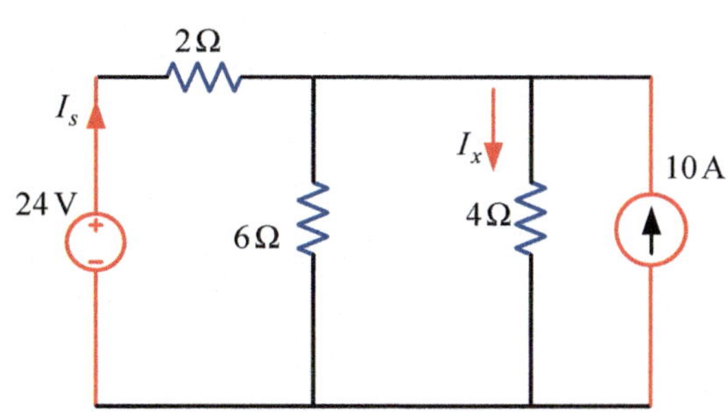

Fig. 4.15 Circuit for Example 4.3

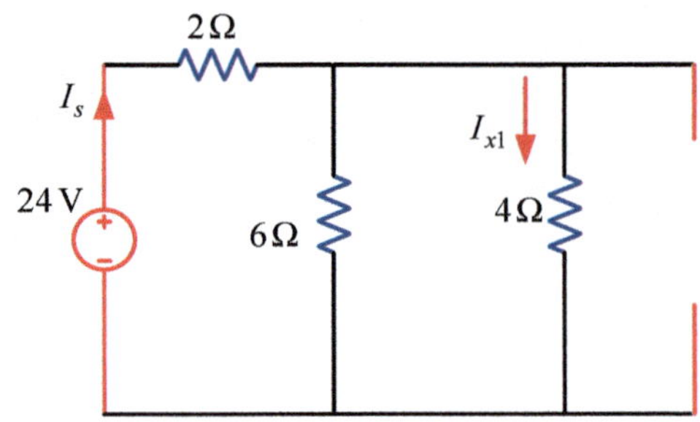

Fig. 4.16 A circuit for Example 4.3 with active voltage source

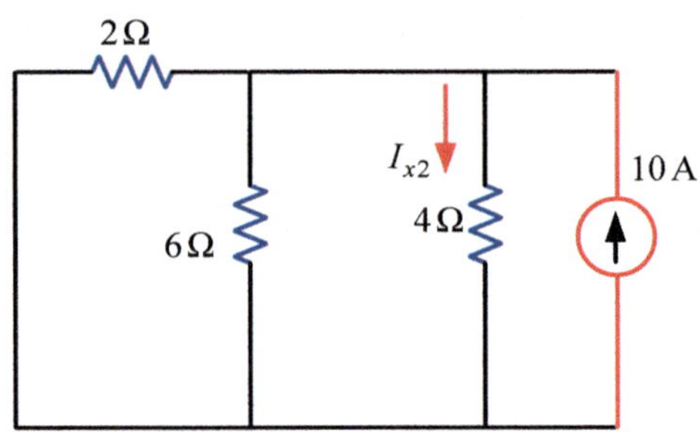

Fig. 4.17 Circuit for Example 4.3 with active current source

The source current is,

$$I_s = \frac{24}{4.4} = 5.45\ \text{A} \tag{4.36}$$

The current in the 4 Ω resistor is,

$$I_{x1} = 5.45 \times \frac{6}{6+4} = 3.27\ \text{A} \tag{4.37}$$

Again, consider that the voltage is short circuited and the current source is active as shown in Fig. 4.17. The 2 and 6 Ω resistors are in parallel and their equivalent resistance is,

$$R_1 = \frac{2 \times 6}{2+6} = 1.5\ \Omega \tag{4.38}$$

The current in the 4 Ω resistor is,

$$I_{x2} = 10 \times \frac{1.5}{1.5+4} = 2.73\ \text{A} \tag{4.39}$$

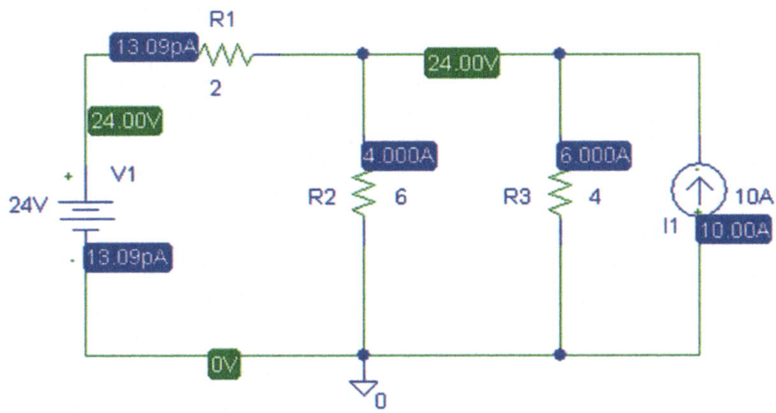

Fig. 4.18 PSpice simulation circuit for Example 4.3

Fig. 4.19 Circuit for Practice Problem 4.3

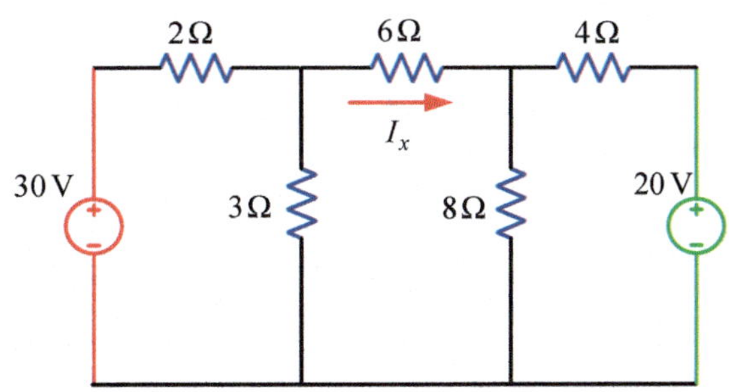

Finally, the current in the 4 Ω resistor can be obtained by adding Eqs. (4.37) and (4.39) as,

$$I_x = I_{x1} + I_{x2} = 3.27 + 2.73 = 6\,\text{A} \tag{4.40}$$

The PSpice simulation circuit is shown in Fig. 4.18. The simulation result is found to be the same as the calculated result.

Practice Problem 4.3

Calculate the current through 6 Ω resistor in the circuit shown in Fig. 4.19 using superposition theorem.

4.4 Analysis of Superposition Theorem with Dependent Source

An electrical circuit with a combination of independent and dependent sources is shown in Fig. 4.20. Here, the expression of voltage V_x will be determined by

Fig. 4.20 Circuit with dependent and independent sources

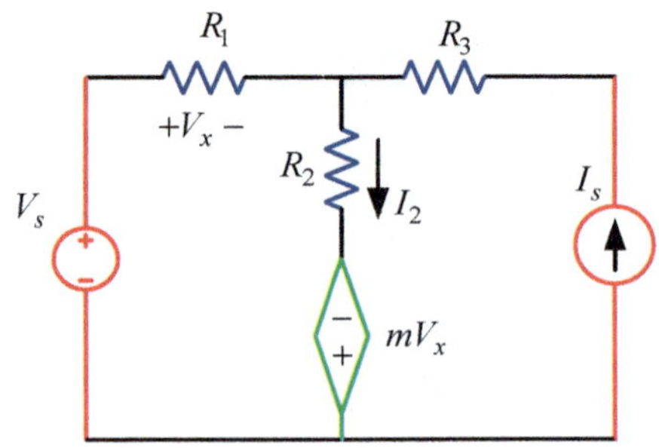

Fig. 4.21 Circuit with active voltage source

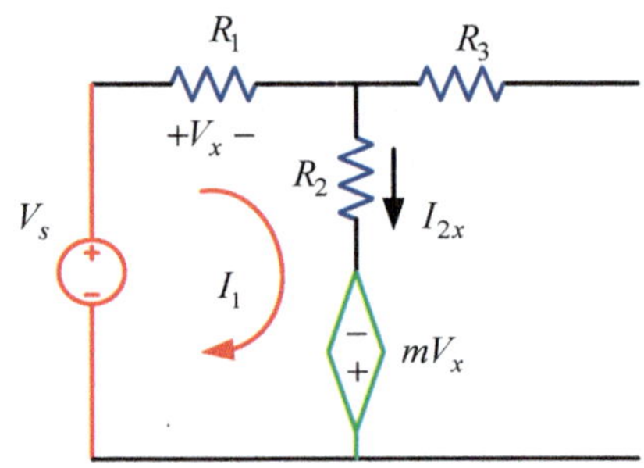

considering the independent sources separately. Initially, consider that the independent voltage source is active and the current source is open circuited as shown in Fig. 4.21.

Applying KVL to the circuit in Fig. 4.21 yields,

$$(R_1 + R_2)I_1 - V_s - mV_x = 0 \tag{4.41}$$

Now, according to the circuit,

$$V_x = I_1 R_1 \tag{4.42}$$

Substituting Eq. (4.42) into Eq. (4.41) yields,

$$(R_1 + R_2)I_1 - V_s - mI_1 R_1 = 0 \tag{4.43}$$

$$I_1 = \frac{V_s}{R_1 + R_2 + mR_1} \tag{4.44}$$

The current in the R_2 resistor is,

$$I_{2x} = \frac{V_s}{R_1 + R_2 + mR_1} \tag{4.45}$$

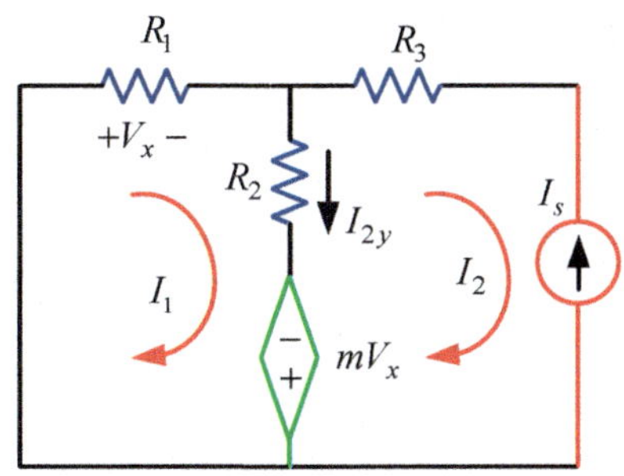

Fig. 4.22 Circuit with active current source

Now, considering the active current source and the inactive (short circuited) voltage source as shown in Fig. 4.22, the application of KVL to the circuit yields,

$$I_2 = -I_s \tag{4.46}$$

$$R_1 I_1 + R_2(I_1 - I_2) - mV_x = 0 \tag{4.47}$$

Substituting Eqs. (4.42) and (4.46) into Eq. (4.47) yields,

$$R_1 I_1 + R_2(I_1 + I_s) - mR_1 I_1 = 0 \tag{4.48}$$

$$I_1 = \frac{R_2 I_s}{R_1 + R_2 + mR_1} \tag{4.49}$$

The current in the R_2 resistor is,

$$I_{2y} = I_1 - I_2 \tag{4.50}$$

Substituting Eqs. (4.46) and (4.49) into Eq. (4.50) yields,

$$I_{2y} = \frac{R_2 I_s}{R_1 + R_2 + mR_1} + I_s \tag{4.51}$$

Finally, the current in the R_2 resistor is,

$$I_2 = I_{2x} + I_{2y} \tag{4.52}$$

Substituting Eqs. (4.45) and (4.51) into Eq. (4.52) yields,

$$I_2 = \frac{V_s}{R_1 + R_2 + mR_1} + \frac{R_2 I_s}{R_1 + R_2 + mR_1} + I_s \tag{4.53}$$

Example 4.4 Find the power absorbed by the 6 Ω resistor using superposition theorem for the circuit in Fig. 4.23. Apply PSpice simulation to find the current in the 6 Ω resistor.

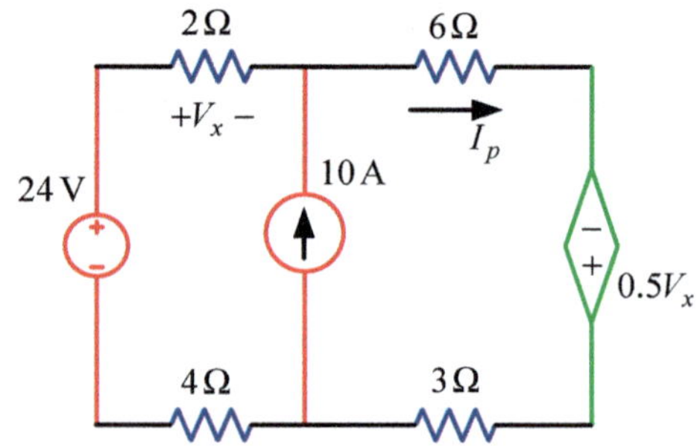

Fig. 4.23 Circuit for Example 4.4

Solution:
Consider that the voltage source is active and the current source is open circuited as shown in Fig. 4.23. Then, applying KVL to the circuit in Fig. 4.24 yields,

$$(2+3+4+6)I_1 - 0.5V_x = 24 \tag{4.54}$$

Now, according to the circuit,

$$V_x = 2I_1 \tag{4.55}$$

Substituting Eq. (4.55) into Eq. (4.54) yields,

$$15I_1 - 1I_1 = 24 \tag{4.56}$$

$$I_1 = \frac{24}{14} = 1.71\,\text{A} \tag{4.57}$$

The current in the 6 Ω resistor is,

$$I_{2x} = I_1 = 1.71\,\text{A} \tag{4.58}$$

Again, consider that the voltage source is short circuited and the current source is active as shown in Fig. 4.25.

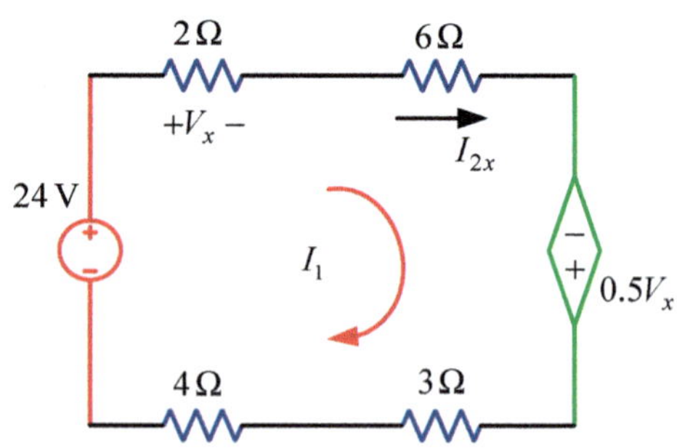

Fig. 4.24 Circuit for Example 4.4 with active voltage source

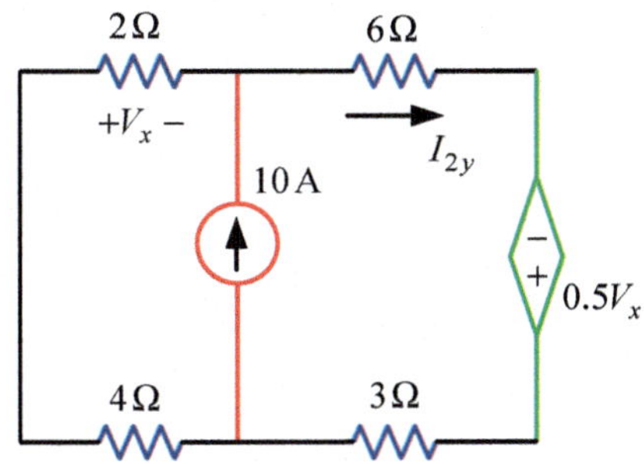

Fig. 4.25 Circuit for Example 4.4 with active current source

In this case, the current in Fig. 4.25 represents the supermesh and it can be redrawn as shown in Fig. 4.26. According to the circuit shown in Fig. 4.26,

$$I_1 = I_2 - 10 \tag{4.59}$$

Applying KVL to the supermesh circuit in Fig. 4.26 yields,

$$6I_1 + 9I_2 = 0.5V_x \tag{4.60}$$

Substituting Eqs. (4.55) and (4.59) into Eq. (4.60) yields,

$$6(I_2 - 10) + 9I_2 = 0.5 \times 2(I_2 - 10) \tag{4.61}$$

$$14I_2 = 50 \tag{4.62}$$

$$I_2 = \frac{50}{14} = 3.57\,\text{A} \tag{4.63}$$

In this case, the current in the 6 Ω resistor is,

$$I_{2y} = I_2 = 3.57\,\text{A} \tag{4.64}$$

Finally, the current in the 6 Ω resistor is,

$$I_p = I_{2x} + I_{2y} = 3.57 + 1.71 = 5.28\,\text{A} \tag{4.65}$$

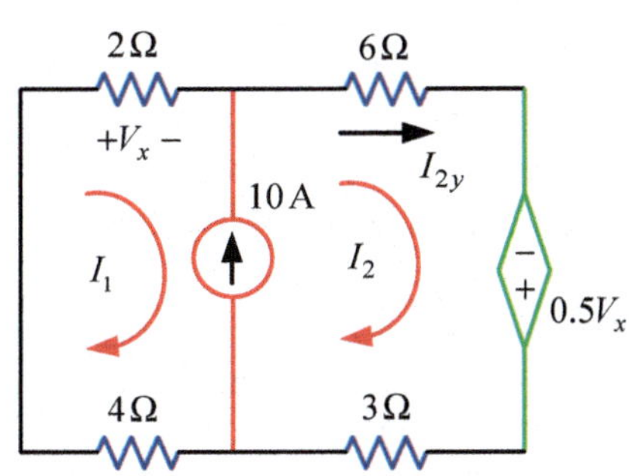

Fig. 4.26 Circuit for Example 4.4 with supermesh

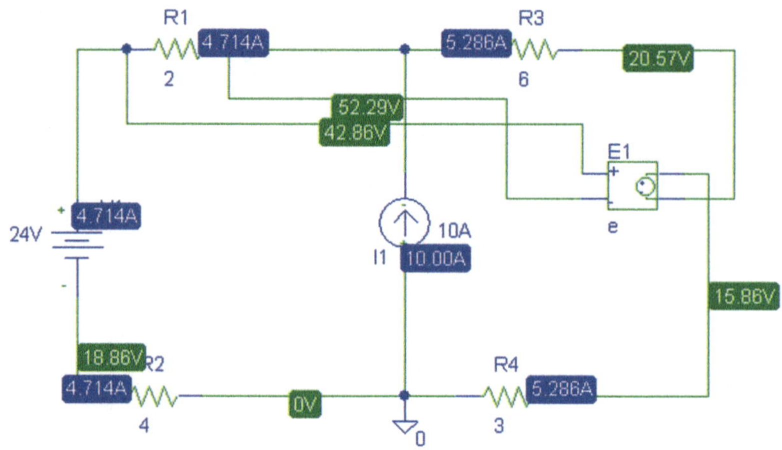

Fig. 4.27 PSpice simulation circuit for Example 4.4

Fig. 4.28 Circuit for Practice Problem 4.4

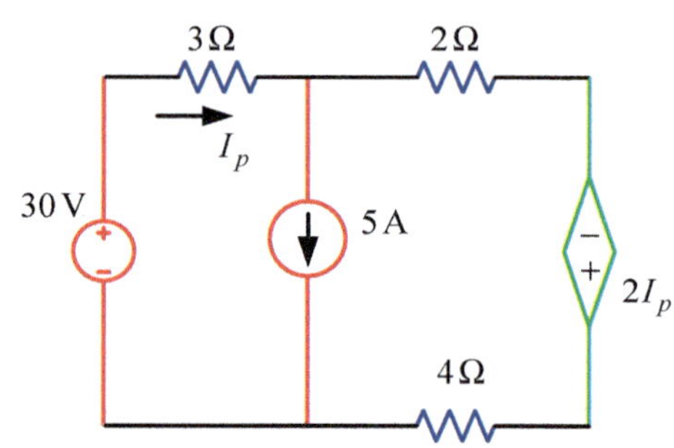

The PSpice simulation result is shown in Fig. 4.27, which shows that the calculated current in the 6 Ω resistor is same as the simulated result.

The power absorbed by the 6 Ω resistor is,

$$P_{6\Omega} = I_2^2 \times 6 = 5.28^2 \times 6 = 167.27\,\text{W} \tag{4.66}$$

Practice Problem 4.4

Determine the current in the 3 Ω resistor using superposition theorem for the circuit shown in Fig. 4.28. Verify the result by PSpice simulation.

4.5 Thevenin's Theorem

Thevenin's theorem is a useful theorem to analyse equivalent circuit of a three-phase induction motor, low-frequency hybrid model and amplifier model of a transistor. In 1883, a French telegraph engineer M. Leon Thevenin introduced this theorem to reduce a complex two-terminal linear circuit into a simple circuit. Thevenin's theorem states that any two-terminal linear circuit consisting of sources and resistors can be replaced by a simple circuit consisting of a Thevenin's equivalent voltage (open-circuit voltage) source V_{Th} in series with a Thevenin's equivalent resistance R_{Th}. Thevenin's equivalent voltage (open-circuit voltage) source is calculated from the same terminals by turning off the independent sources in the circuit. Let us explain the theorem with that aid of the circuit shown in Fig. 4.29.

Initially, open the resistance R_x, mark the terminals and draw the circuit as shown in Fig. 4.30. Thevenin's resistance R_{Th} can be calculated from Fig. 4.30, by short circuiting the voltage source as shown in Fig. 4.31.

The expression of the Thevenin resistance is,

$$R_{\mathrm{Th}} = R_2 + \frac{R_1 R_3}{R_1 + R_3} \tag{4.67}$$

Now, let us redraw the circuit as shown in Fig. 4.32 and calculate the open-circuit or Thevenin's voltage. The source current in the circuit of Fig. 4.32 is,

$$I_s = \frac{V_s}{R_1 + R_3} \tag{4.68}$$

The expression of the Thevenin's (open-circuit) voltage is,

$$V_{\mathrm{Th}} = I_s R_3 \tag{4.69}$$

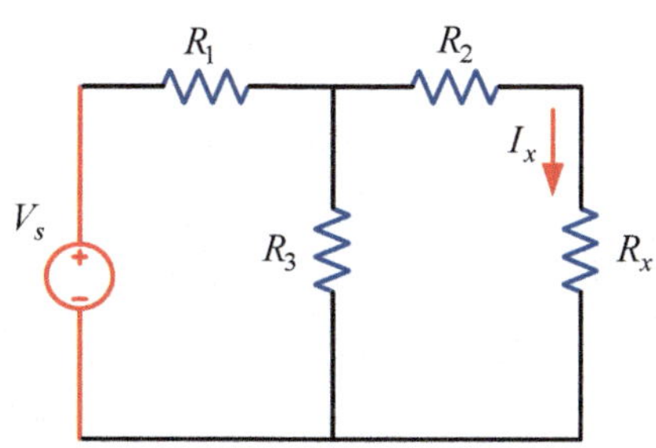

Fig. 4.29 A circuit for Thevenin's theorem

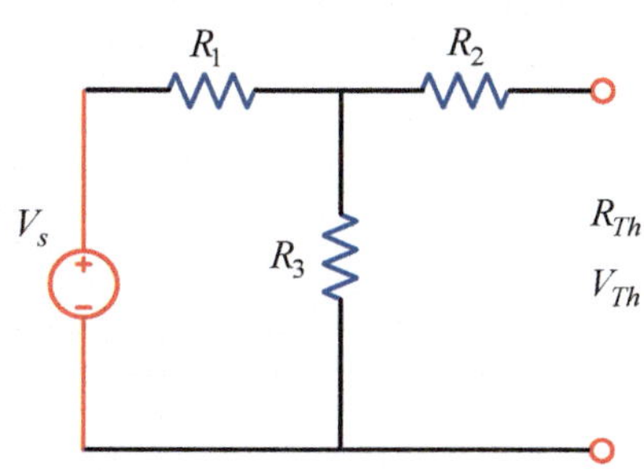

Fig. 4.30 Open terminals for Thevenin's theorem

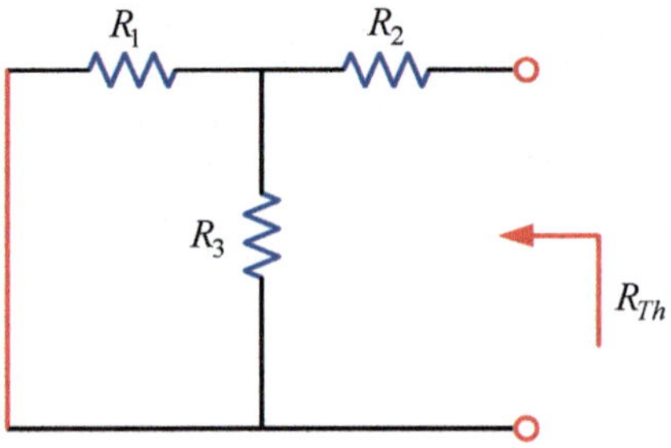

Fig. 4.31 Circuit for calculating Thevenin's resistance

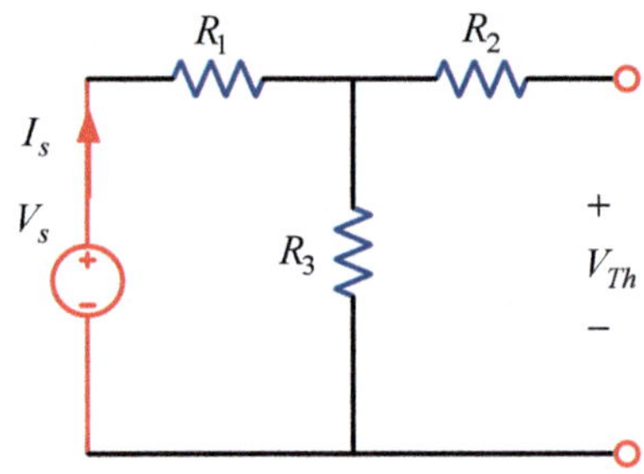

Fig. 4.32 Circuit for calculating Thevenin's voltage

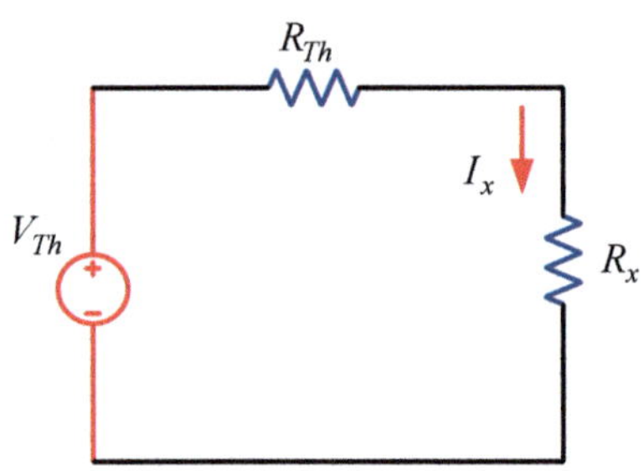

Fig. 4.33 Equivalent Thevenin's circuit

Substituting Eq. (4.68) into Eq. (4.69) yields,

$$V_{\text{Th}} = \frac{V_s}{R_1 + R_3} R_3 \tag{4.70}$$

Finally, the Thevenin's equivalent circuit is shown in Fig. 4.33. The current through the R_x resistor is,

$$I_x = \frac{V_{\text{Th}}}{R_{\text{Th}} + R_x} \tag{4.71}$$

Example 4.5 Use Thevenin's theorem to calculate the current in the 12 Ω resistor of the circuit in Fig. 4.34. Verify the result by the PSpice simulation.

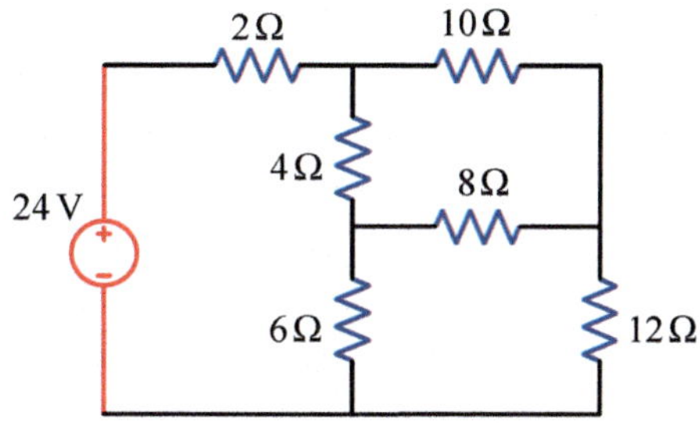

Fig. 4.34 Circuit for Example 4.5

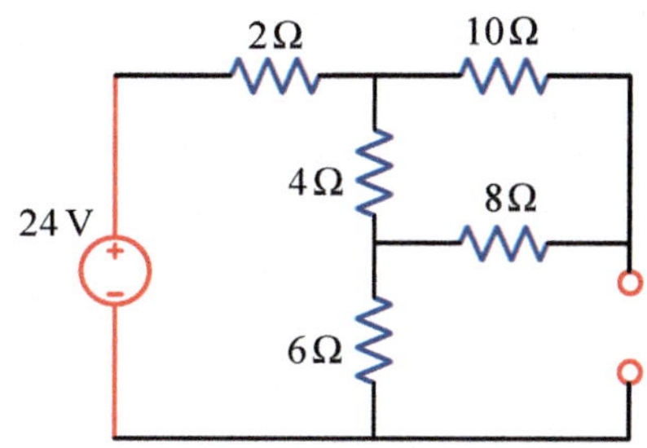

Fig. 4.35 Circuit for Example 4.5 without 12 Ω resistor

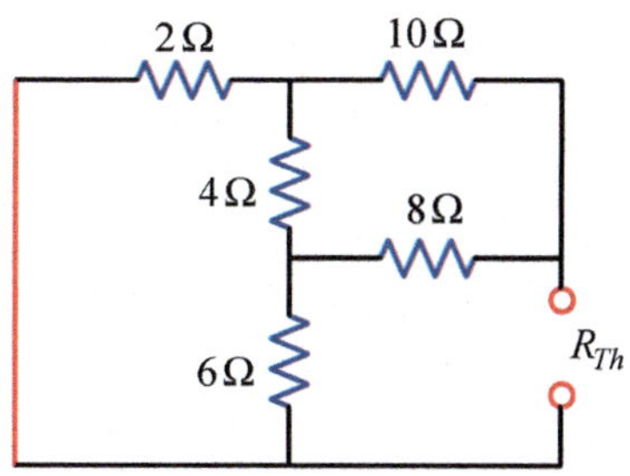

Fig. 4.36 Circuit for Example 4.5 for calculating Thevenin's resistance

Solution:
Since, the point of interest is the current through 12 Ω resistor. Let us open that resistor (12 Ω) from the circuit, and redraw the circuit as shown in Fig. 4.35. Short circuit (turn off) the voltage source to calculate the Thevenin's resistance with the aid of the redrawn circuit in Fig. 4.36.

The left delta circuit of Fig. 4.36 is converted to a wye circuit as shown in Fig. 4.37. The resistances are,

$$R_1 = \frac{4 \times 6}{4+6+2} = 2\,\Omega \tag{4.72}$$

$$R_2 = \frac{6 \times 2}{4+6+2} = 1\,\Omega \tag{4.73}$$

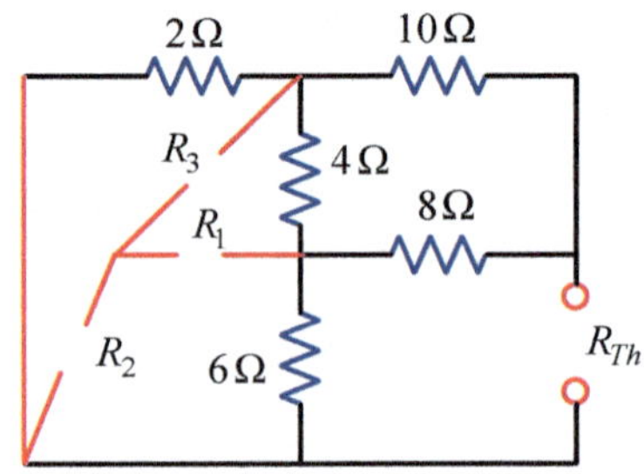

Fig. 4.37 Circuit for Example 4.5 for delta to wye conversion

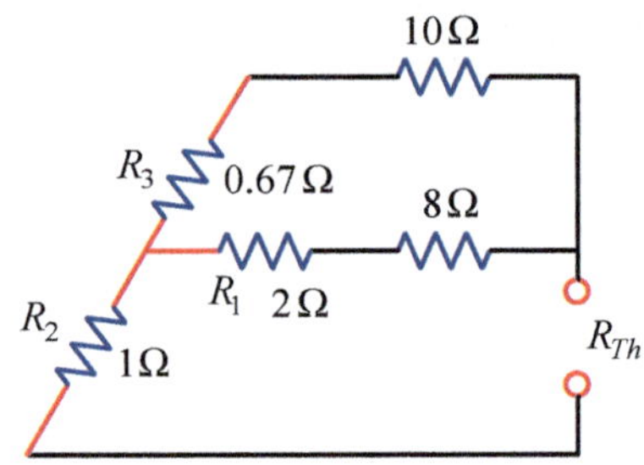

Fig. 4.38 Circuit for calculating Thevenin's resistance for Example 4.5

$$R_3 = \frac{4 \times 2}{4 + 6 + 2} = 0.67\,\Omega \tag{4.74}$$

After conversion from delta to wye, the circuit can be redrawn as shown in Fig. 4.38. The resistor 0.67 Ω is in series with 10 Ω resistor. Again, the resistor 2 Ω is in series with 8 Ω resistor. Therefore, the Thevenin's resistance can be calculated as,

$$R_{\mathrm{Th}} = 1 + \frac{(8 + 2)(10 + 0.67)}{10 + 10.67} = 6.16\,\Omega \tag{4.75}$$

To calculate the open-circuit voltage, use the circuit in Fig. 4.35. The 8 Ω resistor is in series with 10 Ω resistor, and in parallel with 4 Ω resistor. In this case, the equivalent circuit resistance is,

$$R_{\mathrm{eq}} = 2 + \frac{(4)(10 + 8)}{4 + 18} + 6 = 11.27\,\Omega \tag{4.76}$$

The source current is,

$$I_s = \frac{24}{11.27} = 2.13\,\mathrm{A} \tag{4.77}$$

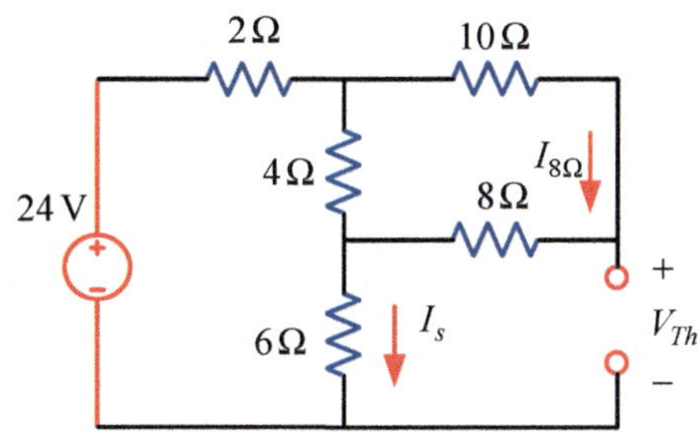

Fig. 4.39 Circuit for calculating Thevenin's voltage

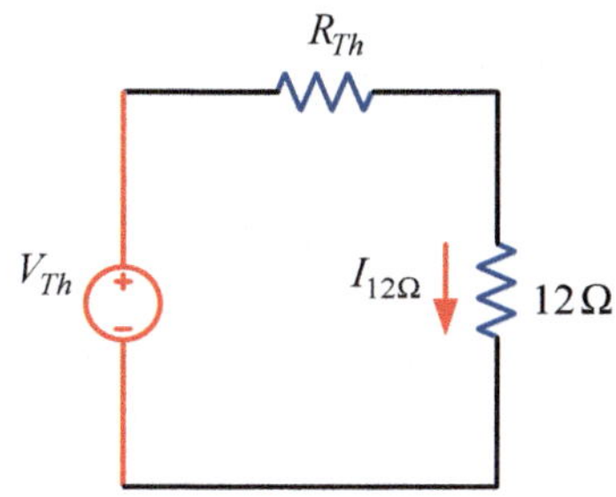

Fig. 4.40 Final circuit for Example 4.5

The current in the 8 Ω resistor is,

$$I_{8\Omega} = 2.13 \times \frac{4}{10+8+4} = 0.39\,\Omega \tag{4.78}$$

Applying KVL to the circuit in Fig. 4.39 yields,

$$-V_{\text{Th}} + 8I_{8\Omega} + 6I_s = 0 \tag{4.79}$$

$$V_{\text{Th}} = 8 \times 0.39 + 6 \times 2.13 = 15.9\,\text{V} \tag{4.80}$$

Thevenin's equivalent circuit is shown in Fig. 4.40. The current in the 12 Ω resistor is,

$$I_{12\Omega} = \frac{15.9}{6.16+12} = 0.875\,\text{A} \tag{4.81}$$

The PSpice simulation results are shown in Fig. 4.41. The simulation current is same as the calculated result.

Practice Problem 4.5

Calculate the current in the 8 Ω resistor of the circuit in Fig. 4.42 using Thevenin's theorem. Verify the results using PSpice simulation.

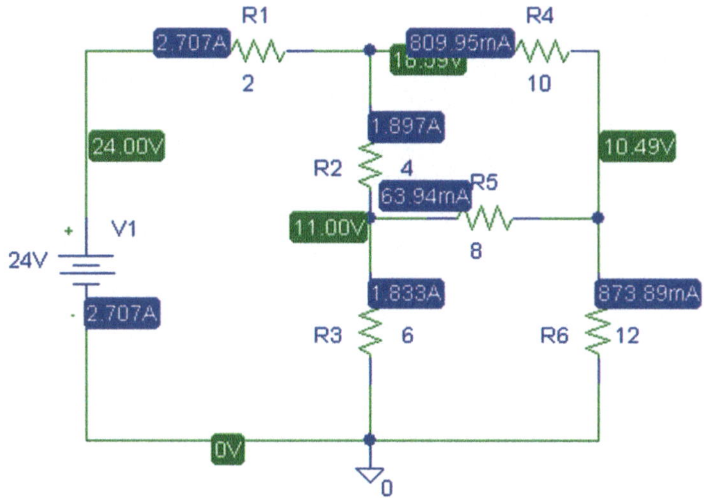

Fig. 4.41 PSpice simulation circuit for Example 4.5

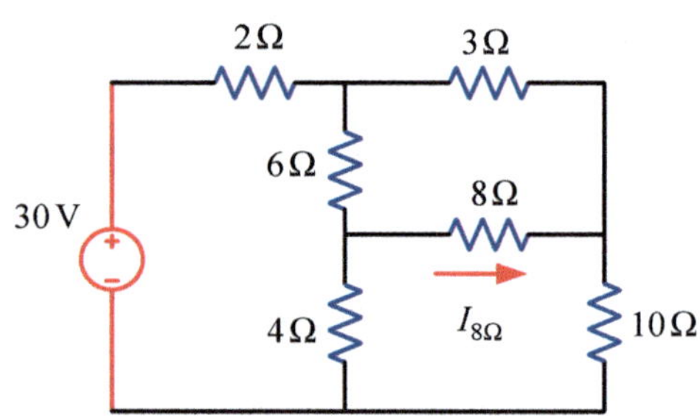

Fig. 4.42 Circuit for Practice Problem 4.5

4.6 Thevenin's Theorem with Dependent Source

Thevenin's theorem with a dependent source requires little more work. This idea can be explained with the help of the circuit as shown in Fig. 4.43. Here, current I_p through resistor R_3 needs to be evaluated using Thevenin's theorem. Initially, open the resistor R_3 from the circuit and calculate the Thevenin's (open-circuit) voltage from the circuit as shown in Fig. 4.44.

In this case, the Thevenin's voltage will be the sum of the voltage drop across R_2 and the dependent source voltage, which is,

$$V_{\text{Th}} = \frac{V_s}{R_1 + R_2} R_2 + mV_x \tag{4.82}$$

Now to calculate the Thevenin's resistance, turn off (short circuited) the voltage source and connect a voltage or current source with a unity magnitude across the

Fig. 4.43 Circuit with a dependent voltage source

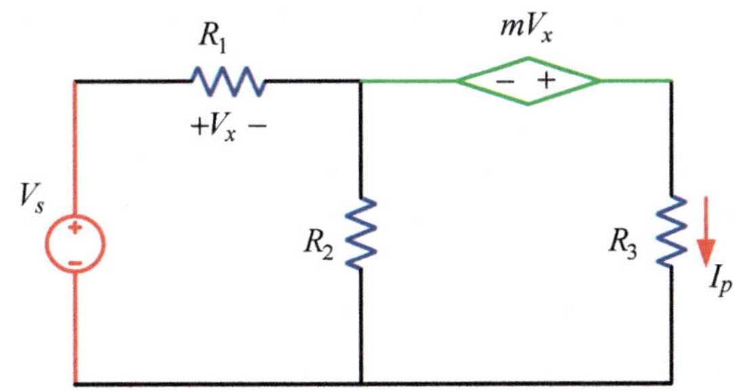

Fig. 4.44 Circuit for calculating Thevenin's voltage

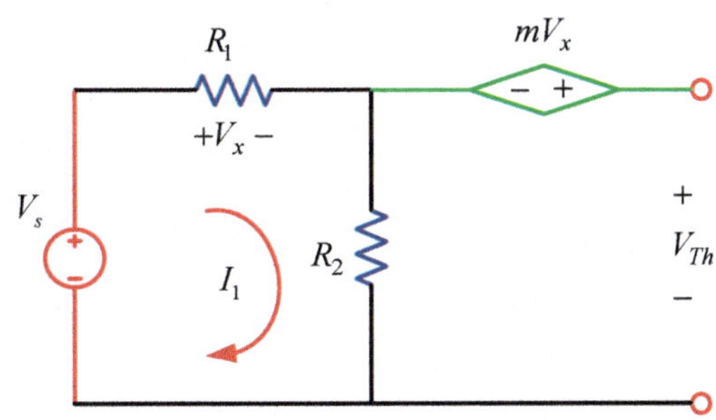

Fig. 4.45 Circuit with an additional voltage source

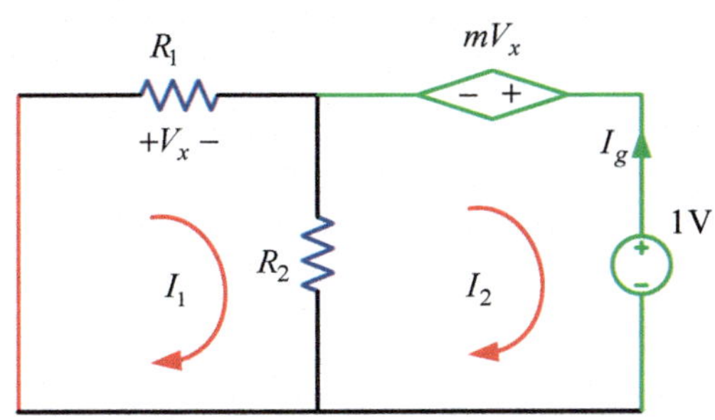

open terminals. Insert the voltage source with a magnitude of 1 V and find the current I_g to calculate the Thevenin's resistance as shown in Fig. 4.45. Applying KVL to the circuit in Fig. 4.45 yields,

$$I_1R_1 + R_2(I_1 - I_2) = 0 \tag{4.83}$$

$$I_1(R_1 + R_2) - I_2R_2 = 0 \tag{4.84}$$

$$R_2(I_2 - I_1) - mV_x + 1 = 0 \tag{4.85}$$

However, according to the circuit, the following equation can be written:

$$V_x = I_1 R_1 \tag{4.86}$$

Substituting Eq. (4.86) into Eq. (4.85) yields,

$$R_2(I_2 - I_1) - mI_1R_1 + 1 = 0 \tag{4.87}$$

$$I_1(mR_1 + R_2) - I_2R_2 = 1 \tag{4.88}$$

From Eq. (4.84) and (4.88), the current I_2 can be determined as,

$$I_2 = \frac{\begin{vmatrix} (R_1 + R_2) & 0 \\ (mR_1 + R_2) & 1 \end{vmatrix}}{\begin{vmatrix} (R_1 + R_2) & -R_2 \\ (mR_1 + R_2) & -R_2 \end{vmatrix}} \tag{4.89}$$

$$I_g = -I_2 = \frac{\begin{vmatrix} (R_1 + R_2) & 0 \\ (mR_1 + R_2) & 1 \end{vmatrix}}{\begin{vmatrix} (R_1 + R_2) & -R_2 \\ (mR_1 + R_2) & -R_2 \end{vmatrix}} \tag{4.90}$$

Finally, Thevenin's resistance can be determined as,

$$R_{\text{Th}} = \frac{1}{I_g} \tag{4.91}$$

Example 4.6 Determine the current in the $10\,\Omega$ resistor of the circuit in Fig. 4.46 using Thevenin's theorem. Verify the results using PSpice simulation.

Solution:
Initially, open the $10\,\Omega$ resistor and calculate the Thevenin's voltage as shown in Fig. 4.47. In this case, the source current is,

$$I_s = \frac{30}{2+4} = 5\,\text{A} \tag{4.92}$$

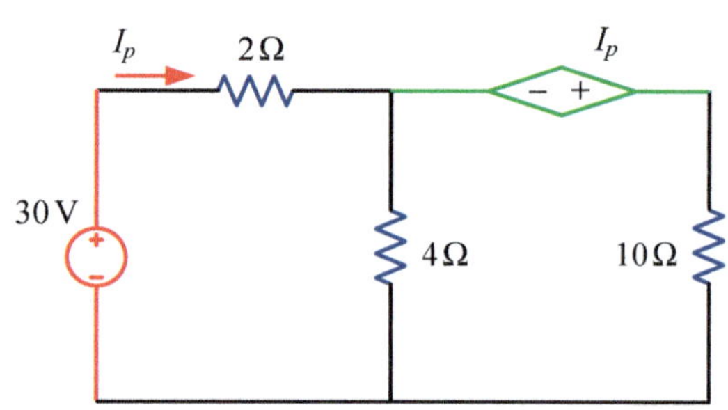

Fig. 4.46 Circuit for Example 4.6

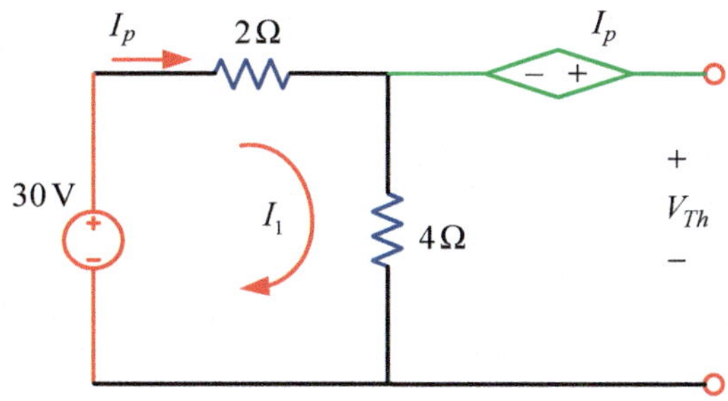

Fig. 4.47 Circuit for calculating Thevenin's voltage for Example 4.6

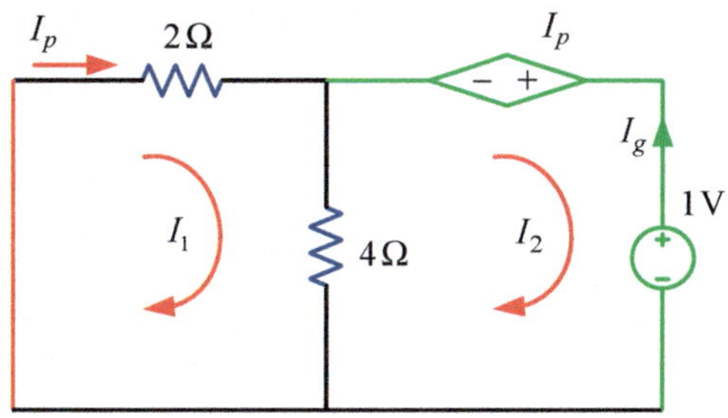

Fig. 4.48 Circuit for calculating Thevenin's resistance for Example 4.6

Thevenin's voltage can be calculated as,

$$V_{\text{Th}} = 4 \times 5 + 5 = 25\,\text{V} \tag{4.93}$$

Insert a unity voltage source to the circuit in Fig. 4.48 to calculate the Thevenin's resistance. Applying KVL to the circuit in Fig. 4.48 yields,

$$6I_1 - 4I_2 = 0 \tag{4.94}$$

$$I_1 = \frac{2}{3} I_2 \tag{4.95}$$

$$-4I_1 + 4I_2 - I_p + 1 = 0 \tag{4.96}$$

$$-4I_1 + 4I_2 - I_1 + 1 = 0 \tag{4.97}$$

$$5I_1 - 4I_2 = 1 \tag{4.98}$$

Substituting Eq. (4.95) into Eq. (4.98) yields,

$$5 \times \frac{2}{3} I_2 - 4I_2 = 1 \tag{4.99}$$

$$3.33I_2 - 4I_2 = 1 \tag{4.100}$$

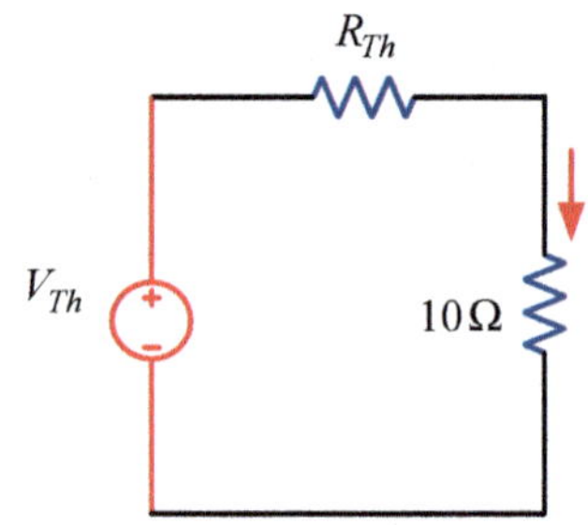

Fig. 4.49 Final circuit for Example 4.6

$$-0.67I_2 = 1 \tag{4.101}$$

$$I_2 = -\frac{1}{0.67} = -1.49\,\text{A} \tag{4.102}$$

$$I_g = -I_2 = 1.49\,\text{A} \tag{4.103}$$

Thevenin's resistance can be determined by omitting negative sign as,

$$R_{\text{Th}} = \frac{1}{I_g} = \frac{1}{1.49} = 0.67\,\Omega \tag{4.104}$$

Thevenin's equivalent circuit is shown in Fig. 4.49. The current in the $10\,\Omega$ resistor is,

$$I_{10\Omega} = \frac{25}{10 + 0.67} = 2.34\,\text{A} \tag{4.105}$$

The PSpice simulation circuit is shown in Fig. 4.50. The simulation current is found to be the same as the calculated current.

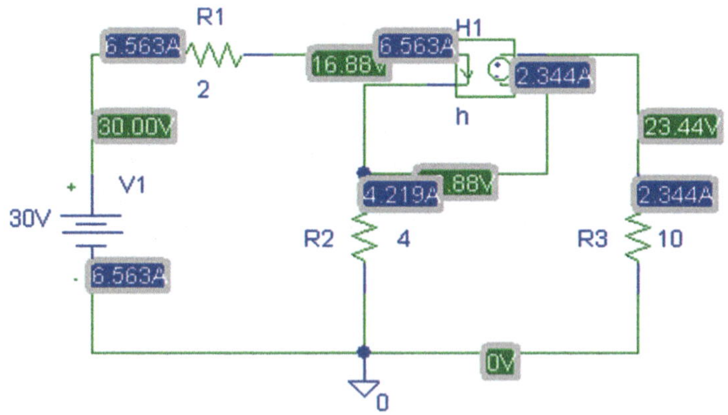

Fig. 4.50 PSpice simulation circuit for Example 4.6

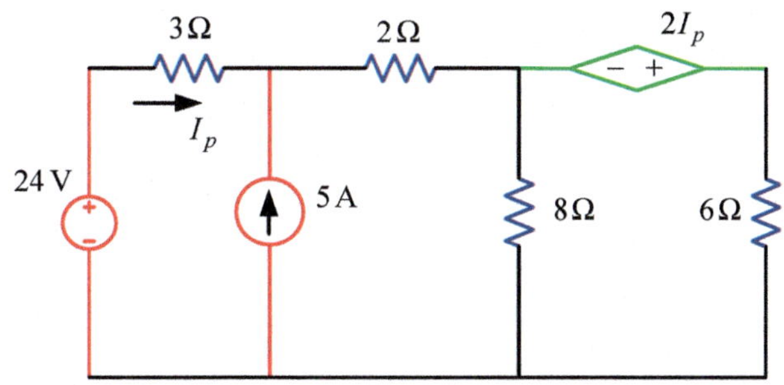

Fig. 4.51 Circuit for Practice Problem 4.6

Practice Problem 4.6

Find the current in the 6 Ω resistor of the circuit in Fig. 4.51 using Thevenin's theorem. Verify the results using PSpice simulation.

4.7 Norton's Theorem

Edward Lawry Norton (1898–1983), a Bell-Lab engineer, derived this theorem in 1926. Norton's theorem states that any linear circuit containing sources and resistors can be replaced by an equivalent circuit that consists of a current source in parallel with an equivalent resistor at a given pair of terminals.

This Norton's equivalent current at a given pair of terminals can be determined by short circuiting the terminals. The Norton's equivalent resistance can be determined from the opening terminals of a specific circuit by turning off the sources, that is, by short circuiting the voltage source and open circuiting the current source.

Figure 4.52 shows a circuit to calculate the current in the resistor R_4 by Norton's theorem. At the first step, the circuit is redrawn in Fig. 4.53 by removing the resistor R_4 to create the open terminals. Now, by turning off (short circuiting) the voltage source as shown in Fig. 4.54, the Norton's resistance is calculated as,

$$R_N = R_3 + \frac{R_1 R_2}{R_1 + R_2} \tag{4.106}$$

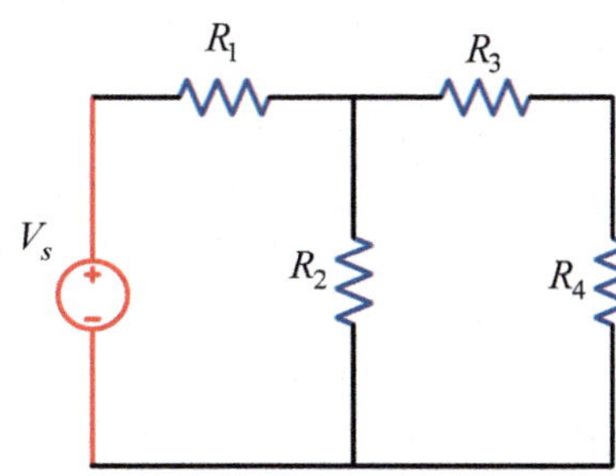

Fig. 4.52 Circuit for Norton's theorem

Fig. 4.53 Circuit for Norton's theorem with opening resistance

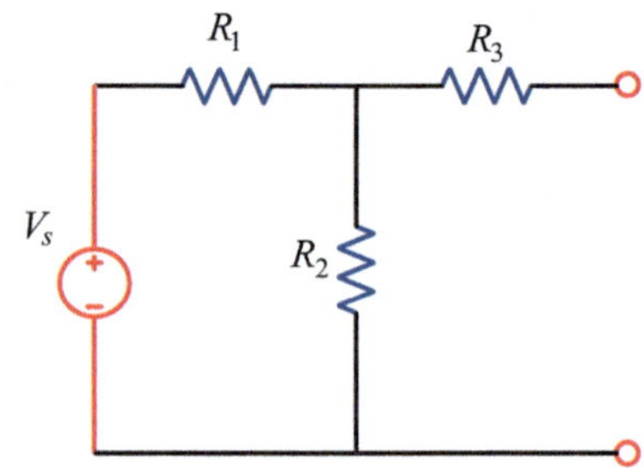

Fig. 4.54 Circuit for calculating Norton's resistance

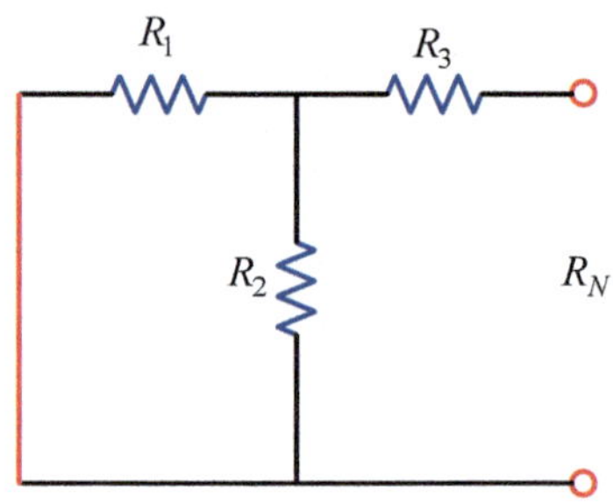

Fig. 4.55 Circuit for calculating Norton's current

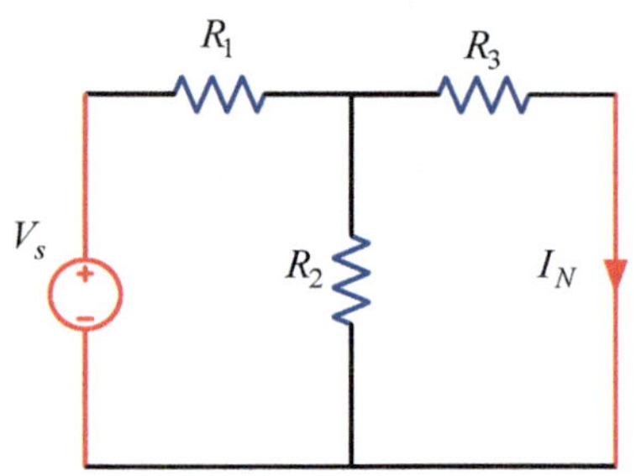

The Norton's or short-circuit current is calculated from Fig. 4.55 as,

$$R_t = R_1 + \frac{R_2 R_3}{R_2 + R_3} \tag{4.107}$$

The source current is,

$$I_s = \frac{V_s}{R_t} = \frac{V_s}{R_1 + \frac{R_2 R_3}{R_2 + R_3}} \tag{4.108}$$

The Norton's current can be determined as,

$$I_N = I_s \frac{R_2}{R_2 + R_3} \tag{4.109}$$

Finally, the current in the R_4 resistor can be determined from the Norton's equivalent circuit in Fig. 4.56 as,

$$I_4 = I_N \frac{R_N}{R_N + R_4} \tag{4.110}$$

Example 4.7 Find the current in the 1 Ω resistor of the circuit in Fig. 4.57 using Norton's theorem. Apply PSpice simulation to compare the result.

Solution:
Open the 1 Ω resistance and redraw the circuit as shown in Fig. 4.58. To calculate the Norton's resistance, open the current source and short circuit the voltage source as shown in Fig. 4.59.

Fig. 4.56 Norton's equivalent circuit

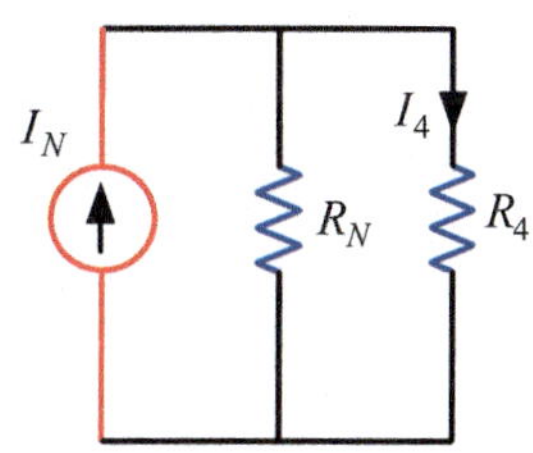

Fig. 4.57 Circuit for Example 4.7

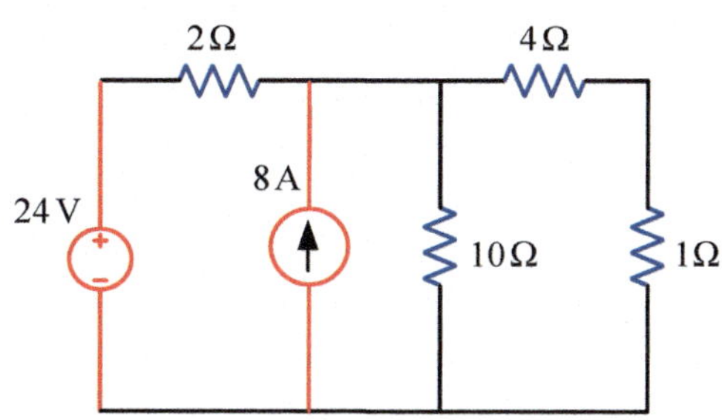

Fig. 4.58 Circuit for Example 4.7 with opening resistance

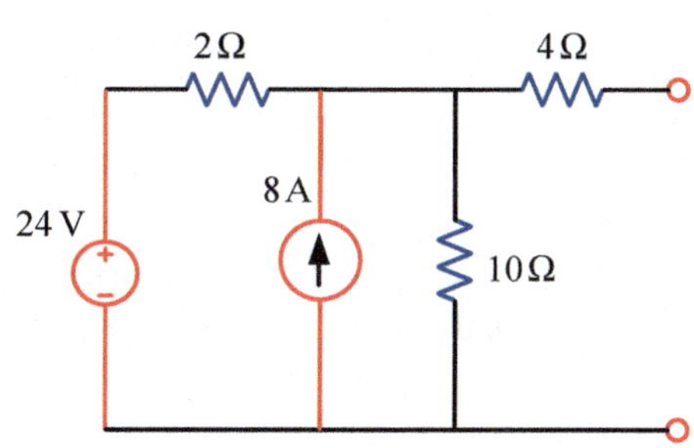

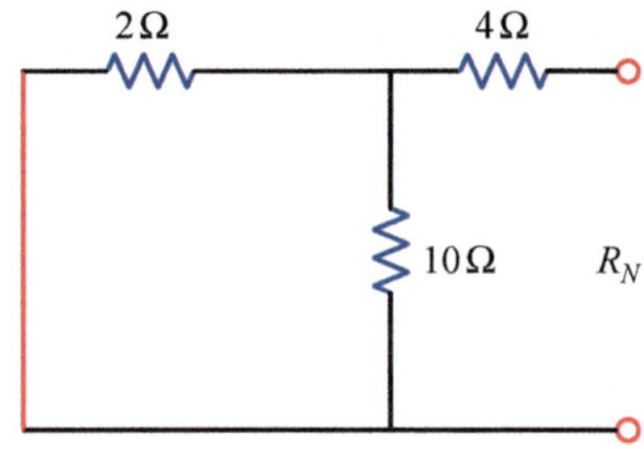

Fig. 4.59 Circuit for Example 4.7 for calculating Norton's resistance

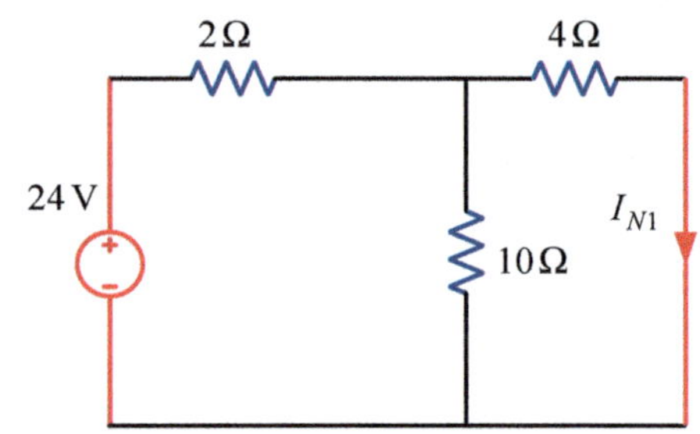

Fig. 4.60 Norton's current calculation with voltage source for Example 4.7

The Norton's resistance can be determined as,

$$R_N = 4 + \frac{2 \times 10}{2 + 10} = 5.67\,\Omega \tag{4.111}$$

First, consider that the voltage source is in active condition and the current source is open circuited (turned off) to find the short-circuit current as shown in Fig. 4.60.

The total circuit resistance is,

$$R_t = 2 + \frac{4 \times 10}{4 + 10} = 4.86\,\Omega \tag{4.112}$$

The source current is,

$$I_s = \frac{24}{4.86} = 4.94\,\text{A} \tag{4.113}$$

With the active voltage source, the value of the short-circuit current is,

$$I_{N1} = 4.94 \times \frac{10}{10 + 4} = 3.53\,\text{A} \tag{4.114}$$

Again, considering the current source is in active condition and voltage source is short circuited (turned off) to find the short-circuit current as shown in Fig. 4.61.

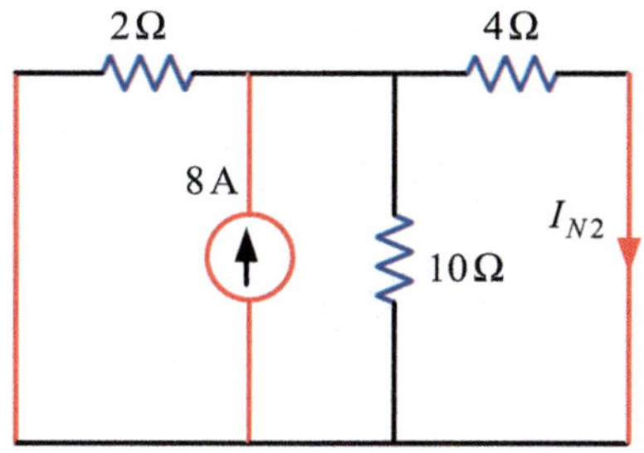

Fig. 4.61 Norton's current calculation with current source for Example 4.7

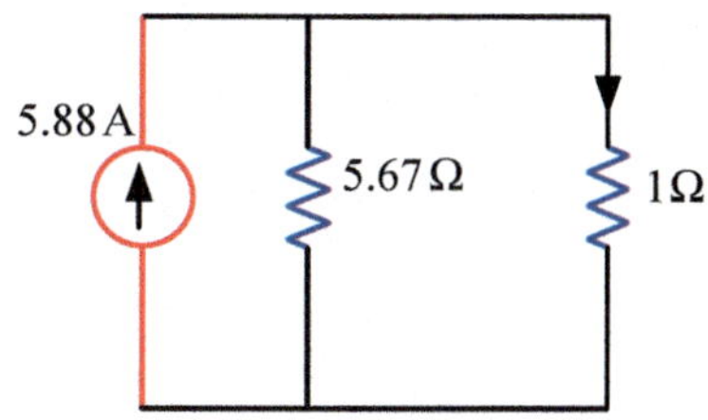

Fig. 4.62 Norton's equivalent circuit for Example 4.7

The following current can be calculated:

$$I_{4\Omega,10\Omega} = 8 \times \frac{2}{2 + \frac{4\times10}{4+10}} = 3.29\,\Omega \tag{4.115}$$

With the active current source, the value of the Norton's current is,

$$I_{N2} = 3.29 \times \frac{10}{10+4} = 2.35\,\Omega \tag{4.116}$$

Finally, the Norton's current for both active sources can be calculated as,

$$I_N = I_{N1} + I_{N2} = 3.53 + 2.35 = 5.88\,\text{A} \tag{4.117}$$

The Norton's equivalent circuit is shown in Fig. 4.62.
The current in the 1 Ω resistor is,

$$I_{1\Omega} = 5.88 \times \frac{5.67}{5.67+1} = 5\,\text{A} \tag{4.118}$$

The PSpice simulation current is found to be the same as the calculated current as can be seen in Fig. 4.63.

Practice Problem 4.7
An electrical circuit is shown in Fig. 4.64. Use Norton's theorem to find the current in the 10 Ω resistor. Verify the result by PSpice simulation.

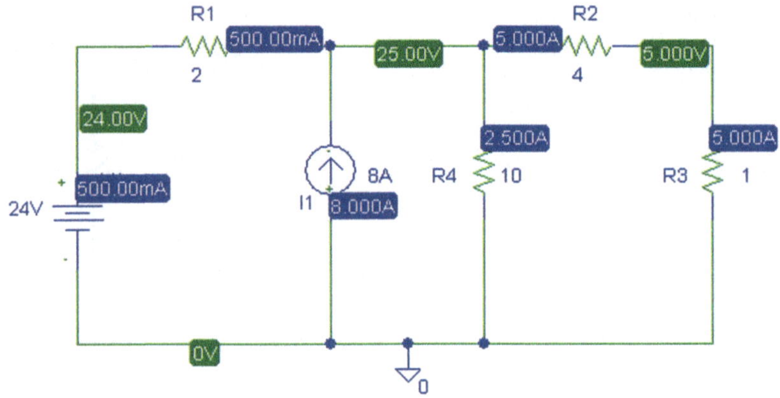

Fig. 4.63 PSpice simulation circuit for Example 4.7

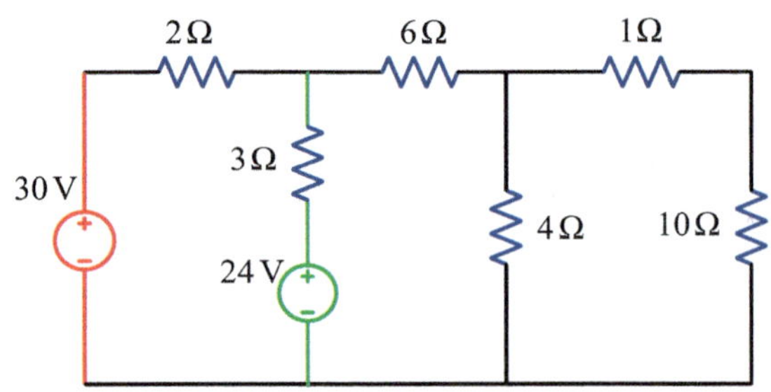

Fig. 4.64 Circuit for Practice Problem 4.7

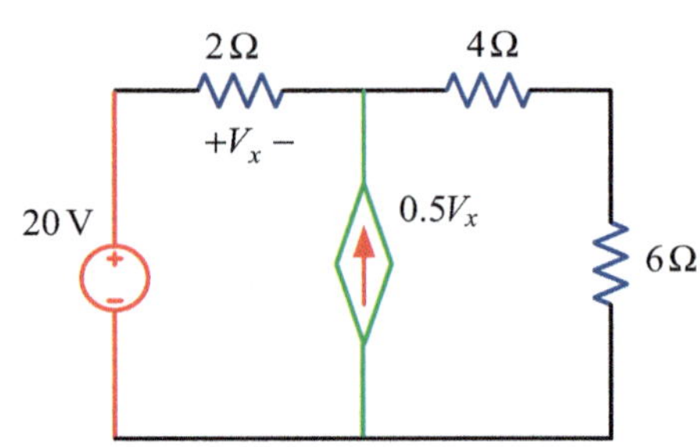

Fig. 4.65 Circuit for Example 4.8

Example 4.8 An electrical circuit is shown in Fig. 4.65. Use Norton's theorem to find the current in the $6\,\Omega$ resistor. Verify the result by PSpice simulation.

Solution:
Open the $6\,\Omega$ resistor and redraw the circuit as shown in Fig. 4.66. Applying KCL at the node a yields,

$$\frac{20 - V_{\mathrm{Th}}}{2} + 0.5V_x = 0 \tag{4.119}$$

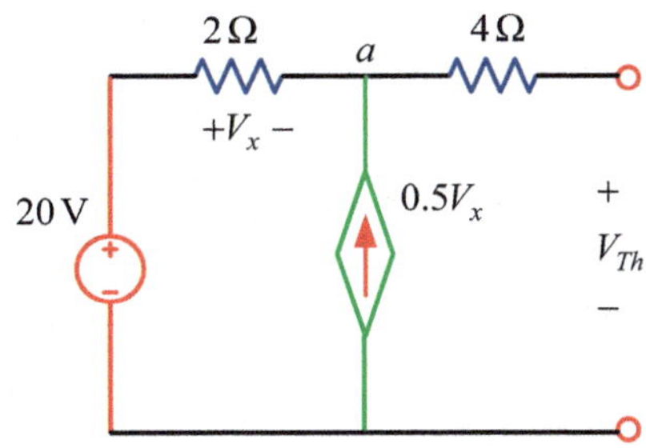

Fig. 4.66 Open-circuit voltage calculation for Example 4.8

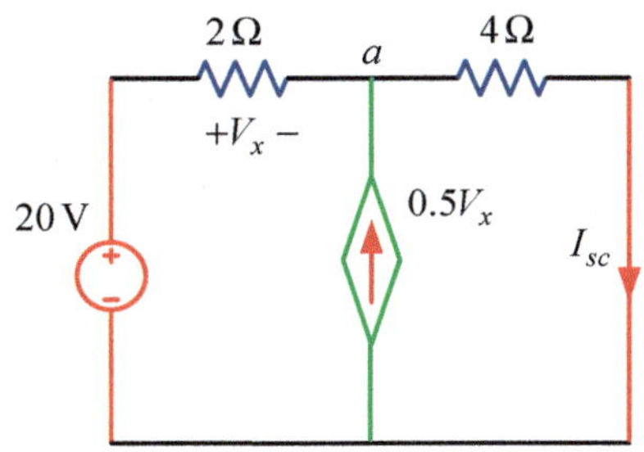

Fig. 4.67 Short-circuit current calculation for Example 4.8

Again, applying KVL to the circuit in Fig. 4.66 yields,

$$-20 + V_x + V_{\text{Th}} = 0 \tag{4.120}$$

$$V_x = 20 - V_{\text{Th}} \tag{4.121}$$

Substituting Eq. (4.121) into Eq. (4.119) yields,

$$\frac{20 - V_{\text{Th}}}{2} + 0.5(20 - V_{\text{Th}}) = 0 \tag{4.122}$$

$$V_{\text{Th}} = 20\,\text{V} \tag{4.123}$$

Again short circuit the open terminals as shown in Fig. 4.67. Applying KCL at the node a yields,

$$\frac{20 - V_a}{2} + 0.5V_x = I_{\text{sc}} = \frac{V_a}{4} \tag{4.124}$$

According to the circuit, the following equation can be written:

$$V_x = 20 - V_a \tag{4.125}$$

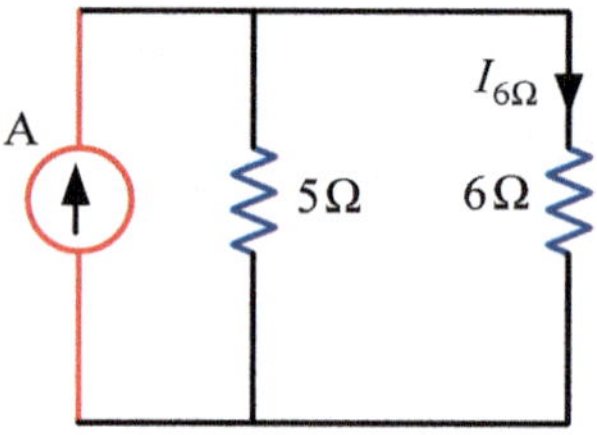

Fig. 4.68 Final Norton's equivalent circuit for Example 4.8

Substituting Eq. (4.125) into Eq. (4.124) yields,

$$\frac{20 - V_a}{2} + 10 - 0.5V_a = \frac{V_a}{4} \tag{4.126}$$

$$V_a = \frac{20}{1.25} = 16\,\text{V} \tag{4.127}$$

Then, the short-circuit current is,

$$I_{\text{sc}} = \frac{16}{4} = 4\,\text{A} \tag{4.128}$$

The Norton's resistance is calculated as,

$$R_N = \frac{V_{\text{Th}}}{I_{\text{sc}}} = \frac{20}{4} = 5\,\Omega \tag{4.129}$$

The final Norton's equivalent circuit is shown in Fig. 4.68. The current in the $6\,\Omega$ resistor is calculated as,

$$I_{6\Omega} = \frac{4 \times 5}{5 + 6} = 1.82\,\text{A} \tag{4.130}$$

The calculated current is found to be the same as the PSpice simulated current as shown in Fig. 4.69.

Practice Problem 4.8
Using Norton's theorem determine the current in the $6\,\Omega$ resistor of the circuit in Fig. 4.70. Apply PSpice simulation to verify the result.

4.8 Maximum Power Transfer Theorem

The prime importance of power utility companies to transfer power from the source to the load through transmission lines without a loss. In this case, the source and the load impedance are required to be matched with the characteristic impedance of the transmission line. In another case, the maximum power point tracking (MPPT) of

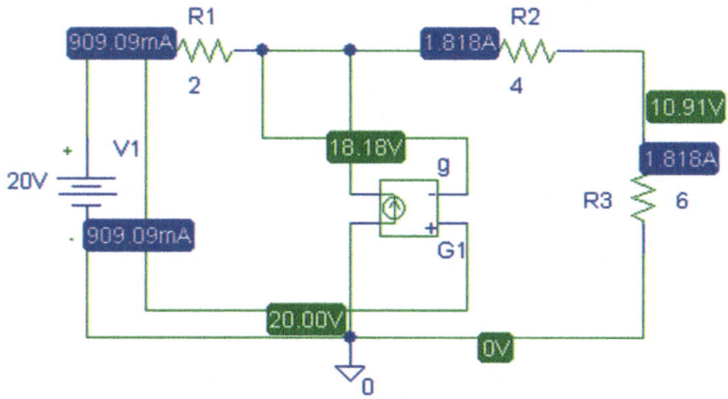

Fig. 4.69 PSpice simulation circuit for Example 4.8

Fig. 4.70 Circuit for Example 4.8

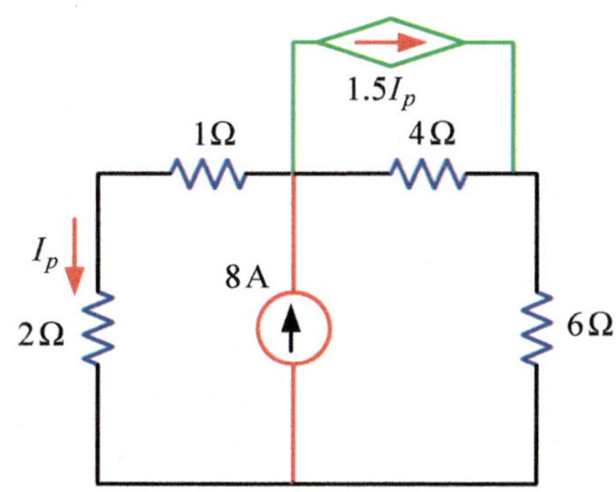

the photovoltaic system is obtained by the incremental conductance method when the load resistance is equal to the resistance of a photovoltaic system. Therefore, there is an importance of the maximum power transfer theorem. Maximum power transfer theorem states 'in any linear network, maximum power will be transferred from the source to the load when the load resistance is equal to the internal resistance of the source'. This internal resistance is basically equal to the Thevenin's resistance seen from the open terminals. Let us explain this theorem with the aid of the given circuit in Fig. 4.71.

Fig. 4.71 Circuit for maximum power transfer theorem

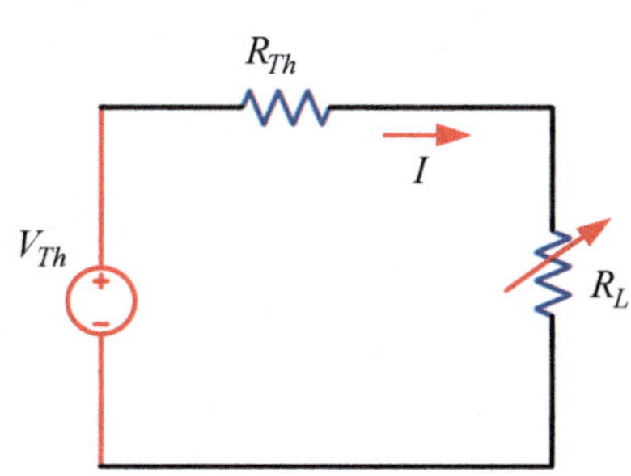

The load current is,

$$I = \frac{V_{\mathrm{Th}}}{R_{\mathrm{Th}} + R_L} \tag{4.131}$$

The power absorbed by the load is,

$$P = I^2 R_L \tag{4.132}$$

Substituting Eq. (4.131) into Eq. (4.132) yields,

$$P = \frac{V_{\mathrm{Th}}^2}{(R_{\mathrm{Th}} + R_L)^2} R_L \tag{4.133}$$

In Eq. (4.133), the load resistance is a variable parameter. Therefore, differentiate Eq. (4.133) with respect to the load resistance and set it to zero for finding maximum power. These steps are as follows:

$$\frac{\mathrm{d}P}{\mathrm{d}R_L} = V_{\mathrm{Th}}^2 \left\{ \frac{(R_{\mathrm{Th}} + R_L)^2 \frac{\mathrm{d}}{\mathrm{d}R_L}(R_L) - R_L \frac{\mathrm{d}}{\mathrm{d}R_L}(R_{\mathrm{Th}} + R_L)^2}{(R_{\mathrm{Th}} + R_L)^4} \right\} \tag{4.134}$$

$$\frac{\mathrm{d}P}{\mathrm{d}R_L} = V_{\mathrm{Th}}^2 \left\{ \frac{(R_{\mathrm{Th}} + R_L)^2 - 2R_L(R_{\mathrm{Th}} + R_L)}{(R_{\mathrm{Th}} + R_L)^4} \right\} = 0 \tag{4.135}$$

$$(R_{\mathrm{Th}} + R_L)^2 - 2R_L(R_{\mathrm{Th}} + R_L) = 0 \tag{4.136}$$

$$R_{\mathrm{Th}}^2 + R_L^2 + 2R_{\mathrm{Th}}R_L - 2R_L R_{\mathrm{Th}} - 2R_L^2 = 0 \tag{4.137}$$

$$R_{\mathrm{Th}}^2 = R_L^2 \tag{4.138}$$

$$R_{\mathrm{Th}} = R_L \tag{4.139}$$

From Eq. (4.138), it is concluded that maximum power will be transferred from the source to the load when the Thevenin's resistance is equal to the load resistance.

Substituting Eq. (4.139) into Eq. (4.133) yields,

$$P_{\mathrm{mxm}} = \frac{V_{\mathrm{Th}}^2}{(R_{\mathrm{Th}} + R_{\mathrm{Th}})^2} R_{\mathrm{Th}} \tag{4.140}$$

$$P_{\mathrm{mxm}} = \frac{V_{\mathrm{Th}}^2}{4R_{\mathrm{Th}}} \tag{4.141}$$

Again, substituting Eq. (4.139) into Eq. (4.131) yields,

$$I = \frac{V_{\mathrm{Th}}}{2R_{\mathrm{Th}}} \tag{4.142}$$

The expression of input power can be determined as,

$$P_{\mathrm{in}} = V_{\mathrm{Th}} I \tag{4.143}$$

Substituting Eq. (4.142) into Eq. (4.143) yields,

$$P_{\mathrm{in}} = V_{\mathrm{Th}} \frac{V_{\mathrm{Th}}}{2R_{\mathrm{Th}}} = \frac{V_{\mathrm{Th}}^2}{2R_{\mathrm{Th}}} \tag{4.144}$$

Now, the efficiency can be determined as,

$$\eta = \frac{P_{\mathrm{mxm}}}{P_{\mathrm{in}}} \tag{4.145}$$

Substituting Eqs. (4.141) and (4.144) into Eq. (4.145) yields,

$$\eta = \frac{\frac{V_{\mathrm{Th}}^2}{4R_{\mathrm{Th}}}}{\frac{V_{\mathrm{Th}}^2}{2R_{\mathrm{Th}}}} \times 100 \tag{4.146}$$

$$\eta = 50\% \tag{4.147}$$

From Eq. (4.147), it is seen that the efficiency will be 50% under maximum power transfer condition.

Example 4.9 Find the value of the load resistance and the maximum power absorbed by the load, if the circuit in Fig. 4.72 transfers maximum power.

Solution:
Open the load resistance and redraw the circuit as shown in Fig. 4.73. Find the Thevenin's resistance by opening the current source as shown in Fig. 4.74.

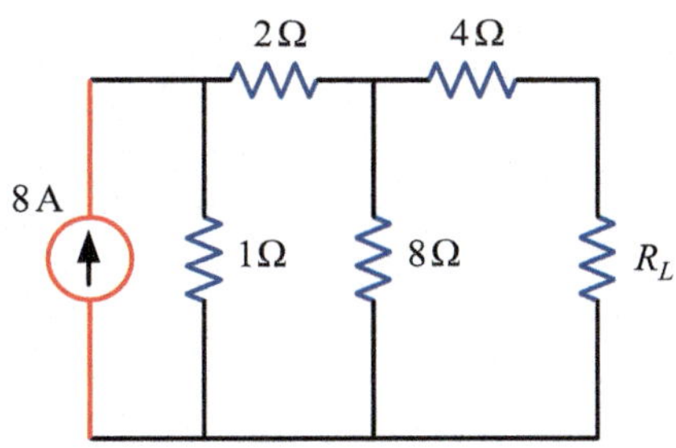

Fig. 4.72 Circuit for Example 4.9

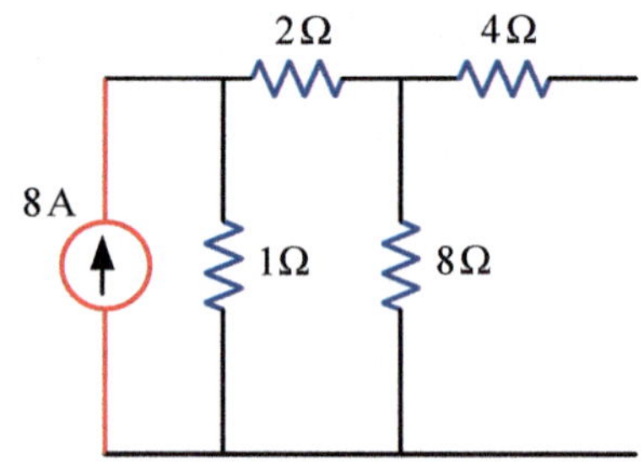

Fig. 4.73 Circuit after removing the load resistance for Example 4.9

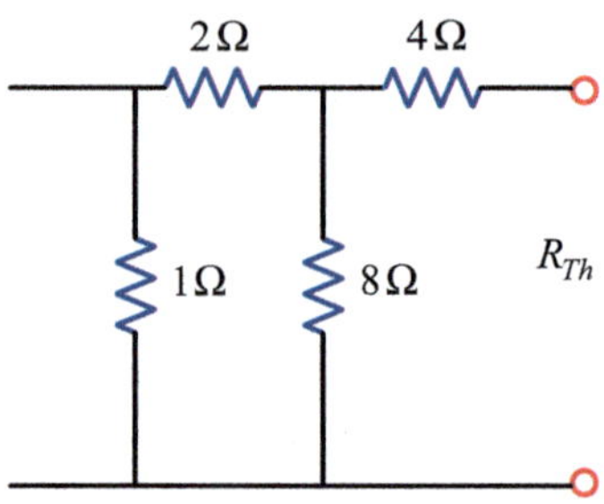

Fig. 4.74 Circuit for calculating Thevenin's resistance for Example 4.9

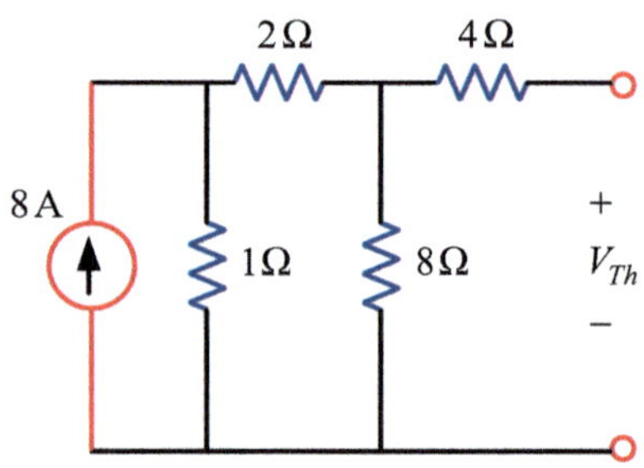

Fig. 4.75 Circuit for calculating Thevenin's voltage for Example 4.9

The Thevenin's resistance is calculated as,

$$R_{\mathrm{Th}} = 4 + \frac{(2+1)\times 8}{3+8} = 6.18\,\Omega \tag{4.148}$$

Thevenin's voltage can be determined from the circuit in Fig. 4.75 as,

$$I_{8\Omega} = 8 \times \frac{1}{8+2+1} = 0.73\,\mathrm{A} \tag{4.149}$$

$$V_{\mathrm{Th}} = I_{8\Omega} \times 8 = 0.73 \times 8 = 5.84\,\mathrm{V} \tag{4.150}$$

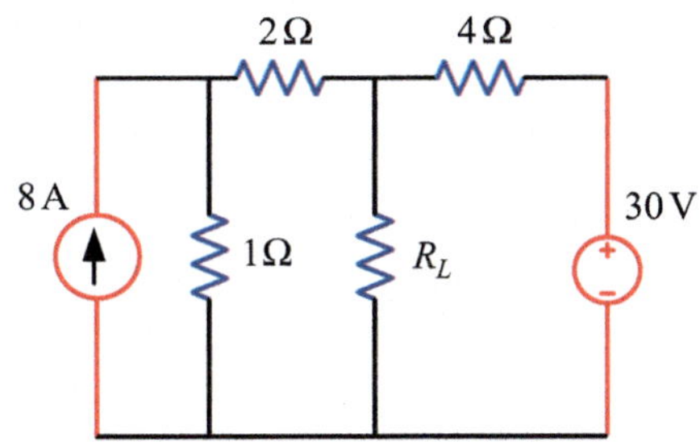

Fig. 4.76 Circuit for Practice Problem 4.9

The maximum power is,

$$P_{\mathrm{mxm}} = \frac{V_{\mathrm{Th}}^2}{4R_L} = \frac{5.84^2}{4 \times 6.18} = 1.38\,\mathrm{W} \tag{4.151}$$

Practice Problem 4.9
Determine the value of the load resistance and the maximum power absorbed by the load, if the circuit in Fig. 4.76 transfers maximum power.

Exercise Problems

4.1 An electrical circuit is shown in Fig. 4.77. Assume $I_p = 1$ A. By using the linearity property find the actual value of I_p.
4.2 Figure 4.78 shows an electrical circuit with the source voltage assigned to $V_s = 16$ V. Calculate the actual value of I_0.
4.3 Use linearity property to calculate the actual value of the voltage V_0 of the circuit in Fig. 4.79. Assume $V_0 = 1$ V.
4.4 Using superposition theorem determine the current in the 4 Ω resistor of the circuit in Fig. 4.80. Compare the result by PSpice simulation.
4.5 Using superposition theorem calculate the current in the 6 Ω resistor of the circuit in Fig. 4.81. Compare the result by PSpice simulation.

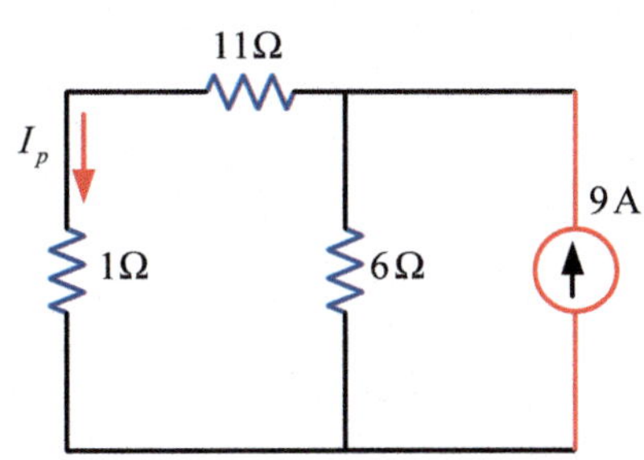

Fig. 4.77 Circuit for Exercise Problem 4.1

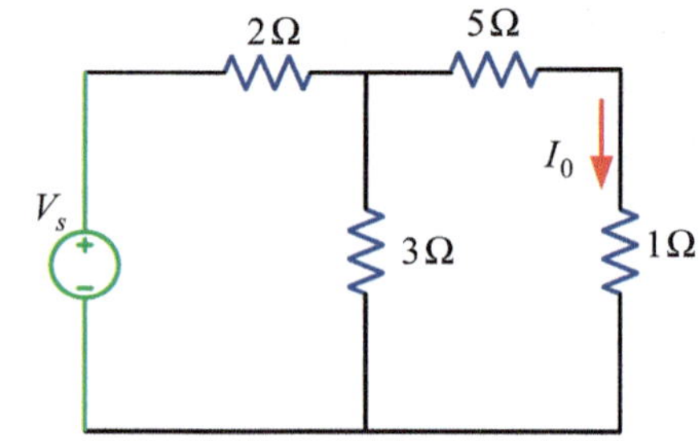

Fig. 4.78 Circuit for Exercise Problem 4.2

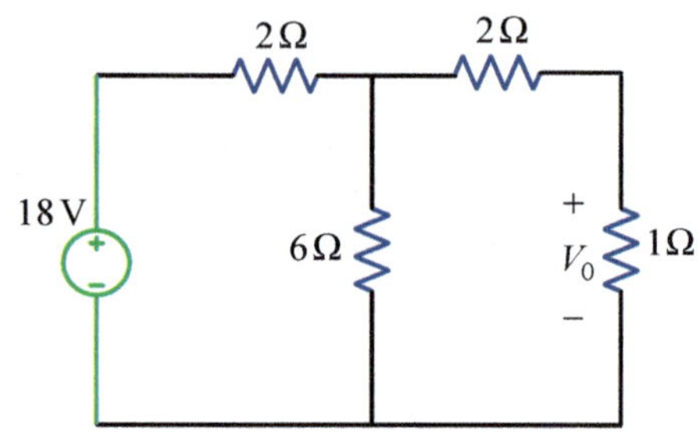

Fig. 4.79 Circuit for Exercise Problem 4.3

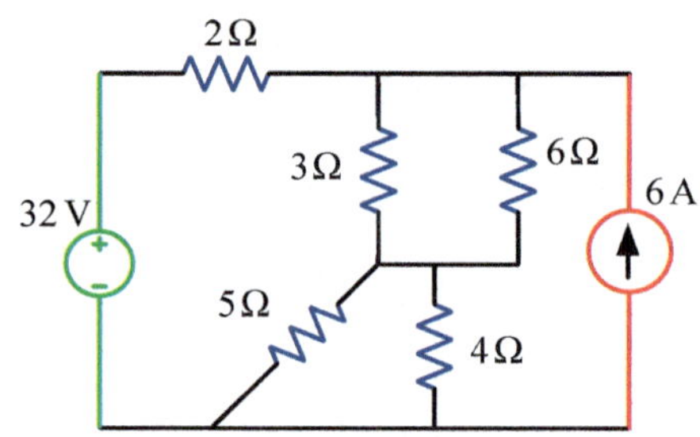

Fig. 4.80 Circuit for Exercise Problem 4.4

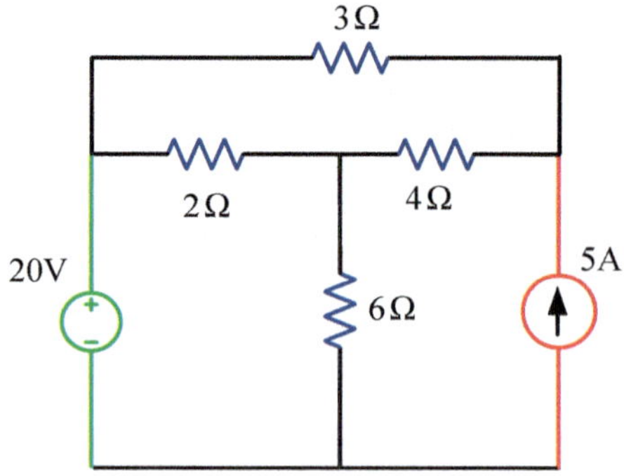

Fig. 4.81 Circuit for Exercise Problem 4.5

4.6 Using superposition theorem find the current in the 6 Ω resistor of the circuit in Fig. 4.82. Verify the result by PSpice simulation.

4.7 Using superposition theorem calculate the current in the 3 Ω resistor of the circuit in Fig. 4.83. Verify the result by PSpice simulation.

4.8 Use superposition theorem to calculate the current in the 4 Ω resistor of the circuit in Fig. 4.84. Verify the result by PSpice simulation.

4.9 Using superposition theorem calculate the current in the 2 Ω resistor of the circuit in Fig. 4.85. Verify the result by PSpice simulation.

Fig. 4.82 Circuit for Exercise Problem 4.6

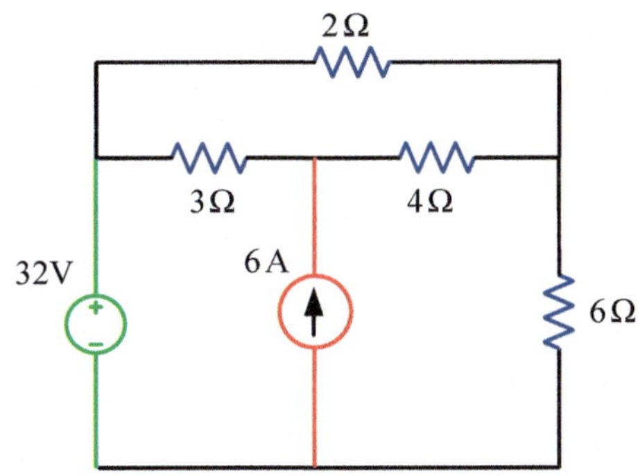

Fig. 4.83 Circuit for Exercise Problem 4.7

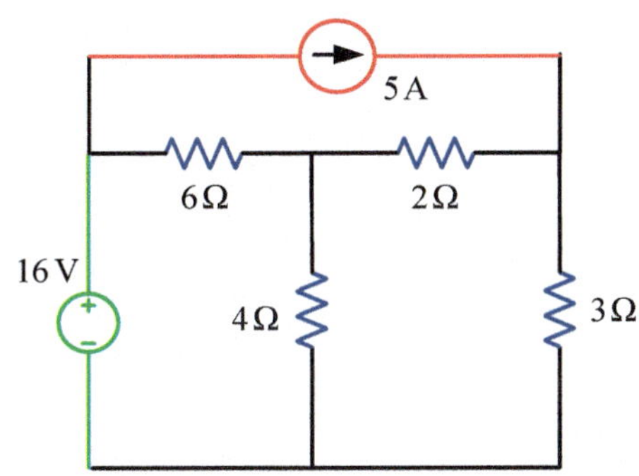

Fig. 4.84 Circuit for Exercise Problem 4.8

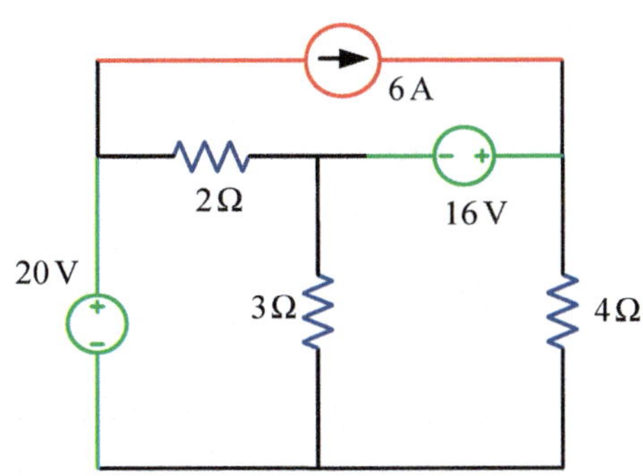

4.10 Using superposition theorem determine the current in the 4 Ω resistor of the circuit in Fig. 4.86. Verify the result by PSpice simulation.

4.11 Using superposition theorem find the current in the 8 Ω resistor of the circuit in Fig. 4.87. Verify the result by PSpice simulation.

4.12 Using superposition theorem find the current in the 4 Ω resistor of the circuit in Fig. 4.88. Verify the result by PSpice simulation.

Fig. 4.85 Circuit for Exercise Problem 4.9

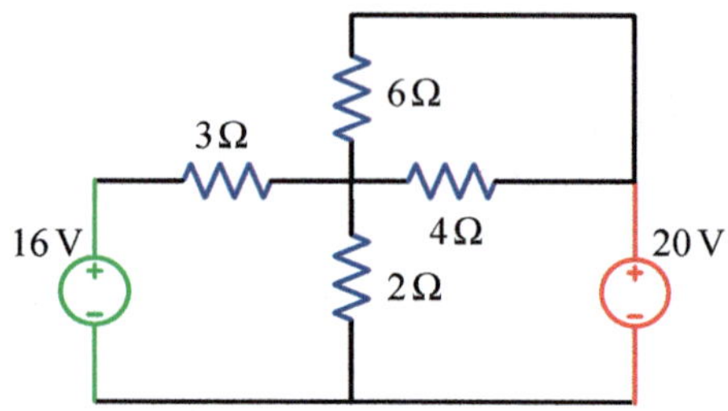

Fig. 4.86 Circuit for Exercise Problem 4.10

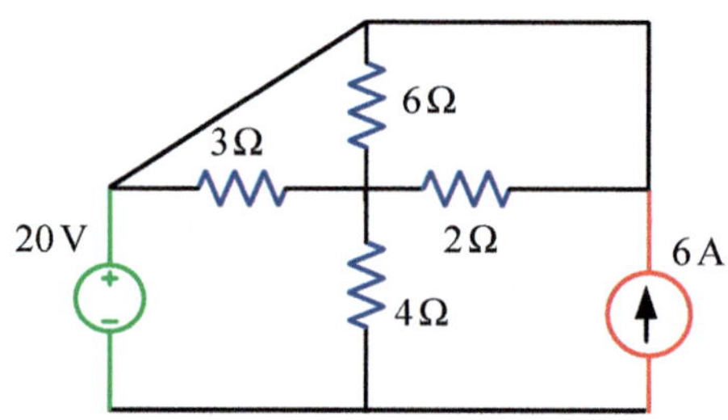

Fig. 4.87 Circuit for Exercise Problem 4.11

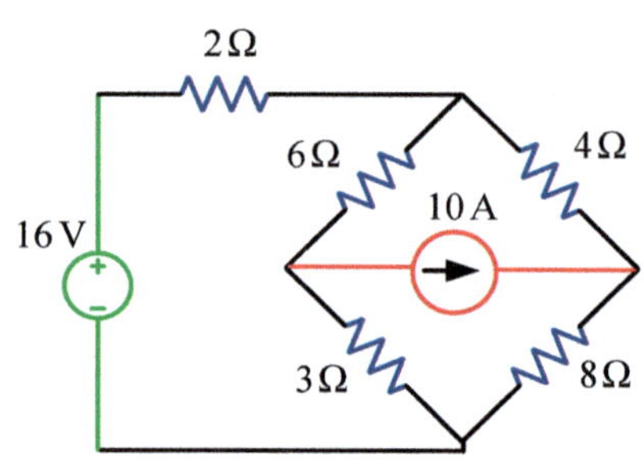

Fig. 4.88 Circuit for Exercise Problem 4.12

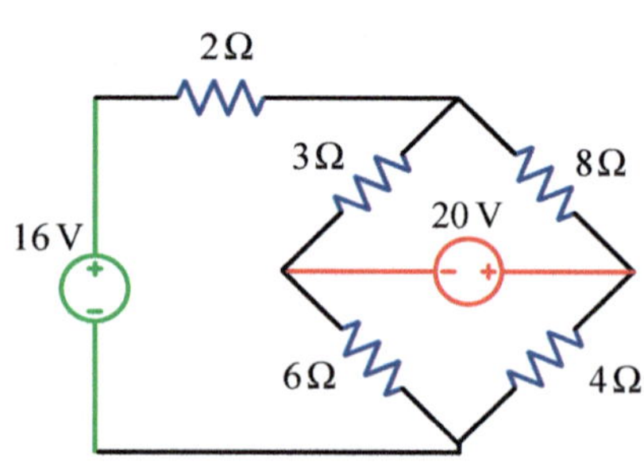

4.13 Using superposition theorem determine the current in the 6 Ω resistor of the circuit in Fig. 4.89. Verify the result by PSpice simulation.

4.14 Use superposition theorem to calculate the current in the 3 Ω resistor of the circuit in Fig. 4.90. Verify the result by PSpice simulation.

4.15 Using superposition theorem determine the current in the 5 Ω resistor of the circuit in Fig. 4.91. Verify the result by PSpice simulation.

4.16 Using superposition theorem find the current in the 1 Ω resistor of the circuit in Fig. 4.92. Verify the result by PSpice simulation.

4.17 Using superposition theorem calculate the current in the 8 Ω resistor of the circuit in Fig. 4.93. Verify the result by PSpice simulation.

4.18 Using superposition theorem determine the current in the 5 Ω resistor of the circuit in Fig. 4.94. Verify the result by PSpice simulation.

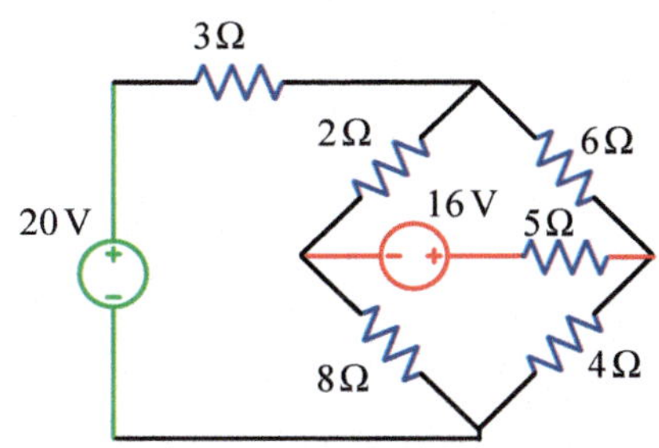

Fig. 4.89 Circuit for Exercise Problem 4.13

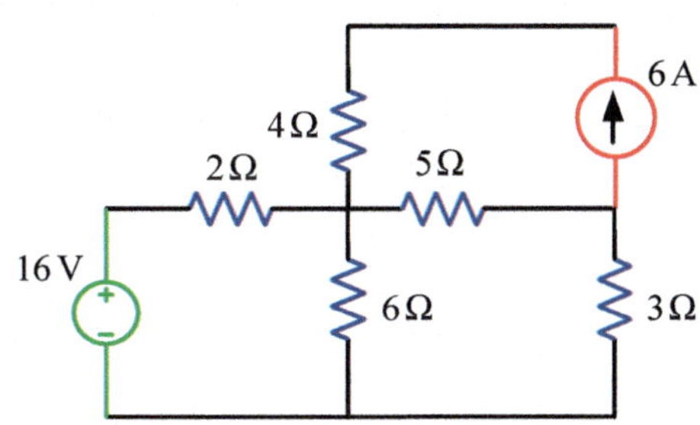

Fig. 4.90 Circuit for Exercise Problem 4.14

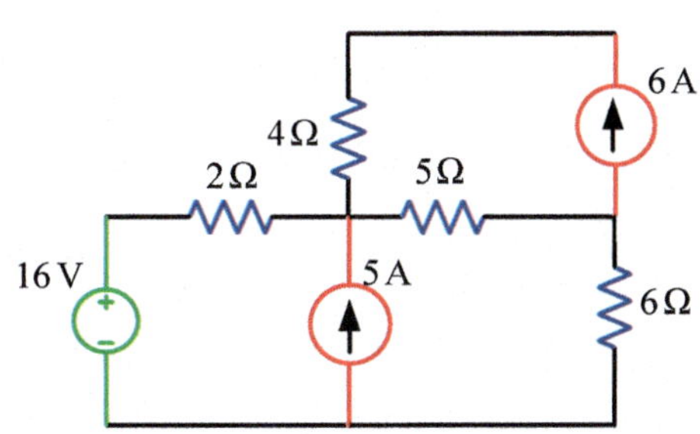

Fig. 4.91 Circuit for Exercise Problem 4.15

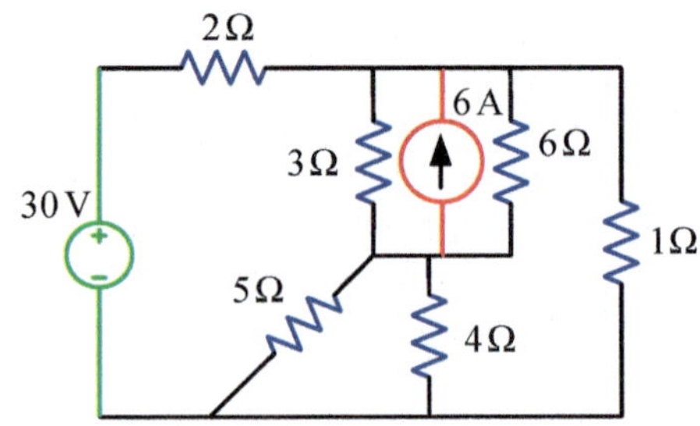

Fig. 4.92 Circuit for Exercise Problem 4.16

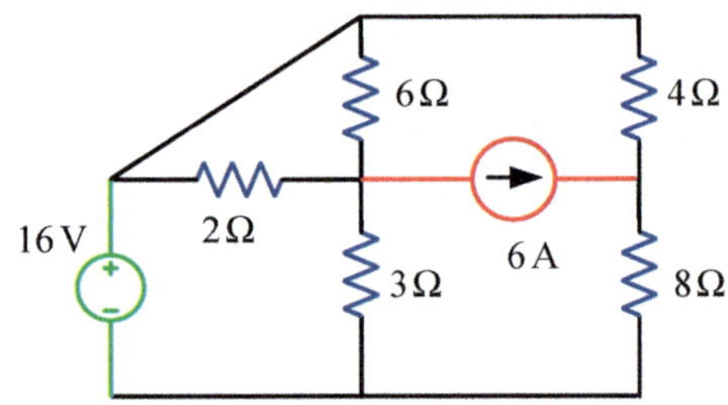

Fig. 4.93 Circuit for Exercise Problem 4.17

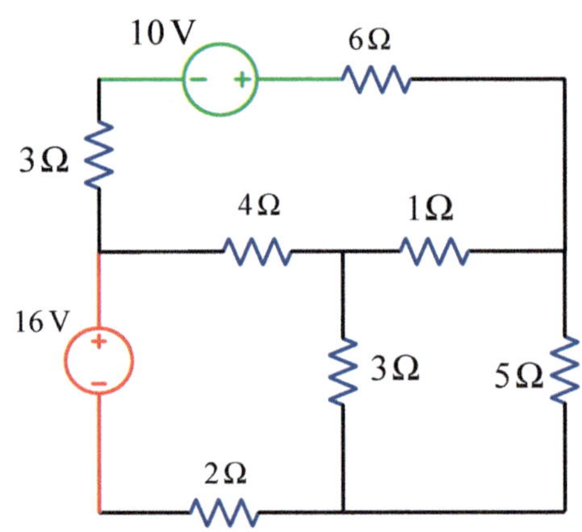

Fig. 4.94 Circuit for Exercise Problem 4.18

4.19 Use superposition theorem to determine the current in the 3 Ω resistor of the circuit in Fig. 4.95. Verify the result by PSpice simulation.

4.20 Using superposition theorem calculate the through the 4 Ω resistor of the circuit in Fig. 4.96 and verify the result by PSpice simulation.

4.21 Figure 4.97 shows an electrical circuit. Use superposition theorem to calculate the current in the 6 Ω resistor and verify the result by PSpice simulation.

4.22 Using superposition theorem determine the current in the 5 Ω resistor of the circuit in Fig. 4.98. Verify the result by PSpice simulation.

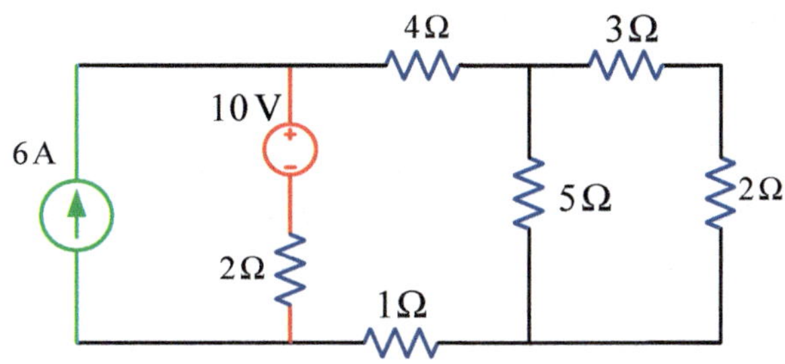

Fig. 4.95 Circuit for Exercise Problem 4.19

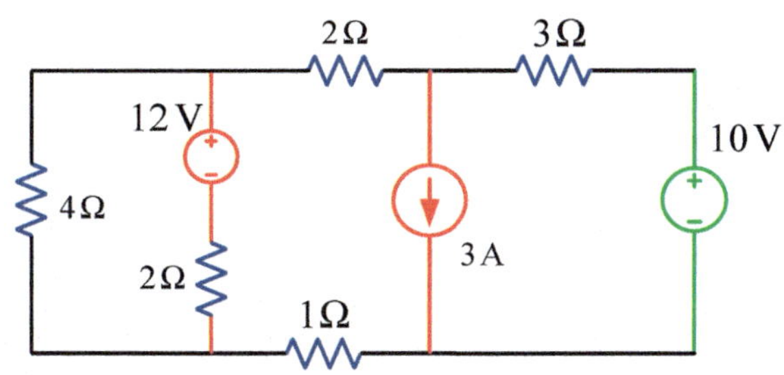

Fig. 4.96 Circuit for Exercise Problem 4.20

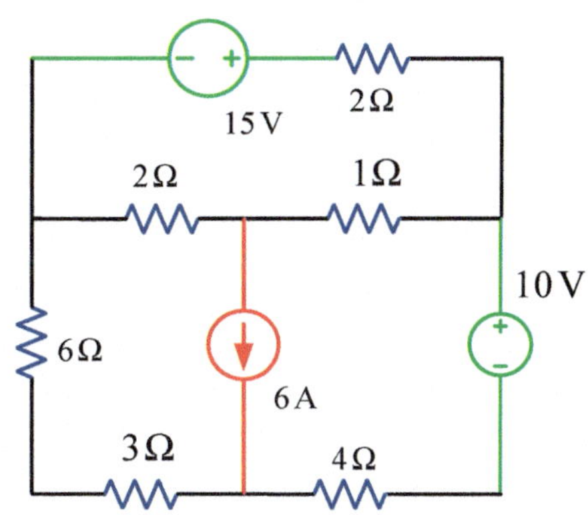

Fig. 4.97 Circuit for Exercise Problem 4.21

4.23 Using superposition theorem calculate the voltage across the 12 Ω resistor of the circuit in Fig. 4.99. Verify the result by PSpice simulation.
4.24 Using superposition theorem find the voltage across the 4 Ω resistor of the circuit in Fig. 4.100. Verify the result by PSpice simulation.
4.25 Using superposition theorem calculate the voltage across the 6 Ω resistor of the circuit in Fig. 4.101. Verify the result by PSpice simulation.
4.26 Using superposition theorem find the voltage across the 3 Ω resistor of the circuit in Fig. 4.102. Verify the result by PSpice simulation.

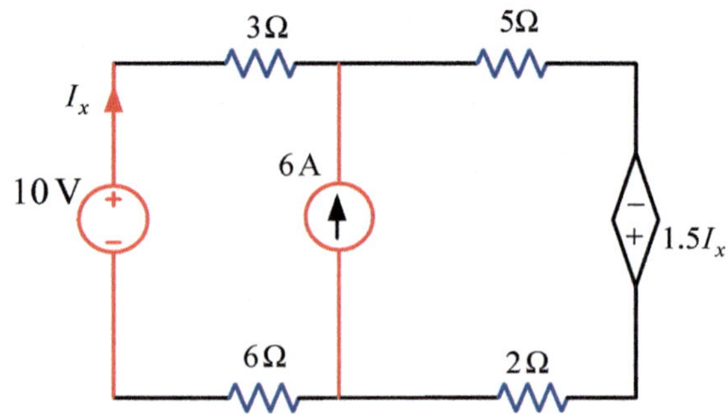

Fig. 4.98 Circuit for Exercise Problem 4.22

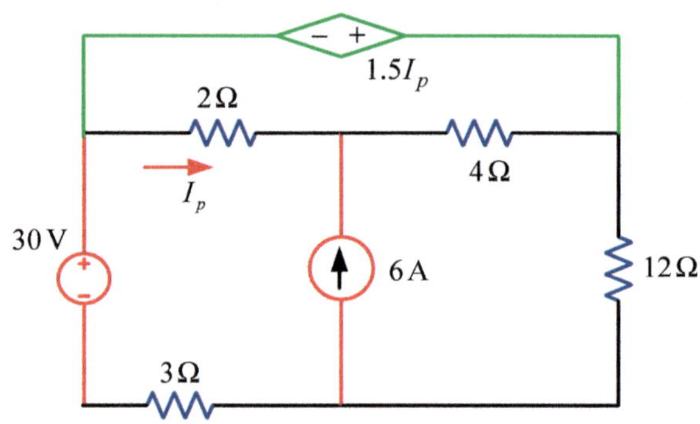

Fig. 4.99 Circuit for Exercise Problem 4.23

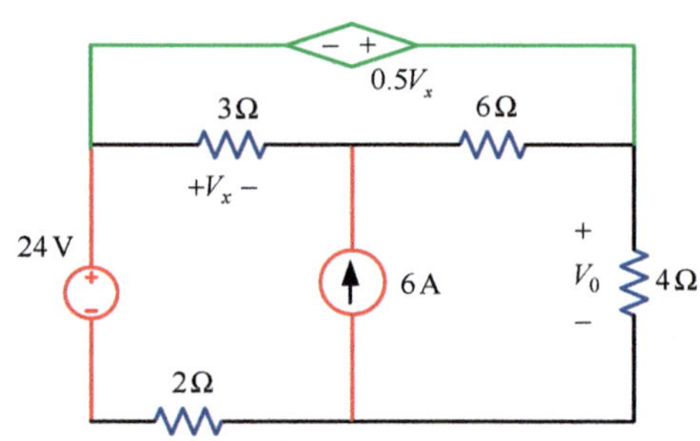

Fig. 4.100 Circuit for Exercise Problem 4.24

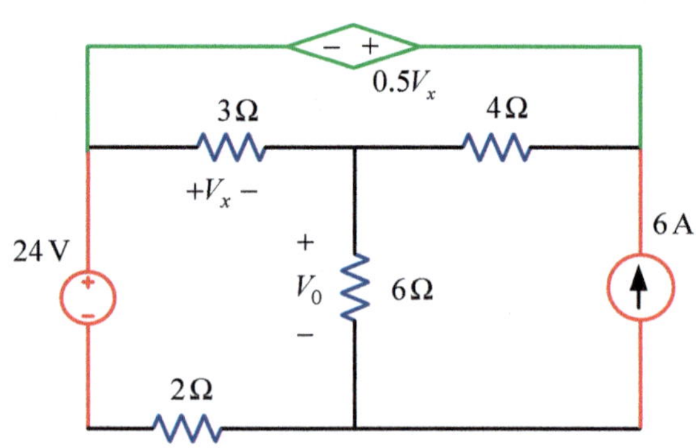

Fig. 4.101 Circuit for Exercise Problem 4.25

4.27 Using Thevenin's theorem find the current in the 5 Ω resistor of the circuit in Fig. 4.103. Verify the result by PSpice simulation.
4.28 Use Thevenin's theorem to determine the current in the 6 Ω resistor of the circuit in Fig. 4.104. Verify the result by PSpice simulation.
4.29 Using Thevenin's theorem find the current in the 4 Ω resistor of the circuit in Fig. 4.105. Verify the result by PSpice simulation.
4.30 Use Thevenin's theorem to determine the current in the 3 Ω resistor of the circuit in Fig. 4.106. Verify the result by PSpice simulation.
4.31 Using Thevenin's theorem calculate the current in the 8 Ω resistor of the circuit in Fig. 4.107. Verify the result by PSpice simulation.
4.32 Using Thevenin's theorem determine the current in the 10 Ω resistor of the circuit in Fig. 4.108. Verify the result by PSpice simulation.
4.33 Use Thevenin's theorem to find the current in the 6 Ω resistor of the circuit in Fig. 4.109. Verify the result by PSpice simulation.

Fig. 4.102 Circuit for Exercise Problem 4.26

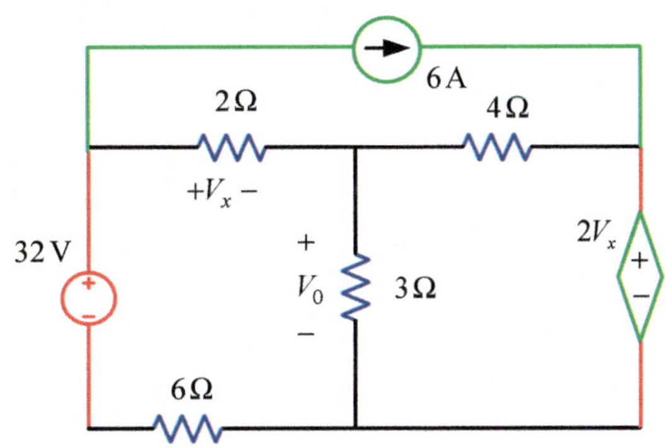

Fig. 4.103 Circuit for Exercise Problem 4.27

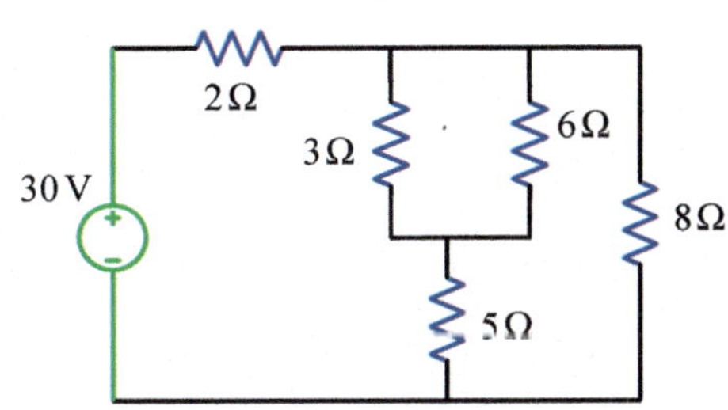

Fig. 4.104 Circuit for Exercise Problem 4.28

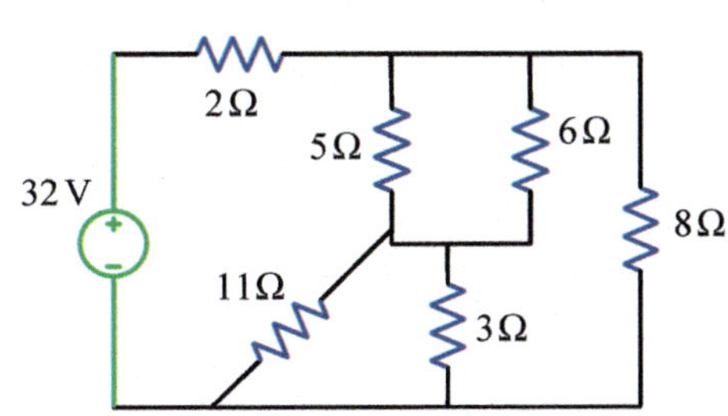

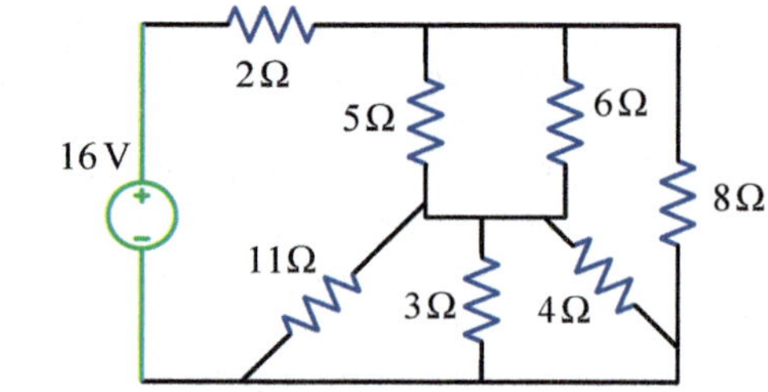

Fig. 4.105 Circuit for Exercise Problem 4.29

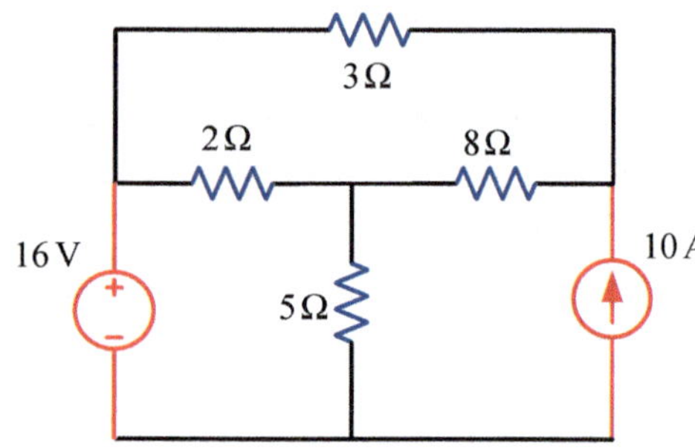

Fig. 4.106 Circuit for Exercise Problem 4.30

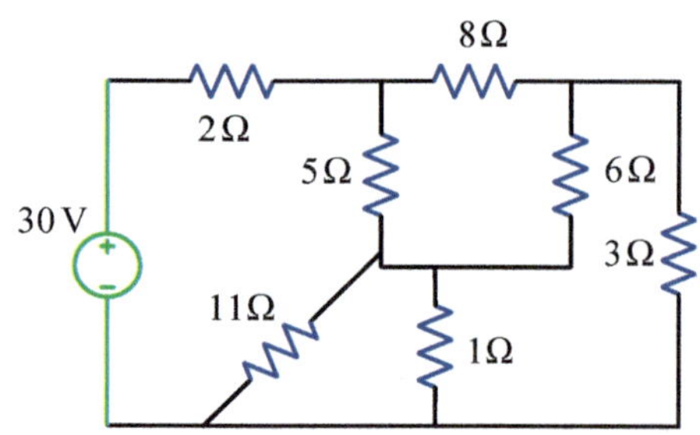

Fig. 4.107 Circuit for Exercise Problem 4.31

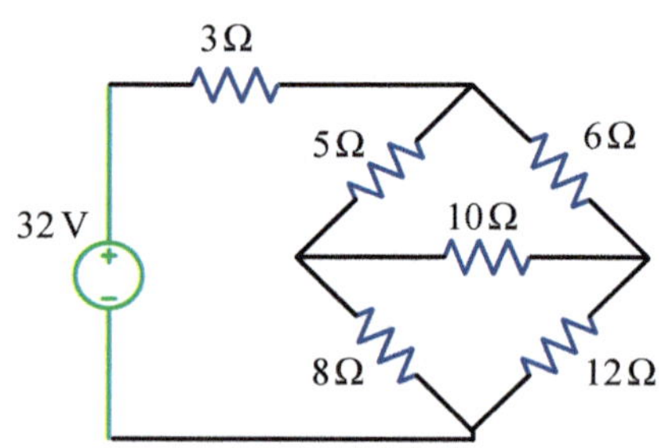

Fig. 4.108 Circuit for Exercise Problem 4.32

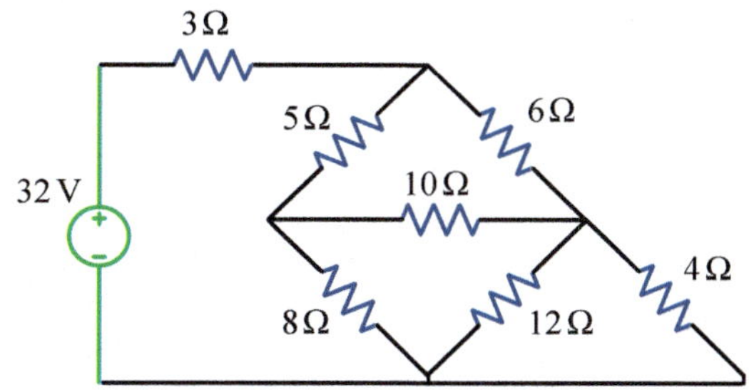

Fig. 4.109 Circuit for Exercise Problem 4.33

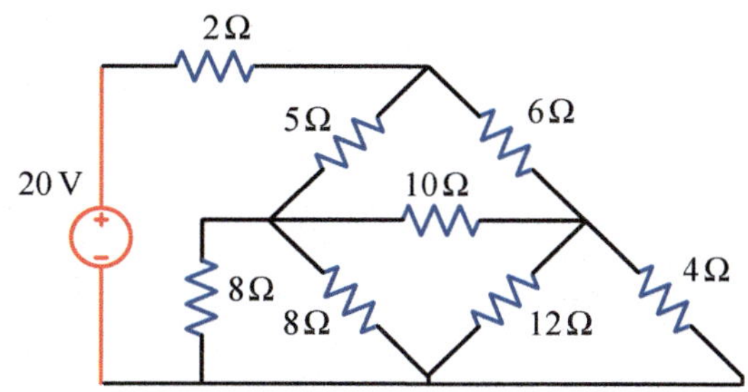

Fig. 4.110 Circuit for Exercise Problem 4.34

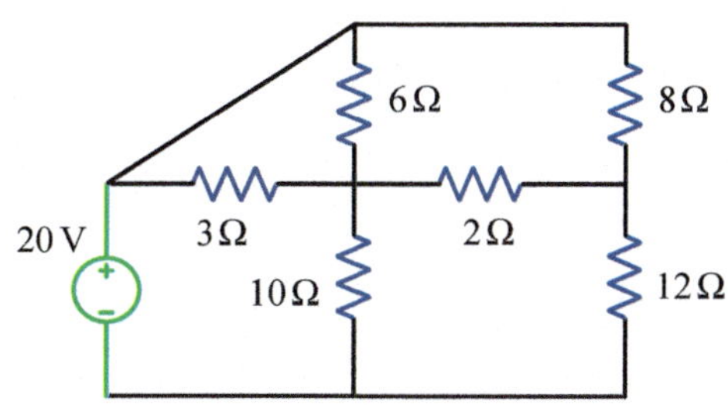

Fig. 4.111 Circuit for Exercise Problem 4.35

4.34 Using Thevenin's theorem calculate the current in the 5 Ω resistor of the circuit in Fig. 4.110. Verify the result by PSpice simulation.

4.35 Use Thevenin's theorem to find the current in the 10 Ω resistor of the circuit in Fig. 4.111. Verify the result by PSpice simulation.

4.36 Using Thevenin's theorem determine the current in the 12 Ω resistor of the circuit in Fig. 4.112. Verify the result by PSpice simulation.

4.37 Use Thevenin's theorem to determine the current in the 6 Ω resistor of the circuit in Fig. 4.113. Verify the result by PSpice simulation.

4.38 Using Thevenin's theorem calculate the voltage across the 5 Ω resistor of the circuit in Fig. 4.114. Verify the result by PSpice simulation.

4.39 Use Thevenin's theorem to determine the current in the 2 Ω resistor of the circuit in Fig. 4.115. Verify the result by PSpice simulation.

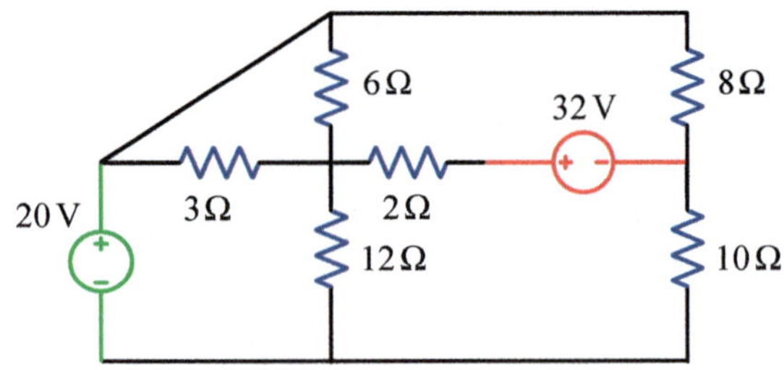

Fig. 4.112 Circuit for Exercise Problem 4.36

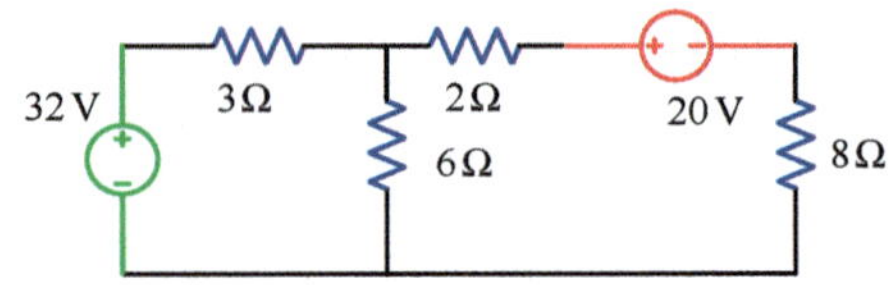

Fig. 4.113 Circuit for Exercise Problem 4.37

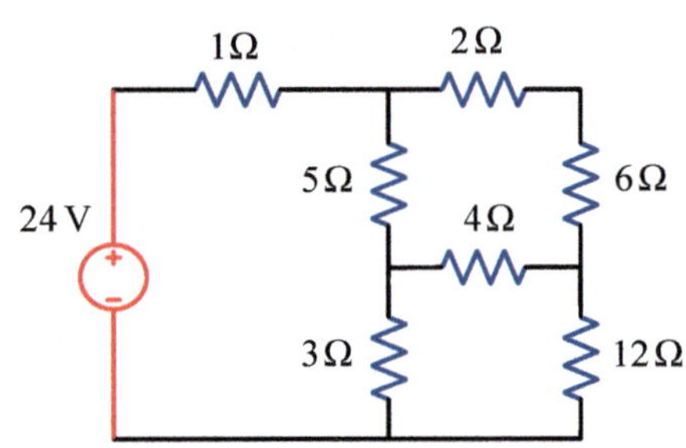

Fig. 4.114 Circuit for Exercise Problem 4.38

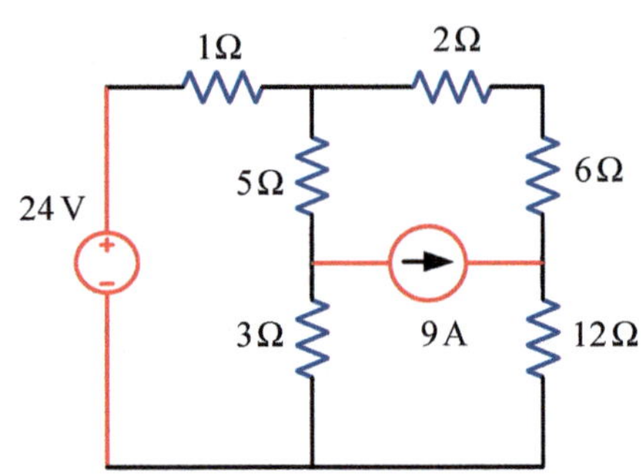

Fig. 4.115 Circuit for Exercise Problem 4.39

4.40 Using Thevenin's theorem calculate the power absorbed by the 5 Ω resistor of the circuit in Fig. 4.116. Also, find the current by PSpice simulation.

4.41 Using Thevenin's theorem determine the voltage across the 3 Ω resistor of the circuit in Fig. 4.117. Verify the result by PSpice simulation.

4.42 Use Thevenin's theorem to find the current in the 11 Ω resistor of the circuit in Fig. 4.118. Verify the result by PSpice simulation.

Fig. 4.116 Circuit for Exercise Problem 4.40

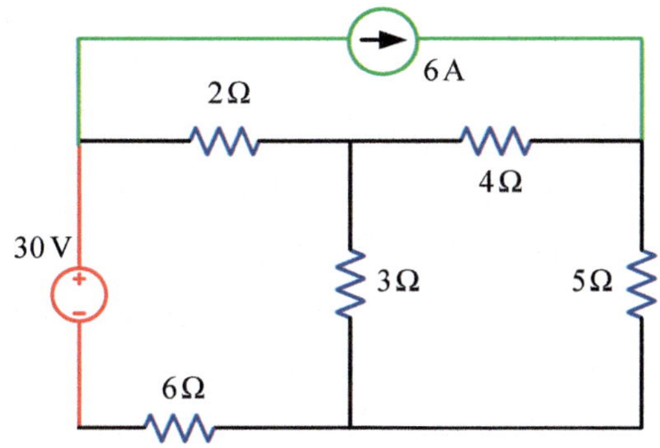

Fig. 4.117 Circuit for Exercise Problem 4.41

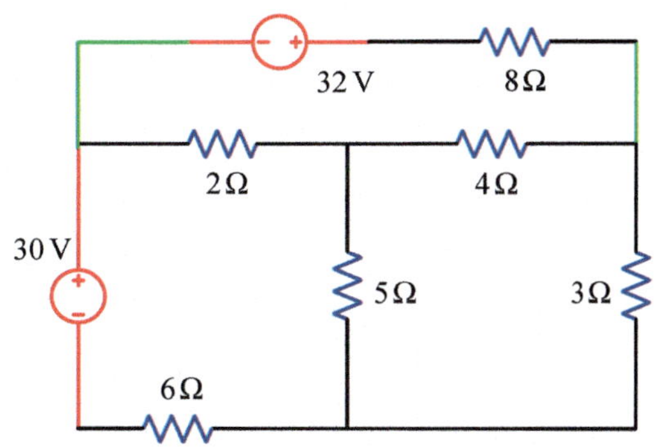

Fig. 4.118 Circuit for Exercise Problem 4.42

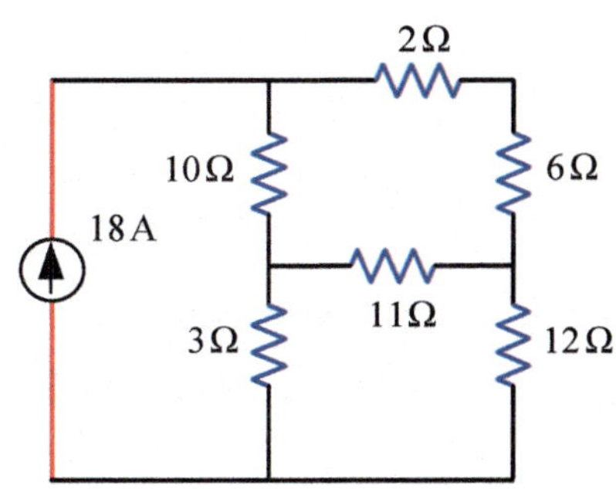

4.43 Using Thevenin's theorem determine the current in the 8 Ω resistor of the circuit in Fig. 4.119. Verify the result by PSpice simulation.

4.44 Using Thevenin's theorem find the current in the 4 Ω resistor of the circuit in Fig. 4.120. Verify the result by PSpice simulation.

4.45 Use Thevenin's theorem to determine the current in the 3 Ω resistor of the circuit in Fig. 4.121. Verify the result by PSpice simulation.

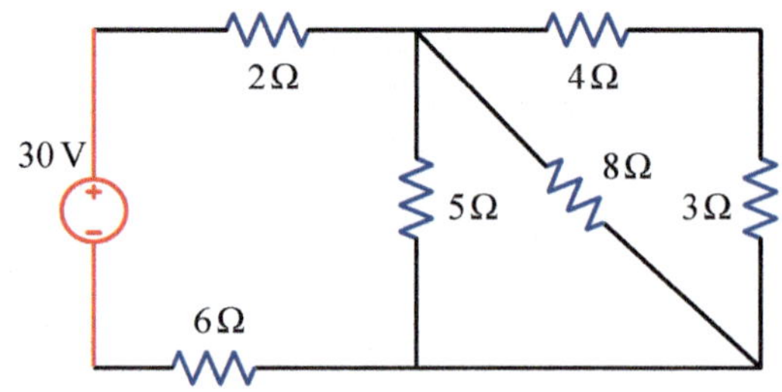

Fig. 4.119 Circuit for Exercise Problem 4.43

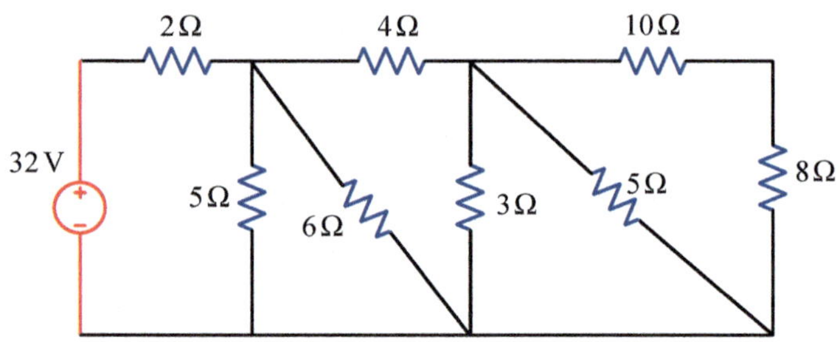

Fig. 4.120 Circuit for Exercise Problem 4.44

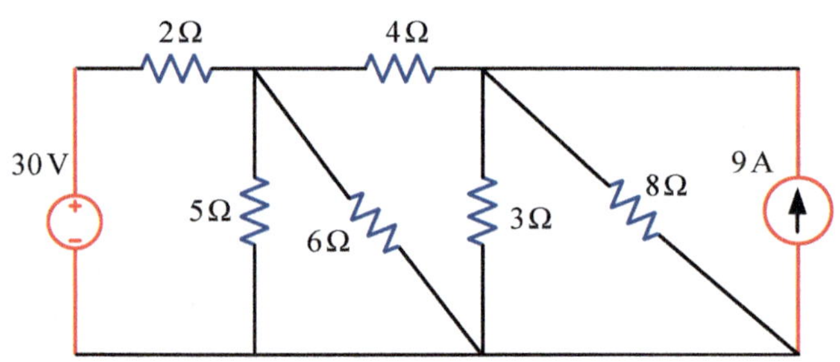

Fig. 4.121 Circuit for Exercise Problem 4.45

4.46 Using Thevenin's theorem calculate the power absorbed by the 12 Ω resistor of the circuit in Fig. 4.122.

4.47 Using Thevenin's theorem calculate the current in the 5 Ω resistor of the circuit in Fig. 4.123. Verify the result by PSpice simulation.

4.48 Using Thevenin's theorem find the current in the 6 Ω resistor of the circuit in Fig. 4.124. Verify the result by PSpice simulation.

4.49 Using Thevenin's theorem find the current in 2 Ω resistor of the circuit in Fig. 4.125.

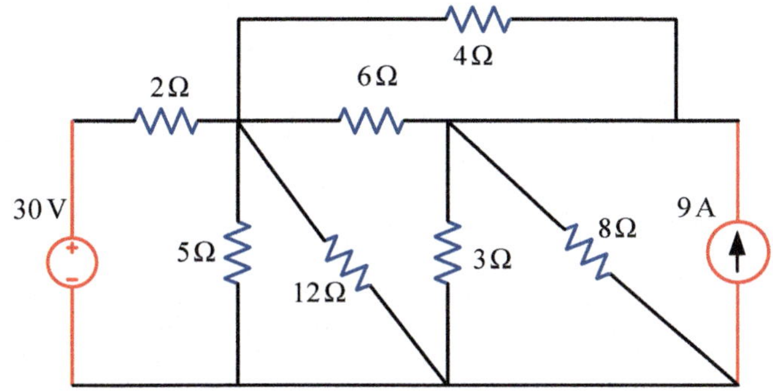

Fig. 4.122 Circuit for Exercise Problem 4.46

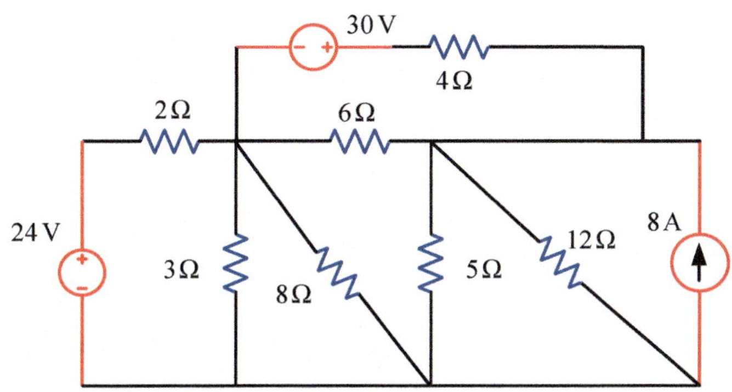

Fig. 4.123 Circuit for Exercise Problem 4.47

Fig. 4.124 Circuit for Exercise Problem 4.48

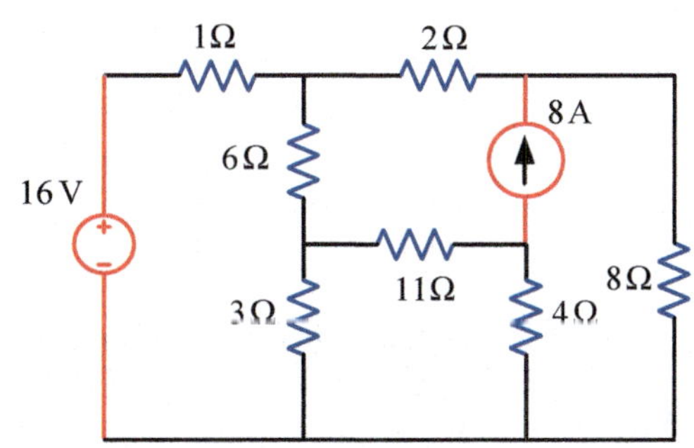

4.50 Using Thevenin's theorem determine the current in the 12 Ω resistor of the circuit in Fig. 4.126. Verify the result by PSpice simulation.

4.51 Using Thevenin's theorem calculate the current in the 8 Ω resistor of the circuit in Fig. 4.127. Verify the result by PSpice simulation.

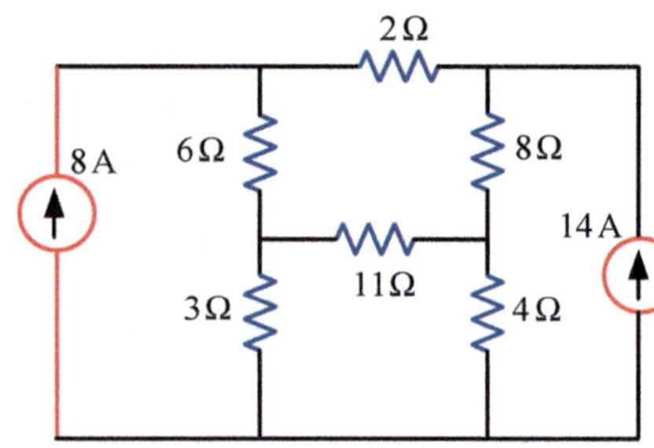

Fig. 4.125 Circuit for Exercise Problem 4.49

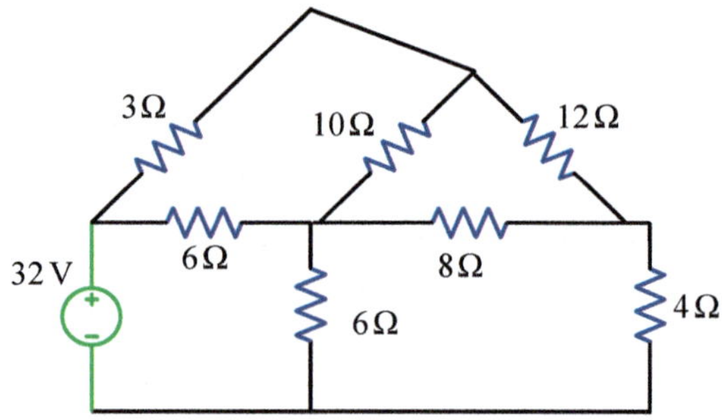

Fig. 4.126 Circuit for Exercise Problem 4.50

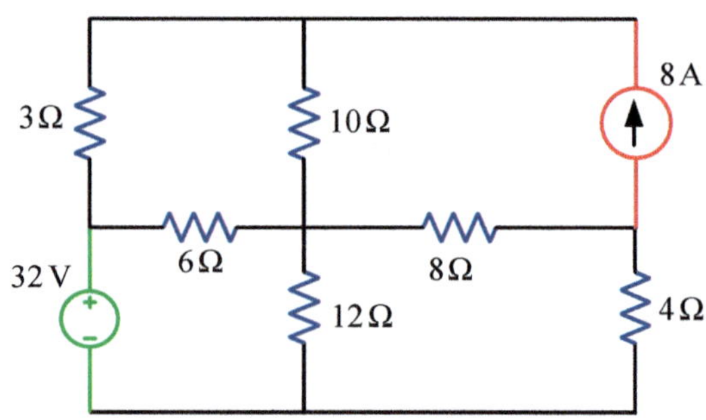

Fig. 4.127 Circuit for Exercise Problem 4.51

4.52 Using Thevenin's theorem find the current in the 10 Ω resistor of the circuit in Fig. 4.128. Verify the result by PSpice simulation.

4.53 Using Thevenin's theorem determine the current in the 12 Ω resistor of the circuit in Fig. 4.129. Verify the result by PSpice simulation.

4.54 Using Thevenin's theorem calculate the current in the 6 Ω resistor of the circuit in Fig. 4.130. Verify the result by PSpice simulation.

4.55 Using Thevenin's theorem find the current in the 3 Ω resistor of the circuit in Fig. 4.131. Verify the result by PSpice simulation.

4.56 Using Thevenin's theorem determine the current in the 8 Ω resistor of the circuit in Fig. 4.132. Verify the result by PSpice simulation.

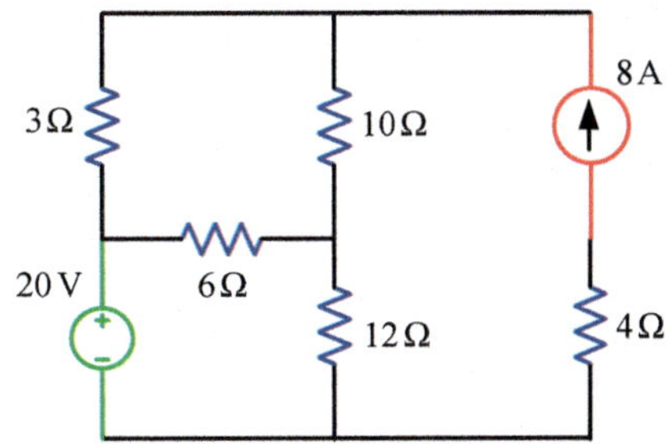

Fig. 4.128 Circuit for Exercise Problem 4.52

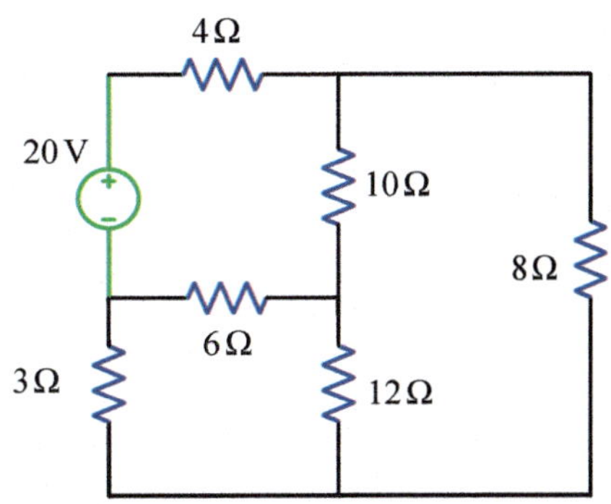

Fig. 4.129 Circuit for Exercise Problem 4.53

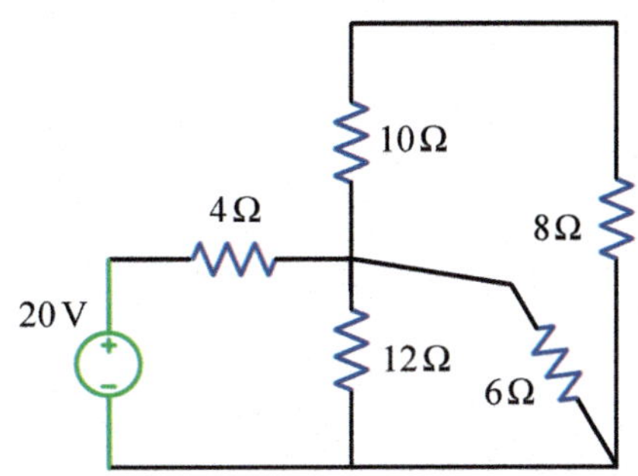

Fig. 4.130 Circuit for Exercise Problem 4.54

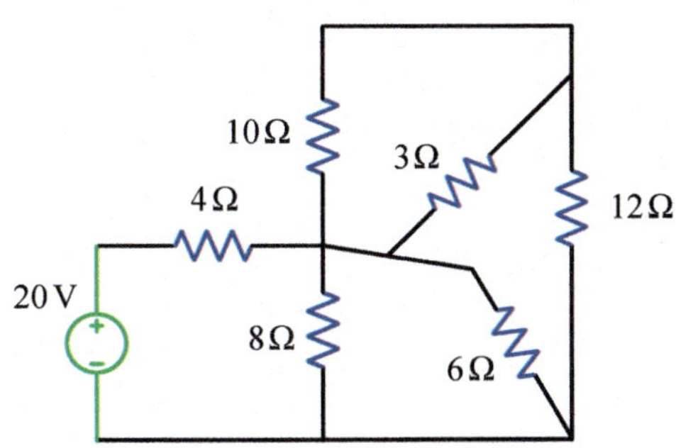

Fig. 4.131 Circuit for Exercise Problem 4.55

Fig. 4.132 Circuit for Exercise Problem 4.56

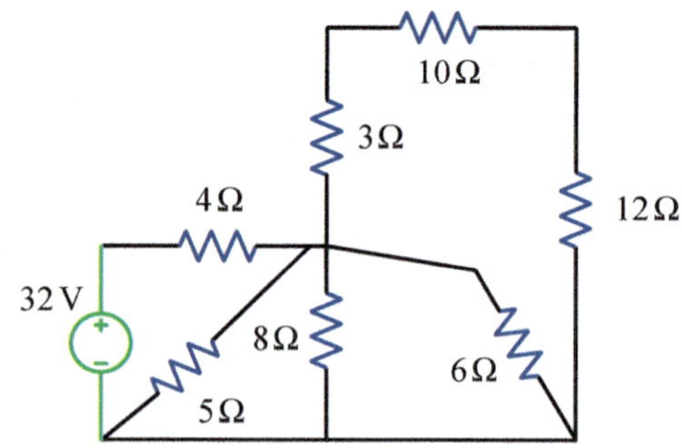

Fig. 4.133 Circuit for Exercise Problem 4.57

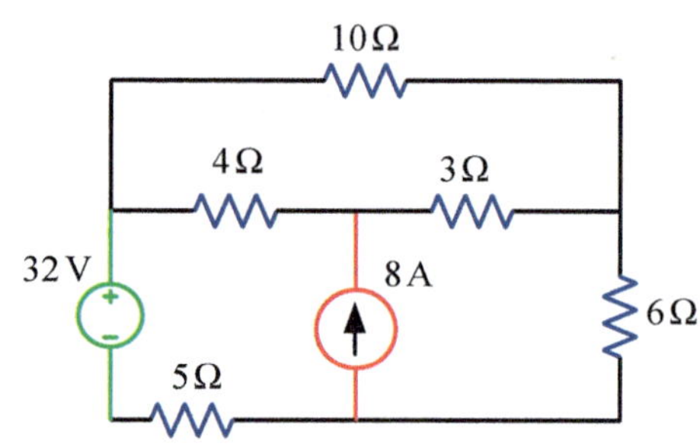

Fig. 4.134 Circuit for Exercise Problem 4.58

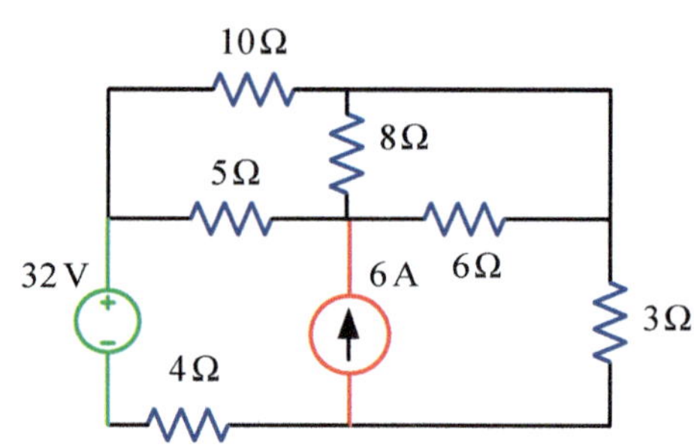

4.57 Use Thevenin's theorem to calculate the current in the 6 Ω resistor of the circuit in Fig. 4.133. Verify the result by PSpice simulation.

4.58 Use Thevenin's theorem to calculate the current in the 10 Ω resistor of the circuit in Fig. 4.134. Verify the result by PSpice simulation.

4.59 Using Thevenin's theorem determine the current in the 10 Ω resistor of the circuit in Fig. 4.135. Verify the result by PSpice simulation.

4.60 Using Thevenin's theorem calculate the current in the 8 Ω resistor of the circuit in Fig. 4.136. Verify the result by PSpice simulation.

4.61 Using Thevenin's theorem calculate the current in the 4 Ω resistor of the circuit in Fig. 4.137. Verify the result by PSpice simulation.

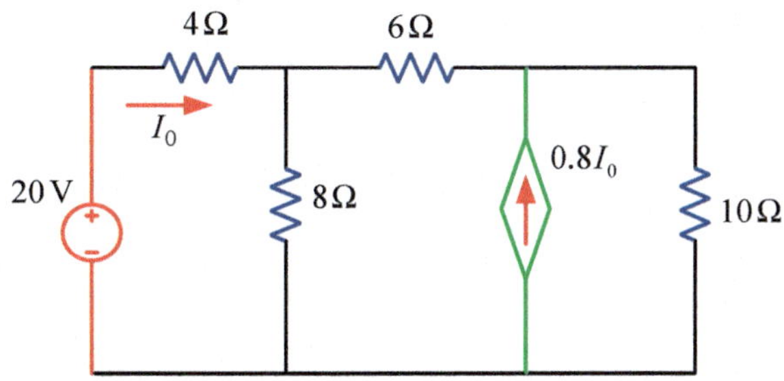

Fig. 4.135 Circuit for Exercise Problem 4.59

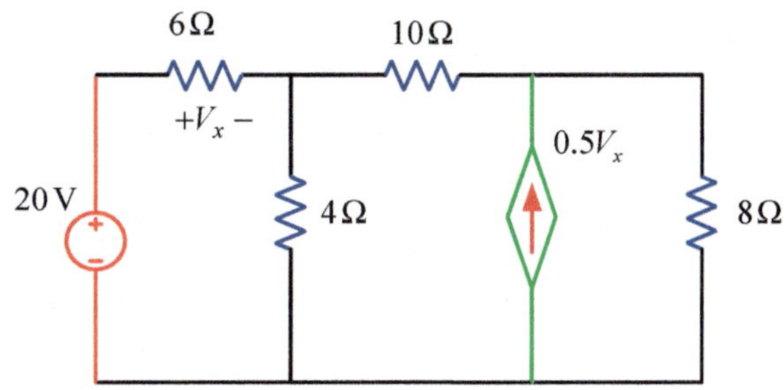

Fig. 4.136 Circuit for Exercise Problem 4.60

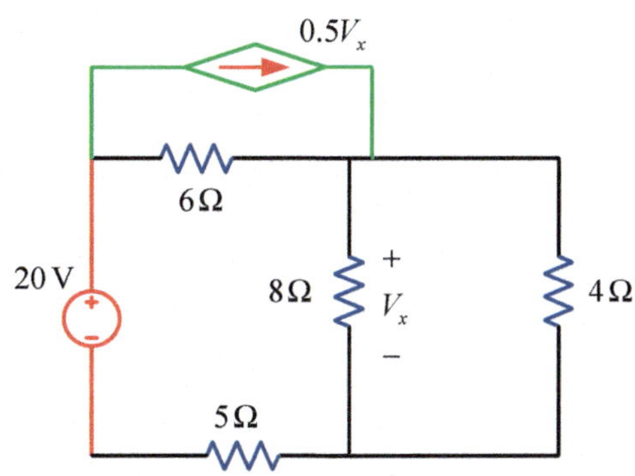

Fig. 4.137 Circuit for Exercise Problem 4.61

4.62 Use Thevenin's theorem to find the current in the 5 Ω resistor of the circuit in Fig. 4.138. Verify the result by PSpice simulation.

4.63 Use Thevenin's theorem to find the current in the 4 Ω resistor of the circuit is shown in Fig. 4.139.

4.64 Use Thevenin's theorem to find the current in the 6 Ω resistor of the circuit in Fig. 4.140.

4.65 Using Norton's theorem calculate the current in the 8 Ω resistor of the circuit in Fig. 4.141. Verify the result by PSpice simulation.

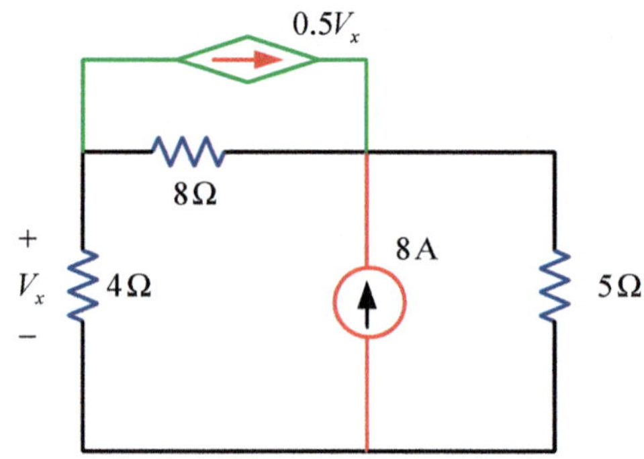

Fig. 4.138 Circuit for Exercise Problem 4.62

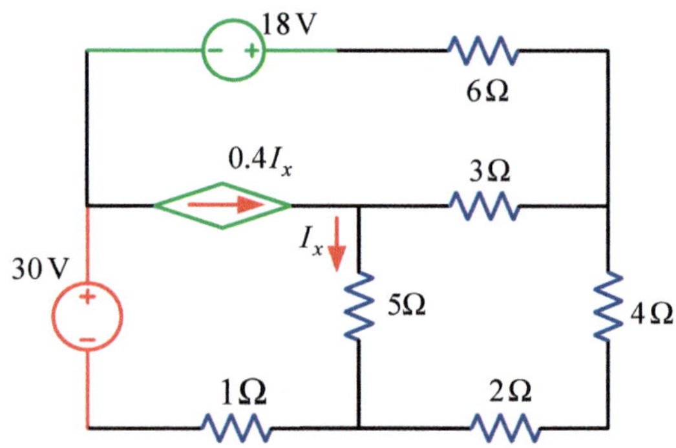

Fig. 4.139 Circuit for Exercise Problem 4.63

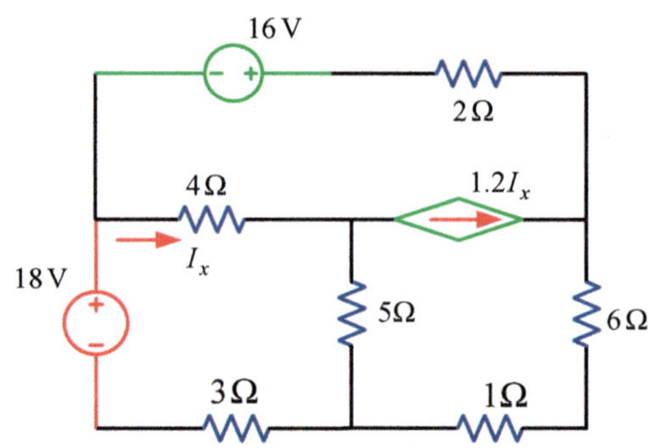

Fig. 4.140 Circuit for Exercise Problem 4.64

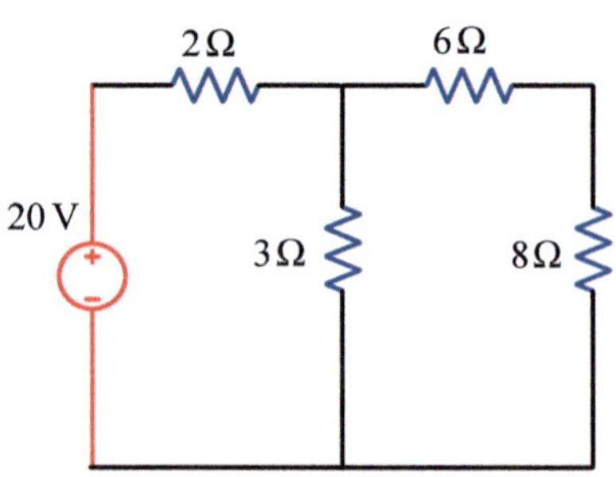

Fig. 4.141 Circuit for Exercise Problem 4.65

4.66 Using Norton's theorem find the current in the 4 Ω resistor of the circuit in Fig. 4.142. Verify the result by PSpice simulation.

4.67 Using Norton's theorem calculate the current in the 5 Ω resistor of the circuit in Fig. 4.143. Verify the result by PSpice simulation.

4.68 Using Norton's theorem determine the current in the 8 Ω resistor of the circuit in Fig. 4.144. Verify the result by PSpice simulation.

4.69 Using Norton's theorem find the current in the 12 Ω resistor of the circuit in Fig. 4.145. Verify the result by PSpice simulation.

Fig. 4.142 Circuit for Exercise Problem 4.66

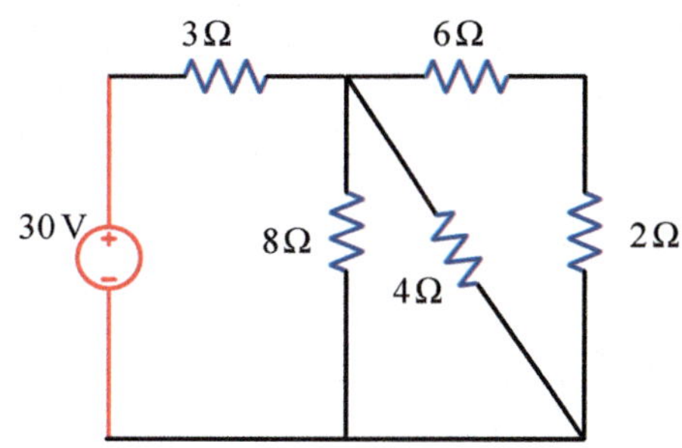

Fig. 4.143 Circuit for Exercise Problem 4.67

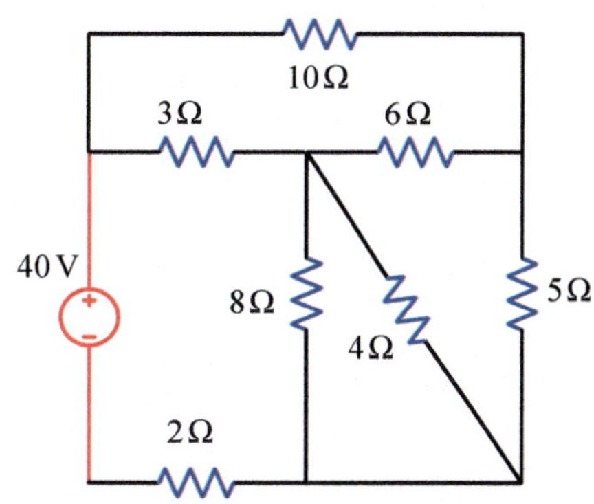

Fig. 4.144 Circuit for Exercise Problem 4.68

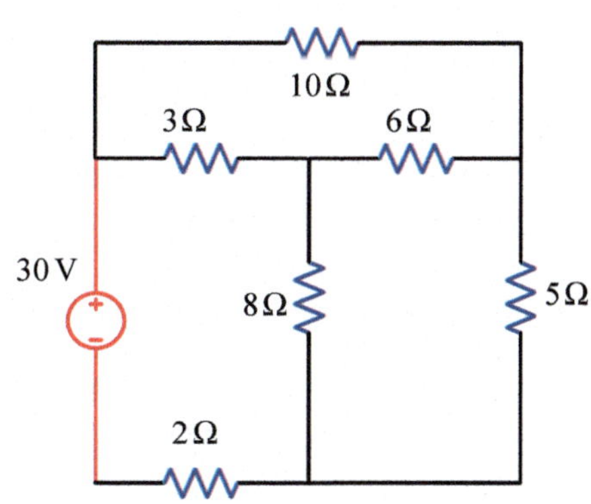

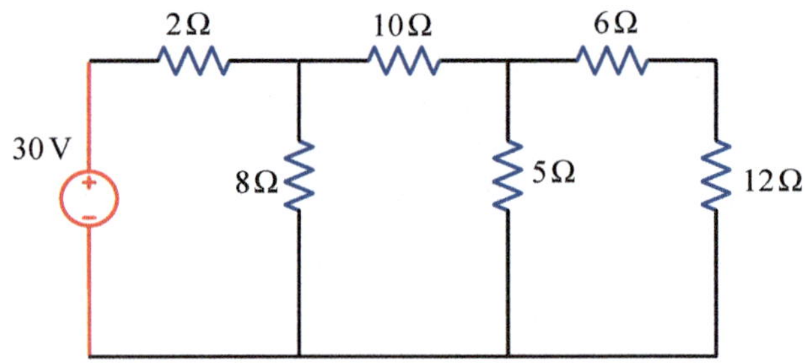

Fig. 4.145 Circuit for Exercise Problem 4.69

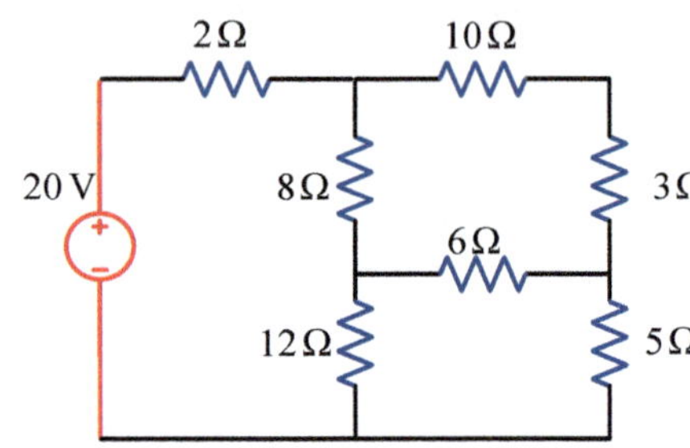

Fig. 4.146 Circuit for Exercise Problem 4.70

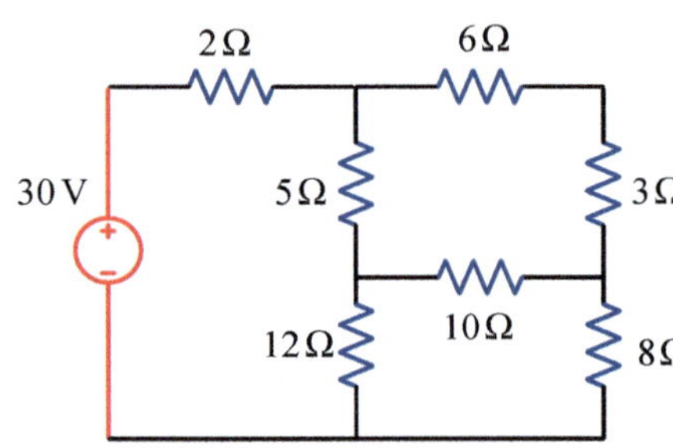

Fig. 4.147 Circuit for Exercise Problem 4.71

4.70 Using Norton's theorem calculate the current in the 5 Ω resistor of the circuit in Fig. 4.146. Verify the result by PSpice simulation.

4.71 Using Norton's theorem determine the current in the 3 Ω resistor of the circuit in Fig. 4.147. Verify the result by PSpice simulation.

4.72 Using Norton's theorem calculate the current in the 6 Ω resistor of the circuit in Fig. 4.148. Verify the result by PSpice simulation.

4.73 Use Norton's theorem to find the current in the 4 Ω resistor of the circuit in Fig. 4.149. Verify the result by PSpice simulation.

4.74 An electrical circuit is shown in Fig. 4.150. Use Norton's theorem to find the current in the 6 Ω resistor and compare the result with PSpice simulation.

4.75 Fig. 4.151 shows an electrical circuit. Calculate the current in the 5 Ω resistor by using Norton's theorem and verify the result by PSpice simulation.

Fig. 4.148 Circuit for Exercise Problem 4.72

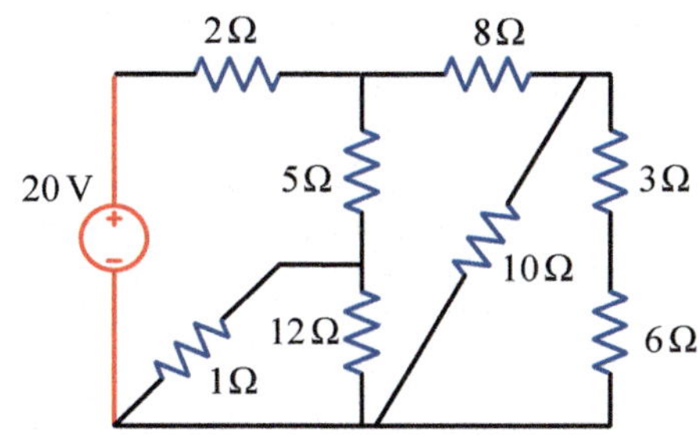

Fig. 4.149 Circuit for Exercise Problem 4.73

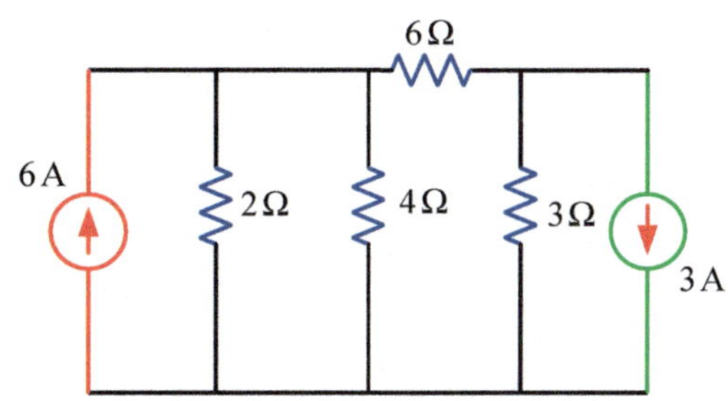

Fig. 4.150 Circuit for Exercise Problem 4.74

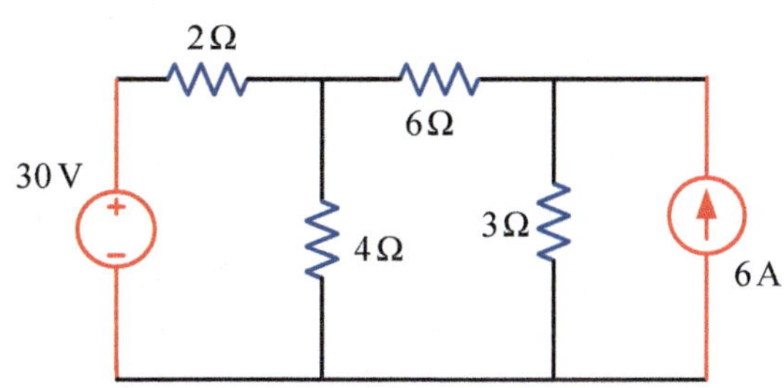

Fig. 4.151 Circuit for Exercise Problem 4.75

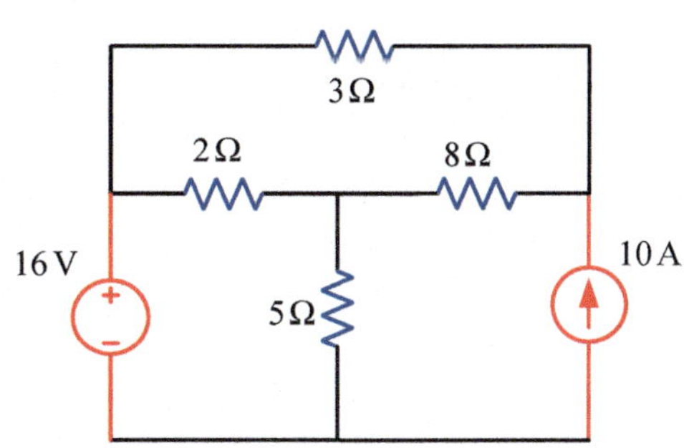

4.76 An electrical circuit is shown in Fig. 4.152. Determine the current in the 3 Ω resistor by using Norton's theorem and verify the result by PSpice simulation.

4.77 Use Norton's theorem to calculate the current in the 3 Ω resistor of the circuit in Fig. 4.153. Verify the result by PSpice simulation.

4.78 An electrical circuit is shown in Fig. 4.154. Use Norton's theorem to calculate the current in the 5 Ω resistor. Verify the result by PSpice simulation.

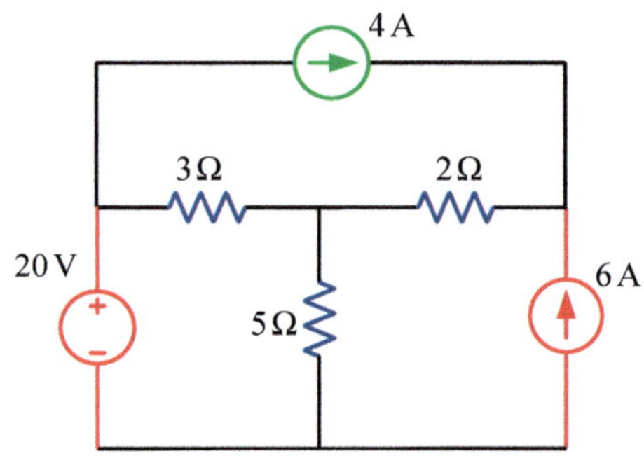

Fig. 4.152 Circuit for Exercise Problem 4.76

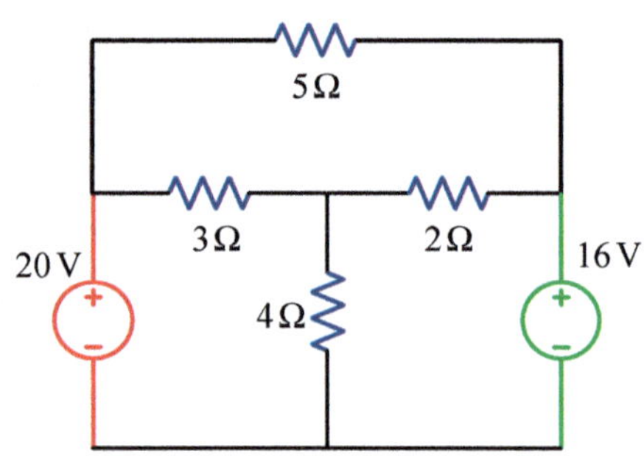

Fig. 4.153 Circuit for Exercise Problem 4.77

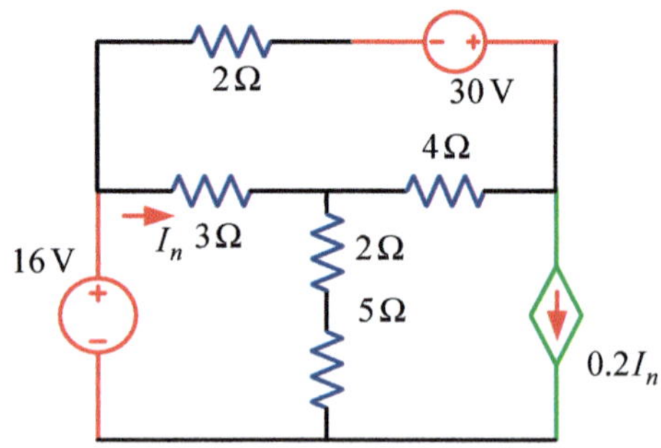

Fig. 4.154 Circuit for Exercise Problem 4.78

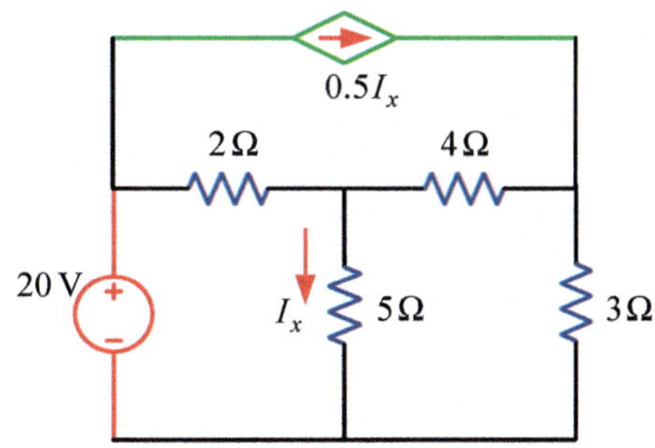

Fig. 4.155 Circuit for Exercise Problem 4.79

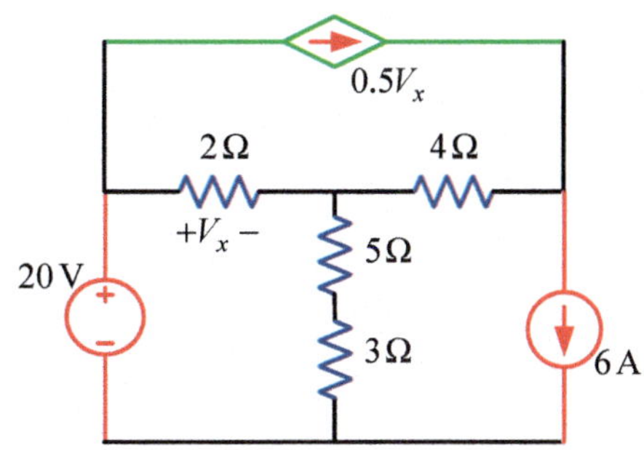

Fig. 4.156 Circuit for Exercise Problem 4.80

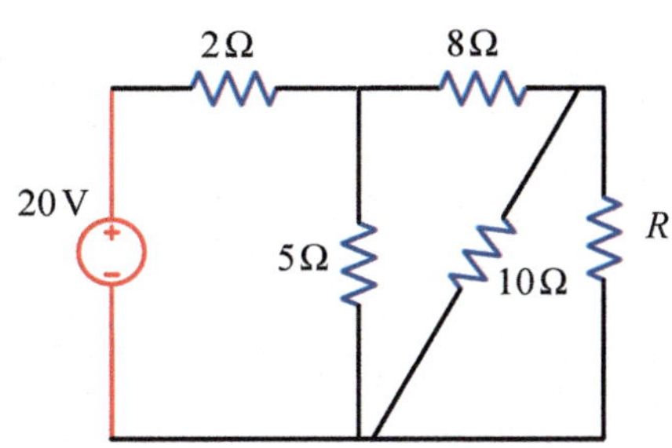

Fig. 4.157 Circuit for Exercise Problem 4.81

4.79 Figure 4.155 shows an electrical circuit. Use Norton's theorem to calculate the current in the 3 Ω resistor. Verify the result by PSpice simulation.

4.80 Figure 4.156 shows an electrical circuit. Use Norton's theorem to calculate the current in the 3 Ω resistor. Verify the result by PSpice simulation.

4.81 Using maximum power transfer theorem calculate the load resistance R of the circuit in Fig. 4.157, and also find the maximum power.

4.82 Using maximum power transfer theorem determine the load resistance *R* of the circuit in Fig. 4.158, and also find the maximum power.

4.83 Using maximum power transfer theorem calculate the load resistance *R* of the circuit in Fig. 4.159, and also find the maximum power.

4.84 Using maximum power transfer theorem calculate the load resistance *R* of the circuit in Fig. 4.160, and also find the maximum power.

4.85 Using maximum power transfer theorem calculate the load resistance *R* of the circuit in Fig. 4.161, and also find the maximum power.

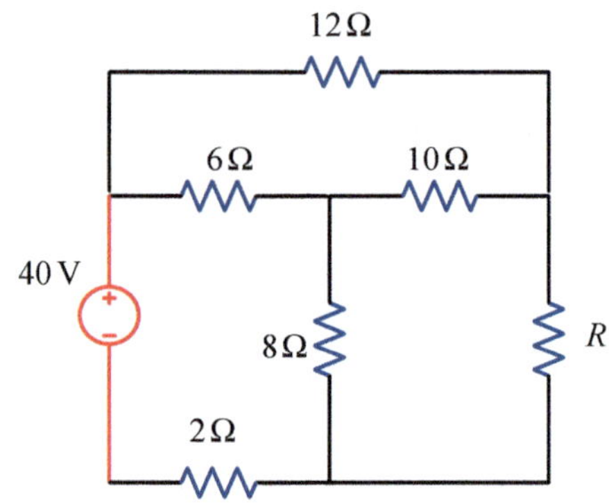

Fig. 4.158 Circuit for Exercise Problem 4.82

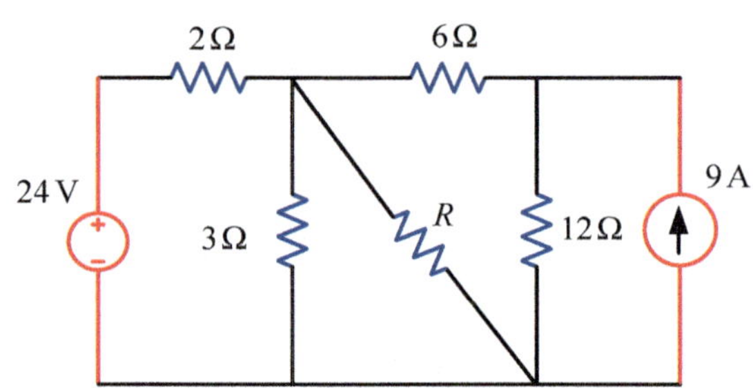

Fig. 4.159 Circuit for Exercise Problem 4.83

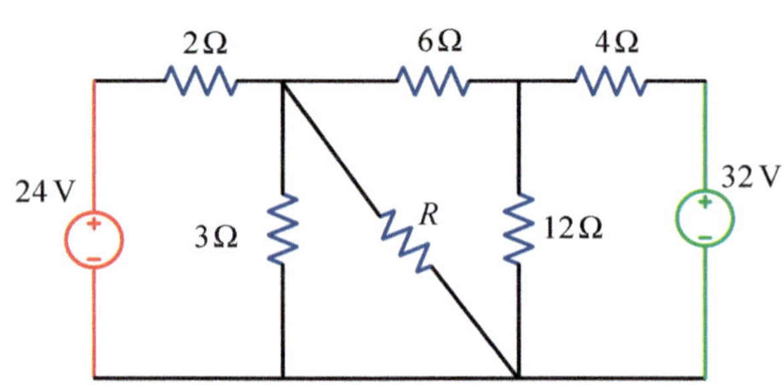

Fig. 4.160 Circuit for Exercise Problem 4.84

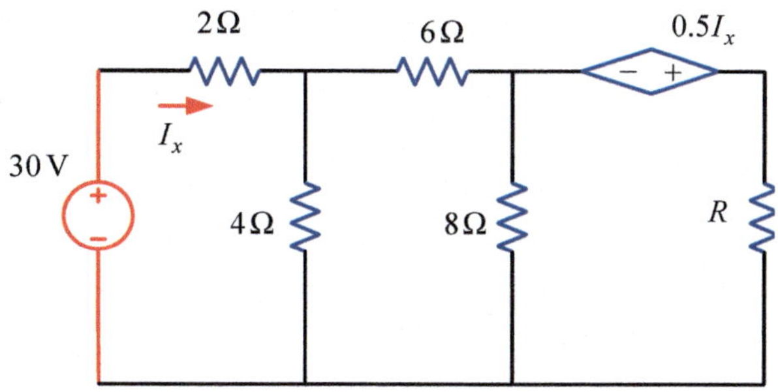

Fig. 4.161 Circuit for Exercise Problem 4.85

Chapter 5
Capacitors and Inductors

5.1 Introduction

Resistor, with different laws and network theorems, has already been discussed and analysed in the previous chapters. Like a resistor, capacitor and inductor are also important linear circuit elements. Capacitor and inductor do not dissipate energy like resistor, but store energy when these elements are connected to energy source. Later on, this stored energy can be used for other applications. A capacitor finds its application in power factor improvement, single-phase induction motor starting, filter circuit, laser, electronic flash on a camera, power systems, microphones, tuning circuit of a radio receiver and memory elements in a computer. Similarly, an inductor finds its applications in many areas, namely, in the transformer, power electronic circuit, electrical machines, radio, television and communication system. This chapter presents different types of analysis on electric circuits containing capacitors and inductors.

5.2 Capacitors

Capacitor is a passive two-terminal electrical component that stores electrical energy, and it consists of one or more pairs of parallel plate conductors separated by an insulating material, called dielectric. Different insulating materials, also known as dielectric materials, such as mica, glass, air, paper, ceramic and electrolytic, are used as in a capacitor. The capacitor, made of mica, glass or ceramic dielectric materials, is suitable for the high-frequency applications. A typical parallel plate capacitor and its symbol are shown in Fig. 5.1.

A capacitor is connected across a voltage source as shown in Fig. 5.2. In this case, the positive charge accumulates in one plate and the negative charge accumulates in the other plate. The resultant charge between these two plates creates the electric field

Md. A. Salam and Q. Mehbubar Rahman, *Fundamentals of Electrical Circuit Analysis*,
https://doi.org/10.1007/978-981-10-8624-3_5

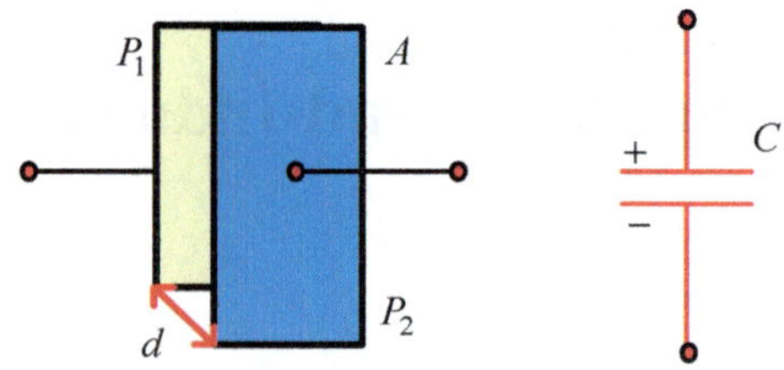

Fig. 5.1 Parallel plate capacitor and symbol

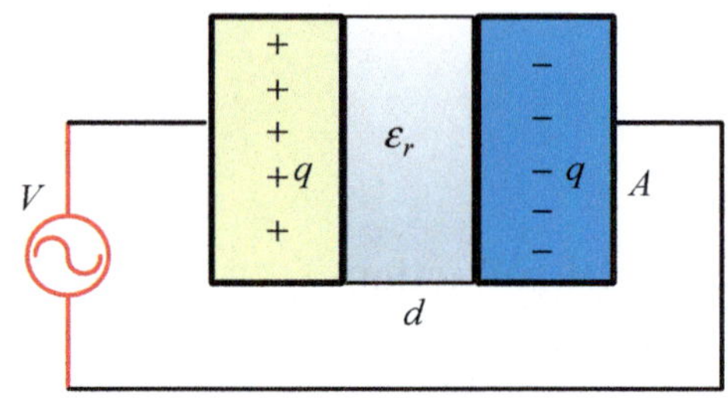

Fig. 5.2 Parallel plates with voltage source

which stores energy. This ability of a capacitor to store charge is known as capacitance. The capacitance is represented by the letter C, and its unit is coulombs/volts or farad (F), in honour of English Physicist Michael Faraday (1791–1867).

It has been established experimentally that the stored charge in a capacitor is directly proportional to the applied voltage, and it can be expressed as [1–3],

$$q \propto V \tag{5.1}$$

$$q = CV \tag{5.2}$$

$$C = \frac{q}{V} \tag{5.3}$$

where C is the proportionality constant known as capacitance. The SI unit of the capacitance can be derived from Eq. (5.3) as,

$$C = \frac{q}{V}\left(\frac{\text{Coulomb}}{\text{Volts}} = \text{Farad (F)}\right) \tag{5.4}$$

Farad is the largest unit of the capacitor. Generally, the capacitors with smaller values in micro-farad (μF) to pico-farad (pF) range are widely in use in practice.

5.3 Parallel Plate Capacitor with Uniform Medium

Consider a capacitor whose plates are separated by a uniform dielectric of thickness d metre, where the area of each plate is A square metre as shown in Fig. 5.2. Let E and D be the electric field intensity and flux density, respectively, between the plates of the capacitor. The electric field is defined as the voltage per unit length and it can be written as,

$$E = \frac{V}{d} \tag{5.5}$$

The electric flux density is defined as the charge per unit area and it can be written as,

$$D = \frac{q}{A} \tag{5.6}$$

The electric flux density is directly proportional to the electric field intensity and it can be written as,

$$D = \varepsilon E \tag{5.7}$$

where ε is the proportionality constant, known as the permittivity of the dielectric medium. Substituting Eq. (5.7) into Eq. (5.6) yields,

$$\varepsilon E = \frac{q}{A} \tag{5.8}$$

$$E = \frac{q}{\varepsilon A} \tag{5.9}$$

Substituting Eq. (5.9) into the Eq. (5.5) yields,

$$\frac{q}{\varepsilon A} = \frac{V}{d} \tag{5.10}$$

$$\frac{q}{V} = \frac{\varepsilon A}{d} \tag{5.11}$$

Substituting Eq. (5.3) into Eq. (5.11) yields,

$$C = \frac{\varepsilon A}{d} \tag{5.12}$$

$$C = \frac{\varepsilon_0 \varepsilon_r A}{d} \tag{5.13}$$

where ε_r the relative permittivity of the dielectric medium and ε_0 is the permittivity of free space.

From Eq. (5.12), it is concluded that the capacitance is directly proportional to the relative permittivity of the dielectric medium and the area of the plate. It is also inversely proportional to the distance between the plates.

Example 5.1 The dielectric thickness between the parallel plate capacitor is 0.5 cm and its relative permittivity is 3. The capacitor consists of a rectangular plate having dimensions 10 cm × 15 cm. The electric field strength of the dielectric medium is found to be 30 per cm. Determine the capacitance, and total charge on each plate.

Solution:
The area of the plates is,

$$A = 10 \times 15 = 15\,\text{cm}^2 \tag{5.14}$$

The capacitance is calculated as,

$$C = \frac{\varepsilon_0 \varepsilon_r A}{d} = \frac{8.854 \times 10^{-12} \times 3 \times 150}{0.5} = 7.97 \times 10^{-9}\text{F} = 0.0079\,\mu\text{F} \tag{5.15}$$

The voltage across the plates is,

$$V = E \times d = 30 \times 0.5 = 15\,\text{V} \tag{5.16}$$

The charge is calculated as,

$$q = CV = 0.0079 \times 10^{-6} \times 15 = 0.12\,\mu\text{C} \tag{5.17}$$

Practice Problem 5.1
The positive and the negative plates of a capacitor are separated by 0.3 cm solid dielectric medium. The relative permittivity of the dielectric medium is 8, and the area of the plate is 95 cm^2. The electric field strength of the dielectric medium is 280 V per cm. Calculate the capacitance and total charge on each plate.

5.4 Current–Voltage Terminal Characteristics of a Capacitor

As presented in Chap. 1, the current is expressed as [4–6],

$$i = \frac{\mathrm{d}q}{\mathrm{d}t} \tag{5.18}$$

Substituting Eq. (5.2) into Eq. (5.18) yields,

$$i = \frac{\mathrm{d}}{\mathrm{d}t}(Cv) \tag{5.19}$$

$$i = C\frac{\mathrm{d}v}{\mathrm{d}t} \tag{5.20}$$

Rearrangement of Eq. (5.20) yields,

$$dv = \frac{1}{C} i\,dt \tag{5.21}$$

Integration of Eq. (5.21) with respect to time (t) yields,

$$v = \frac{1}{C} \int_{-\infty}^{t} i dt \tag{5.22}$$

$$v = \frac{1}{C} \int_{0}^{t} i dt + v(0) \tag{5.23}$$

where $v(0)$ is the voltage across the capacitor at $t = 0$, which provides the memory property of the capacitor.

From Eq. (5.20), the following points can be summarized:

- If the rate of change of voltage across the capacitor is zero (i.e. DC voltage), the current through the capacitor is zero. Therefore, the capacitor acts as an open circuit to DC source.
- For an alternating voltage, which changes with time across the capacitor, it appears that the current flows through the capacitor.
- Capacitor is usually hold charge without a current flowing through it.
- Capacitor can store energy, and this provides the characteristic of a fundamental memory element.

Example 5.2 A 6 mF capacitor is connected across the voltage source of $v = 20 \sin 377t$ V. Calculate the expression of the current in the capacitor.

Solution:
The current in the capacitor can be calculated as,

$$i = C\frac{dv}{dt} = 6 \times 10^{-3} \times \frac{d}{dt}(20 \sin 377t) = 45.24 \cos 377t \text{ A} \tag{5.24}$$

Practice Problem 5.2
The current in a 3 μF capacitor is $i = 14e^{-1000t}$ mA. Determine the voltage across the capacitor if the initial voltage across the capacitor is zero.

Example 5.3 Figure 5.3 shows a voltage waveform for a 5 μF capacitor. Calculate the current.

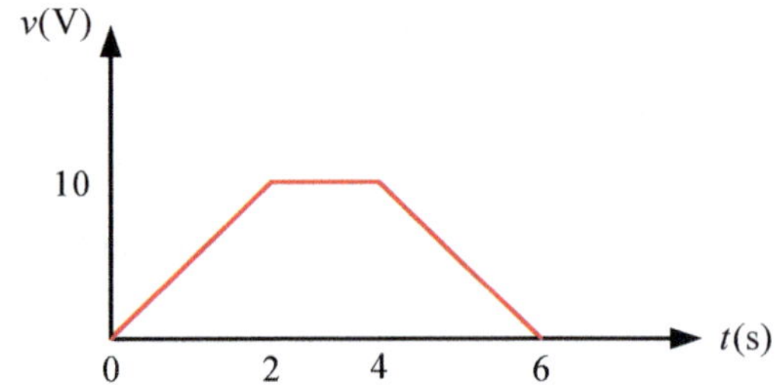

Fig. 5.3 Voltage waveform for Example 5.3

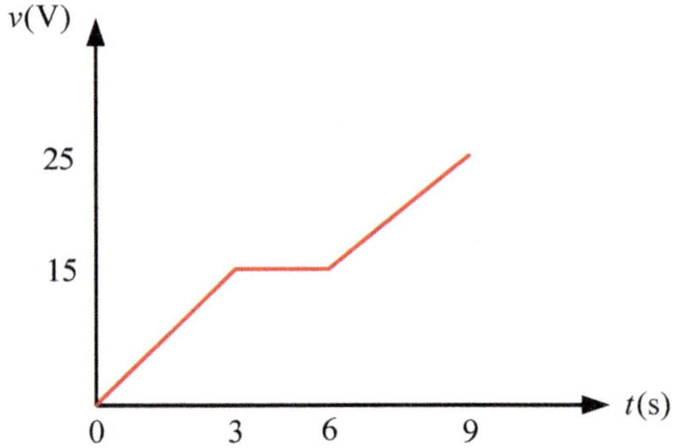

Fig. 5.4 Voltage waveform for Practice Problem 5.3

Solution

The equations of the voltage are,

$$v(t) = \frac{10 - 0}{2 - 0} t = 5t \text{ V} \qquad 0 \leq t \leq 2 \tag{5.25}$$

$$v(t) = 10 \text{ V} \qquad 2 \leq t \leq 4 \tag{5.26}$$

$$\frac{v(t) - 0}{0 - 10} = \frac{t - 6}{6 - 4} \tag{5.27}$$

$$v(t) = -5t + 30 \text{ V} \qquad 4 < t < 6 \tag{5.28}$$

The current is determined as,

$$i = C\frac{dv}{dt} = 5 \times 10^{-6}\left\{\frac{d}{dt}(5t) + \frac{d}{dt}(10) - \frac{d}{dt}(5t) + \frac{d}{dt}(30)\right\} = 0 \text{ A} \tag{5.29}$$

Practice Problem 5.3

Figure 5.4 shows a voltage waveform for a 4 μF capacitor. Find the current.

5.5 Energy Stored in a Capacitor

When a source is connected across a capacitor, the positive charges are transferred to one plate while the negative charges are transferred to the other plate. Thus, the energy is stored between the plates of a capacitor. The power p delivered to the capacitor from the source is,

$$p = vi \tag{5.30}$$

where v is the source voltage and i is the resultant current. Power is also defined as the rate of receiving or delivering energy (w), which is expressed as,

$$p = \frac{\mathrm{d}w}{\mathrm{d}t} \tag{5.31}$$

$$\mathrm{d}w = p\,\mathrm{d}t \tag{5.32}$$

Substituting Eq. (5.30) into (5.32) yields,

$$\mathrm{d}w = vi\,\mathrm{d}t \tag{5.33}$$

Substituting Eq. (5.20) into Eq. (5.33) yields,

$$\mathrm{d}w = vC\frac{\mathrm{d}v}{\mathrm{d}t}\,\mathrm{d}t \tag{5.34}$$

$$\mathrm{d}w = C\,v\,\mathrm{d}v \tag{5.35}$$

Integrating Eq. (5.35) from the negative infinite to time t yields,

$$w = C\int_{t=-\infty}^{t} v(t)\mathrm{d}v(t) \tag{5.36}$$

$$w = \frac{C}{2}\left[v^2(t) - v^2(-\infty)\right] \tag{5.37}$$

The capacitor does not store any energy at $t = -\infty$ i.e. $v(-\infty) = 0$. Therefore, Eq. (5.37) can be modified to,

$$w = \frac{1}{2}Cv^2 \tag{5.38}$$

Substituting Eq. (5.2) into Eq. (5.38) yields,

$$w = \frac{1}{2}Cv^2 = \frac{1}{2}qv = \frac{1}{2}\frac{q^2}{C} \tag{5.39}$$

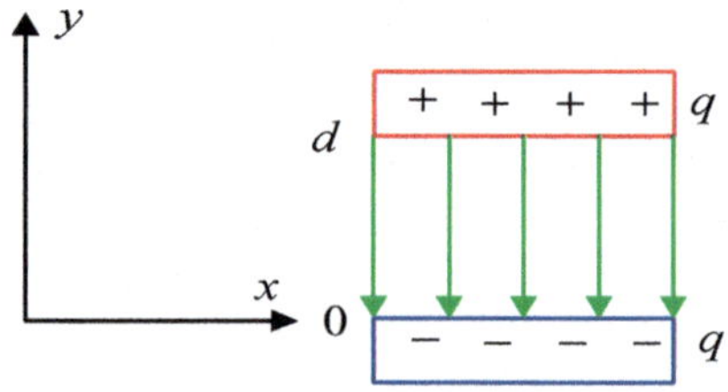

Fig. 5.5 Plates with separation distance

Equation (5.39) can be derived in an alternative way. In this case, consider a parallel plate capacitor as shown in Fig. 5.5. The distance between the plates is *d,* and the area of each plate is *A*. The space in between the plates is filled with a dielectric whose permittivity is ε. The positive and negative charges (q) are accumulated on the upper and lower plates, respectively. The expression of surface charge density (ρ_s) can be written as [7],

$$\rho_s = \frac{q}{A} \tag{5.40}$$

Considering electric flux density is zero, the following equation can be written:

$$\boldsymbol{E} = -\frac{\rho_s}{\varepsilon}\mathbf{a}_y \tag{5.41}$$

where $\boldsymbol{E}$ is the electric field intensity vector and $\mathbf{a}_y$ is the unity vector in y-direction. The expression of the voltage can be calculated as,

$$V = -\int_0^d \mathbf{E} \, . \, \mathrm{d}\mathbf{l} \tag{5.42}$$

Substituting Eq. (5.41) and the expression of differential length in Cartesian coordinates (only y component) into Eq. (5.42) provides,

$$V = -\int_0^d \left(-\frac{\rho_s}{\varepsilon}\mathbf{a}_y\right) . \left(dy\mathbf{a}_y\right) \tag{5.43}$$

$$V = \int_0^d \frac{\rho_s}{\varepsilon}\mathrm{d}y \tag{5.44}$$

$$V = \frac{\rho_s}{\varepsilon}d \tag{5.45}$$

Substituting Eq. (5.40) into Eq. (5.45) yields,

$$V = \frac{q}{A\varepsilon} d \tag{5.46}$$

Again, consider a total charge of q is transferred from the positive to the negative plates. Then, the potential difference between the plates can be written as,

$$V = \frac{\mathrm{d}w}{\mathrm{d}q} \tag{5.47}$$

Substituting Eq. (5.3) into Eq. (5.47) yields,

$$\frac{q}{C} = \frac{\mathrm{d}w}{\mathrm{d}q} \tag{5.48}$$

$$\mathrm{d}w = \frac{q}{C} \mathrm{d}q \tag{5.49}$$

Integrating Eq. (5.49) from 0 to q yields,

$$w = \int_0^q \frac{q}{C} \mathrm{d}q = \frac{1}{2} \frac{q^2}{C} \tag{5.50}$$

Again, substituting Eq. (5.3) into Eq. (5.50) yields,

$$w = \frac{q^2}{2C} = \frac{CV^2}{2} = \frac{CV}{2} \tag{5.51}$$

The electric field in between the plates is uniform, so the potential is written as,

$$V = Ed \tag{5.52}$$

Substituting Eqs. (5.12) and (5.52) into Eq. (5.51) yields,

$$w = \frac{\varepsilon A}{d} \frac{E^2 d^2}{2} \tag{5.53}$$

$$w = \frac{\varepsilon E^2 A d}{2} \tag{5.54}$$

In Eq. (5.54), the product Ad is the volume of the field in between the plates. Then, the energy density w_d or energy per unit volume is,

$$w_d = \frac{1}{2} \varepsilon E^2 \tag{5.55}$$

From Eq. (5.55), it is seen that the energy density is equal to the half of the product of the permittivity and the square of the electric field.

Example 5.4 A 20 μF capacitor is connected across a 120 V source. Calculate the charge and energy stored in the capacitor.

Solution:
The charge stored in the capacitor is,

$$q = Cv = 20 \times 10^{-6} \times 120 = 2.4\,\text{mC} \tag{5.56}$$

The energy stored is,

$$w = \frac{1}{2}Cv^2 = \frac{1}{2} \times 20 \times 10^{-6} \times 120^2 = 0.14\ \text{J} \tag{5.57}$$

Example 5.5 Two capacitors are connected to a voltage source as shown in Fig. 5.6. Determine the energy stored in the capacitors under DC condition.

Solution:
Under DC condition, capacitors are open circuited and the circuit is redrawn as shown in Fig. 5.7. The source current is,

$$I_s = \frac{32}{2+4+2} = 4\,\text{A} \tag{5.58}$$

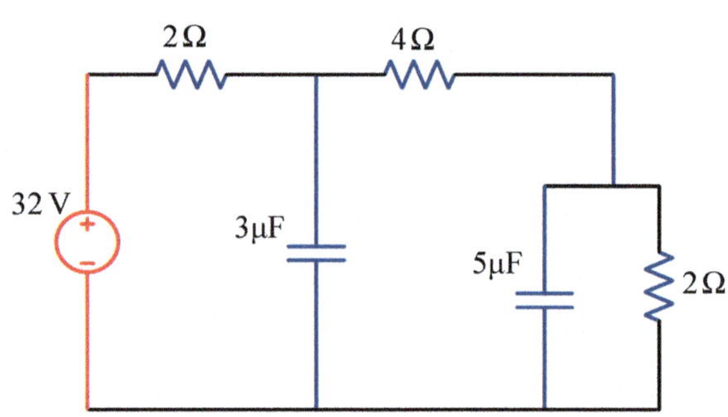

Fig. 5.6 Circuit for Example 5.5

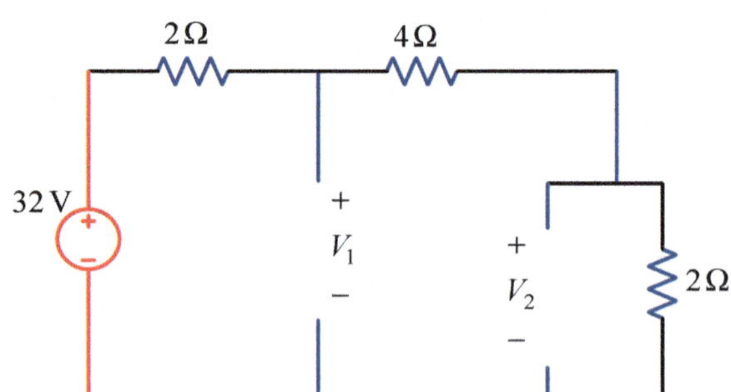

Fig. 5.7 Circuit for Example 5.5 with opening capacitors

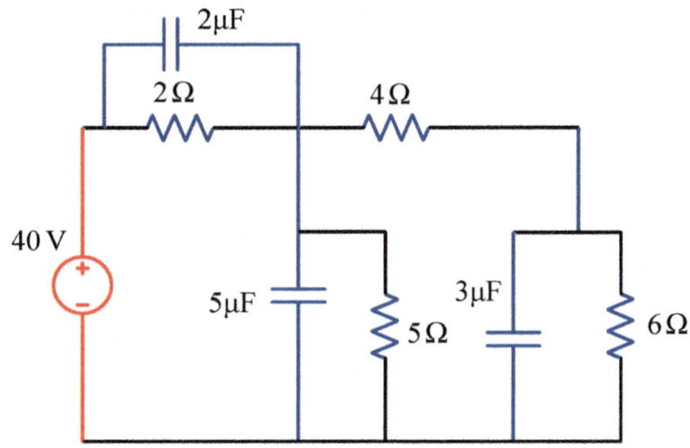

Fig. 5.8 Circuit for Practice Problem 5.5

The voltages across the two capacitors are calculated as,

$$V_1 = 32 - 2 \times 4 = 24 \text{ V} \tag{5.59}$$

$$V_2 = 2 \times 4 = 8 \text{ V} \tag{5.60}$$

The energies stored by the capacitors are,

$$w_1 = \frac{1}{2} C V_1^2 = \frac{1}{2} \times 3 \times 10^{-6} \times 24^2 = 0.86 \text{ mJ} \tag{5.61}$$

$$w_2 = \frac{1}{2} C V_2^2 = \frac{1}{2} \times 5 \times 10^{-6} \times 8^2 = 0.16 \text{ mJ} \tag{5.62}$$

Practice Problem 5.4
Determine the charge and energy stored when a 30 μF capacitor is connected across a 220 V source.

Practice Problem 5.5
Three capacitors are connected to a voltage source as shown in Fig. 5.8. Calculate the energy stored in the capacitors under DC condition.

5.6 Series and Parallel Capacitors

Capacitors are connected either in series or in parallel in an electrical circuit. Here, three capacitors are connected in series as shown in Fig. 5.9.

The same current will flow through all three capacitors as they are connected in series. According to Eq. (5.23), the voltages across each capacitor can be written as,

$$v_1 = \frac{1}{C_1} \int_0^t i dt \tag{5.63}$$

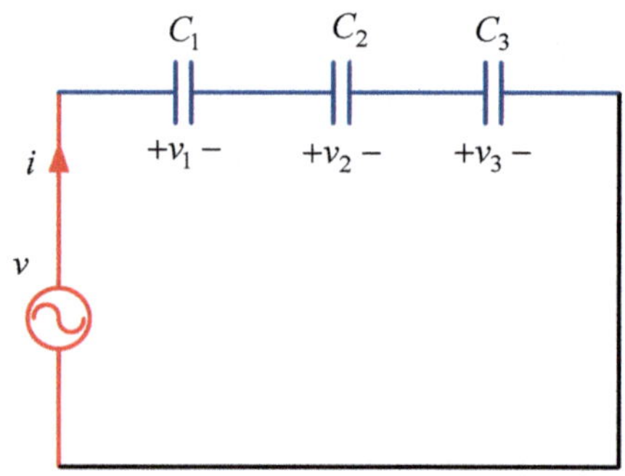

Fig. 5.9 Capacitors in series

$$v_2 = \frac{1}{C_2}\int_0^t i\mathrm{d}t \tag{5.64}$$

$$v_3 = \frac{1}{C_3}\int_0^t i\mathrm{d}t \tag{5.65}$$

Applying KVL to the circuit in Fig. 5.9 yields,

$$v = v_1 + v_2 + v_3 \tag{5.66}$$

Substituting Eqs. (5.63) to (5.65) into Eq. (5.66) yields,

$$v = \left(\frac{1}{C_1} + \frac{1}{C_2} + \frac{1}{C_3}\right)\int_0^t i\mathrm{d}t \tag{5.67}$$

However, the expression of the total voltage is calculated when $v(0) = 0$ as,

$$v = \frac{1}{C_{eq}}\int_0^t i\mathrm{d}t \tag{5.68}$$

Substituting Eq. (5.68) into Eq. (5.67) yields,

$$\frac{1}{C_{\mathrm{eq}}}\int_0^t i\mathrm{d}t = \left(\frac{1}{C_1} + \frac{1}{C_2} + \frac{1}{C_3}\right)\int_0^t i\mathrm{d}t \tag{5.69}$$

$$\frac{1}{C_{\mathrm{eq}}} = \frac{1}{C_1} + \frac{1}{C_2} + \frac{1}{C_3} \tag{5.70}$$

If two capacitors are connected in series, then Eq. (5.70) can be modified as,

$$\frac{1}{C_{eq}} = \frac{1}{C_1} + \frac{1}{C_2} \tag{5.71}$$

$$C_{eq} = \frac{C_1 C_2}{C_1 + C_2} \tag{5.72}$$

From Eq. (5.72), it is concluded that the equivalent capacitance for two capacitors connected in series is equal to the product of the two capacitances divided by their sum.

Again, considering three capacitors are connected in parallel as shown in Fig. 5.10. The voltage across each capacitor will be the same as the source voltage, but the currents will be different. According to Eq. (5.20), the currents flow in the capacitors are,

$$i_1 = C_1 \frac{dv}{dt} \tag{5.73}$$

$$i_2 = C_2 \frac{dv}{dt} \tag{5.74}$$

$$i_3 = C_3 \frac{dv}{dt} \tag{5.75}$$

The total current can also be expressed as,

$$i = C_{eq} \frac{dv}{dt} \tag{5.76}$$

where C_{eq} is the equivalent capacitance of the circuit. Applying KCL to the circuit in Fig. 5.10 yields,

$$i = i_1 + i_2 + i_3 \tag{5.77}$$

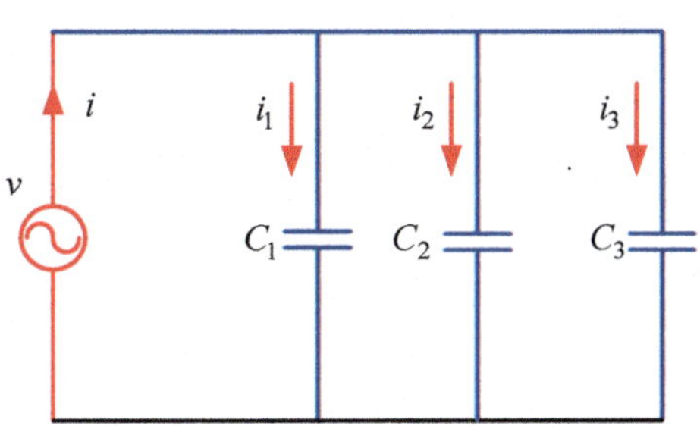

Fig. 5.10 Capacitors in parallel

Substituting Eqs. (5.73) to (5.76) into Eq. (5.77) yields,

$$C_{\text{eq}}\frac{\mathrm{d}v}{\mathrm{d}t} = C_1\frac{\mathrm{d}v}{\mathrm{d}t} + C_2\frac{\mathrm{d}v}{\mathrm{d}t} + C_3\frac{\mathrm{d}v}{\mathrm{d}t} \tag{5.78}$$

$$C = C_1 + C_2 + C_3 \tag{5.79}$$

From Eq. (5.79), it can be concluded that the equivalent capacitance in a parallel connection is equal to the sum of their individual capacitance.

Example 5.6 Capacitors are connected in a circuit as shown in Fig. 5.11. Determine the equivalent circuit capacitance.

Solution:
Capacitors 3 and 5 μF are connected in series. Here, the total capacitance is,

$$C_1 = \frac{3 \times 5}{3 + 5} = 1.88\ \mu\text{F} \tag{5.80}$$

The capacitors 1.88 and 8 μF are connected in parallel, and the total capacitance is,

$$C_2 = 8 + 1.88 = 9.88\ \mu\text{F} \tag{5.81}$$

Again, 9.88 and 2 μF are connected in series and the equivalent circuit capacitance is,

$$C_{\text{eq}} = \frac{9.88 \times 2}{9.88 + 2} = 1.66\ \mu\text{F} \tag{5.82}$$

Practice Problem 5.6
Figure 5.12 shows a circuit where capacitors are connected in a series and parallel connection. Calculate the equivalent circuit capacitance.

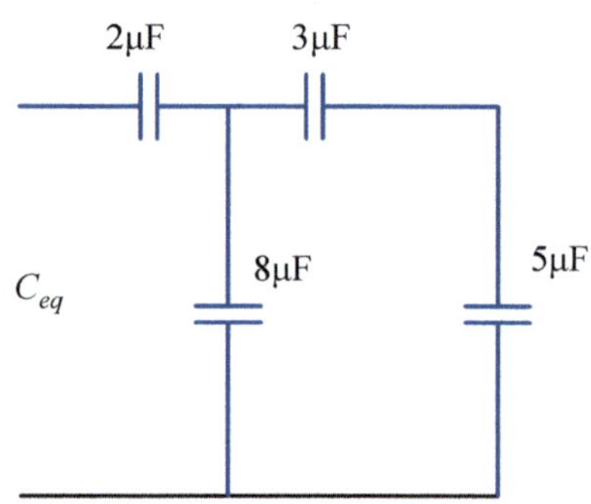

Fig. 5.11 Circuit for Example 5.6

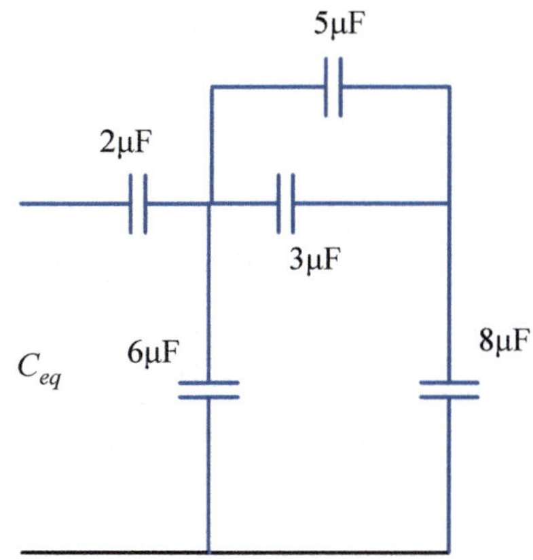

Fig. 5.12 Circuit for Practice Problem 5.6

5.7 Current and Voltage Divider Rules for Capacitor

Consider that two capacitors are connected in parallel with a voltage source as shown in Fig. 5.13.

Applying KCL to the circuit in Fig. 5.13 yields,

$$i = i_1 + i_2 \tag{5.83}$$

Substituting Eqs. (5.73) and (5.74) into Eq. (5.83) yields,

$$i = C_1 \frac{dv}{dt} + C_2 \frac{dv}{dt} \tag{5.84}$$

$$i = (C_1 + C_2) \frac{dv}{dt} \tag{5.85}$$

$$\frac{dv}{dt} = \frac{i}{C_1 + C_2} \tag{5.86}$$

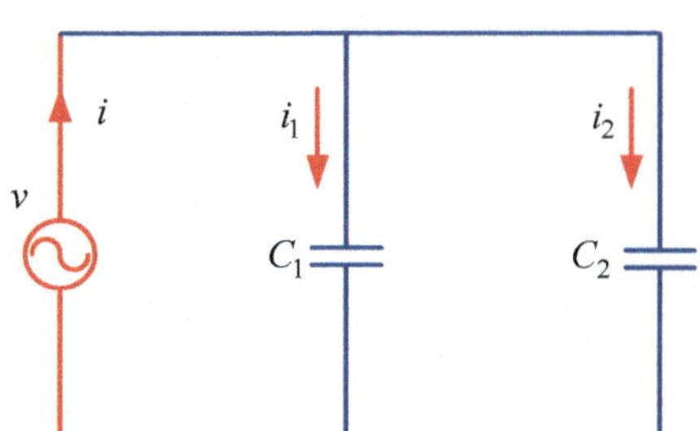

Fig. 5.13 Two capacitors are in parallel

Substituting Eq. (5.86) into Eqs. (5.73) and (5.74) yields,

$$i_1 = \frac{i}{C_1 + C_2} C_1 \tag{5.87}$$

$$i_2 = \frac{i}{C_1 + C_2} C_2 \tag{5.88}$$

Equations (5.87) and (5.88) present the current division rule for the two capacitors connected in parallel.

Again, consider that two capacitors are connected in series as shown in Fig. 5.14. Assume both capacitors are uncharged initially. Applying KVL to the circuit in Fig. 5.14 yields,

$$v = v_1 + v_2 \tag{5.89}$$

Substituting Eqs. (5.63) and (5.64) into Eq. (5.89) yields,

$$v = \frac{1}{C_1} \int_0^t i dt + \frac{1}{C_2} \int_0^t i dt \tag{5.90}$$

$$v = \left(\frac{C_1 + C_2}{C_1 C_2} \right) \int_0^t i dt \tag{5.91}$$

$$\int_0^t i dt = \frac{C_1 C_2}{C_1 + C_2} v \tag{5.92}$$

Substituting Eq. (5.92) into Eqs. (5.63) and (5.64) yields,

$$v_1 = \frac{C_2}{C_1 + C_2} v \tag{5.93}$$

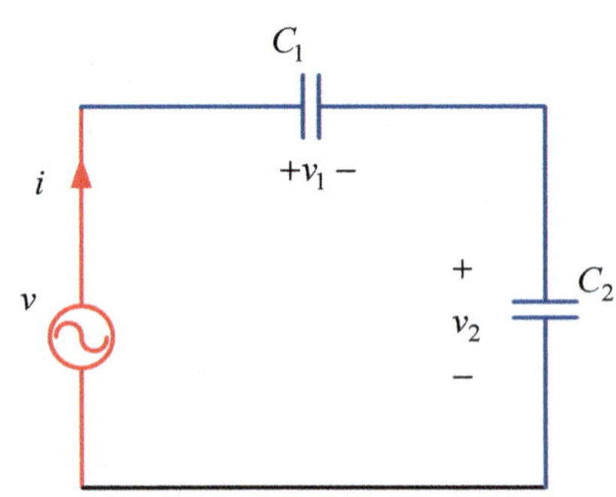

Fig. 5.14 Two capacitors are in series

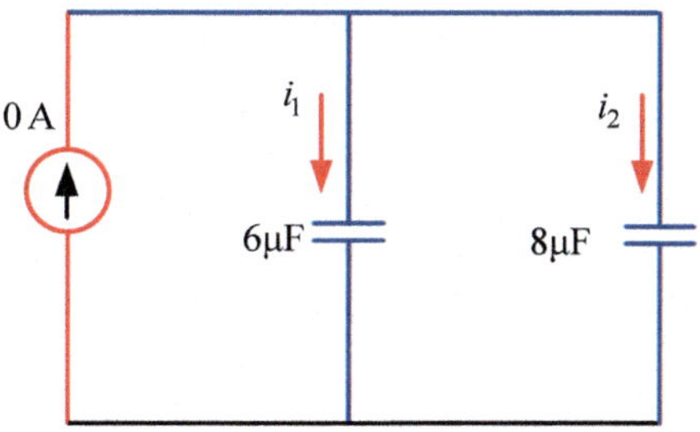

Fig. 5.15 Circuit for Example 5.7

$$v_2 = \frac{C_1}{C_1 + C_2} v \tag{5.94}$$

Equations (5.93) and (5.94) present the voltage division rule for the two capacitors connected in series.

Example 5.7 Two capacitors are connected in parallel as shown in Fig. 5.15. Determine the current in each branch.

Solution:
The currents in the branches are calculated as,

$$i_1 = \frac{i}{C_1 + C_2} C_1 = 10 \times \frac{6}{6+8} = 4.29\,\text{A} \tag{5.95}$$

$$i_2 = \frac{i}{C_1 + C_2} C_2 = 10 \times \frac{8}{6+8} = 5.71\,\text{A} \tag{5.96}$$

Practice Problem 5.7
Two capacitors are connected in series as shown in Fig. 5.16. Find the voltage across each capacitor.

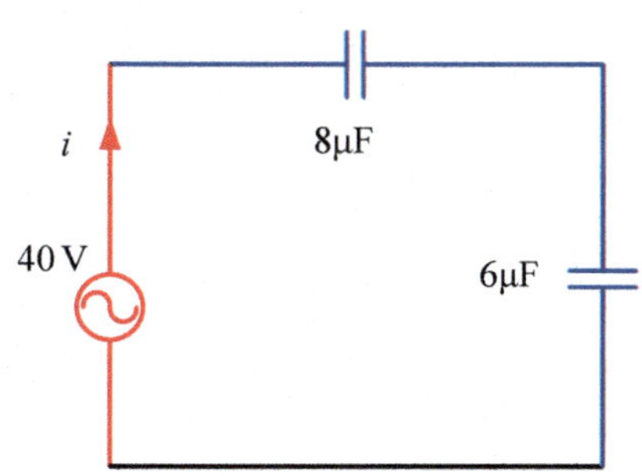

Fig. 5.16 Circuit for Practice Problem 5.7

5.8 Coaxial Capacitor

Consider a charged coaxial cable with a length of l as shown in Fig. 5.17. The inner and outer radii of the cable are a and b, respectively. The respective charges are $+q$ and $-q$. The space between the conductors is filled with a homogeneous dielectric material whose permittivity is ε. Then, the total charge can be determined as [7],

$$q = \int \mathbf{D}.d\mathbf{S} \tag{5.97}$$

$$q = \varepsilon \int \mathbf{E}.d\mathbf{S} \tag{5.98}$$

where $\mathbf{D}$ is the electric flux density vector. The electric field intensity vector ($\mathbf{E}$) is usually normal to the cylindrical Gaussian surface and it can be represented as,

$$\mathbf{E} = E_\rho \mathbf{a}_\rho \tag{5.99}$$

The term $\int d\mathbf{S}$ is the surface area of the Gaussian surface and it can be represented as,

$$\int d\mathbf{S} = 2\pi\rho l \tag{5.100}$$

Here, l is the arbitrary length of the cylinder, and ρ is the surface charge density. Substituting Eqs. (5.99) and (5.100) into Eq. (5.98) yields,

$$q = \varepsilon E_\rho 2\pi\rho l \tag{5.101}$$

$$E_\rho = \frac{q}{2\varepsilon\pi\rho l} \tag{5.102}$$

Again, substituting Eq. (5.102) into Eq. (5.99) yields,

$$\mathbf{E} = \frac{q}{2\varepsilon\pi\rho l}\mathbf{a}_\rho \tag{5.103}$$

Fig. 5.17 Schematic of coaxial cable

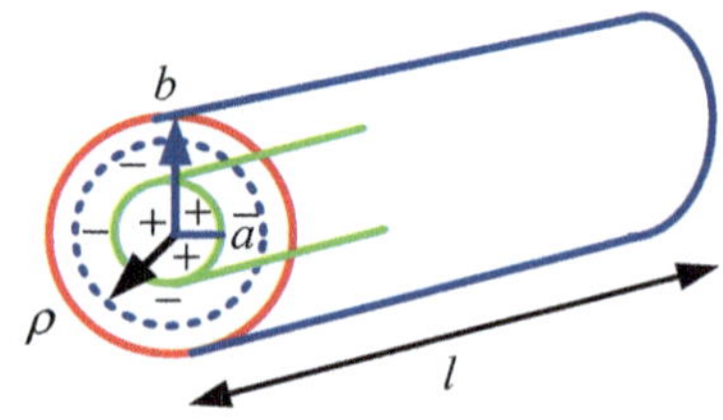

The voltage difference between two conductors can be determined as,

$$V = -\int_{b}^{a} \mathbf{E}\,.\,\mathrm{dl} \tag{5.104}$$

Substituting Eq. (5.103) and differential length in cylindrical coordinates into Eq. (5.104) yields,

$$V = -\int_{b}^{a} \frac{q}{2\varepsilon\pi\rho l}\mathbf{a}_{\rho}.(d\rho\mathbf{a}_{\rho}) \tag{5.105}$$

$$V = -\frac{q}{2\varepsilon\pi l}[\ln \rho]_{b}^{a} \tag{5.106}$$

$$V = -\frac{q}{2\varepsilon\pi l}\ln\frac{a}{b} \tag{5.107}$$

$$V = \frac{q}{2\varepsilon\pi l}\ln\frac{b}{a} \tag{5.108}$$

Substituting Eq. (5.108) into Eq. (5.3) yields,

$$C = \frac{2\varepsilon\pi l}{\ln\frac{b}{a}} \tag{5.109}$$

Equation (5.109) represents the expression of capacitance for coaxial cable.

5.9 Spherical Capacitor

Consider two concentric spheres whose radii are a and b as shown in Fig. 5.18. The radius a is greater than the radius b. As shown in Fig. 5.18, the inner and outer spheres contain $+q$ and $-q$ charges, respectively. Applying Gauss' law to an arbitrary Gaussian spherical surface with radius $(a < r < b)$, the following analysis can be presented.

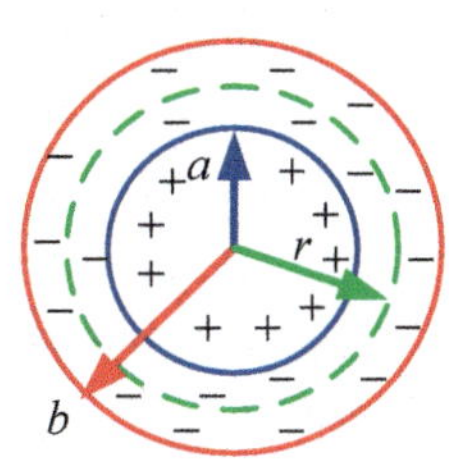

Fig. 5.18 Schematic of spheres

The electric field intensity is normal to the spherical Gaussian surface and it can be represented as [7],

$$\mathbf{E} = E_r \mathbf{a}_r \tag{5.110}$$

where E_r is the electric field at radius r.

The term $\int d\mathbf{S}$ is the surface area of the Gaussian surface and it can be represented as,

$$\int d\mathbf{S} = 4\pi r^2 \tag{5.111}$$

Substituting Eqs. (5.110) and (5.111) into Eq. (5.98) yields,

$$q = \varepsilon E_r 4\pi r^2 \tag{5.112}$$

$$E_r = \frac{q}{4\varepsilon\pi r^2} \tag{5.113}$$

Again, substituting Eq. (5.113) into Eq. (5.110) yields,

$$\mathbf{E} = \frac{q}{4\varepsilon\pi r^2}\mathbf{a}_r \tag{5.114}$$

Substituting Eq. (5.114) and differential length in spherical coordinates into Eq. (5.014) yields,

$$V = -\int_b^a \frac{q}{4\varepsilon\pi r^2}\mathbf{a}_r .\,(dr\mathbf{a}_r) \tag{5.115}$$

$$V = -\frac{q}{4\varepsilon\pi}\left[\frac{r^{-2+1}}{-2+1}\right]_b^a \tag{5.116}$$

$$V = \frac{q}{4\varepsilon\pi}\left[\frac{1}{r}\right]_b^a \tag{5.117}$$

$$V = \frac{q}{4\varepsilon\pi}\left[\frac{1}{a} - \frac{1}{b}\right] \tag{5.118}$$

Substituting Eq. (5.118) into Eq. (5.3) yields,

$$C = \frac{4\varepsilon\pi}{\frac{1}{a} - \frac{1}{b}} \tag{5.119}$$

Equation (5.119) represents the expression of capacitance for two concentric spheres.

Example 5.8 The inner and outer radii of a 2 km coaxial cable are 3 mm and 5 mm, respectively. Determine the capacitance of the cable.

Solution:
The capacitance can be determined as,

$$C = \frac{2\pi\varepsilon l}{\ln\frac{b}{a}} = \frac{2\pi \times 8.854 \times 10^{-12} \times 2 \times 10^3}{\ln\frac{5}{3}} = 0.218\ \mu\text{F} \tag{5.120}$$

Practice Problem 5.8
The outer radius of a 5 km coaxial cable is 5 mm. Calculate the inner radius of the cable if the capacitance of the cable is 0.3 μF.

5.10 Parallel Plate Capacitor with Two Dielectric Slabs

Consider that two conductors a and b are separated by two different dielectric slabs as shown in Fig. 5.19. The lengths of the dielectric slabs are l_1 and l_2, respectively. The electric fields for two dielectric slabs in the z-direction can be written as [7],

$$\mathbf{E}_1 = -\frac{\rho_s}{\varepsilon_1}\mathbf{a}_z \tag{5.121}$$

$$\mathbf{E}_2 = -\frac{\rho_s}{\varepsilon_2}\mathbf{a}_z \tag{5.122}$$

The voltage difference between two conductors can be determined as,

$$V = -\int \mathbf{E}.d\mathbf{l} = -\left[\int_0^{l_1} \mathbf{E}_1.d\mathbf{l} + \int_{l_1}^{l_1+l_1} \mathbf{E}_2.d\mathbf{l}\right] \tag{5.123}$$

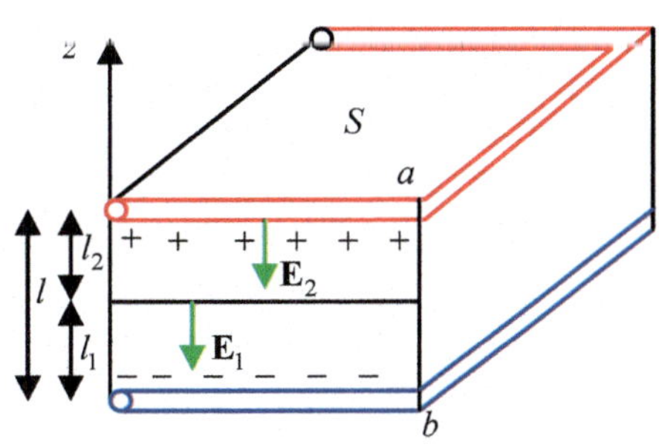

Fig. 5.19 Capacitor with two dielectric slabs

Substituting Eqs. (5.121), (5.122) and the differential length in z-component into Eq. (5.104) yields,

$$V = \int_{0}^{l_1} \frac{\rho_s}{\varepsilon_1} \mathbf{a}_z . \mathrm{d}z\mathbf{a}_z + \int_{l_1}^{l_1+l_1} \frac{\rho_s}{\varepsilon_2} \mathbf{a}_z . \mathrm{d}z\mathbf{a}_z \tag{5.124}$$

$$V = \int_{0}^{l_1} \frac{\rho_s}{\varepsilon_1} \mathrm{d}z + \int_{l_1}^{l_1+l_1} \frac{\rho_s}{\varepsilon_2} \mathrm{d}z \tag{5.125}$$

$$V = \frac{\rho_s}{\varepsilon_1} l_1 + \frac{\rho_s}{\varepsilon_2} (l_1 + l_2 - l_1) \tag{5.126}$$

$$V = \frac{\rho_s}{\varepsilon_1} l_1 + \frac{\rho_s}{\varepsilon_2} l_2 \tag{5.127}$$

Substituting Eq. (5.40) into Eq. (5.127) yields,

$$V = \frac{q}{A} \left(\frac{1}{\varepsilon_1} l_1 + \frac{1}{\varepsilon_2} l_2 \right) \tag{5.128}$$

Substituting Eq. (5.128) into Eq. (5.3) yields,

$$C = \frac{1}{\frac{l_1}{\varepsilon_1 A} + \frac{l_2}{\varepsilon_2 A}} \tag{5.129}$$

$$C = \frac{1}{\frac{1}{\frac{\varepsilon_1 A}{l_1}} + \frac{1}{\frac{\varepsilon_2 A}{l_2}}} \tag{5.130}$$

According to the expression of capacitance [see Eq. (5.12)], the terms $\frac{\varepsilon_1 A}{l_1}$ and $\frac{\varepsilon_2 A}{l_2}$ can be defined as the capacitance C_1 and C_2 of the lower and upper dielectric slabs. Then, Eq. (5.130) can be modified as,

$$C = \frac{1}{\frac{1}{C_1} + \frac{1}{C_2}} \tag{5.131}$$

$$C = \frac{C_1 C_2}{C_1 + C_2} \tag{5.132}$$

Equation (5.132) represents the equivalent capacitance when two capacitors are in series.

Example 5.9 In a parallel plate capacitor with two dielectric slabs, the lengths of the dielectric slabs are 3 mm and 5 mm, respectively, while their associated permittivities are $\varepsilon_1 = 2$ and $\varepsilon_2 = 4$, respectively. A voltage of 100 V is applied across the parallel plates, where the area of each plate is $0.56 \times 10^{-6}\,\text{m}^2$. Determine the (i) capacitance, (ii) q, (iii) D, (iv) E_1, E_2 and (v) $V_1,\ V_2$.

Solution:

(i) The capacitance can be determined as,

$$C = \frac{\varepsilon_0 S}{\frac{l_1}{\varepsilon_1} + \frac{l_2}{\varepsilon_2}} = \frac{8.854 \times 10^{-12} \times 0.56 \times 10^{-6}}{\frac{0.003}{2} + \frac{0.005}{4}} = 1803 \times 10^{-18}\,\text{F} \tag{5.133}$$

(ii) The charge is,

$$q = CV = 0.0018 \times 100 = 0.18\ \text{pC} \tag{5.134}$$

(iii) The electric flux density can be determined as,

$$D = \frac{q}{S} = \frac{0.18 \times 10^{-12}}{0.56 \times 10^{-6}} = 0.321 \times 10^{-6}\ \text{C/m}^2 \tag{5.135}$$

(iv) The electric field intensities are,

$$E_1 = \frac{D}{\varepsilon_0 \varepsilon_1} = \frac{0.321 \times 10^{-6}}{2 \times 8.854 \times 10^{-12}} = 18127.4\ \text{V/m} \tag{5.136}$$

$$E_2 = \frac{D}{\varepsilon_0 \varepsilon_2} = \frac{0.321 \times 10^{-6}}{4 \times 8.854 \times 10^{-12}} = 9063.7\ \text{V/m} \tag{5.137}$$

(v) The voltages can be determined as,

$$V_1 = E_1 l_1 = 18127.4 \times 003 = 54.38\ \text{V} \tag{5.138}$$

$$V_2 = E_2 l_2 = 9063.7 \times 0.005 = 45.32\ \text{V} \tag{5.139}$$

Practice Problem 5.9

In a parallel plate capacitor with two dielectric slabs, the lengths of the dielectric slabs are 2 and 3.5 mm, respectively, while their associated permittivities are $\varepsilon_1 = 3$ and $\varepsilon_2 = 5$, respectively. An 80 V source is connected across the parallel plates, where the area of each plate is $0.026 \times 10^{-7}\,\text{m}^2$. Calculate the (i) capacitance, (ii) q, (iii) D, (iv) E_1, E_2 and (v) $V_1,\ V_2$.

5.11 Inductor and Inductance

Inductor is a passive two-terminal electrical component that stores electrical energy in the form of magnetic field. An inductor is made of conducting wire with a suitable number of turns wound around a magnetic or non-magnetic core as shown in Fig. 5.20. An American scientist and engineer Joseph Henry (1797–1878) invented the concept of self-inductance and mutual-inductance. When current flows through a conductor, it results in a magnetic field or flux around the conductor. The flux linkage per unit current is known as inductance. The inductance is also defined as the ability of an electrical conductor to induce a voltage by the time-varying current that flows through it. The unit of inductance is henry (H) in honour of Joseph Henry. Mathematically, the inductance L of an inductor can be expressed as,

$$L = \frac{\psi}{i} \tag{5.140}$$

where

ψ is the flux linkage,
i is the current flowing through the inductor.

Consider that a current i is flowing through a coil as shown in Fig. 5.21. Because of this current, the flux ϕ is associated with the inductor and the direction of this flux is determined by the Fleming's right-hand rule. In this case, the total flux ϕ is directly proportional to the current as,

$$\phi \propto i \tag{5.141}$$

$$\phi = Li \tag{5.142}$$

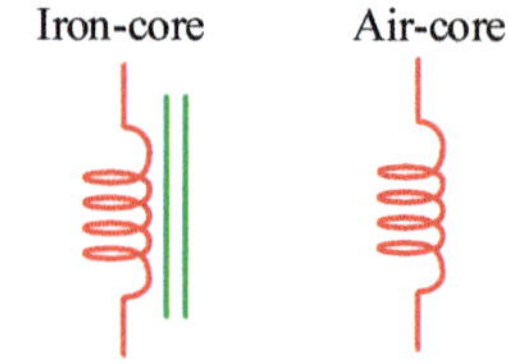

Fig. 5.20 Symbols of inductor

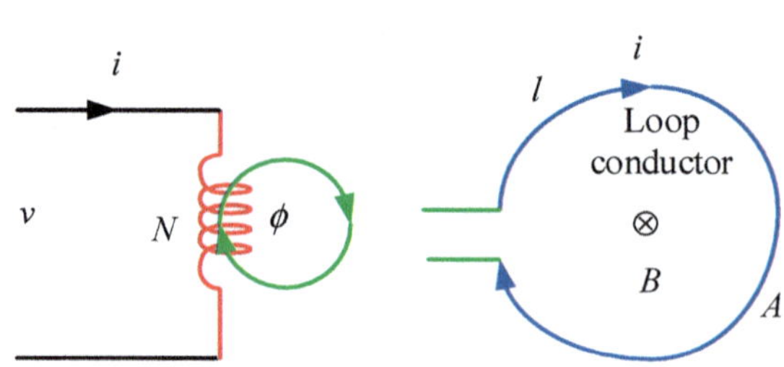

Fig. 5.21 Inductor associated with flux

The current flows through the loop conductor and generates the magnetic field. This magnetic field is expressed as,

$$B = \mu \frac{i}{l} \tag{5.143}$$

where μ is the permeability and l is the length of the core. If the cross-sectional area of the core is A, the magnetic flux can be written as,

$$\phi = BA \tag{5.144}$$

Substituting Eq. (5.143) into Eq. (5.144) yields

$$\phi = \mu \frac{i}{l} A \tag{5.145}$$

$$\phi = Li$$

where the expression of inductance is,

$$L = \frac{\mu A}{l} \tag{5.146}$$

A solenoid with a suitable number of turns of a coil is shown in Fig. 5.22. The mean length and the cross-sectional area of the core are l and A, respectively. In this case, the current in the coil will create a flux. According to the Faraday's laws of electromagnetic induction, the voltage induced across the coil is,

$$v = N \frac{\mathrm{d}\phi}{\mathrm{d}t} \tag{5.147}$$

where N is the number of turns in the coil. Equation (5.147) can be rearranged as,

$$v = N \left[\frac{\mathrm{d}\phi}{\mathrm{d}i} \frac{\mathrm{d}i}{\mathrm{d}t} \right] = L \frac{\mathrm{d}i}{\mathrm{d}t} \tag{5.148}$$

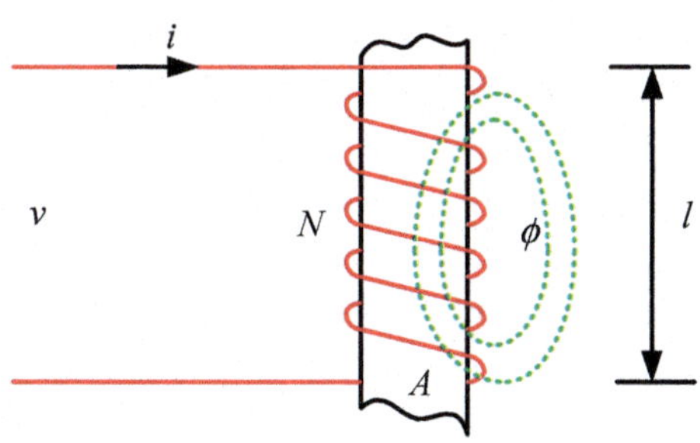

Fig. 5.22 Inductor with a core

where the expression of self-inductance (induction in a coil due to change in the current through this coil) is,

$$L = N\frac{\mathrm{d}\phi}{\mathrm{d}i} \tag{5.149}$$

The flux can be expressed in terms of magnetomotive force ($\Im$) [magnetomotive force is the magnetizing force that represents the magnetic intensity] analogues to electromagnetic force (EMF) and reluctance ($\Re$) [reluctance is the magnetic resistance that opposes the passages of magnetic flux in a magnetic circuit; electric flux is analogues to electric current] as,

$$\phi = \frac{\Im}{\Re} \tag{5.150}$$

But, the expressions of magnetomotive force and reluctance are,

$$\Im = Ni \tag{5.151}$$

$$\Re = \frac{l}{\mu A} \tag{5.152}$$

Substituting Eqs. (5.151) and (5.152) into Eq. (5.150) yields,

$$\phi = \frac{N\mu A}{l}i \tag{5.153}$$

Differentiating Eq. (5.153) with respect to i yields,

$$\frac{\mathrm{d}\phi}{\mathrm{d}i} = \frac{N\mu A}{l} \tag{5.154}$$

Substituting Eq. (5.154) into Eq. (5.149) yields,

$$L = \frac{N^2\mu A}{l} \tag{5.155}$$

From Eq. (5.155), it is seen that inductance is inversely proportional to the mean length of the core, and directly proportional to the product of the permeability of core, square of the number of turns in the coil and the cross-sectional area of the core.

Example 5.10 An air-core solenoid is wound by 40 turns of copper coil whose mean length and cross-sectional area are 5 m and 24 cm^2, respectively. If the permeability of free space is $4\pi \times 10^{-7}\ \mathrm{H/m^2}$, determine the inductance.

Solution:
The inductance is calculated as,

$$L = \frac{N^2 \mu A}{l} = \frac{40^2 \times 1 \times 4\pi \times 10^{-7} \times 24 \times 10^{-4}}{5} = 0.96\,\mu\text{H} \tag{5.156}$$

Practice Problem 5.10
The inductance of an air-core solenoid is found to be 1.5 μH. The core is wound by 100 turns of copper coil whose cross-sectional area is 0.00009 m^2. If the permeability of free space is $4\pi \times 10^{-7}$ H/m^2, calculate the mean length of the core.

5.12 Current–Voltage Terminal Characteristics of an Inductor

Figure 5.21 shows an inductor which carries a current of i. In this case, the flux linkage through the coil will change if the current through the coil changes with time. According to Faraday's laws, for a single-turn of the coil, a voltage will be induced due to this time-varying flux, and the expression of this induced voltage (v) will be,

$$v = \frac{\mathrm{d}\phi}{\mathrm{d}t} \tag{5.157}$$

Substituting Eq. (5.142) into the Eq. (5.157) yields,

$$v = \frac{\mathrm{d}}{\mathrm{d}t}(Li) = L\frac{\mathrm{d}i}{\mathrm{d}t} + i\frac{\mathrm{d}L}{\mathrm{d}t} \tag{5.158}$$

Usually, a fixed value is assigned to an inductor and the rate of change of fixed value of inductance is zero. Then, Eq. (5.158) can be modified as,

$$v = L\frac{\mathrm{d}i}{\mathrm{d}t} + 0 \tag{5.159}$$

$$v = L\frac{\mathrm{d}i}{\mathrm{d}t} \tag{5.160}$$

From Eq. (5.160), the expression of the current can be derived as,

$$\mathrm{d}i = \frac{v}{L}\mathrm{d}t \tag{5.161}$$

Integrating Eq. (5.161) with respect to time yields,

$$i = \frac{1}{L}\int_{-\infty}^{t} v\,\mathrm{d}t \tag{5.162}$$

$$i = \frac{1}{L}\int_{0}^{t} v\,\mathrm{d}t + i(0) \tag{5.163}$$

From Eq. (5.160), it is concluded that an inductor acts as a short circuit under DC condition, when the current through the inductor does not vary with time.

Example 5.11 Figure 5.23 shows a voltage waveform of a 0.5 H inductor. Determine the current in the inductor.

Solution:

$$v(t) = \frac{10}{2}t = 5t \text{ V} \quad 0<t<2 \tag{5.164}$$

$$v(t) = 10 \text{ V} \quad 2<t<4 \tag{5.165}$$

$$i = \frac{1}{L}\int v\,\mathrm{d}t = \frac{1}{0.5}\left[\int_{0}^{2} 5t\,\mathrm{d}t + \int_{2}^{4} 10\mathrm{d}t\right] \tag{5.166}$$

$$i = \frac{1}{0.5}\left[\frac{5\times(4-0)}{2} + 10(4-2)\right] = 60 \text{ A} \tag{5.167}$$

Practice Problem 5.11
The voltage waveform of a 3 H inductor is shown in Fig. 5.24. Calculate the current in the inductor.

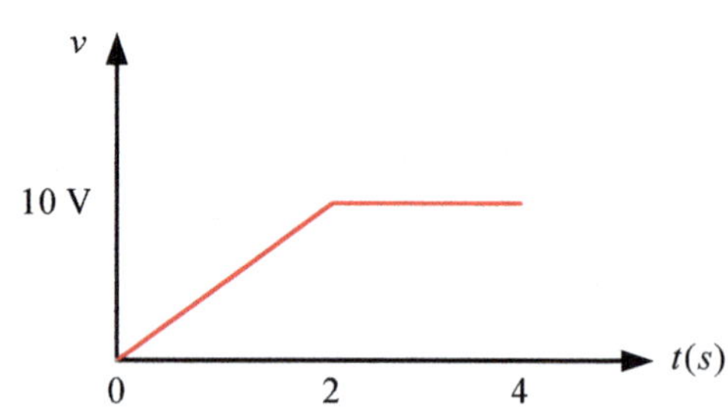

Fig. 5.23 Voltage waveform for Example 5.11

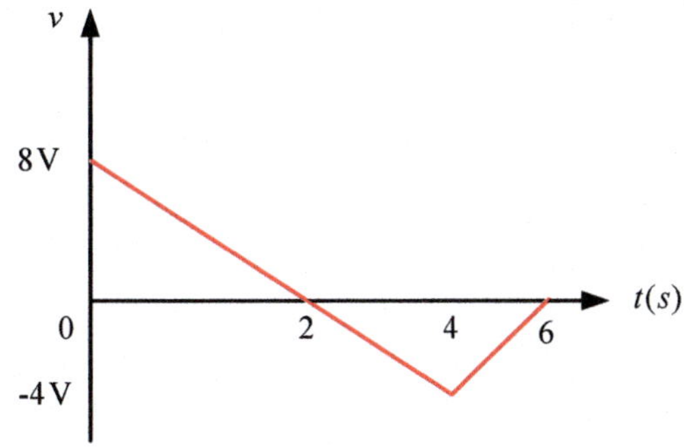

Fig. 5.24 Voltage waveform for Practice Problem 5.11

5.13 Energy Stored in an Inductor

Energy is stored in an inductor when current flows through it. The instantaneous power can be expressed as,

$$p = vi \tag{5.168}$$

Substituting the Eq. (5.160) into Eq. (5.168) yields,

$$p = Li\frac{di}{dt} \tag{5.169}$$

The rate of receiving or delivering energy is known as power and it is expressed as,

$$p = \frac{dw}{dt} \tag{5.170}$$

Substituting Eq. (5.169) into Eq. (5.170) yields,

$$\frac{dw}{dt} = Li\frac{di}{dt} \tag{5.171}$$

$$dw = L\,i\,di \tag{5.172}$$

Integrating the Eq. (5.172) yields,

$$w = L\int_{-\infty}^{t} i\,(t)di\,(t) \tag{5.173}$$

$$w = \frac{L}{2}\left[i^2(t) - i^2(-\infty)\right] \tag{5.174}$$

Since $i(-\infty) = 0$, then Eq. (5.174) becomes,

$$w = \frac{1}{2}Li^2 \tag{5.175}$$

From Eq. (5.175), it is concluded that the energy stored in an inductor is equal to the half of the inductance times the square of the current.

Example 5.12 The length and diameter of an air-cored solenoid are 20 cm and 4 cm, respectively. This solenoid is wound with 600 turns of copper wire and the coil carries a current of 4 A. Calculate the inductance and energy stored in the inductor.

Solution:
Area of solenoid is calculated as,

$$A = \pi r^2 = \pi \times \left(\frac{4 \times 10^{-2}}{2}\right)^2 = 1.26 \times 10^{-3}\,\text{m}^2 \tag{5.176}$$

The inductance is,

$$L = \frac{N^2 \mu A}{l} = \frac{600^2 \times 4\pi \times 10^{-7} \times 1 \times 1.26 \times 10^{-3}}{0.2} = 2.85\ \text{mH} \tag{5.177}$$

The energy stored is calculated as,

$$w = \frac{1}{2}Li^2 = \frac{1}{2} \times 2.85 \times 10^{-3} \times 4^2 = 22.8\ \text{mJ} \tag{5.178}$$

Practice problem 5.12
The energy stored in a coil is found to be 30 mJ. An air-cored solenoid with a specific length and a diameter of 6 cm is wound with 800 turns of copper wire. Determine the length of the solenoid if the coil carries a current of 5 A.

5.14 Series Inductors

Figure 5.25 shows a circuit where three inductors are connected in series with a voltage source. In this circuit, the same current will flow through each inductor. However, the voltage across each inductor will be different. Applying KVL to the circuit in Fig. 5.25 yields,

$$v = v_1 + v_2 + v_3 \tag{5.179}$$

Fig. 5.25 Inductors are in series

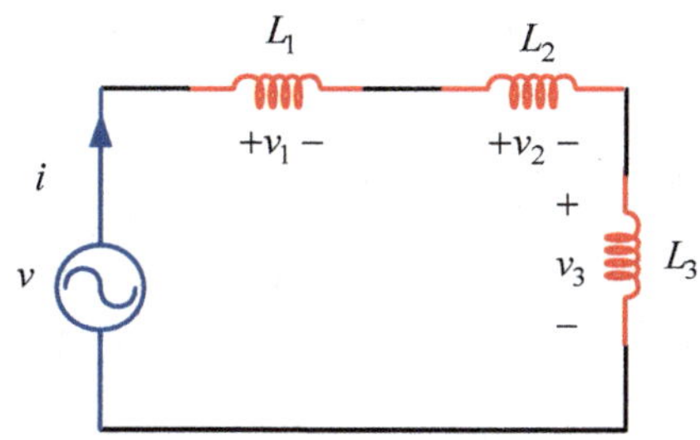

According to Eq. (5.160), Eq. (5.179) can be expressed as,

$$v = L_1 \frac{\mathrm{d}i}{\mathrm{d}t} + L_2 \frac{\mathrm{d}i}{\mathrm{d}t} + L_3 \frac{\mathrm{d}i}{\mathrm{d}t} \tag{5.180}$$

$$v = (L_1 + L_2 + L_3)\frac{\mathrm{d}i}{\mathrm{d}t} = L_S \frac{\mathrm{d}i}{\mathrm{d}t} \tag{5.181}$$

From Eq. (5.181), the expression for the total inductance can be written as,

$$L_S = L_1 + L_2 + L_3 \tag{5.182}$$

When two inductors are connected in series, then Eq. (5.182) becomes,

$$L_S = L_1 + L_2 \tag{5.183}$$

From Eq. (5.182), it is concluded that the total inductance of the inductors connected in series is equal to the sum of individual inductances.

Again, consider that the two inductors are connected in series with a voltage source as shown in Fig. 5.26a whose equivalent circuit is shown in Fig. 5.26b. In this case, the expression of the source voltage v can be written as,

$$v = L_s \frac{\mathrm{d}i}{\mathrm{d}t} \tag{5.184}$$

Substituting Eq. (5.183) into Eq. (5.184) yields,

$$\frac{\mathrm{d}i}{\mathrm{d}t} = \frac{v}{L_s} = \frac{v}{L_1 + L_2} \tag{5.185}$$

Fig. 5.26 **a** Two inductors are in series and **b** equivalent circuit

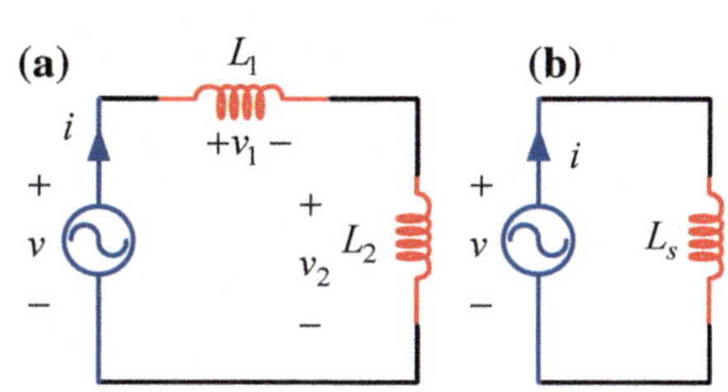

The voltage across the first inductor is,

$$v_1 = L_1 \frac{\mathrm{d}i}{\mathrm{d}t} \tag{5.186}$$

Substituting Eq. (5.185) into Eq. (5.186) yields,

$$v_1 = \frac{L_1}{L_1 + L_2} v \tag{5.187}$$

The voltage across the second inductor is,

$$v_2 = L_2 \frac{\mathrm{d}i}{\mathrm{d}t} \tag{5.188}$$

Substituting Eq. (5.185) into Eq. (5.188) yields,

$$v_2 = \frac{L_2}{L_1 + L_2} v \tag{5.189}$$

Equations (5.187) and (5.189) present the voltage division rule for the two inductors connected in series.

5.15 Parallel Inductors

Figure 5.27 shows a circuit where the two inductors are connected in parallel across a voltage source. Here, the voltage across each capacitor will be the same as the source voltage, but the currents will be different. Applying KCL to the circuit of Fig. 5.27 yields,

$$i = i_1 + i_2 \tag{5.190}$$

Differentiating Eq. (5.190) with respect to time gives,

$$\frac{\mathrm{d}i}{\mathrm{d}t} = \frac{\mathrm{d}i_1}{\mathrm{d}t} + \frac{\mathrm{d}i_2}{\mathrm{d}t} \tag{5.191}$$

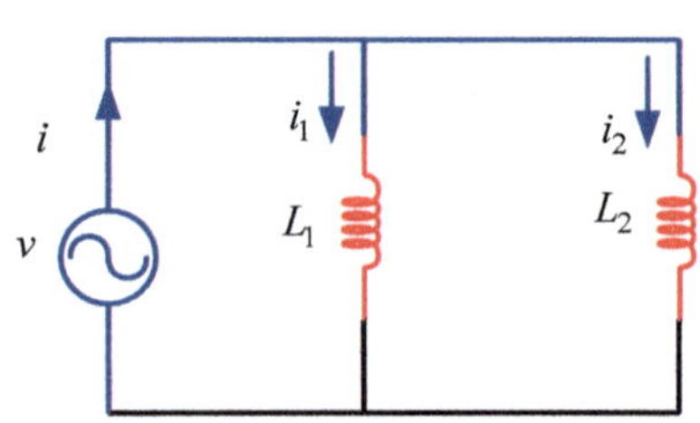

Fig. 5.27 Inductors are in series

The expressions of voltage across different inductances can be written as,

$$v = L_p \frac{di}{dt} \tag{5.192}$$

where L_p is the equivalent inductance of the circuit in Fig. 5.27.

$$\frac{di}{dt} = \frac{v}{L_p} \tag{5.193}$$

$$v = L_1 \frac{di_1}{dt} \tag{5.194}$$

$$\frac{di_1}{dt} = \frac{v}{L_1} \tag{5.195}$$

$$v = L_2 \frac{di_2}{dt} \tag{5.196}$$

$$\frac{di_2}{dt} = \frac{v}{L_2} \tag{5.197}$$

Substituting Eqs. (5.193), (5.195) and (5.197) into Eq. (5.191) yields,

$$\frac{v}{L_p} = \frac{v}{L_1} + \frac{v}{L_2} \tag{5.198}$$

$$L_p = \frac{L_1 L_2}{L_1 + L_2} \tag{5.199}$$

From Eq. (5.199), it is concluded that the equivalent inductance for two inductors connected in parallel is equal to the product of the two inductances divided by their sum.

For inductors with single-turn coils, the voltage across each inductor is the same for the circuit shown in Fig. 5.27, and in terms of flux, it can be written as,

$$v = \frac{d\phi}{dt} = \frac{d\phi_1}{dt} = \frac{d\phi_2}{dt} \tag{5.200}$$

From Eq. (5.200), it is seen that the derivatives of the fluxes with respect to time are the same. In this case, the following relationship can be written:

$$\phi = \phi_1 = \phi_2 \tag{5.201}$$

From the circuit in Fig. 5.27, the expression of the total flux can be written as,

$$\phi = L_p i \tag{5.202}$$

Substituting Eq. (5.199) into Eq. (5.202) yields,

$$\phi = \frac{L_1 L_2}{L_1 + L_2} i \tag{5.203}$$

As per Eq. (5.202), the current through inductors can be determined as,

$$i_1 = \frac{\phi_1}{L_1} = \frac{\phi}{L_1} \tag{5.204}$$

$$i_2 = \frac{\phi_2}{L_2} = \frac{\phi}{L_2} \tag{5.205}$$

Substituting Eq. (5.203) into Eqs. (5.204) and (5.205) yields,

$$i_1 = \frac{1}{L_1} \frac{L_1 L_2}{L_1 + L_2} i \tag{5.206}$$

$$i_1 = \frac{L_2}{L_1 + L_2} i \tag{5.207}$$

$$i_2 = \frac{1}{L_2} \frac{L_1 L_2}{L_1 + L_2} i \tag{5.208}$$

$$i_2 = \frac{L_1}{L_1 + L_2} i \tag{5.209}$$

Equations (5.207) and (5.209) present the current division rule for the two inductors connected in parallel.

Example 5.13 A current of $i = 6(3 - e^{-2t})$A flows in the circuit as shown in Fig. 5.28. Given, $i_2(0) = 4\,\text{A}$, calculate $i_1(0)$, i_1, i_2 and the voltage across the 3 H inductor.

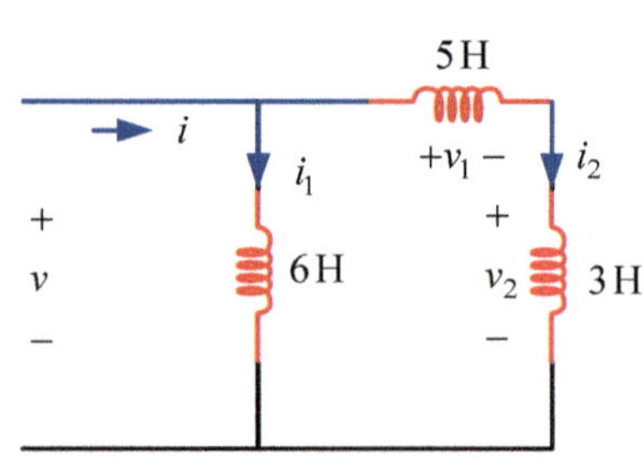

Fig. 5.28 Circuit for Example 5.13

Solution:

$$\text{At } t = 0, i(0) = 6(3 - 1) = 12 \text{ A} \tag{5.210}$$

$$i_1(0) = i(0) - i_2(0) = 12 - 4 = 8 \text{ A} \tag{5.211}$$

The equivalent inductance is,

$$L_{\text{eq}} = \frac{(5+3) \times 6}{8+6} = 3.43 \text{ H} \tag{5.212}$$

The source voltage is determined as,

$$v = L_{\text{eq}} \frac{\mathrm{d}i}{\mathrm{d}t} = 3.43 \frac{\mathrm{d}}{\mathrm{d}t} 6\left(3 - \mathrm{e}^{-2t}\right) = 41.14\mathrm{e}^{-2t} \text{ V} \tag{5.213}$$

Again, according to the circuit,

$$v_{6\text{H}} = v = 6 \frac{\mathrm{d}i_1}{\mathrm{d}t} \tag{5.214}$$

$$i_1 = \frac{41.14}{6} \int_0^t \mathrm{e}^{-2t} \mathrm{d}t + i_1(0) \tag{5.215}$$

$$i_1 = -3.43\mathrm{e}^{-2t} + 8 \text{ A} \tag{5.216}$$

The current i_2 can be determined as,

$$i_2 = i - i_1 = 6\left(3 - \mathrm{e}^{-2t}\right) + 3.43\mathrm{e}^{-2t} - 8 = 10 - 2.57\mathrm{e}^{-2t} \text{ A} \tag{5.217}$$

The voltage across 3 H inductor is,

$$v_2 = L_{3\text{H}} \frac{\mathrm{d}i_2}{\mathrm{d}t} = 3 \frac{\mathrm{d}}{\mathrm{d}t}\left(10 - 2.57\mathrm{e}^{-2t}\right) = 15.42\mathrm{e}^{-2t} \text{ V} \tag{5.218}$$

Practice Problem 5.13

Figure 5.29 shows a circuit where a current of $i = 1 + \mathrm{e}^{-0.02t}$ A flows. Given, $i_2(0) = -2$ A, determine the $i_2(0)$, i_1, i_2 and the voltage across the 2 H inductor.

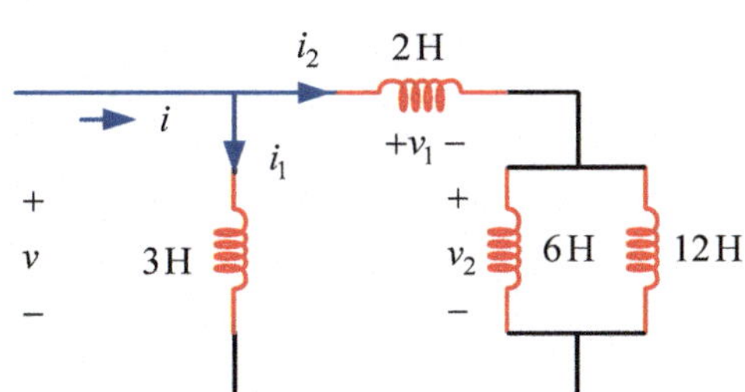

Fig. 5.29 Circuit for Practice Problem 5.13

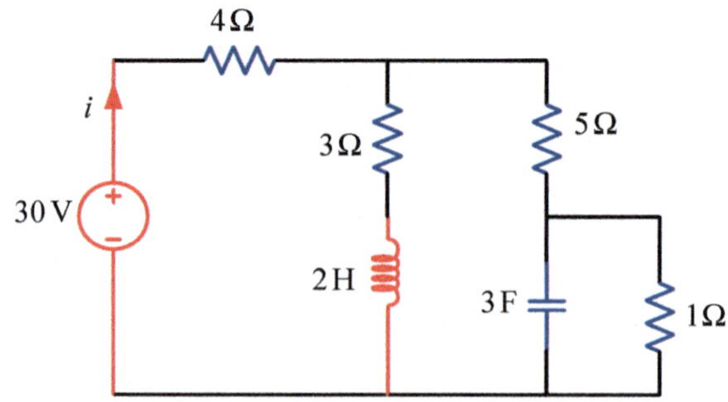

Fig. 5.30 Circuit for Example 5.14

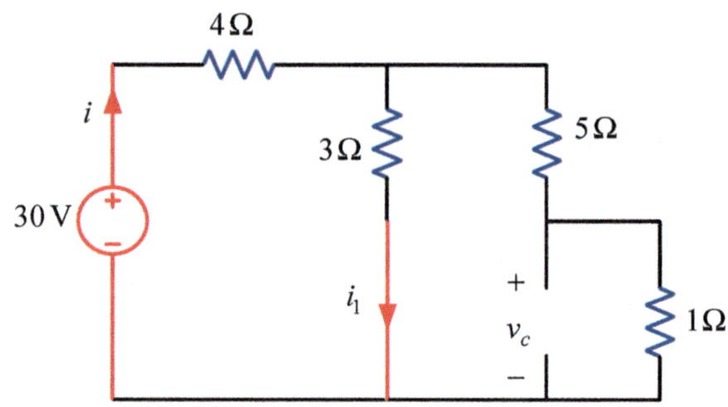

Fig. 5.31 Circuit for Example 5.14 with opening capacitor and short the inductor

Example 5.14 A circuit with resistance, inductance and capacitance is shown in Fig. 5.30. Calculate the source current and energy stored by the inductor and capacitor under DC condition.

Solution:
Under DC condition, the inductor is short circuited and the capacitor is open circuited. In this case, the circuit is shown in Fig. 5.31. The equivalent circuit resistance is,

$$R_{\text{eq}} = 4 + \frac{3 \times (5+1)}{3+6} = 6\,\Omega \tag{5.219}$$

The source current is,

$$i = \frac{30}{6} = 5\,\text{A} \tag{5.220}$$

The branch currents are,

$$i_1 = 5\frac{6}{6+3} = 3.33\,\text{A} \tag{5.221}$$

$$i_{1\Omega} = 5 - 3.33 = 1.67\,\text{A} \tag{5.222}$$

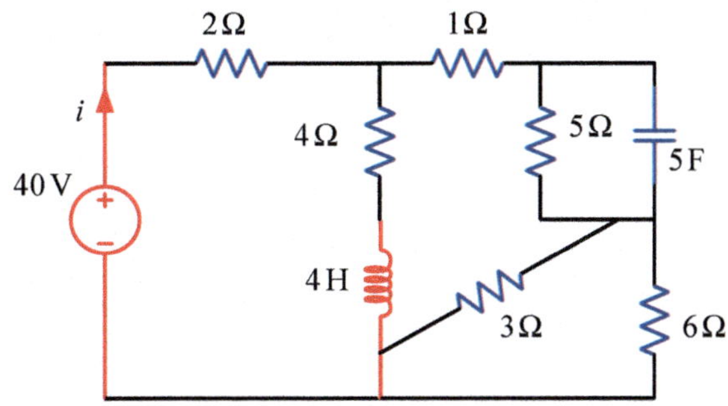

Fig. 5.32 Circuit for Practice Problem 5.14

The voltage drop across the open circuit is,

$$v_c = 1.67 \times 1 = 1.67 \text{ V} \tag{5.223}$$

The energies stored by the inductor and capacitor are,

$$w_l = \frac{1}{2} \times 2 \times 3.33^2 = 11.09 \text{ J} \tag{5.224}$$

$$w_c = \frac{1}{2} \times 3 \times 1.67^2 = 4.18 \text{ J} \tag{5.225}$$

Practice Problem 5.14
Determine the source current and energy stored by the inductor and capacitor under DC condition for the circuit shown in Fig. 5.32.

5.16 RC Circuit Analysis

An uncharged capacitor is connected to the source through a resistance and switch as shown in the circuit in Fig. 5.33. The capacitor will be charged when the switch

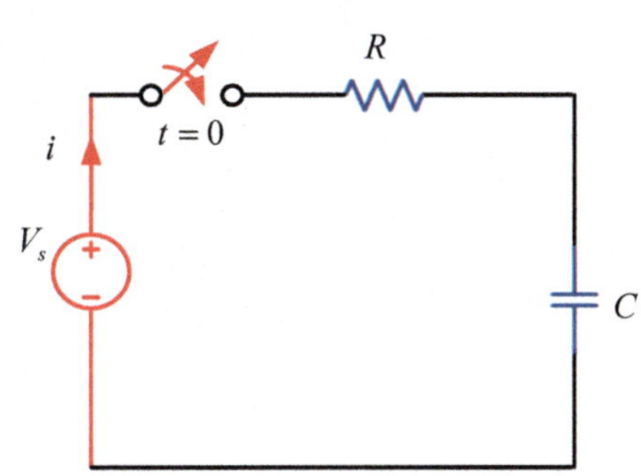

Fig. 5.33 Circuit with switch, resistance and capacitance

is closed at $t = 0$. Under this condition, applying KVL to the circuit in Fig. 5.33 yields,

$$Ri + \frac{1}{C}\int_0^t i\mathrm{d}t = V_s \tag{5.226}$$

Differentiating Eq. (5.226) with respect to time yields,

$$R\frac{\mathrm{d}i}{\mathrm{d}t} + \frac{1}{C}i = 0 \tag{5.227}$$

$$\frac{\mathrm{d}i}{i} = -\frac{1}{RC}\mathrm{d}t \tag{5.228}$$

Integrating Eq. (5.228) yields,

$$\ln i = -\frac{1}{RC}t + \ln k \tag{5.229}$$

$$\ln\left(\frac{i}{k}\right) = -\frac{1}{RC}t \tag{5.230}$$

$$\frac{i}{k} = \mathrm{e}^{-\frac{1}{RC}t} \tag{5.231}$$

$$i = k\mathrm{e}^{-\frac{1}{RC}t} \tag{5.232}$$

At $t = 0^+$, the capacitor is short circuited and the voltage drop across the capacitor is zero. Then, Eq. (5.226) becomes,

$$i = \frac{V_s}{R} \tag{5.233}$$

At $t = 0$, Eq. (5.232) becomes,

$$i = k = \frac{V_s}{R} \tag{5.234}$$

Substituting Eq. (5.234) into Eq. (5.232) yields the final expression of charging current,

$$i(t) = \frac{V_s}{R}\mathrm{e}^{-\frac{1}{RC}t} \tag{5.235}$$

From the circuit in Fig. 5.33, the charging voltage of the capacitor can be written as,

$$v_C = V_s - Ri \tag{5.236}$$

Substituting Eq. (5.236) into Eq. (5.235) yields,

$$v_C = V_s - R \times \frac{V_s}{R} \mathrm{e}^{-\frac{1}{RC}t} \tag{5.237}$$

$$v_C = V_s\left(1 - \mathrm{e}^{-\frac{1}{RC}t}\right) \tag{5.238}$$

The charging voltage and current in the capacitor are shown in Fig. 5.34.

Here, RC is the time constant. Using basic definition of resistance and capacitance, it can be demonstrated as follows that the quantity RC has a unit of time,

$$RC = \frac{V}{I} \times \frac{q}{V} = \frac{q}{I} = t = \lambda \tag{5.239}$$

At $t = \lambda$, the charging current and voltage from Eqs. (5.235) and (5.238) can be determined as,

$$i(t) = \frac{V_s}{R} \mathrm{e}^{-1} = 0.37 I_m \tag{5.240}$$

$$v_C = V_s\left(1 - \mathrm{e}^{-1}\right) = 0.63\, V_s \tag{5.241}$$

From Eq. (5.240), it can be concluded that one time constant marks the current through the RC series circuit that is 37% of its maximum current, while from Eq. (5.241), it can be observed that the voltage drop across the capacitor is 63% of the maximum source voltage at the same time mark.

Now to observe the discharging phenomenon of a fully charged capacitor in an RC circuit, it is considered that the capacitor is connected to the resistance without the source as shown in Fig. 5.35. Applying KCL to the circuit in Fig. 5.35 yields,

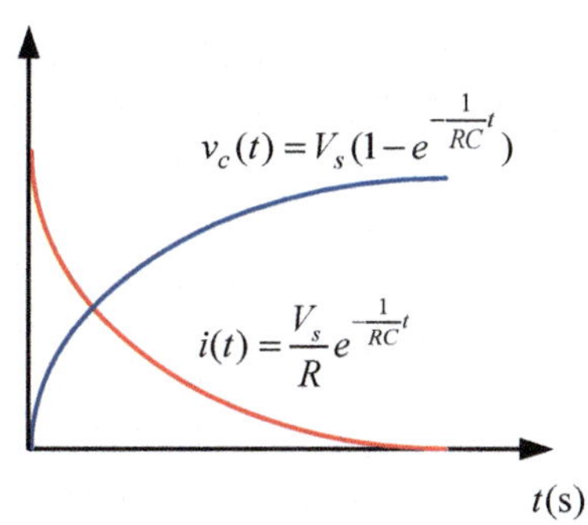

Fig. 5.34 Charging current and voltage of a capacitor

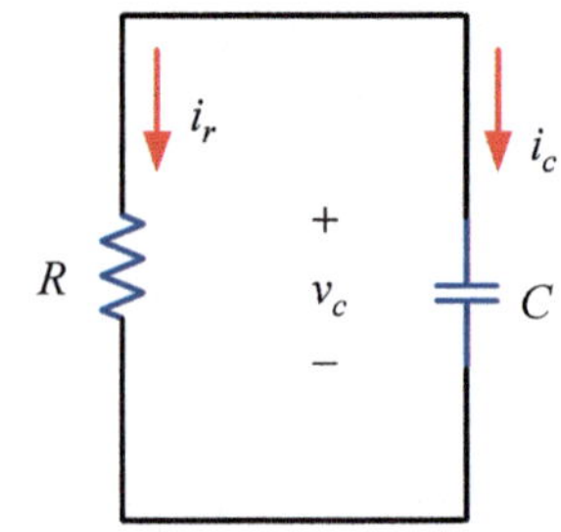

Fig. 5.35 RC circuit for discharge

$$i_c = -i_r \tag{5.242}$$

The currents in the capacitor and resistance are,

$$i_c = C\frac{\mathrm{d}v_c}{\mathrm{d}t} \tag{5.243}$$

$$i_r = \frac{v_c}{R} \tag{5.244}$$

Substituting Eqs. (5.243) and (5.244) into Eq. (5.242) yields,

$$C\frac{\mathrm{d}v_c}{\mathrm{d}t} = -\frac{v_c}{R} \tag{5.245}$$

$$\frac{\mathrm{d}v_c}{v_c} = -\frac{1}{RC}\mathrm{d}t \tag{5.246}$$

Integrating Eq. (5.246) yields,

$$\ln v_c = -\frac{1}{RC}t + \ln A \tag{5.247}$$

$$\ln \frac{v_c}{A} = -\frac{1}{RC}t \tag{5.248}$$

$$v_c(t) = A\mathrm{e}^{-\frac{1}{RC}t} \tag{5.249}$$

Initially, the capacitor was fully charged. Therefore, at $t = 0$, the voltage $v(0) = V_s$. Equation (5.245) becomes,

$$A = V_s \tag{5.250}$$

Substituting Eq. (5.250) into Eq. (5.249) yields,

$$v_c(t) = V_s\mathrm{e}^{-\frac{1}{RC}t} \tag{5.251}$$

Substituting Eq. (5.251) into Eq. (5.243) yields,

$$i_c = C\frac{\mathrm{d}}{\mathrm{d}t}\left(V_0 \mathrm{e}^{-\frac{1}{RC}t}\right) \tag{5.252}$$

$$i_c = -\frac{V_s}{R}\mathrm{e}^{-\frac{1}{RC}t} \tag{5.253}$$

Equations (5.251) and (5.253) present the expressions for discharging capacitor voltage and current, respectively; Fig. 5.36 depicts these phenomena.

Example 5.15 A circuit with resistance and capacitance is shown in Fig. 5.37. Find the charging voltage of the capacitor when the switch is closed at $t = 0$.

Solution:
At $t = 0$, the capacitor will be open circuited. In this condition, the Thevenin resistance and voltage can be determined as,

$$R_{\mathrm{Th}} = 1 + \left[4 + \left(\frac{3 \times 6}{9}\right)\right] \parallel 2 = 1 + \frac{2 \times 6}{2 + 6} = 2.5\,\Omega \tag{2.254}$$

$$V_{\mathrm{Th}} = \frac{30}{4 + 2 + 2} \times (4 + 2) = 22.5\,\mathrm{V} \tag{2.255}$$

The charging voltage is,

$$v_c = V_{\mathrm{Th}}\left(1 - \mathrm{e}^{-\frac{1}{R_{\mathrm{Th}}C}t}\right) = 22.5\left(1 - \mathrm{e}^{-\frac{1}{2.5 \times 5}t}\right) = 22.5(1 - \mathrm{e}^{-0.08t})\,\mathrm{V} \tag{5.256}$$

Example 5.16 Figure 5.38 shows a circuit with resistance, capacitance and switch. Determine the voltage across the capacitor when the switch is closed at $t<0$.

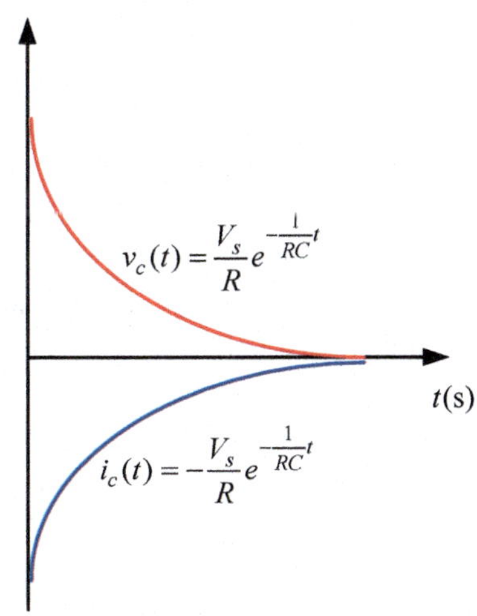

Fig. 5.36 Discharging current and voltage of a capacitor

Fig. 5.37 Circuit for Example 5.15

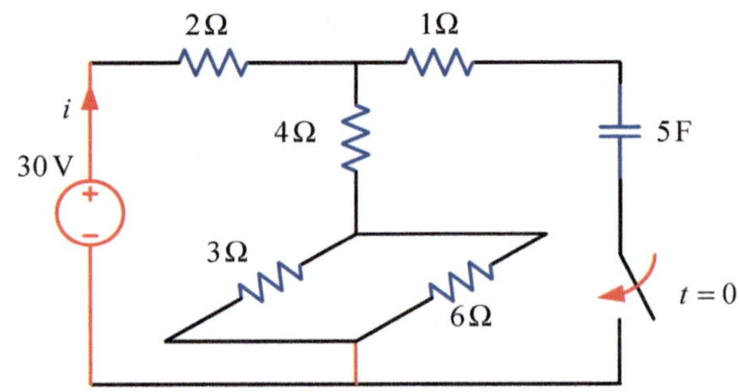

Fig. 5.38 Circuit for Example 5.16

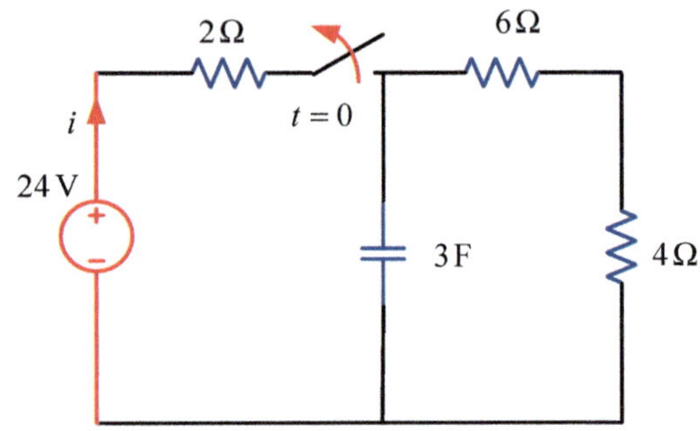

Fig. 5.39 Circuit for Example 5.16 at $t<0$

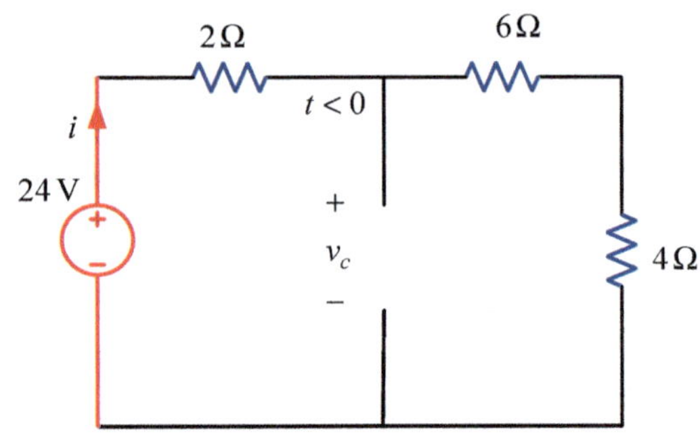

Solution:
At $t<0$, the switch is closed and the capacitor is open circuited under DC condition as shown in Fig. 5.39. In this case, the circuit is shown in Fig. 5.40. Here, the source current is calculated as,

$$i = \frac{24}{2+6+4} = 2\text{ A} \tag{5.257}$$

The voltage across the capacitor is,

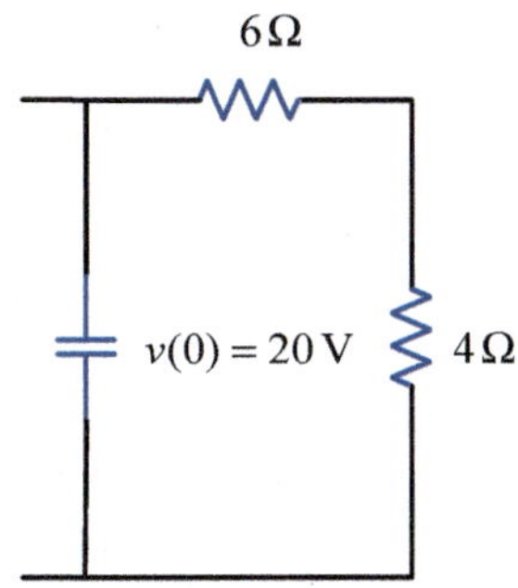

Fig. 5.40 Circuit for Example 5.16 at $t \geq 0$

$$v_c = 2(4+6) = 20\,\text{V} \tag{5.258}$$

The voltage across the capacitor does not change instantaneously. Therefore, the voltage at $t = 0^+$ will be equal to the voltage at $t = 0^-$.

$$v(0) = v_c = 20\,\text{V} \tag{5.259}$$

At $t \geq 0$, the source is disconnected as shown in Fig. 5.40. The equivalent circuit resistance is,

$$R_x = 4 + 6 = 10\,\Omega \tag{5.260}$$

The time constant is,

$$R_x C = 10 \times 3 = 30 \tag{2.261}$$

The voltage across the capacitor at $t \geq 0$ is,

$$v_c = v(0)\mathrm{e}^{-\frac{1}{R_x C}t} = 20\mathrm{e}^{-\frac{1}{30}t} = 20\mathrm{e}^{-0.03t}\,\text{V} \tag{2.262}$$

Practice Problem 5.15

Figure 5.41 shows a circuit with resistance and capacitance. Determine the charging voltage of the capacitor when the switch is closed at $t = 0$.

Practice Problem 5.16

Figure 5.42 shows a circuit with resistance, capacitance and switch. Find the voltage across the capacitor when the switch is closed at $t = 0$.

5.17 RL Circuit Analysis

An inductor is connected in series with a resistance as shown in Fig. 5.43. Applying KVL to the circuit in Fig. 5.43 yields,

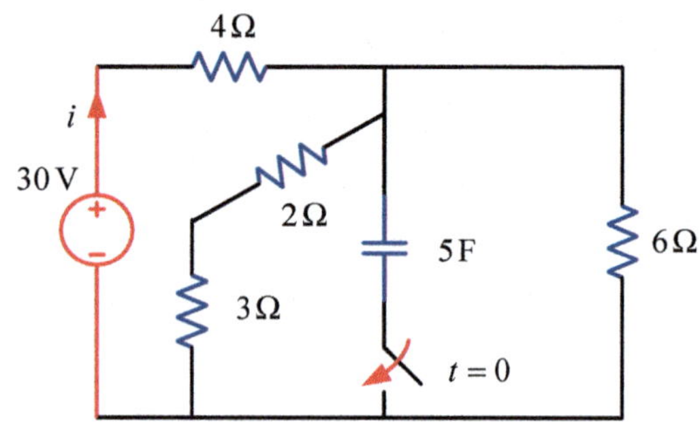

Fig. 5.41 Circuit for Practice Problem 5.15

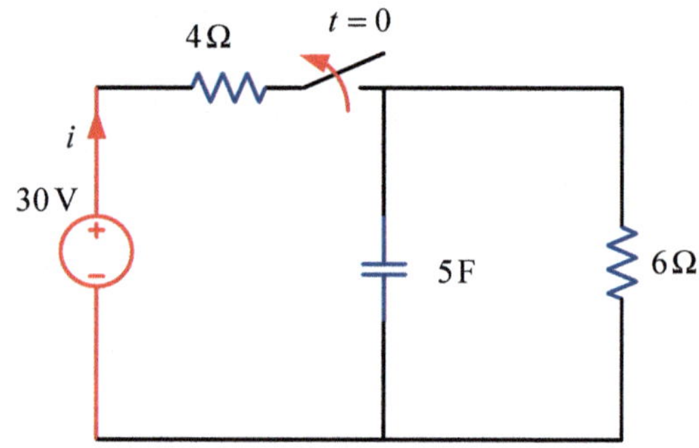

Fig. 5.42 Circuit for Practice Problem 5.16

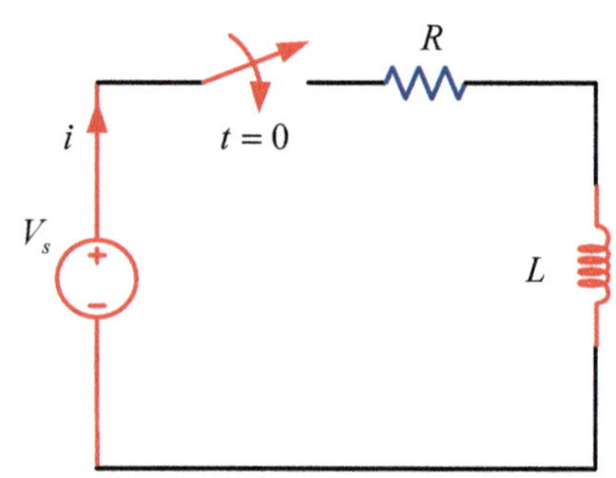

Fig. 5.43 A RL series circuit

$$Ri + L\frac{\mathrm{d}i}{\mathrm{d}t} = V_s \tag{5.263}$$

An auxiliary solution of Eq. (5.263) is,

$$\frac{R}{L}i_a + \frac{\mathrm{d}i_a}{\mathrm{d}t} = 0 \tag{5.264}$$

$$i_a = k\mathrm{e}^{-\frac{R}{L}t} \tag{5.265}$$

A particular solution of Eq. (5.263) is,

$$Ri_p + L\frac{di_p}{dt} = V_s \tag{5.266}$$

According to the excitation, consider the particular solution $i_p = A_p$, then Eq. (5.266) becomes,

$$RA_p + L \times 0 = V_s \tag{5.267}$$

$$A_p = \frac{V_s}{R} \tag{5.268}$$

The complete solution of Eq. (5.263) is,

$$i(t) = i_a + i_p = \frac{V_s}{R} + ke^{-\frac{R}{L}t} \tag{5.269}$$

At $t = 0^+$, the inductor is open circuited, the current in the inductor is zero, and then Eq. (5.269) becomes,

$$0 = \frac{V_s}{R} + k \tag{5.270}$$

$$k = -\frac{V_s}{R} \tag{5.271}$$

The current in the inductor is,

$$i(t) = \frac{V_s}{R}\left(1 - e^{-\frac{R}{L}t}\right) \tag{5.272}$$

Again, consider the inductor is connected in series with the resistance without the source as shown in the circuit in Fig. 5.44. Applying KVL in the circuit in Fig. 5.44 yields,

$$Ri + L\frac{di}{dt} = 0 \tag{5.273}$$

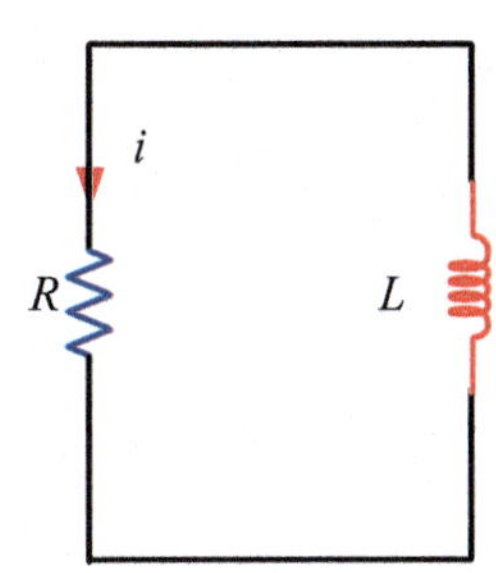

Fig. 5.44 A RL series circuit without a source

The general solution of Eq. (5.273) is,

$$i(t) = k_1 \mathrm{e}^{-\frac{R}{L}t} \tag{5.274}$$

At $t = 0$, initial current in the inductor is,

$$i(0) = I_0 \tag{5.275}$$

At $t = 0$, Eq. (5.274) becomes,

$$i(0) = I_0 = k_1 \tag{5.276}$$

Substituting Eq. (5.276) into Eq. (5.274) yields,

$$i(t) = I_0 \mathrm{e}^{-\frac{R}{L}t} \tag{5.277}$$

The rising and falling of the current are shown in Fig. 5.45. The ratio of inductance to resistance is defined as time constant and it can be derived as follows:

$$v = L\frac{\mathrm{d}i}{\mathrm{d}t} \tag{5.278}$$

$$\mathrm{V} = L\frac{\mathrm{A}}{\mathrm{s}} \tag{5.279}$$

$$\frac{\text{Volt.Sec}}{\text{Amp}} = L \tag{5.280}$$

The unit of the resistance is,

$$\frac{\text{Volt}}{\text{Amp}} = R \tag{5.281}$$

From Eqs. (5.280) and (5.281), it can be written as,

$$\frac{L}{R} = \lambda \tag{5.282}$$

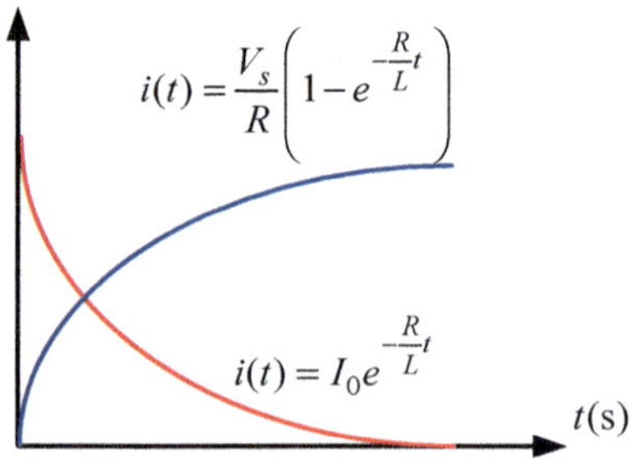

Fig. 5.45 Rising and decay of current

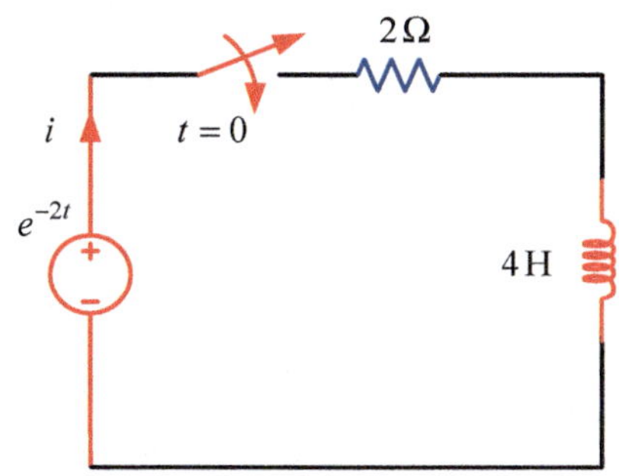

Fig. 5.46 Circuit for Example 5.17

where λ has a unit of second.

Example 5.17 Figure 5.46 shows a circuit with resistance, capacitance and switch. Calculate the current in the circuit when the switch is closed at $t = 0$.

Solution:
At $t = 0$, the current in the circuit is,

$$2i + 4\frac{\mathrm{d}i}{\mathrm{d}t} = \mathrm{e}^{-2t} \tag{5.283}$$

$$\frac{1}{4}i + \frac{\mathrm{d}i}{\mathrm{d}t} = \frac{1}{4}\mathrm{e}^{-2t} \tag{5.284}$$

The auxiliary solution is,

$$i_a = k_1\mathrm{e}^{-0.25t} \tag{5.285}$$

Considering the particular solution $i_p = A\mathrm{e}^{-2t}$ and Eq. (5.284) becomes,

$$\frac{1}{4}A\mathrm{e}^{-2t} + A(-2)\mathrm{e}^{-2t} = \frac{1}{4}\mathrm{e}^{-2t} \tag{5.286}$$

$$A = -\frac{1}{7} \tag{5.287}$$

The complete solution is,

$$i(t) = k_1\mathrm{e}^{-0.25t} - \frac{1}{7}\mathrm{e}^{-2t} \tag{5.288}$$

At $t = 0$, the initial current in an inductor is zero, i.e. $i(0) = 0$ and Eq. (5.288) becomes,

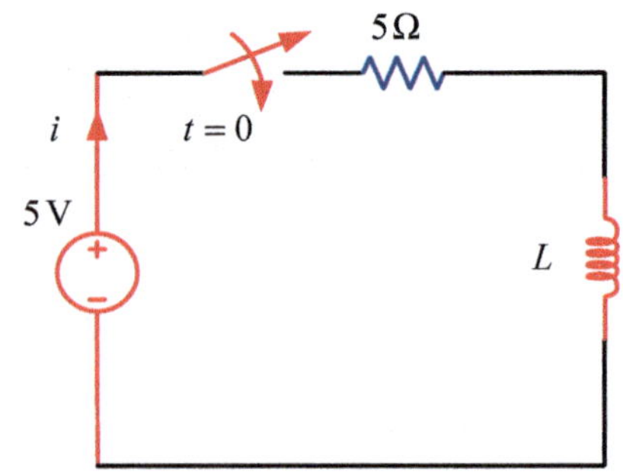

Fig. 5.47 Circuit for Practice Problem 5.17

$$i(0) = 0 = k_1 - \frac{1}{7} \tag{5.289}$$

$$k_1 = \frac{1}{7} \tag{5.290}$$

The final expression of the current is,

$$i(t) = \frac{1}{7}\left(e^{-0.25t} - e^{-2t}\right)\text{ A} \tag{5.291}$$

Practice Problem 5.17
An RL series circuit is shown in Fig. 5.47. Given, $i(-0.1) = 0.6$, calculate the inductance when the switch is closed at $t = 0$.

5.18 RC Circuit with Step Response

Step response of any circuit is the time-dependent behaviour of the output of the circuit when its inputs change from zero to no-zero value in a very short time between $t = 0^-$ and $t = 0^+$. The step response can be observed in a series RC or RL circuit when any switching takes place in the circuit. In an RC series circuit, the voltage across the capacitor does not change instantaneously, but the current changes suddenly. In these cases, the relationships can be written as,

$$v_c(0^-) = v_c(0^+) \tag{5.292}$$

$$i_c(0^-) \neq i_c(0^+) \tag{5.293}$$

where

$v_c(0^-)$ is the voltage across the capacitor just prior to switching,
$v_c(0^+)$ is the capacitor voltage just immediately after switching.

The RC series circuit with a step input voltage is shown in Fig. 5.48. In Fig. 5.48, $u(t)$ is a unit step function which is 0 at $t = 0^-$, and 1 at $t = 0^+$. Applying KVL to the circuit in Fig. 5.48 yields,

$$V_s u(t) = Ri + \frac{1}{C}\int_0^t i\mathrm{d}t \tag{5.294}$$

The voltage across the capacitor is,

$$v_c = \frac{1}{C}\int_0^t i\mathrm{d}t \tag{5.295}$$

$$\frac{\mathrm{d}v_c}{\mathrm{d}t} = \frac{1}{C}i \tag{5.296}$$

Substituting Eqs. (5.295) and (5.296) in Eq. (5.294) yields,

$$V_s u(t) = RC\frac{\mathrm{d}v_c}{\mathrm{d}t} + v_c \tag{5.297}$$

At $t > 0$, Eq. (5.297) becomes,

$$V_s = RC\frac{\mathrm{d}v_c}{\mathrm{d}t} + v_c \tag{5.298}$$

$$\frac{\mathrm{d}v_c}{v_c - V_s} = -\frac{\mathrm{d}t}{RC} \tag{5.299}$$

Integrating Eq. (5.299) yields,

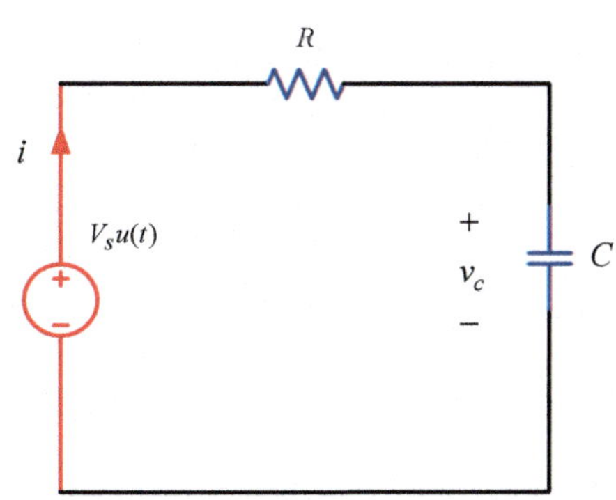

Fig. 5.48 A series RC circuit with step response

$$\int_{V_0}^{v_c} \frac{\mathrm{d}v_c}{v_c - V_s} = -\int_0^t \frac{\mathrm{d}t}{RC} \tag{5.300}$$

$$\ln\left(\frac{v_c - V_s}{V_0 - V_s}\right) = -\frac{t}{RC} \tag{5.301}$$

$$v_c - V_s = (V_0 - V_s)\mathrm{e}^{\frac{-t}{RC}} \tag{5.302}$$

$$v_c(t) = V_s + (V_0 - V_s)\mathrm{e}^{\frac{-t}{RC}} \tag{5.303}$$

Generally, Eq. (5.303) can be expressed as,

$$v_c(t) = V_{ss,C} + \left\{v_c(0^+) - V_{ss,C}\right\}\mathrm{e}^{\frac{-t}{R_{th}C}} \tag{5.304}$$

where

$v_c(0^+)$ is the voltage across the capacitor at $t = 0^+$,
$V_{ss,C}$ is the steady-state capacitor voltage in the circuit at $t > 0$,
R_{Th} is the Thevenin resistance seen by the capacitor terminals in the circuit at $t > 0$,
$R_{\mathrm{Th}}C$ is the circuit time constant.

Example 5.18 The switch is opened at $t = 0$ in the circuit as shown in Fig. 5.49. Calculate the voltage across the capacitor at $t > 0$ condition.

Solution:
At $t = 0^-$, the circuit is shown in Fig. 5.50. The source and branch currents are calculated as,

$$i = \frac{24}{4 + \frac{3\times 6}{9}} = 4\ \mathrm{A} \tag{5.305}$$

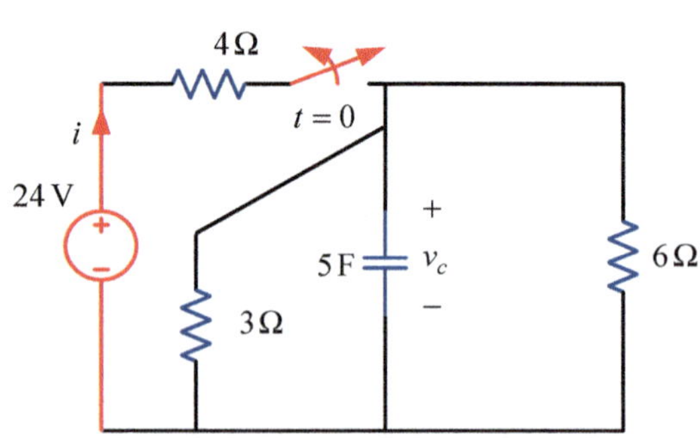

Fig. 5.49 Circuit for Example 5.18

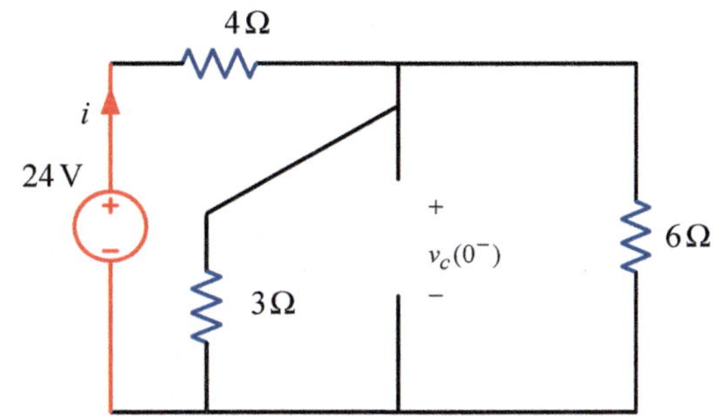

Fig. 5.50 Circuit for Example 5.18 at $t < 0$

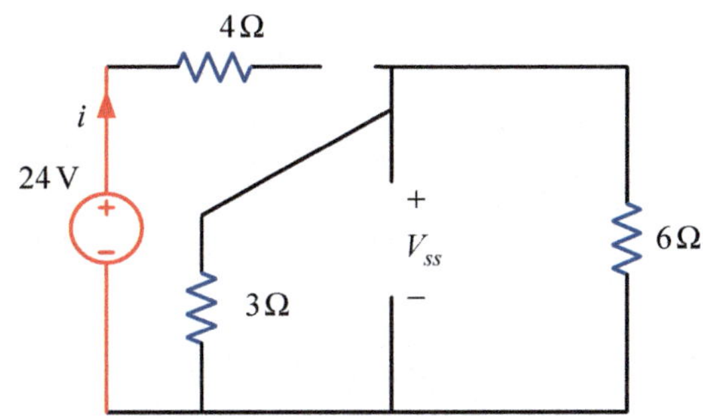

Fig. 5.51 Circuit for Example 5.18 at $t > 0$

$$i_{6\Omega} = 4 \times \frac{3}{3+6} = 1.33 \text{ A} \tag{5.306}$$

The voltage across the capacitor is,

$$v_c(0^-) = 6 \times 1.33 = 7.98 \text{ V} \tag{5.307}$$

$$v_c(0^-) = v_c(0^+) = 7.98 \text{ V} \tag{5.308}$$

At $t > 0$, the circuit is shown in Fig. 5.51 and the steady-state voltage across the capacitor is,

$$V_{ss} = 0 \tag{5.309}$$

Thevenin resistance and time constant are,

$$R_{\text{Th}} = \frac{3 \times 6}{3+6} = 2\,\Omega \tag{5.310}$$

$$R_{\text{Th}}C = 5 \times 2 = 10\,\text{s} \tag{5.311}$$

At $t > 0$, the voltage across the capacitor is,

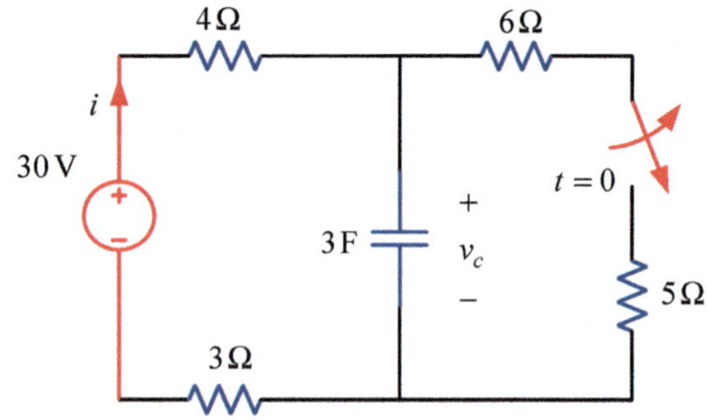

Fig. 5.52 Circuit for Practice Problem 5.18

$$v_c(t) = V_{ss} + [v_c(0^+) - V_{ss}]e^{\frac{-t}{R_{Th}C}} = 0 + [7.98 - 0]e^{\frac{-t}{10}} = 7.98e^{-0.1t}\ \mathrm{V} \tag{5.312}$$

Practice Problem 5.18

Figure 5.52 shows a circuit where the switch is closed at $t = 0$. Find the voltage across the capacitor at $t > 0$.

5.19 RL Circuit with Step Response

Figure 5.53 shows an RL series circuit with a step input. Applying KVL to the circuit as shown in Fig. 5.53 yields,

$$L\frac{di}{dt} + Ri = V_s \tag{5.313}$$

The complete solution of Eq. (5.313) is,

$$i(t) = Be^{\frac{-R}{L}t} + \frac{V_s}{R} \tag{5.314}$$

The current in the inductor does not change instantaneously, whereas voltage changes abruptly and it can be expressed as,

$$i_L(0^-) = i_L(0^+) \tag{5.315}$$

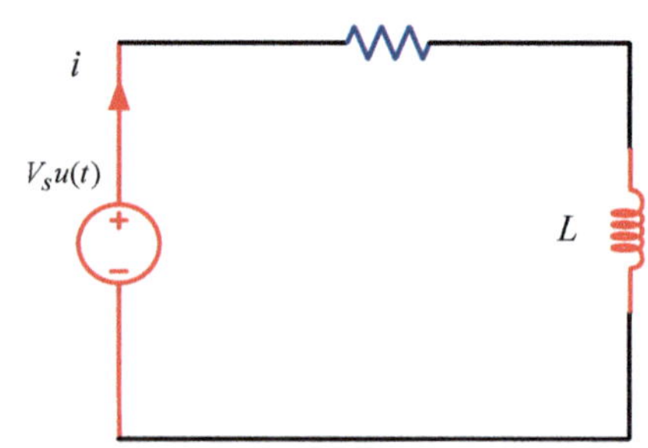

Fig. 5.53 A RL series circuit with step response

$$v_L(0^-) \neq v_L(0^+) \tag{5.316}$$

Consider the initial current in the inductor is,

$$i_L(0^-) = i_L(0^+) = I_0 \tag{5.317}$$

At $t = 0^+$, Eq. (5.314) can be modified as,

$$i_L(0^+) = B + \frac{V_s}{R} \tag{5.318}$$

Substituting Eq. (5.317) into Eq. (5.318) yields,

$$B = i_L(0^+) - \frac{V_s}{R} \tag{5.319}$$

Substituting Eq. (5.319) into Eq. (5.314) yields,

$$i(t) = \left[i_L(0^+) - \frac{V_s}{R}\right] e^{\frac{-R}{L}t} + \frac{V_s}{R} \tag{5.320}$$

In general, Eq. (5.320) can be expressed as,

$$i(t) = [i_L(0^+) - i_{ss,L}] e^{\frac{-R_{\mathrm{Th}}}{L}t} + i_{ss,L} \tag{5.321}$$

where

$i_L(0^+)$ is the current in the inductor at $t = 0^+$,
$i_{ss,L}$ is the steady-state inductor current in the circuit at $t > 0$,
R_{Th} is the Thevenin resistance seen by the inductor terminals in the circuit at $t > 0$,
$\frac{L}{R_{\mathrm{Th}}}$ is the circuit time constant.

Example 5.19 Figure 5.54 shows a circuit where the switch is opened for a long time. Determine the current $i(t)$ for $t > 0$.

Solution:
At $t = 0^-$, the switch is still open circuited and the inductor is short circuited. This circuit in this condition is shown in Fig. 5.55 and in this case, the current is,

$$i_L(0^-) = i_L(0^+) = \frac{16}{3+4} = 2.29\,\mathrm{A} \tag{5.322}$$

At $t = 0^+$, i.e. $t > 0$, the switch is closed and the circuit is shown in Fig. 5.56. The total circuit resistance is,

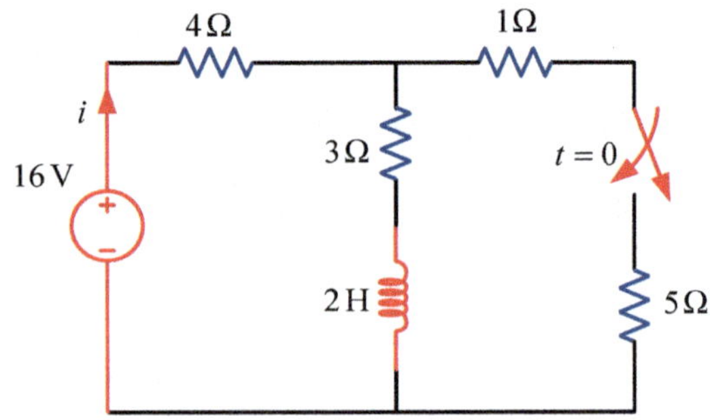

Fig. 5.54 Circuit for Example 5.19

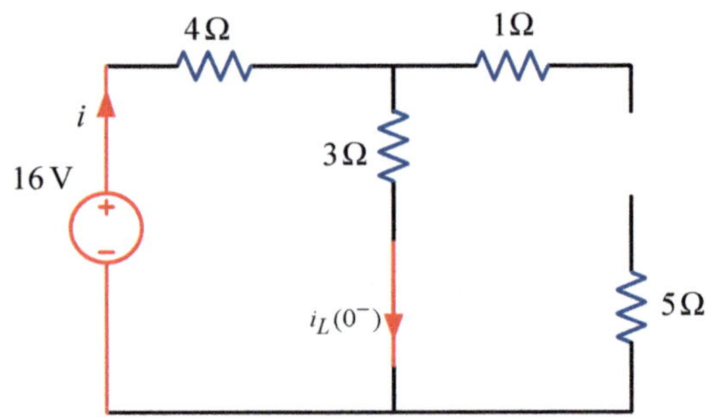

Fig. 5.55 Circuit for Example 5.19 at $t < 0$

$$i = \frac{16}{4 + 3 \parallel (1 + 5)} = 2.67\,\text{A} \tag{5.323}$$

$$i_{ss,L} = 2.67 \times \frac{(1 + 5)}{6 + 3} = 1.78\,\text{A} \tag{5.324}$$

The Thevenin resistance can be calculated from the circuit as shown in Fig. 5.56 as,

$$R_{\text{Th}} = 3 + \frac{4 \times 6}{10} = 5.4\,\Omega \tag{5.325}$$

The final solution is,

$$i(t) = 1.78 + (2.29 - 1.78)\text{e}^{-\frac{5.4}{2}t} = 1.78 + 0.51\text{e}^{-2.7t}\ \text{A} \tag{5.326}$$

Practice Problem 5.19
The switch in the circuit is opened for a long time as shown in Fig. 5.57. Calculate the expression of the current $i(t)$ for $t > 0$.

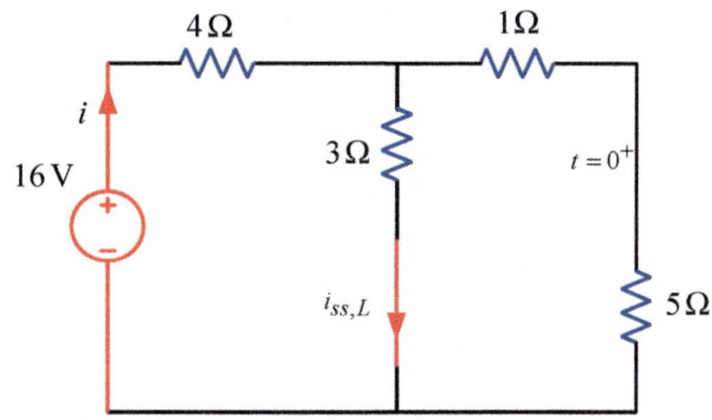

Fig. 5.56 Circuit for Example 5.19 for $t > 0$

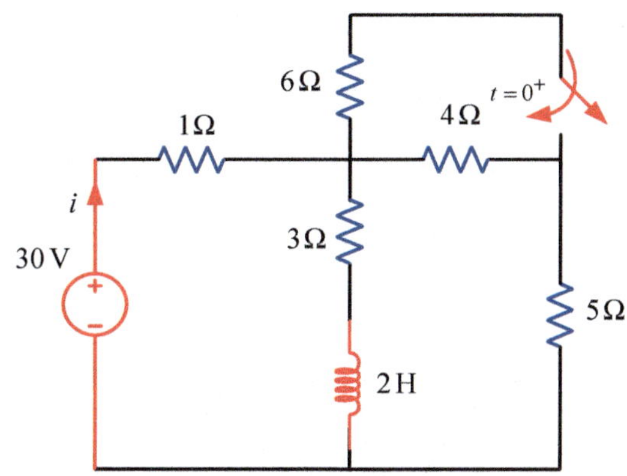

Fig. 5.57 Circuit for Practice Problem 5.19

Exercise Problems

5.1 The distance between the positive and the negative plates of a capacitor is 0.5 cm. The solid dielectric medium is placed between the plates. The relative permittivity of the dielectric is 2 and the area of each plate is 105 cm^2. Determine the capacitance and total charge on each plate if the electric field strength of the dielectric medium is 200 V per cm.

5.2 A voltage source of $v = 5\sin(377t - 20°)$ V is connected across a 3 mF capacitor. Find the expression for the current in the capacitor.

5.3 A current of $i = 8e^{-1500t}$ mA flows in a 5 μF capacitor. Calculate the voltage across the capacitor if the initial voltage across the capacitor is zero.

5.4 Figure 5.58 shows a voltage waveform of a 10 μF capacitor. Determine the current.

5.5 Figure 5.59 shows a voltage waveform of a 15 μF capacitor. Find the current.

5.6 Calculate the charge and energy stored when a 12 μF capacitor is connected across a 240 V source.

5.7 Resistor and capacitor are connected with a voltage source as shown in Fig. 5.60. Determine the energy stored in the capacitor under DC condition

5.8 Resistor and capacitor are connected with a voltage source as shown in Fig. 5.61. Calculate the energy stored in the capacitor under DC condition.

5.9 Capacitors are connected in a circuit as shown in Fig. 5.62. Determine the equivalent circuit capacitance.

5.10 The outer radius of a 10 km coaxial cable is 4 mm. Calculate the value of the inner radius of the cable if the value of the capacitance is 1.5 μF.

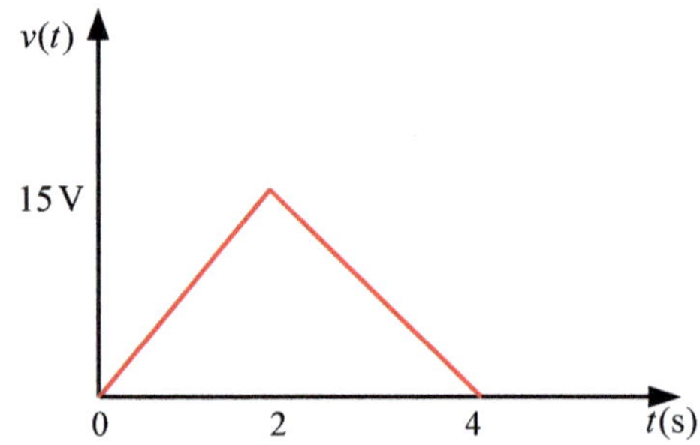

Fig. 5.58 Voltage waveform for Problem 5.4

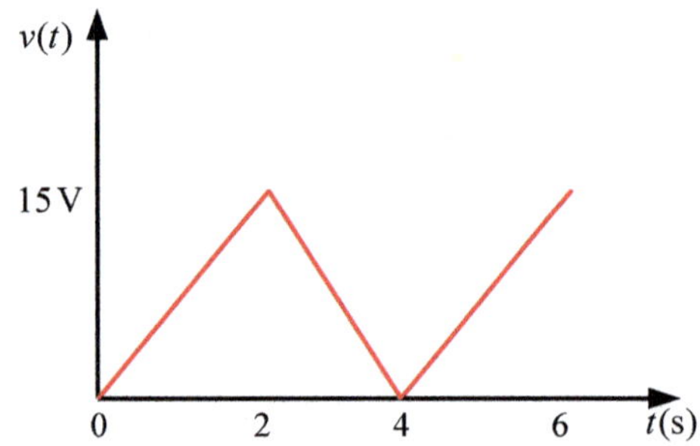

Fig. 5.59 Current waveform with times

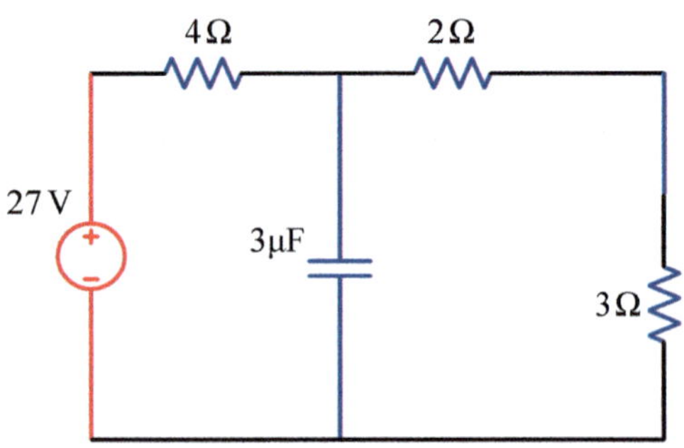

Fig. 5.60 Circuit for Problem 5.7

Fig. 5.61 Circuit for Problem 5.8

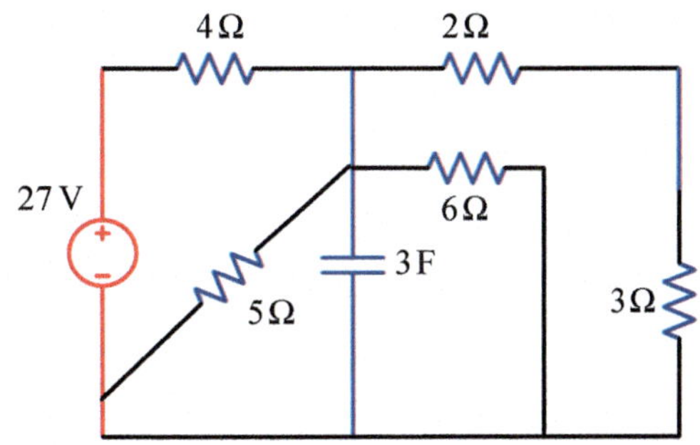

Fig. 5.62 Circuit for Problem 5.9

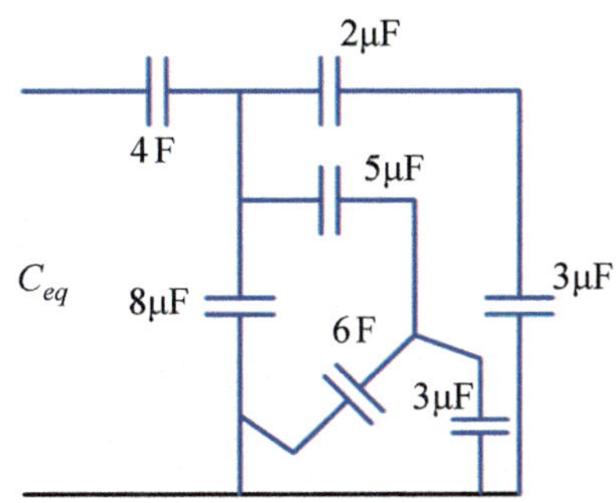

5.11 An air-core solenoid is wound by 120-turn copper coil whose mean length and cross-sectional area are 2 m and 16 cm^2, respectively. Find the inductance.

5.12 The inductance of an air-core solenoid is found to be 3 H. The core is wound by 500 turns copper coil whose cross-sectional area is 0.4 m^2. Calculate the mean length of the core.

5.13 A current of $i = 1.5\left(4 - e^{-1.5t}\right)$ A flows in the circuit as shown in Fig. 5.63. Calculate $i_1(0)$, i_1, i_2 and the voltage across the 6 H inductor if $i_2(0) = 2$ A.

5.14 Figure 5.64 shows a circuit with resistance, inductance and capacitance. Find the source current and energy stored by the inductor and capacitor under DC condition.

5.15 Figure 5.65 shows a circuit where a switch is opened at $t = 0$. Determine the voltage across the capacitor at $t > 0$ condition.

5.16 A switch is opened for a long time in the circuit in Fig. 5.66. Calculate the current $i(t)$ for $t > 0$.

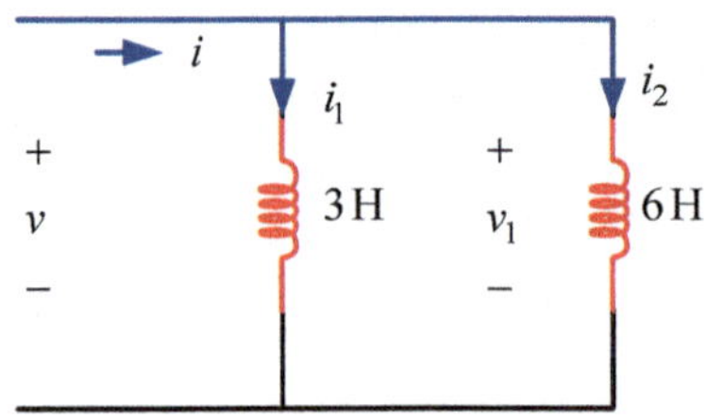

Fig. 5.63 Circuit for Problem 5.13

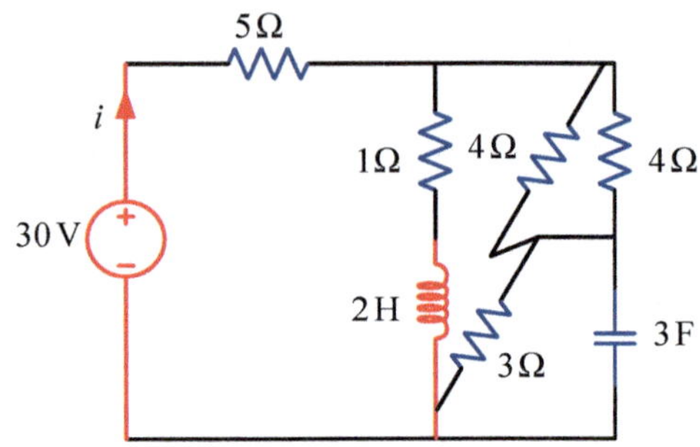

Fig. 5.64 Circuit for Problem 5.14

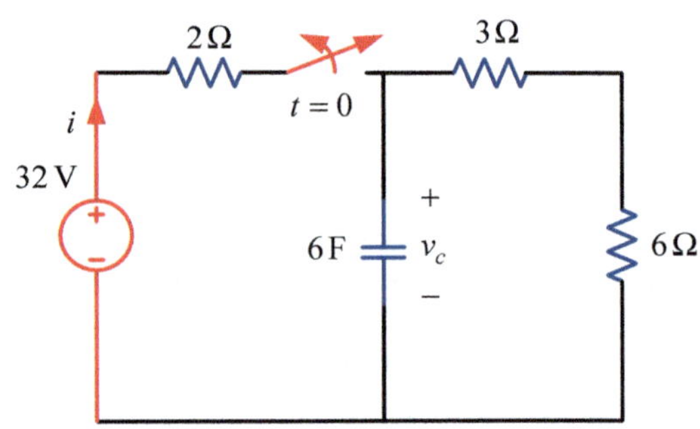

Fig. 5.65 Circuit for Problem 5.15

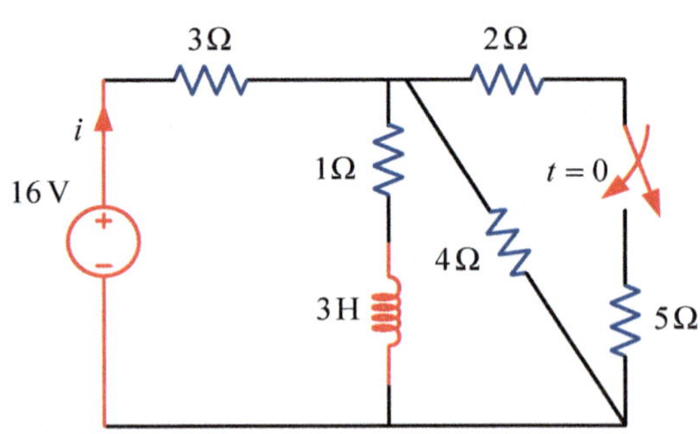

Fig. 5.66 Circuit for Problem 5.16

References

1. R.L. Boylestad, *Introductory Circuit Analysis*, 13th edn. (Pearson, 2016)
2. J.D. Irwin, R.M. Nelms, *Basic Engineering Circuit Analysis*, 11th edn. (Wiley, USA, 2015)
3. C.K. Alexander, M.N.O. Sadiku, *Fundamentals of Electric Circuits*, 6th edn. (McGraw-Hill Higher Education, January 2016)
4. J.W. Nilsson, S.A. Riedel, *Electric Circuits*, 10th edn. (Prentice Hall International Edition, 2015)
5. M.A. Salam, *Basic Electrical Circuits*, 2nd edn. (Shroff Publishers & Distributors Pvt. Ltd, India, 2007)
6. J.R. Cogdell, *Foundation of Electric Circuits*. (Pearson Higher Education, 1999)
7. M.A. Salam, *Electromagnetic Field Theories for Engineering*, 1st edn. (Springer, 2014)

Chapter 6
Alternating Current

6.1 Introduction

Up until now, the analysis has been limited to the electrical circuit networks with time-invariant sources, also known as DC source. At this point, the focus of the circuit analysis will be shifted towards time-varying sources. In electrical domain, the magnitude of a time-varying source varies between two preset levels with time. In this domain, a sinusoidal time-varying source is generally known as AC (derived from the name, Alternating Current) source. Circuits, driven by AC sources (current or voltage), are known as AC circuits. Generation, transmission and distribution of AC sources are more efficient than time-invariant sources. AC signals in any electrical circuit are also easy to analyse with diverse mathematical tools. The development of AC sources along with the formulation of associated mathematical theories has expanded the electrical power industries in an enormous rate in the whole world. This chapter presents all the fundamental concepts on AC sources along with some relevant circuit analysis.

6.2 Alternating Current Parameters

Sinusoidal source generates voltage or current that varies with time. A sinusoidal signal has the form of either sine or cosine function [1, 2], as shown in Fig. 6.1 (in the form of a sine function). Other than sinusoidal signals, AC circuits deal with few more time-varying signalling waveforms. These are triangular, sawtooth and pulse waveforms. The waveform whose rate of rise and the rate of fall are equal and constant is known as triangular waveform. The waveform whose magnitude varies between two constant positive and negative magnitudes is known as pulse or square waveform. The waveform whose rate of rise is different from the rate of fall is known as sawtooth waveform (looks like sawtooth). All these waveforms are shown in Fig. 6.1.

Md. A. Salam and Q. Mehbubar Rahman, *Fundamentals of Electrical Circuit Analysis*,
https://doi.org/10.1007/978-981-10-8624-3_6

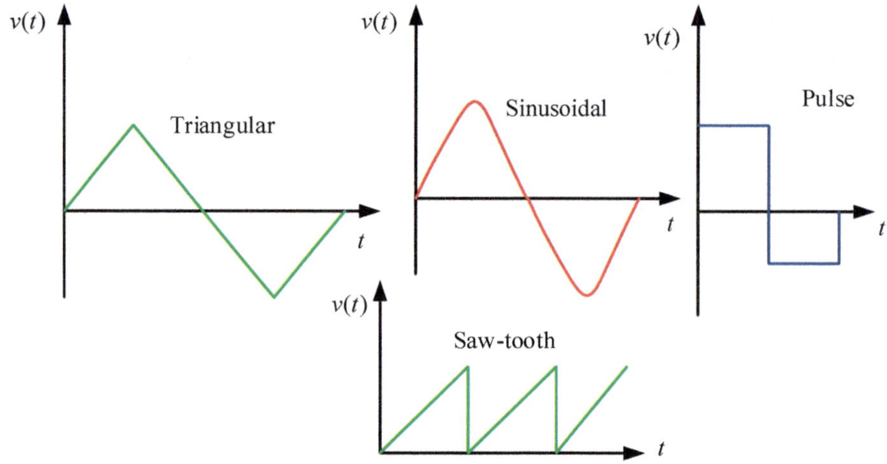

Fig. 6.1 Different waveforms

The value of the waveform at any instant of time is known as the instantaneous value. One complete set of positive and negative values of an alternating waveform is known as cycle. The number of cycles that occur in one second is known as the frequency. It is represented by the letter f and its unit is cycles per second or hertz (Hz) in honour of the German physicist Heinrich Hertz (1857–1894). The total time required to complete one cycle of a waveform is known as the time period, which is denoted by the capital letter T. The time period is inversely proportional to the frequency and it is written as,

$$T = \frac{1}{f} \tag{6.1}$$

From Fig. 6.2, it is seen that an angular distance (ωt) of 2π is traversed by the sinusoidal signal over a time period of T. In other words, the angular distance at $t = T$ is,

$$\omega T = 2\pi \tag{6.2}$$

Substituting Eq. (6.1) into Eq. (6.2) yields,

$$\omega \frac{1}{f} = 2\pi \tag{6.3}$$

$$\omega = 2\pi f \tag{6.4}$$

In Eq. (6.4), ω is known as the angular frequency.

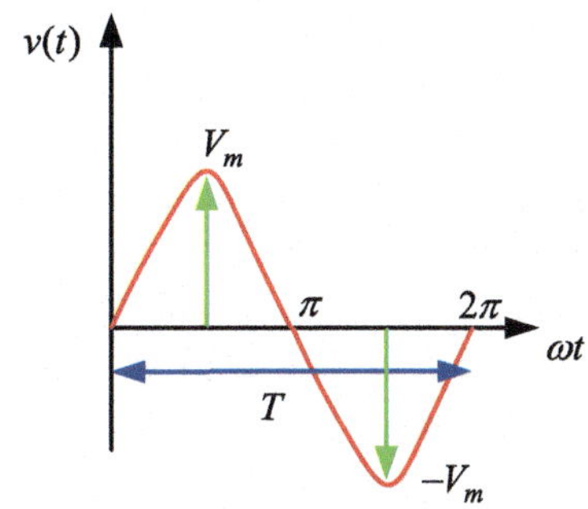

Fig. 6.2 Sinusoidal waveform with different parameters

The highest magnitude of any waveform is known as the peak value or the maximum value. The maximum value of voltage and current is represented by $V_{\rm m}$ for a voltage waveform and by $I_{\rm m}$ for current waveform. The sum of the positive and the negative peaks of a voltage waveform is known as the peak-to-peak value. From Fig. 6.2, the peak-to-peak value of the voltage waveform is written as,

$$V_{\rm pp} = |V_{\rm m}| + |-V_{\rm m}| = 2V_{\rm m} \tag{6.5}$$

The phase and phase difference between the two waveforms are very important to identify the terms lag and lead.

The phase (also known as phase angle) is the initial angle of a sinusoidal function at its origin or at a reference point. The phase difference is the difference in phase angles between the two waveforms as shown in Fig. 6.3.

From Fig. 6.3, the expressions of the two voltage waveforms can be written as,

$$v_1(t) = V_{\rm m}\sin(\omega t + \theta) \tag{6.6}$$

$$v_2(t) = V_{\rm m}\sin(\omega t + 0^\circ) \tag{6.7}$$

The phase difference between the two waveforms can be written as,

$$\theta_{\rm pd} = (\omega t + \theta) - (\omega t + 0^\circ) = \theta \tag{6.8}$$

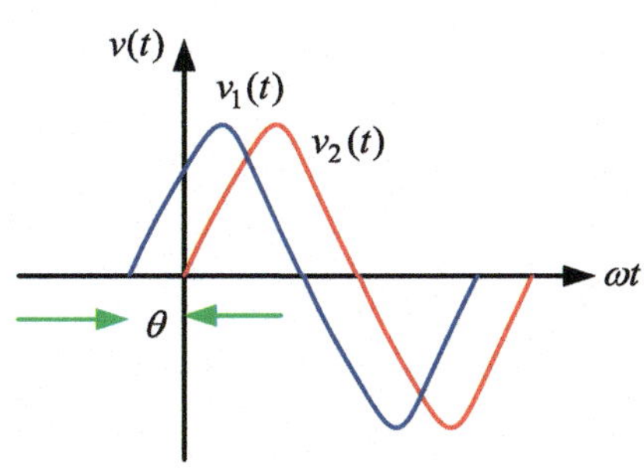

Fig. 6.3 Two sinusoidal voltage waveforms

In an inductor, current through the inductor lags the voltage across it by 90°, while in a capacitor the current leads the voltage by 90°. The lagging and leading phenomenon can be best understood by Figs. 6.4 and 6.5, respectively. In Fig. 6.4, the current waveform, i, lags the voltage waveform v, while in Fig. 6.5, the current leads the voltage.

Example 6.1 The expression of an alternating voltage waveform is given by $v(t) = 50\sin(314t - 35°)\,\text{V}$. Determine the maximum value of the voltage, frequency and time period.

Solution:
The maximum value of the voltage is,

$$V_\text{m} = 50\,\text{V} \tag{6.9}$$

The frequency is calculated as,

$$\omega = 314 \tag{6.10}$$

$$f = \frac{314}{2\pi} = 50\,\text{Hz} \tag{6.11}$$

Practice Problem 6.1
The expression of an alternating current waveform is given by $i(t) = 15\sin 377t$ A. Find the maximum value of the current, frequency, time period and instantaneous current at $t = 0.01$ s.

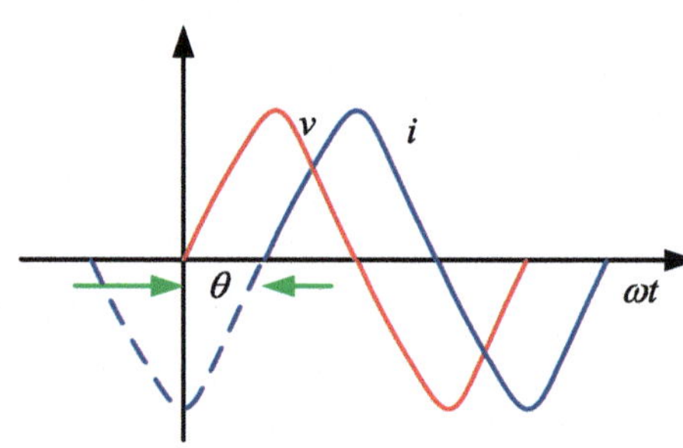

Fig. 6.4 Current lags the voltage

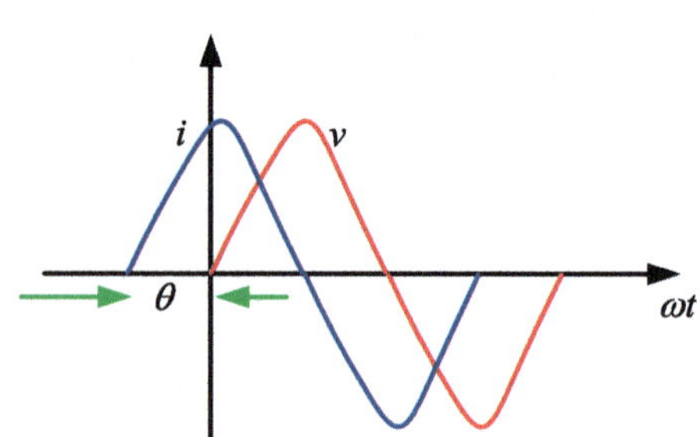

Fig. 6.5 Current leads the voltage

6.3 Root Mean Square Value

The root mean square is an important term which is used in an AC circuit for analysis and it is abbreviated as rms. It is often known as effective value or virtual value. The rms value of an alternating current can be defined as the steady-state (DC) current which, when flows through a given resistance for a given time length, produces the same heat energy that is produced by an alternating current when flows through the same resistance for the same length of time. According to the definition of rms value, the following equation can be written [3, 4]:

$$P_{\mathrm{dc}} = P_{\mathrm{ac}} \tag{6.12}$$

As shown in the circuit in Fig. 6.6, when terminal 'a' connects to terminal 'b', the heat energy produced due to the current I is,

$$P_{\mathrm{dc}} = I^2 R \tag{6.13}$$

Again, for the same circuit, when terminal 'b' connects the terminal 'c', the heat energy produced due to the current i is,

$$P_{\mathrm{ac}} = \frac{1}{T}\int_0^T i^2(t) R \mathrm{d}t \tag{6.14}$$

Substituting Eqs. (6.13) and (6.14) into Eq. (6.12) yields,

$$I^2 R = \frac{1}{T}\int_0^T i^2(t) R \mathrm{d}t \tag{6.15}$$

$$I_{\mathrm{rms}} = \sqrt{\frac{1}{T}\int_0^T i^2(t) \mathrm{d}t} \tag{6.16}$$

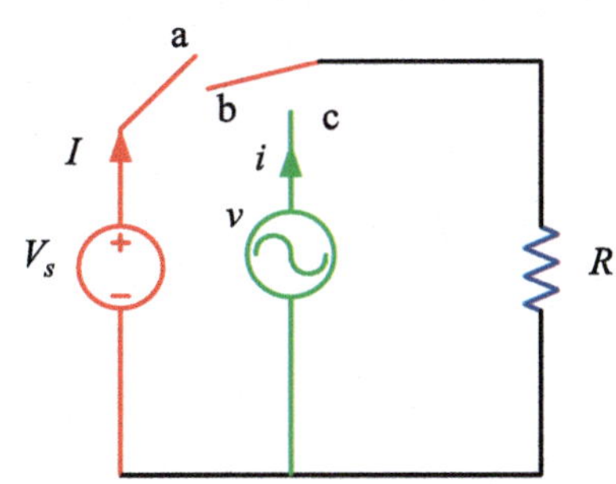

Fig. 6.6 DC and AC sources with resistance

Consider a sinusoidal current waveform as shown in Fig. 6.7. This sinusoidal current is expressed as,

$$i = I_{\mathrm{m}} \sin \omega t \tag{6.17}$$

Substituting Eq. (6.17) into Eq. (6.16) yields,

$$I_{\mathrm{rms}} = \sqrt{\frac{1}{2\pi} \int_0^{2\pi} (I_{\mathrm{m}} \sin \omega t)^2 \mathrm{d}(\omega t)} \tag{6.18}$$

$$I_{\mathrm{rms}} = \sqrt{\frac{I_{\mathrm{m}}^2}{4\pi} \int_0^{2\pi} 2 \sin^2 \omega t \mathrm{d}(\omega t)} \tag{6.19}$$

$$I_{\mathrm{rms}} = \frac{I_{\mathrm{m}}}{2} \sqrt{\frac{1}{\pi} \int_0^{2\pi} (1 - \cos 2\omega t) \mathrm{d}(\omega t)} \tag{6.20}$$

$$I_{\mathrm{rms}} = \frac{I_{\mathrm{m}}}{2} \sqrt{\frac{1}{\pi} [\omega t]_0^{2\pi} - \frac{1}{2\pi} [\sin 2\omega t]_0^{2\pi}} \tag{6.21}$$

$$I_{\mathrm{rms}} = \frac{I_{\mathrm{m}}}{2} \sqrt{\frac{1}{\pi} [2\pi - 0] - 0} \tag{6.22}$$

$$I_{\mathrm{rms}} = \frac{I_{\mathrm{m}}}{2} \times \sqrt{2} \tag{6.23}$$

$$I_{\mathrm{rms}} = \frac{I_{\mathrm{m}}}{\sqrt{2}} = 0.707 I_{\mathrm{m}} \tag{6.24}$$

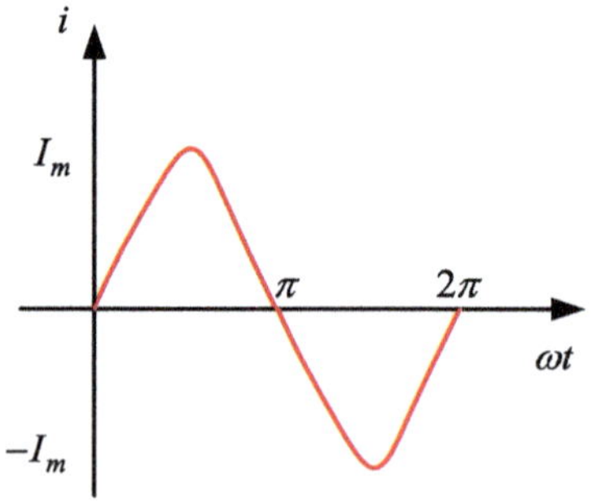

Fig. 6.7 Sinusoidal current waveform

Equation (6.24) provides the expression for the rms current. Following similar approach, the expression for the rms voltage can be derived as,

$$V_{\mathrm{rms}} = \frac{V_{\mathrm{m}}}{\sqrt{2}} = 0.707V_{\mathrm{m}} \tag{6.25}$$

In this case, the power absorbed by any resistance R can be calculated as,

$$P = I_{\mathrm{rms}}^2 R = \frac{V_{\mathrm{rms}}^2}{R} \tag{6.26}$$

Example 6.2 A sawtooth current waveform is shown in Fig. 6.8. Calculate the rms value of this current, and when it flows through a 6 Ω resistor calculate the power absorbed by the resistor.

Solution:
In the sawtooth waveform, the time period is 2 s. The value of the slope is,

$$m = \frac{12}{2} = 6 \tag{6.27}$$

$$i(t) = 6t\,\mathrm{A} \quad 0 < t < 2 \tag{6.28}$$

The rms value of the current waveform is calculated as,

$$I_{\mathrm{rms}}^2 = \frac{1}{2}\int_0^2 (6t)^2 \mathrm{d}t \tag{6.29}$$

$$I_{\mathrm{rms}}^2 = \frac{18}{3}\left[t^3\right]_0^2 = 6 \times 8 = 48 \tag{6.30}$$

$$I_{\mathrm{rms}} = 6.92\ \mathrm{A} \tag{6.31}$$

The power absorbed by the 6 Ω resistor is determined as,

$$P = 6.92^2 \times 6 = 288\ \mathrm{W} \tag{6.32}$$

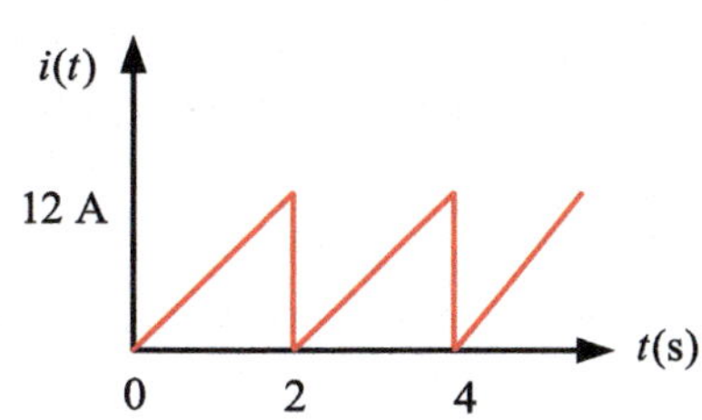

Fig. 6.8 Current waveform for Example 6.2

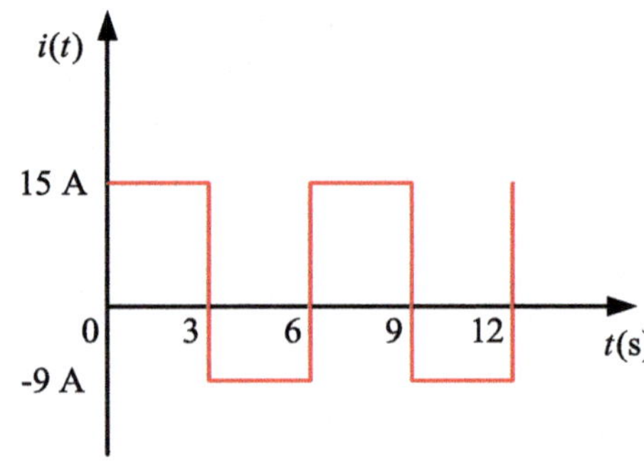

Fig. 6.9 Current waveform for Practice Problem 6.2

Practice Problem 6.2
Figure 6.9 shows a square-wave current waveform. Determine the rms value of the current, and the power absorbed by a 2 Ω resistor when the current flows through this resistor.

6.4 Average Value

The average of all the values of an alternating waveform in one cycle is known as average value. The average value for a non-periodic waveform is expressed as,

$$\text{Average value} = \frac{\text{Area under one cycle}}{\text{base length}} \tag{6.33}$$

In general, the average value for a non-periodic waveform is written as,

$$I_{\text{av}} = \frac{1}{2\pi}\int_0^{2\pi} i d(\omega t) \tag{6.34}$$

For a symmetrical or a periodic waveform, the positive area cancels the negative area. Thus, the average value is zero. In this case, the average value is calculated using the positive half of the waveform. The average value for a periodic waveform is expressed as,

$$\text{Average value} = \frac{\text{Area under half cycle}}{\text{base length}} \tag{6.35}$$

In general, the average value for a periodic waveform is written as,

$$I_{\text{av}} = \frac{1}{\pi}\int_0^{\pi} i d(\omega t) \tag{6.36}$$

The average value of the sinusoidal current waveform (periodic) is calculated as,

$$I_{av} = \frac{1}{\pi}\int_{0}^{\pi} I_m \sin \omega t d(\omega t) \tag{6.37}$$

$$I_{av} = \frac{I_m}{\pi}[-\cos \omega t]_0^{\pi} = -\frac{I_m}{\pi}[\cos \pi - 1] \tag{6.38}$$

$$I_{av} = \frac{2I_m}{\pi} = 0.637 I_m \tag{6.39}$$

Example 6.3 A triangular-wave current waveform is shown in Fig. 6.10. Find the average value of this current waveform.

Solution:
The time period of the waveform is 4 s. For different times, the expressions of voltage are,

$$v(t) = 5t\,\text{V} \quad 0 < t < 1 \tag{6.40}$$

$$\frac{v(t) - 0}{0 - 5} = \frac{t - 2}{2 - 1} \tag{6.41}$$

$$v(t) = -5t + 10 \tag{6.42}$$

$$v(t) = -5t + 10\,\text{V} \quad 1 < t < 2 \tag{6.43}$$

The average value of the voltage waveform (periodic) is calculated as,

$$V_{av} = \frac{1}{2}\left[\int_{0}^{1} 5t\text{d}t + \int_{1}^{2} (-5t + 10)\text{d}t\right] \tag{6.44}$$

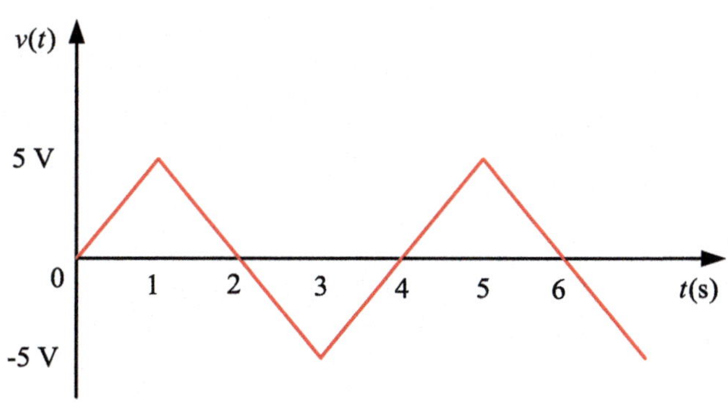

Fig. 6.10 Voltage waveform for Example 6.3

$$V_{\mathrm{av}} = \frac{2.5}{2}\left[t^2\right]_0^1 - \frac{2.5}{2}\left[t^2\right]_1^2 + \frac{10}{2}\left[t\right]_1^2 \tag{6.45}$$

$$V_{\mathrm{av}} = \frac{2.5}{2} - \frac{2.5}{2} \times 3 + 5 = 2.5\,\mathrm{V} \tag{6.46}$$

Practice Problem 6.3
A current waveform is shown in Fig. 6.11. Determine the average value of this current waveform.

6.5 RMS Value for Complex Waveform

Consider that an instantaneous value i of a periodic (with a time period T) complex current waveform that flows through a resistor R is given by,

$$i = I_0 + I_{\mathrm{m1}}\sin(\omega t + \theta_1) + I_{\mathrm{m2}}\sin(2\omega t + \theta_2) + I_{\mathrm{m3}}\sin(3\omega t + \theta_3) \tag{6.47}$$

According to Eq. (6.24), the rms values of $I_{\mathrm{m1}}\sin(\omega t + \theta_1)$, $I_{\mathrm{m2}}\sin(2\omega t + \theta_2)$ and $I_{\mathrm{m3}}\sin(3\omega t + \theta_3)$ are,

$$I_1 = \frac{I_{\mathrm{m1}}}{\sqrt{2}} \tag{6.48}$$

$$I_2 = \frac{I_{\mathrm{m2}}}{\sqrt{2}} \tag{6.49}$$

$$I_3 = \frac{I_{\mathrm{m3}}}{\sqrt{2}} \tag{6.50}$$

Heat produced due to the first component of the current is,

$$H_1 = I_0^2 RT \tag{6.51}$$

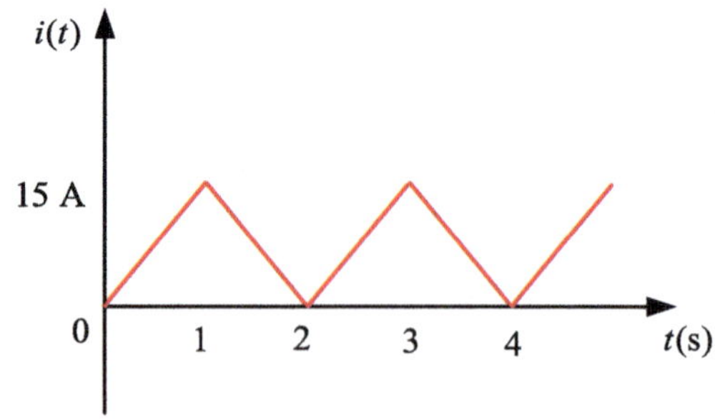

Fig. 6.11 Voltage waveform for Example 6.3

Heat produced due to the second component of the current is,

$$H_2 = \left(\frac{I_{\mathrm{m}1}}{\sqrt{2}}\right)^2 RT \tag{6.52}$$

Heat produced due to the third component of the current is,

$$H_3 = \left(\frac{I_{\mathrm{m}2}}{\sqrt{2}}\right)^2 RT \tag{6.53}$$

Total heat produced by these current components can be expressed as,

$$H_{\mathrm{t}} = I_0^2 RT + \left(\frac{I_{\mathrm{m}1}}{\sqrt{2}}\right)^2 RT + \left(\frac{I_{\mathrm{m}2}}{\sqrt{2}}\right)^2 RT \tag{6.54}$$

According to the definition of rms value, Eq. (6.54) can be modified as,

$$I^2 RT = I_0^2 RT + \left(\frac{I_{\mathrm{m}1}}{\sqrt{2}}\right)^2 RT + \left(\frac{I_{\mathrm{m}2}}{\sqrt{2}}\right)^2 RT \tag{6.55}$$

$$I_{\mathrm{rms}} = \sqrt{I_0^2 + \left(\frac{I_{\mathrm{m}1}}{\sqrt{2}}\right)^2 + \left(\frac{I_{\mathrm{m}2}}{\sqrt{2}}\right)^2} \tag{6.56}$$

Similarly, for instantaneous value of a complex voltage waveform, it can be expressed as,

$$V_{\mathrm{rms}} = \sqrt{V_0^2 + \left(\frac{V_{\mathrm{m}1}}{\sqrt{2}}\right)^2 + \left(\frac{V_{\mathrm{m}2}}{\sqrt{2}}\right)^2} \tag{6.57}$$

Example 6.4 A complex voltage waveform is given by $v(t) = 2 + 3\sin\omega t + 5\sin 2\omega t\,\mathrm{V}$. Calculate the rms value and the power absorbed by a $2\,\Omega$ resistor.

Solution:
The rms value of the complex voltage waveform can be calculated as,

$$V_{\mathrm{rms}} = \sqrt{2^2 + \left(\frac{3}{\sqrt{2}}\right)^2 + \left(\frac{5}{\sqrt{2}}\right)^2} = 4.58\,\mathrm{V} \tag{6.58}$$

The power absorbed by the 2 Ω resistor is calculated as,

$$P = \frac{V_{rms}^2}{R} = \frac{4.58^2}{2} = 10.49\,\text{W} \tag{6.59}$$

Practice Problem 6.4
A complex voltage waveform is given by $v(t) = 1 + 1.5\sin\omega t\,\text{V}$. Determine the rms value and the power absorbed by a 4 Ω resistor.

6.6 Form Factor and Peak Factor

The ratio of rms value to the average value of an alternating voltage or current is known as form factor. Mathematically, the form factor is expressed as,

$$\text{Form factor} = \frac{I_{rms}}{I_{av}} = \frac{V_{rms}}{V_{av}} \tag{6.60}$$

The ratio of maximum value to the rms value of an alternating voltage or current is known as peak factor, which also known as crest factor. Mathematically, it is expressed as,

$$\text{Peak factor} = \frac{I_m}{I_{rms}} = \frac{V_m}{V_{rms}} \tag{6.61}$$

6.7 Concept of Phasor

Charles Proteus Steinmetz, a mathematician and an electrical engineer, established the phasor in an AC circuit in between the years 1865–1923. A phasor is a complex term which represents the magnitude and the phase angle of a sinusoidal waveform. Phasor is often known as vector [5–7]. A definite length of line rotating in the anticlockwise direction with a constant angular speed is also known as phasor. As shown in Fig. 6.12, the phasor ***P*** rotates in the anticlockwise direction with a

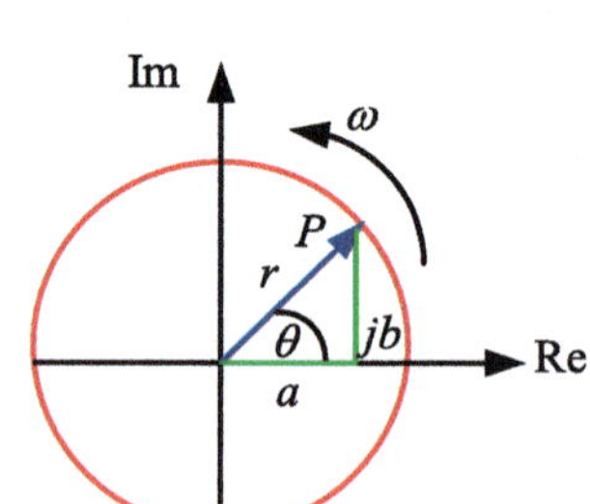

Fig. 6.12 Phasor with angular speed

constant angular speed ω, where the phase angle θ is measured with respect to the real axis.

Here, r is the magnitude of the phasor. In a rectangular form, the phasor can be represented as,

$$P = a + jb \tag{6.62}$$

In a polar form, the phasor is represented as,

$$P = r\angle\theta \tag{6.63}$$

The magnitude and the phase angle of the phasor are calculated as,

$$r = \sqrt{a^2 + b^2} \tag{6.64}$$

$$\theta = \tan^{-1}\left(\frac{b}{a}\right) \tag{6.65}$$

Again, from Fig. 6.12, the following expressions relationship can be written:

$$a = r\cos\theta \tag{6.66}$$

$$b = r\sin\theta \tag{6.67}$$

Substituting Eqs. (6.66) and (6.67) into Eq. (6.62) yields,

$$P = r\cos\theta + jr\sin\theta = r(\cos\theta + j\sin\theta) = r\angle\theta \tag{6.68}$$

The phasor in an exponential form is expressed as,

$$P = re^{j\theta} \tag{6.69}$$

where

$$e^{j\theta} = \cos\theta + j\sin\theta \tag{6.70}$$

A sinusoidal voltage $v(t) = V_{\mathrm{m}}\sin(\omega t + \theta)$ can be expressed as,

$$v(t) = \mathrm{Im}(V_{\mathrm{m}}e^{j(\omega t + \theta)}) = \mathrm{Im}(V_{\mathrm{m}}e^{j\omega t}e^{j\theta}) = \mathrm{Im}(Ve^{j\omega t}) \tag{6.71}$$

where

$$V = V_{\mathrm{m}}e^{j\theta} = V_{\mathrm{m}}\angle\theta \tag{6.72}$$

The derivative of a sinusoidal voltage $v(t) = V_{\mathrm{m}}\sin(\omega t + \theta)$ can be expressed as,

$$\frac{dv(t)}{dt} = V_m\omega\cos(\omega t + \theta) = V_m\omega\sin\{90^\circ + (\omega t + \theta)\} \tag{6.73}$$

$$\frac{dv(t)}{dt} = \omega\,\mathrm{Im}\left(V_m e^{j90^\circ} e^{j\omega t} e^{j\theta}\right) = j\omega\,\mathrm{Im}\left(V e^{j\omega t}\right) \tag{6.74}$$

$$\frac{dv(t)}{dt} = j\omega V \tag{6.75}$$

Similarly, the following expression:

$$\int v(t)dt = \frac{V}{j\omega} \tag{6.76}$$

Example 6.5 Convert the voltage $v(t) = 10\sin(10t - 45^\circ)$ V and current $i(t) = 5\cos(20t - 65^\circ)$ A expressions into their phasor forms.

Solution:
The phasor form of voltage can be written as,

$$V = 10\underline{|-45^\circ}\text{ V} \tag{6.77}$$

The current expression can be rearranged as,

$$i(t) = 5\cos(20t - 65^\circ) = 5\sin(20t - 65^\circ + 90^\circ) = 5\sin(20t + 25^\circ)\text{ A} \tag{6.78}$$

The phasor form of current can be expressed as,

$$I = 5\underline{|25^\circ}\text{ A} \tag{6.79}$$

Practice Problem 6.5
The voltage and the current expressions are given by $v(t) = 2\sin(314t + 25^\circ)$ V and $i(t) = -5\sin(-20t - 95^\circ)$ A, respectively. Convert these expressions into phasor forms.

6.8 The *j*-Operator

The *j*-operator ($j = \sqrt{-1}$) is an important multiplication factor in an AC circuit. It provides a 90° displacement in the anticlockwise direction from the reference axis when multiplying with any phasor. Consider a phasor P works in the positive x-axis as shown in Fig. 6.13. Multiply the phasor P by j and the new phasor jP works in the positive y-axis which is 90° apart from the x-axis. Again, multiplying the phasor jP by j results in the new phasor jjP works in the negative x-axis.

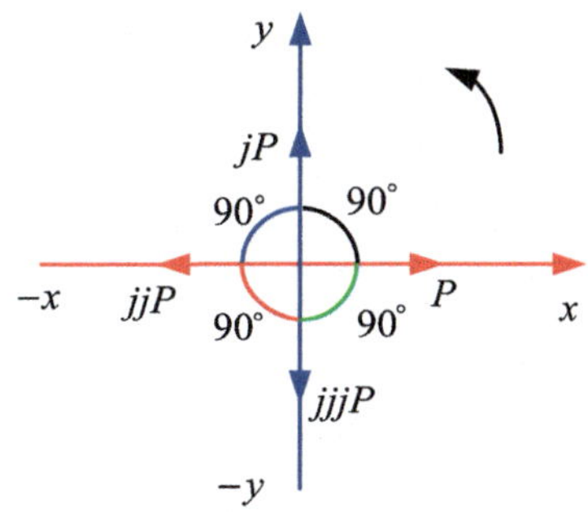

Fig. 6.13 Phasor with angular speed

In the positive and negative *x*-axes, the phasors are equal, but work in the opposite directions and in this case, the associated expression can be written as,

$$-jjP = P \tag{6.80}$$

$$j^2 = -1 \tag{6.81}$$

6.9 Phasor Algebra

Phasor addition, subtraction, multiplication, division, power and conjugate have been discussed here. Consider the following phasors to discuss the phasor algebra.

$$p_1 = a_1 + jb_1 = r_1\underline{|\theta_1} \tag{6.82}$$

$$p_2 = a_2 + jb_2 = r_2\underline{|\theta_2} \tag{6.83}$$

The phasor addition is expressed as,

$$p_a = a_1 + jb_1 + a_2 + jb_2 = (a_1 + a_2) + j(b_1 + b_2) \tag{6.84}$$

The phasor subtraction is expressed as,

$$p_s = a_1 + jb_1 - a_2 - jb_2 = (a_1 - a_2) + j(b_1 - b_2) \tag{6.85}$$

The phasor multiplication is expressed as,

$$p_{\mathrm{m}} = r_1\underline{|\theta_1} \times r_2\underline{|\theta_2} = r_1 r_2\underline{|\theta_1 + \theta_2} \tag{6.86}$$

The phasor division is expressed as,

$$p_d = r_1\underline{|\theta_1} \div r_2\underline{|\theta_2} = \frac{r_1}{r_2}\underline{|\theta_1 - \theta_2} \tag{6.87}$$

The power of the phasor is expressed as,

$$p_p = (r\angle\theta)^n = r^n\angle r\theta \tag{6.88}$$

The conjugate of any phasor is expressed as,

$$p_{conj} = a - jb = r\angle{-\theta} \tag{6.89}$$

Example 6.6 Evaluate the following phasor expression in polar form:

$$A = \frac{(2+j5+3-j2)(3-j2)}{(2+j3)^3}$$

Solution:
The value of the voltage is determined as,

$$A = \frac{(2+j5+3-j2)(3-j2)}{(2+j3)^3} = \frac{21.02\angle{-2.73^\circ}}{46.87\angle 168.93^\circ} = 0.45\angle{-171.66^\circ} \tag{6.90}$$

Practice Problem 6.6
Evaluate the following expression in polar form:

$$B = \frac{(6+j5)(3-j2)}{(5+j3)(-j3)}$$

6.10 Alternating Current Circuit with Resistor

Figure 6.14 shows an AC circuit with a resistor. Consider that the expression of the source voltage in this circuit is,

$$v(t) = V_{\mathrm{m}} \sin \omega t \tag{6.91}$$

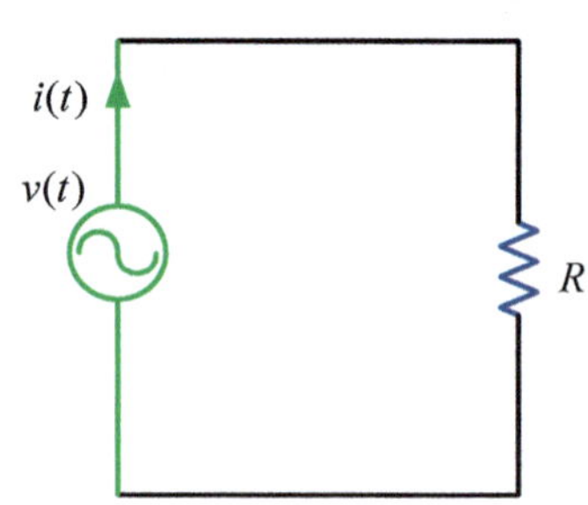

Fig. 6.14 Resistance with a voltage source

The phasor form of this voltage is,

$$V = V_{\mathrm{m}}\underline{|0^{\circ}} \tag{6.92}$$

The current in the resistor R is,

$$i(t) = \frac{v(t)}{R} \tag{6.93}$$

Substituting Eq. (6.90) into Eq. (6.92) yields,

$$i(t) = \frac{V_{\mathrm{m}} \sin \omega t}{R} \tag{6.94}$$

$$i(t) = I_{\mathrm{m}} \sin \omega t \tag{6.95}$$

The phasor form of this current is,

$$I = I_{\mathrm{m}}\underline{|0^{\circ}} \tag{6.96}$$

The voltage and current waveforms are drawn in the same graph as shown in Fig. 6.15. The starting point of both the waveforms is the same. Therefore, the voltage and the current waveforms are in the same phase. The impedance (impedance is the resistance to AC circuit that contains both magnitude and phase; resistance contains magnitude only) due to the resistance is written as,

$$Z_{\mathrm{R}} = R + j0 \tag{6.97}$$

The real part (Re) and the imaginary part (Im) of impedance due to resistance are plotted as shown in Fig. 6.16.

Example 6.7 A 3Ω resistor is connected in series with a voltage source whose expression is given by $v(t) = 6\sin(314t + 25^{\circ})$ V. Calculate the current in the resistor using phasor and draw the phasor diagram for the associated voltage and current.

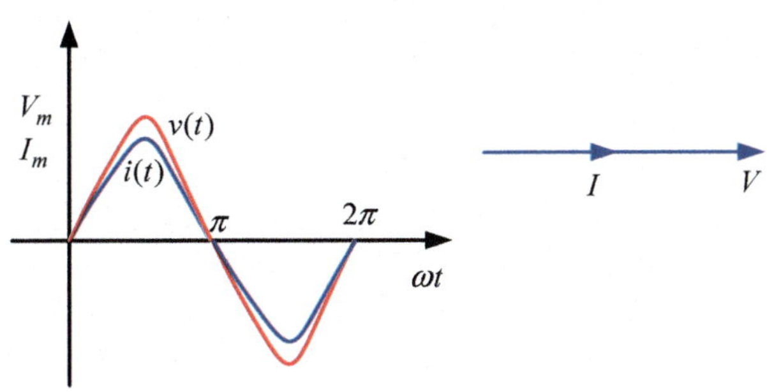

Fig. 6.15 Voltage and current waveforms along with in-phase phasor diagram

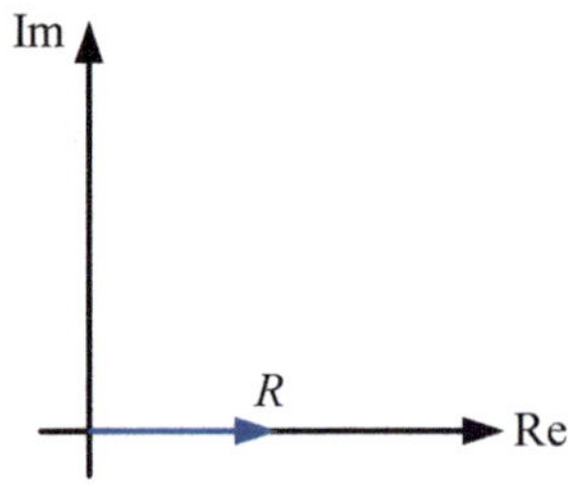

Fig. 6.16 Phasor diagram for impedance due to resistance

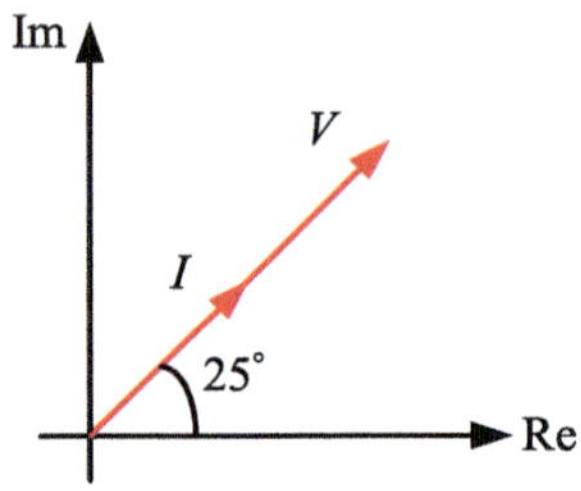

Fig. 6.17 Phasor diagram with voltage and current for Example 6.7

Solution:
The phasor form of the voltage is,

$$V = 6\angle 25^\circ \text{ V} \tag{6.98}$$

The value of the current in the phasor form is,

$$I = \frac{6\angle 25^\circ}{3} = 2\angle 25^\circ \text{ A} \tag{6.99}$$

The current in sinusoidal form is,

$$i(t) = 2\sin(314t + 25^\circ) \text{ A} \tag{6.100}$$

The phasor diagram is shown in Fig. 6.17.

Practice Problem 6.7
A current of $i(t) = 5\sin(377t - 35^\circ)$ A flows through a 5Ω resistor. Use phasor to find the voltage across the resistor and draw the phasor diagram the associated current and voltage.

6.11 Alternating Current Circuit with Inductor

An inductor is connected in series with an alternating voltage source as shown in Fig. 6.18. An alternating current will flow through the inductor. As a result, an emf will be induced in the coil and this induced emf will be equal to the applied voltage.

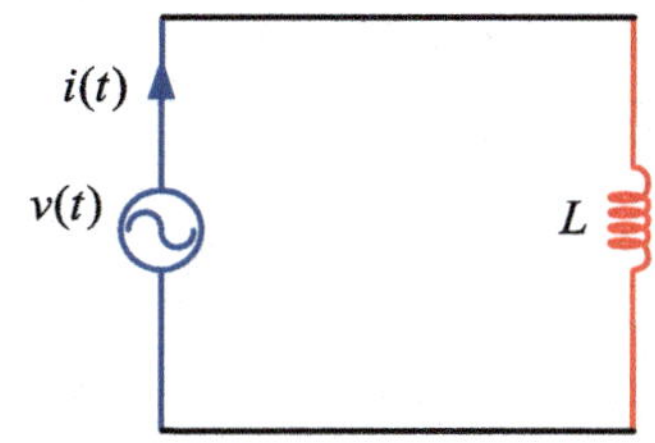

Fig. 6.18 Inductor in series with a voltage source

The induced emf in the inductor is expressed as,

$$v(t) = L\frac{\mathrm{d}i(t)}{\mathrm{d}t} \tag{6.101}$$

Substituting Eq. (6.91) into Eq. (6.101) yields,

$$V_{\mathrm{m}} \sin \omega t = L\frac{\mathrm{d}i(t)}{\mathrm{d}t} \tag{6.102}$$

Rearranging Eq. (6.102) yields,

$$\mathrm{d}i(t) = \frac{V_{\mathrm{m}}}{L} \sin \omega t \, \mathrm{d}t \tag{6.103}$$

Integrating Eq. (6.103) yields,

$$i(t) = -\frac{V_{\mathrm{m}}}{\omega L} \cos \omega t = \frac{V_{\mathrm{m}}}{\omega L} \sin\left(\omega t - \frac{\pi}{2}\right) \tag{6.104}$$

The current in phasor form is,

$$I = I_{\mathrm{m}}\angle -90^\circ \tag{6.105}$$

where the maximum value of the current is,

$$I_{\mathrm{m}} = \frac{V_{\mathrm{m}}}{\omega L} \tag{6.106}$$

The phase difference between voltage and current is $\phi_{\mathrm{pd}} = 0 + 90^\circ = 90^\circ$. Therefore, voltage leads the current by an angle 90° or current lags the voltage waveform by the same angle. The waveforms for the voltage and current are shown in Fig. 6.19.

Impedance due to inductance is defined as the ratio of the phasor form of voltage to the phasor form of current, and it is written as,

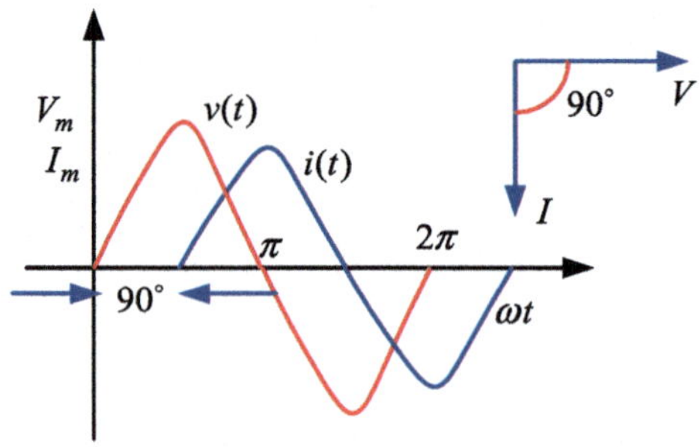

Fig. 6.19 Voltage and current waveforms along with lagging phasor diagram

$$Z_L = \frac{V}{I} \tag{6.107}$$

Substituting Eqs. (6.92) and (6.105) into Eq. (6.107) yields,

$$Z_L = \frac{V_m \angle 0^\circ}{\frac{V_m}{\omega L} \angle -90^\circ} = \omega L \angle 90^\circ = j\omega L \tag{6.108}$$

The magnitude of Z_L is called the inductive reactance and it is represented by X_L. Equation (6.108) can be modified as,

$$Z_L = jX_L \tag{6.109}$$

where the expression of inductive reactance is expressed as,

$$X_L = \omega L \tag{6.110}$$

The phasor representation of impedance due to inductance is shown in Fig. 6.20.

Example 6.8 The voltage across a 2 mH inductor is found to be $v(t) = 5\sin(34t + 35^\circ)$ V. Determine the current in the inductor using phasor and draw the phasor diagram for the associated current and voltage.

Solution:
The phasor form of the voltage is,

$$V = 5\angle 35^\circ \text{ V} \tag{6.111}$$

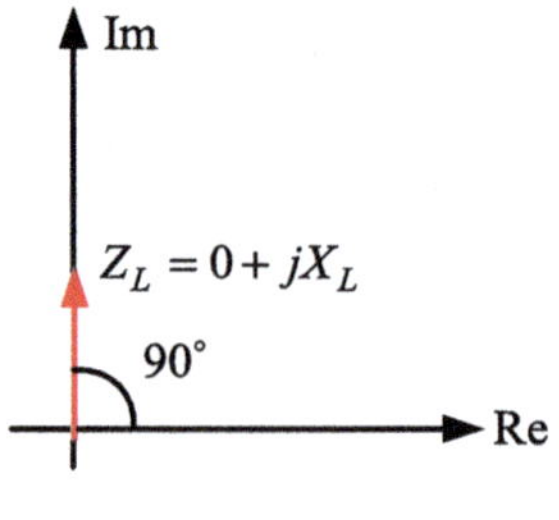

Fig. 6.20 Phasor diagram for impedance due to inductance

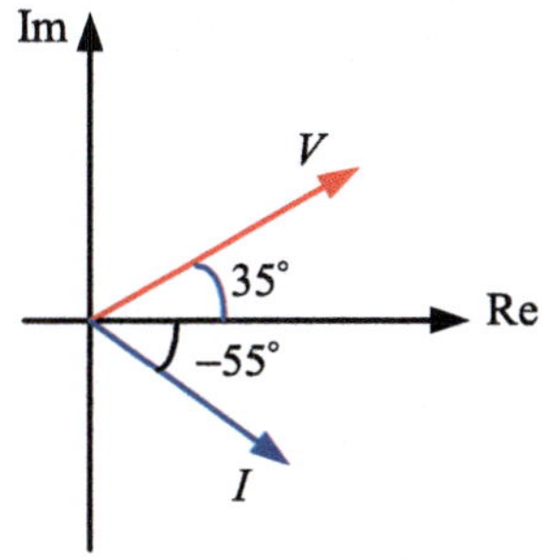

Fig. 6.21 Phasor diagram for Example 6.8

The inductive reactance is,

$$X_{\mathrm{L}} = \omega L = 34 \times 2 \times 10^{-3} = 0.086\,\Omega \tag{6.112}$$

The current is,

$$I = \frac{V}{jX_{\mathrm{L}}} = \frac{5\underline{|35^\circ}}{0.086\underline{|90^\circ}} = 58.14\underline{|-55^\circ}\ \mathrm{A} \tag{6.113}$$

The current in sinusoidal form or in time domain is,

$$i(t) = 58.14\sin(34\,t - 55^\circ)\ \mathrm{A} \tag{6.114}$$

The phasor diagram is shown in Fig. 6.21.

Practice Problem 6.8
The voltage across a 5H inductor is measured as $v(t) = 15\sin(30t - 20^\circ)$ V. Calculate the current in the inductor using phasor and draw the phasor diagram for the associated current and voltage.

6.12 Alternating Current Circuit with Capacitor

Figure 6.22 shows a circuit where a capacitor with C is connected in series with an alternating voltage source v. In this case, when a current flows through it, the instantaneous charge stored in the capacitor will be,

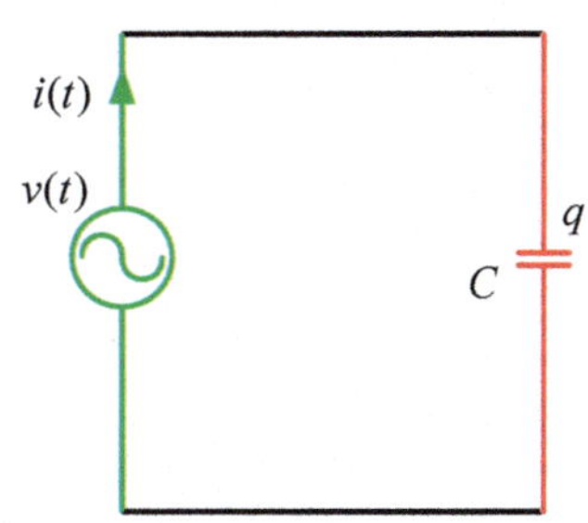

Fig. 6.22 Capacitor in series with a voltage source

$$q = Cv \tag{6.115}$$

Substituting Eq. (6.91) into Eq. (6.115) yields,

$$q = CV_{\mathrm{m}} \sin \omega t \tag{6.116}$$

The current through the capacitor is expressed as,

$$i(t) = \frac{\mathrm{d}q}{\mathrm{d}t} \tag{6.117}$$

Substituting Eq. (6.116) into Eq. (6.117) yields,

$$i(t) = \frac{\mathrm{d}}{\mathrm{d}t}(CV_{\mathrm{m}} \sin \omega t) \tag{6.118}$$

$$i(t) = CV_{\mathrm{m}}\omega \cos \omega t \tag{6.119}$$

$$i(t) = CV_{\mathrm{m}}\omega \sin(90^\circ + \omega t) \tag{6.120}$$

where the maximum value of the current is,

$$I_{\mathrm{m}} = C\omega V_{\mathrm{m}} \tag{6.121}$$

From Eq. (6.120), the expression of the current in phasor form can be written as,

$$I = I_{\mathrm{m}}\underline{|90^\circ} \tag{6.122}$$

In this case, the phase difference between the voltage and the current is,

$$\theta_{\mathrm{pd}} = 0 - 90^\circ = -90^\circ \tag{6.123}$$

From Eq. (6.123), it is observed that the current waveform in a capacitor leads the voltage waveform or the voltage waveform lags the current waveform. The voltage and the current waveforms in a capacitor are shown in Fig. 6.23.

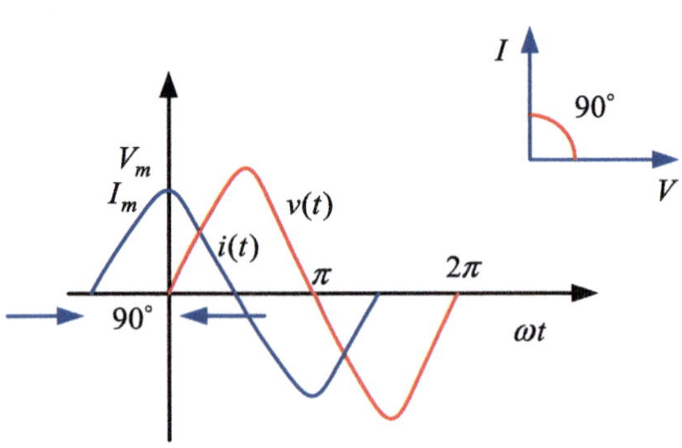

Fig. 6.23 Voltage and current waveforms along with lagging phasor diagram

The impedance due to capacitance is defined as the ratio of the phasor voltage and phasor current. Mathematically, it is expressed as,

$$Z_C = \frac{V}{I} \tag{6.124}$$

Substituting Eqs. (6.92) and (6.122) into Eq. (6.124) yields,

$$Z_C = \frac{V_m\angle 0^\circ}{I_m\angle 90^\circ} \tag{6.125}$$

Substituting Eq. (6.121) into Eq. (6.125) yields,

$$Z_C = \frac{V_m\angle 0^\circ}{CV_m\omega\angle 90^\circ} = \frac{1}{\omega C}\angle -90^\circ \tag{6.126}$$

$$Z_C = \frac{1}{\omega C}\angle -90^\circ = -j\frac{1}{\omega C} \tag{6.127}$$

The magnitude of Z_C is called the capacitive reactance and it is represented by X_C. Equation (6.127) can be rearranged as,

$$Z_C = -jX_C \tag{6.128}$$

where the expression of capacitive reactance can be written as,

$$X_C = \frac{1}{\omega C} \tag{6.129}$$

The phasor representation of impedance due to capacitive reactance is shown in Fig. 6.24.

Example 6.9 Determine the current in both the phasor and sinusoidal forms when a voltage across a 5 μF capacitor is given by $v(t) = 65\sin(700t - 40^\circ)$ V.

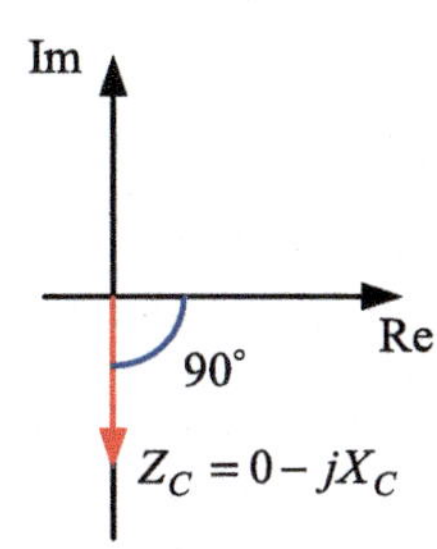

Fig. 6.24 Phasor diagram for impedance due to capacitance

Solution:
The capacitive reactance is,

$$X_C = \frac{1}{\omega C} = \frac{1}{700 \times 5 \times 10^{-6}} = 285.71\,\Omega \tag{6.130}$$

The voltage in phasor form is,

$$V = 65\underline{|-40^\circ}\ \text{V} \tag{6.131}$$

The current in phasor form can be determined as,

$$I = \frac{V}{-jX_C} = \frac{65\underline{|-40^\circ}}{285.71\underline{|-90^\circ}} = 0.23\underline{|50^\circ}\ \text{A} \tag{6.132}$$

The current in sinusoidal form is,

$$i(t) = 0.23\sin(700t + 50^\circ)\ \text{A} \tag{6.133}$$

Practice Problem 6.9
A 3 μF capacitor is connected to a voltage source, and the current through this capacitor is given by $i(t) = 12\sin(377t + 26^\circ)$ A. Determine the voltage across the capacitor in both phasor and sinusoidal forms.

6.13 Impedance and Admittance

The impedance and admittance are considered to be important parameters in analysing AC circuit. The impedance is the characteristic of a circuit or circuit element force that opposes the current flow in an AC circuit. Impedance is equivalent to resistance in a DC circuit. The impedance, which is a vector quality, is also defined as the ratio of the phasor voltage to the phasor current. The impedance is represented by the letter Z, and it is frequency dependent. The reciprocal of impedance is known as admittance. It is represented by the letter Y and its unit is Siemens (S).

The impedance can be written as,

$$Z = \frac{V}{I} \tag{6.134}$$

where the admittance can be written as,

$$Y = \frac{I}{V} = \frac{1}{Z} \tag{6.135}$$

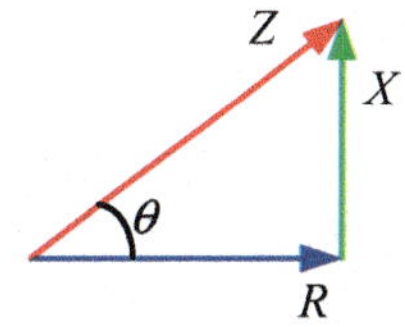

Fig. 6.25 Impedance triangle

Both impedance and admittance are related to resistance and reactance. The impedance triangle is shown in Fig. 6.25. According to Fig. 6.25, the expression of impedance can be written as,

$$Z = R + jX \tag{6.136}$$

The impedance in polar form can be written as,

$$Z = |Z|\underline{|\theta} \tag{6.137}$$

where the magnitude of the impedance is,

$$|Z| = \sqrt{R^2 + X^2} \tag{6.138}$$

The expression of phase angle is,

$$\theta = \tan^{-1}\left(\frac{X}{R}\right) \tag{6.139}$$

In addition, from Fig. 6.25, the following expressions can be written:

$$R = Z\cos\theta \tag{6.140}$$

$$X = Z\sin\theta \tag{6.141}$$

Substituting Eq. (6.136) into Eq. (6.135) yields,

$$Y = \frac{1}{R + jX} \tag{6.142}$$

Multiplying the numerator and denominator of the right-hand side of Eq. (6.142) by $(R - jX)$ yields,

$$Y = \frac{R - jX}{(R + jX)(R - jX)} = \frac{R - jX}{R^2 + X^2} \tag{6.143}$$

$$Y = \frac{R}{R^2 + X^2} - j\frac{X}{R^2 + X^2} \tag{6.144}$$

$$Y = G + jB \tag{6.145}$$

where G and B are known as the conductance and the susceptance, respectively, and their units are Siemens (S). The expressions of these two parameters are written as,

$$G = \frac{R}{R^2 + X^2} \tag{6.146}$$

$$B = -\frac{X}{R^2 + X^2} \tag{6.147}$$

Example 6.10 Determine the inductive reactance and the current in both phasor and sinusoidal forms for the circuit in Fig. 6.26, if the applied voltage is $v(t) = 15\sin(10t - 20°)$ V.

Solution:

The value of the inductive reactance is calculated as,

$$X_L = \omega L = 10 \times 0.5 = 5\,\Omega \tag{6.148}$$

The impedance of the circuit is,

$$Z = 2 + j5 = 3.61\underline{|56.31°}\,\Omega \tag{6.149}$$

The voltage in phasor form is,

$$V = 15\underline{|-20°}\text{ V} \tag{6.150}$$

The current in the phasor form is calculated as,

$$I = \frac{V}{Z} = \frac{15\underline{|-20°}}{3.61\underline{|56.31°}} = 4.16\underline{|-76.31°}\text{ A} \tag{6.151}$$

The expression of the current in the sinusoidal form is,

$$i(t) = 4.16\sin(10t - 76.31°)\text{ A} \tag{6.152}$$

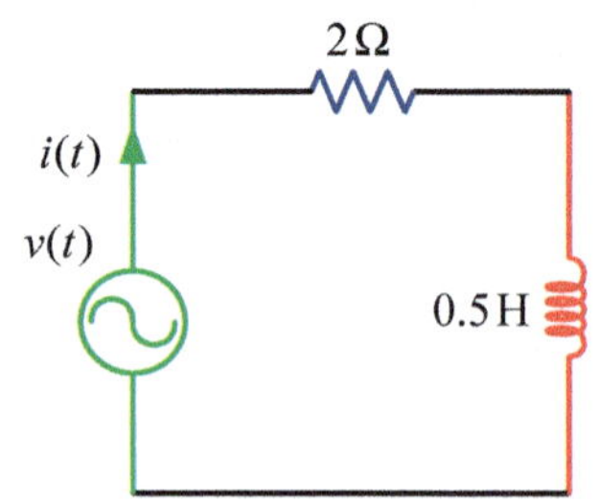

Fig. 6.26 Circuit for Example 6.10

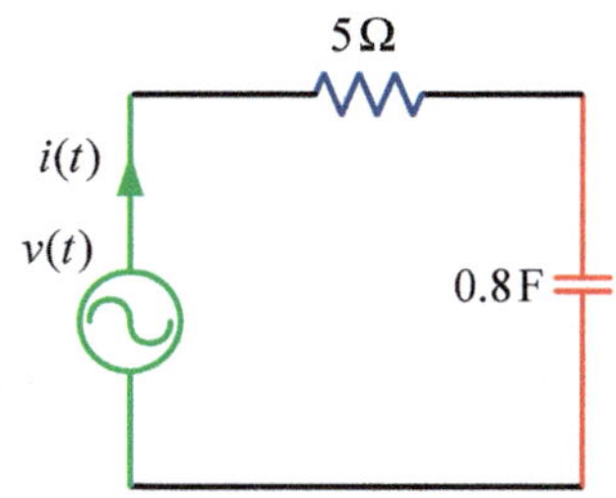

Fig. 6.27 Circuit for Practice Problem 6.10

Example 6.11 Two circuit components with impedances $Z_1 = 2 + j3\,\Omega$ and $Z_2 = 4 - j7\,\Omega$ are connected in parallel to a $V = 220\underline{|35^\circ}$ V AC source. Calculate the total admittance and the current in both phasor and sinusoidal forms.

Solution:
The total admittance is calculated as,

$$Y = Y_1 + Y_2 = \frac{1}{Z_1} + \frac{1}{Z_2} = \frac{1}{2+j3} + \frac{1}{4-j7} = 0.25\underline{|-29.74^\circ}\ \text{S} \tag{6.153}$$

The current is calculated as,

$$I = VY = 220\underline{|35^\circ} \times 0.25\underline{|-29.74^\circ} = 55\underline{|-5.26^\circ}\ \text{A} \tag{6.154}$$

Practice Problem 6.10
Figure 6.27 shows an RC series circuit. Calculate the capacitive reactance and the current in both phasor and sinusoidal forms, if the applied voltage is $v(t) = 5\sin(20t + 35^\circ)$ V.

Practice Problem 6.11
Find the total admittance and the current in both phasor and sinusoidal forms when two circuit components with impedances $Z_1 = 3 + j5\,\Omega$ and $Z_2 = 3 - j4\,\Omega$ are connected in parallel to a $V = 230\underline{|20^\circ}$ V AC source.

6.14 Kirchhoff's Laws in AC Circuit

Consider that n number of time-varying voltage sources are connected in series in a closed circuit as shown in Fig. 6.28. The sum of the time-varying voltages is expressed as,

$$v_1(t) + v_2(t) + v_3(t) + \cdots + v_n(t) = 0 \tag{6.155}$$

Equation (6.155) can be expressed in sinusoidal form as,

$$V_{m1}\sin(\omega t + \theta_1) + V_{m2}\sin(\omega t + \theta_2) + V_{m3}\sin(\omega t + \theta_3) + \cdots + V_{mn}\sin(\omega t + \theta_n) = 0 \tag{6.156}$$

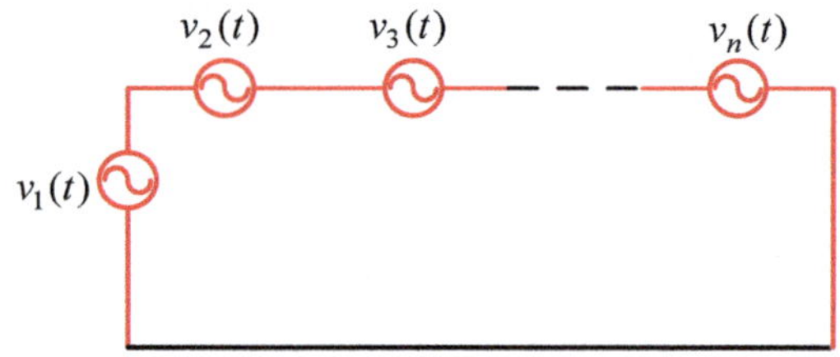

Fig. 6.28 Series circuit with time-varying voltage sources

Again, Eq. (6.156) can be rearranged as,

$$\mathrm{Im}(V_{\mathrm{m1}}e^{j\omega t}e^{j\theta_1}) + \mathrm{Im}(V_{\mathrm{m2}}e^{j\omega t}e^{j\theta_2}) + \mathrm{Im}(V_{\mathrm{m3}}e^{j\omega t}e^{j\theta_3}) + \cdots + \mathrm{Im}(V_{\mathrm{m}n}e^{j\omega t}e^{j\theta_n}) = 0 \tag{6.157}$$

Dividing Eq. (6.157) by the $e^{j\omega t}$ yields,

$$\mathrm{Im}(V_{\mathrm{m1}}e^{j\theta_1} + V_{\mathrm{m2}}e^{j\theta_2} + V_{\mathrm{m3}}e^{j\theta_3} + \cdots + V_{\mathrm{m}n}e^{j\theta_n}) = 0 \tag{6.158}$$

In phasor form, Eq. (6.158) can be written as,

$$V_1 + V_2 + V_3 + \cdots + V_n = 0 \tag{6.159}$$

From Eq. (6.159), the Kirchhoff's voltage law in an AC circuit can be stated as follows: '*The phasor sum of the voltage drop across each element in a closed circuit is equal to zero*'.

Similarly, the Kirchhoff's current law in a phasor form can be written as,

$$I_1 + I_2 + I_3 + \cdots + I_n = 0 \tag{6.160}$$

And this can be stated as follows: 'T*he algebraic sum of the phasor currents in any node of a circuit is equal to zero*'.

6.15 Impedance and Admittance in Series

Consider that three circuit elements with impedances Z_1, Z_2 and Z_3 are connected in a series as shown in the circuit in Fig. 6.29. Applying KVL to the circuit in Fig. 6.29, the expression of the phasor voltage can be written as,

$$V = V_1 + V_2 + V_3 \tag{6.161}$$

According to Ohm's law, with the aid of phasor current I and impedance, the corresponding phasor voltages can be expressed as,

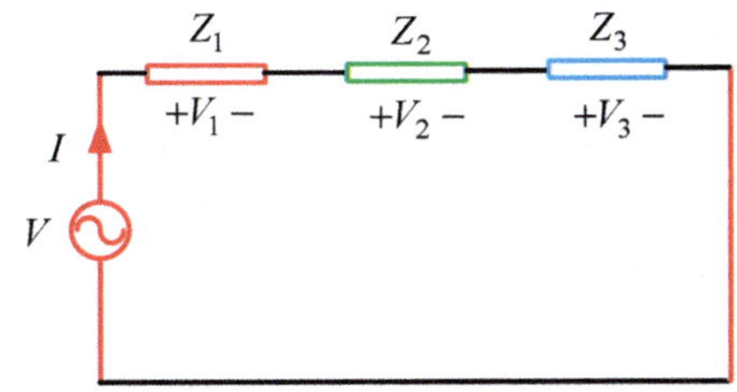

Fig. 6.29 Series impedances with a phasor voltage source

$$V_1 = IZ_1 \tag{6.162}$$

$$V_2 = IZ_2 \tag{6.163}$$

$$V_3 = IZ_3 \tag{6.164}$$

Substituting equations from (6.162) to 6.164) into Eq. (6.161) yields,

$$V = IZ_1 + IZ_2 + IZ_3 \tag{6.165}$$

$$V = I(Z_1 + Z_2 + Z_3) \tag{6.166}$$

$$V = IZ_t \tag{6.167}$$

In this case, Z_t is the total impedance which is given by,

$$Z_t = Z_1 + Z_2 + Z_3 \tag{6.168}$$

In terms of admittance, Eqs. (6.162) to (6.164) can be written as,

$$V_1 = \frac{I}{Y_1} \tag{6.169}$$

$$V_2 = \frac{I}{Y_2} \tag{6.170}$$

$$V_3 = \frac{I}{Y_3} \tag{6.171}$$

Substituting equations from (6.169) to 6.171) into Eq. (6.161) yields,

$$V = I\left(\frac{1}{Y_1} + \frac{1}{Y_2} + \frac{1}{Y_3}\right) \tag{6.172}$$

$$V = I\frac{1}{Y_t} \tag{6.173}$$

where Y_t is the total admittance, which can be equated from the following expression:

$$\frac{1}{Y_t} = \frac{1}{Y_1} + \frac{1}{Y_2} + \frac{1}{Y_3} \tag{6.174}$$

$$Y_t = \frac{Y_1 Y_2 Y_3}{Y_1 Y_2 + Y_2 Y_3 + Y_1 Y_3} \tag{6.175}$$

The equivalent admittance for two circuit elements in series with their admittances (Y_1 and Y_2) can be expressed as,

$$Y_t = \frac{Y_1 Y_2}{Y_1 + Y_2} \tag{6.176}$$

From Eq. (6.176), it is seen that the total admittance for two circuit elements connected in series is equal to the product of the individual admittances over the sum of the corresponding admittances.

6.16 Impedance and Admittance in Parallel Connection

Consider that three circuit elements with associated impedances Z_1, Z_2 and Z_3 are connected in parallel as shown in the circuit in Fig. 6.30. Applying KCL to the circuit in Fig. 6.30 yields,

$$I = I_1 + I_2 + I_3 \tag{6.177}$$

According to Ohm's law, with the aid of phasor voltage V and impedances, the corresponding phasor currents in each parallel branch can be expressed as,

$$I_1 = \frac{V}{Z_1} \tag{6.178}$$

$$I_2 = \frac{V}{Z_2} \tag{6.179}$$

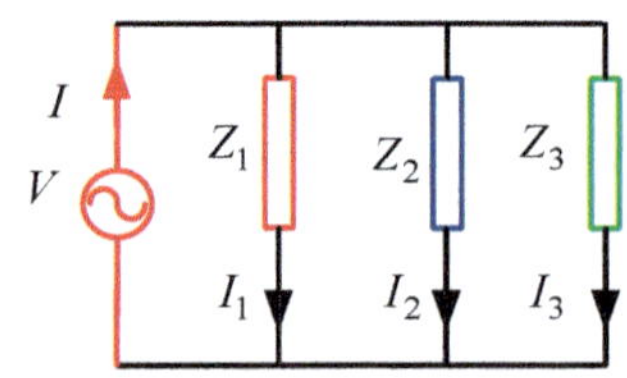

Fig. 6.30 Parallel impedances with a phasor voltage source

$$I_3 = \frac{V}{Z_3} \tag{6.180}$$

Substituting equations from (6.178) to (6.180) into Eq. (6.177) yields,

$$I = \frac{V}{Z_1} + \frac{V}{Z_2} + \frac{V}{Z_3} \tag{6.181}$$

$$\frac{I}{V} = \frac{1}{Z_1} + \frac{1}{Z_2} + \frac{1}{Z_3} \tag{6.182}$$

$$\frac{1}{Z_t} = \frac{1}{Z_1} + \frac{1}{Z_2} + \frac{1}{Z_3} \tag{6.183}$$

When two circuit elements with their corresponding impedances are in parallel, then Eq. (6.183) can be modified as,

$$Z_t = \frac{Z_1 Z_2}{Z_1 + Z_2} \tag{6.184}$$

In case of admittance, Eq. (6.183) can be modified as,

$$Y_t = Y_1 + Y_2 + Y_3 \tag{6.185}$$

From Eq. (6.185), it can be concluded that in a parallel circuit, the total admittance is equal to the sum of the individual admittances.

Example 6.12 A series–parallel AC circuit is shown in Fig. 6.31. Determine the total impedance, total admittance, source current and voltage drop across the inductor.

Solution:
The impedances are calculated as,

$$Z_1 = 5 + j(314 \times 20 \times 10^{-3}) = 5 + j6.28 = 8.03\underline{|51.47^\circ}\ \Omega \tag{6.186}$$

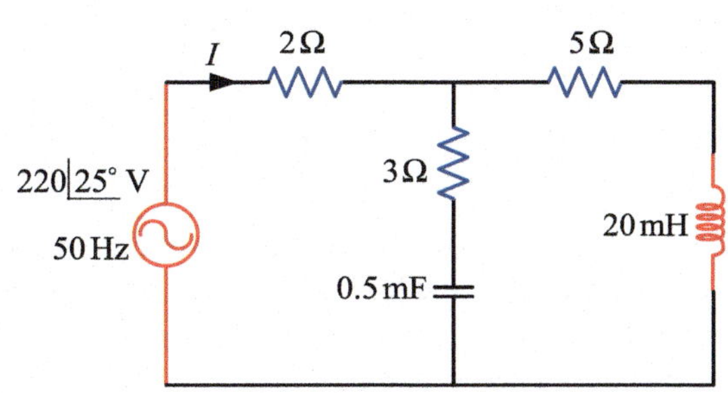

Fig. 6.31 Circuit for Example 6.12

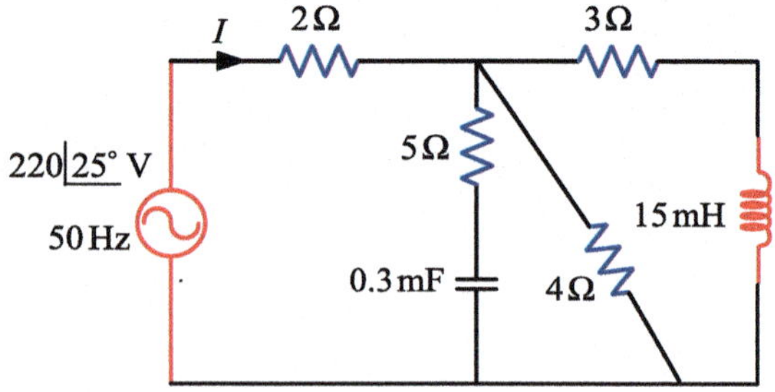

Fig. 6.32 Circuit for Practice Problem 6.12

$$Z_2 = 3 - j\left(\frac{1}{314 \times 0.5 \times 10^{-3}}\right) = 5 - j6.37 = 8.10\underline{|-51.87^\circ}\ \Omega \tag{6.187}$$

The total impedance is,

$$Z_t = 3 + \frac{8.03\underline{|51.47^\circ} \times 8.10\underline{|-51.87^\circ}}{8.03\underline{|51.47^\circ} + 8.10\underline{|-51.87^\circ}} = 9.5\underline{|0.08^\circ}\ \Omega \tag{6.188}$$

The total admittance is,

$$Y_t = \frac{1}{Z_t} = \frac{1}{9.5\underline{|0.08^\circ}} = 0.11\underline{|-0.08^\circ}\ \text{S} \tag{6.189}$$

The source current is calculated as,

$$I = \frac{V}{Z_t} = \frac{220\underline{|25^\circ}}{9.5\underline{|0.08^\circ}} = 23.16\underline{|24.92^\circ}\ \text{A} \tag{6.190}$$

The voltage drop across the inductor is calculated as,

$$I_1 = 23.16\underline{|24.92^\circ} \times \frac{8.10\underline{|-51.87^\circ}}{8.03\underline{|51.47^\circ} + 8.10\underline{|-51.87^\circ}} = 18.75\underline{|-26.44^\circ}\ \text{A} \tag{6.191}$$

$$V_{j6.28} = 6.28\underline{|90^\circ} \times 18.75\underline{|-26.44^\circ} = 117.75\underline{|63.56^\circ}\ \text{V} \tag{6.192}$$

Practice Problem 6.12
Figure 6.32 shows a series–parallel AC circuit. Find the total impedance, total admittance, source current and voltage drop across the capacitor.

6.17 Delta-to-Wye Conversion

Delta-to-wye conversion has already been discussed using DC circuit. In case of AC circuit, same method is followed and finally the same expressions are found using impedances. Based on the AC circuit shown in Fig. 6.33, the delta-to-wye impedances can be written as,

$$Z_a = \frac{Z_3 Z_1}{Z_1 + Z_2 + Z_3} \tag{6.193}$$

$$Z_b = \frac{Z_1 Z_2}{Z_1 + Z_2 + Z_3} \tag{6.194}$$

$$Z_c = \frac{Z_2 Z_3}{Z_1 + Z_2 + Z_3} \tag{6.195}$$

6.18 Wye-to-Delta Conversion

Similar to DC circuit with the aid of the circuit shown in Fig. 6.33, wye-to-delta converted impedances can be written as:

$$Z_1 = \frac{Z_a Z_b + Z_b Z_c + Z_c Z_a}{Z_c} \tag{6.196}$$

$$Z_2 = \frac{Z_a Z_b + Z_b Z_c + Z_c Z_a}{Z_a} \tag{6.197}$$

$$Z_3 = \frac{Z_a Z_b + Z_b Z_c + Z_c Z_a}{Z_b} \tag{6.198}$$

Considering $Z_a = Z_b = Z_c = Z_Y$ and $Z_1 = Z_2 = Z_3 = Z_\Delta$, Eq. (6.195) can be modified as,

$$Z_Y = \frac{Z_\Delta Z_\Delta}{3Z_\Delta} \tag{6.199}$$

$$Z_Y = \frac{Z_\Delta}{3} \tag{6.200}$$

From Eq. (6.200), it can be observed that for a balanced circuit, the wye-connected impedance is equal to one-third of the corresponding delta-connected impedance.

Fig. 6.33 Delta–wye connection

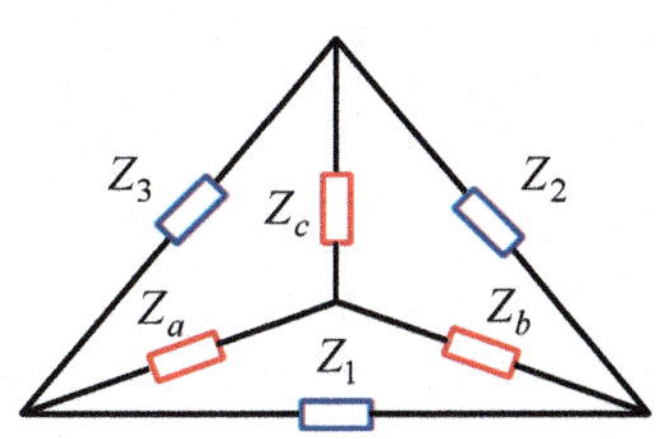

Example 6.13 Calculate the equivalent impedance and the source current of the circuit shown in Fig. 6.34.

Solution:
The inductive and capacitive reactances are,

$$X_L = 314 \times 16 \times 10^{-3} = 5.02\,\Omega \tag{6.201}$$

$$X_C = \frac{1}{314 \times 0.3 \times 10^{-3}} = 10.62\,\Omega \tag{6.202}$$

The delta-to-wye converted impedances are,

$$Z_1 = \frac{5 \times (4 + j5.02)}{12 + j5.02} = 2.47\underline{|28.75^\circ}\,\Omega \tag{6.203}$$

$$Z_2 = \frac{3 \times (4 + j5.02)}{12 + j5.02} = 1.48\underline{|28.75^\circ}\,\Omega \tag{6.204}$$

$$Z_3 = \frac{5 \times 3}{12 + j5.02} = 1.15\underline{|-22.7^\circ}\,\Omega \tag{6.205}$$

The equivalent impedance of the circuit is calculated as,

$$\begin{aligned} Z_t &= 2 + 2.47\underline{|28.75^\circ} + \frac{(1.48\underline{|28.75^\circ} + 6)(1.15\underline{|-22.7^\circ} - j10.62)}{1.48\underline{|28.75^\circ} + 6 + 1.15\underline{|-22.7^\circ} - j10.62} \\ &= 9.73\underline{|-9.91^\circ}\,\Omega \end{aligned} \tag{6.206}$$

The source current is calculated as,

$$I = \frac{220\underline{|25^\circ}}{9.73\underline{|-9.91^\circ}} = 22.61\underline{|34.91^\circ}\,\text{A} \tag{6.207}$$

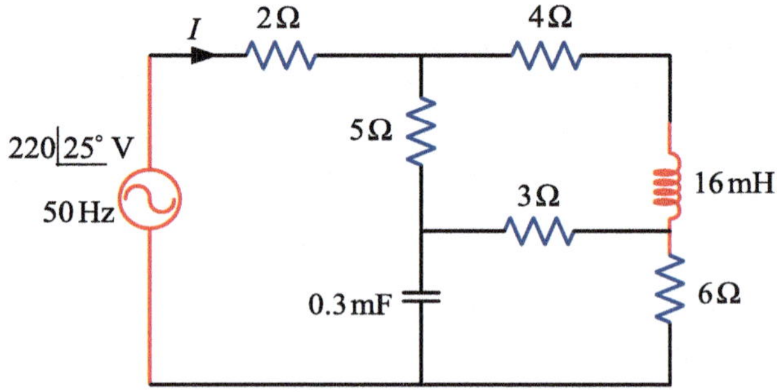

Fig. 6.34 Circuit for Example 6.13

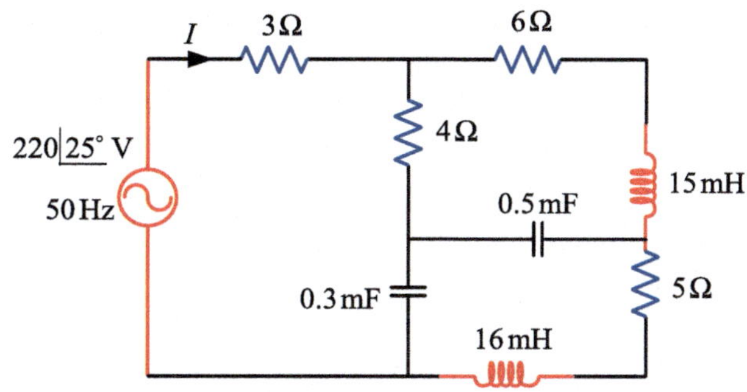

Fig. 6.35 Circuit for Practice Problem 6.13

Practice Problem 6.13
Determine the equivalent impedance and the source current of the circuit as shown in Fig. 6.35.

Exercise Problems

6.1 A sinusoidal current is given by $i(t) = 10\sin 314t$ A. Calculate the maximum value, frequency, time period and the instantaneous value of the current at $t = 1.02$ ms.

6.2 The expression of an alternating voltage is given by $v(t) = 25\sin 377t$ V. Determine the maximum value, frequency, time period of $v(t)$ and the instantaneous value of the voltage at $t = 0.5$ ms.

6.3. A complex voltage waveform is given by $v(t) = 3 + \sin\omega t + 7\sin 3\omega t$ V. Find the rms value of this voltage waveform, and the power absorbed by a 3 Ω resistor when this voltage is applied across this resistor.

6.4. A complex current waveform is given by $i = 5 + 3\sin\phi$ A. Find the rms value of this current waveform.

6.5 Determine the average value, rms value, form factor and peak factor of the voltage waveform shown in Fig. 6.36. Also, calculate the average power absorbed by the 5 Ω resistor, when this current flows through this resistor.

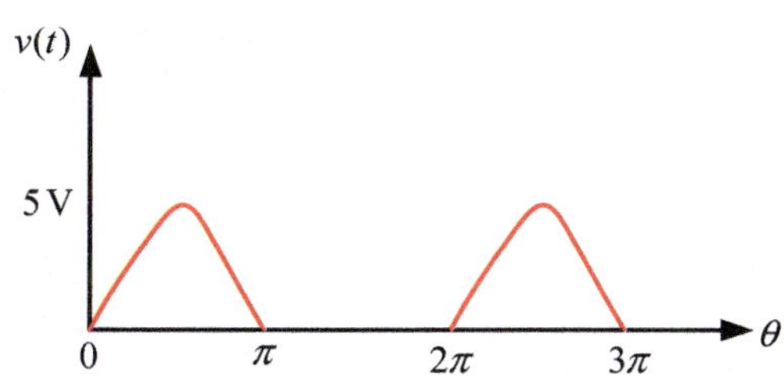

Fig. 6.36 Circuit for Problem 6.5

6.6 Find the average value, rms value, form factor and peak factor of the voltage waveform shown in Fig. 6.37. Also, find average power dissipated by the 9 Ω resistor, when the said voltage waveform is applied across this resistor .

6.7 Determine the average value and rms value of the voltage waveform shown in Fig. 6.38.

6.8 Find the average value and the rms value of the voltage waveform shown in Fig. 6.39.

6.9 Determine the average value and the rms value of the voltage waveform shown in Fig. 6.40.

6.10 For the voltage waveform shown in Fig. 6.41, calculate the average value, rms value and the average power absorbed by a 4 Ω resistor when this voltage is applied across this resistor.

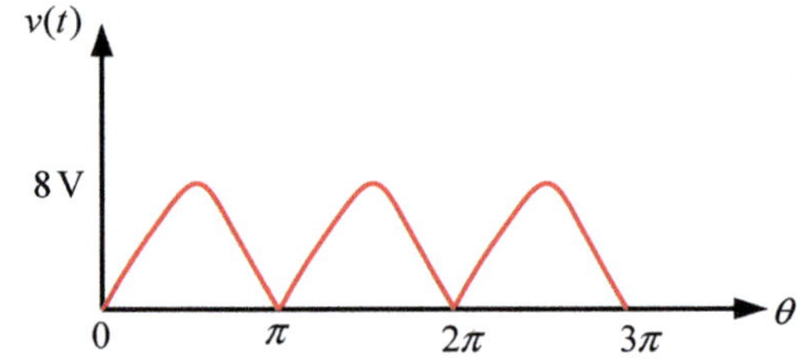

Fig. 6.37 Circuit for Problem 6.6

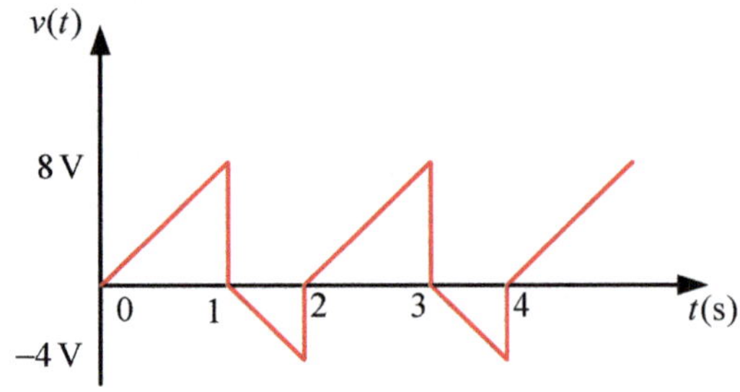

Fig. 6.38 Circuit for Problem 6.7

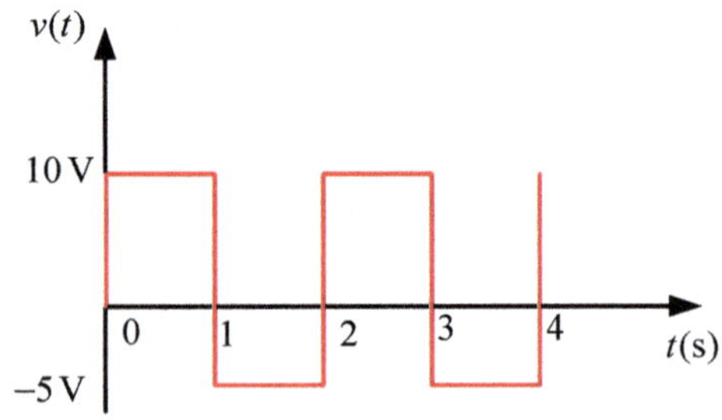

Fig. 6.39 Circuit for Problem 6.8

6.11 For the current waveform shown in Fig. 6.7, determine the average value, rms value and the average power absorbed by a 5 Ω resistor when this current flow through this resistor (Fig. 6.42).

6.12 For the voltage waveform shown in Fig. 6.43, find the average value, rms value and the average power absorbed by an 8 Ω resistor when this voltage is applied across this resistor.

6.13 Find the average value and the rms value of the current waveform shown in Fig. 6.44.

6.14 Determine the rms and average values of the current waveform shown in Fig. 6.45.

6.15 Calculate the rms and average values of the voltage waveform as shown in Fig. 6.11

6.16 Three phasors are given in the forms of $A = 10 - j15$, $B = 8 + j9$ and $C = 3 - j5$, respectively. Find $M = \frac{A+B}{C}$ and $N = AB - C$.

Fig. 6.40 Circuit for Problem 6.9

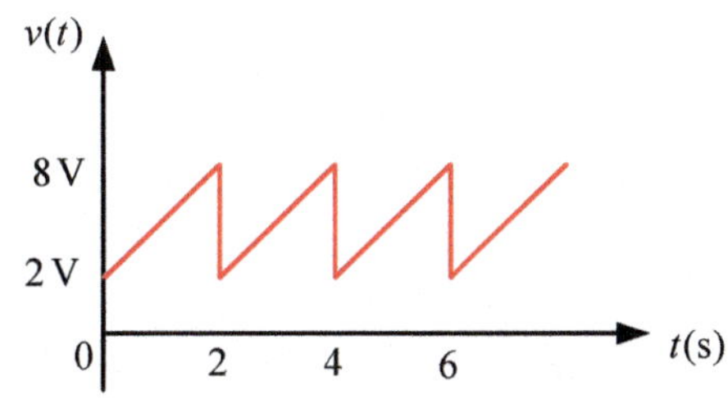

Fig. 6.41 Circuit for Problem 6.10

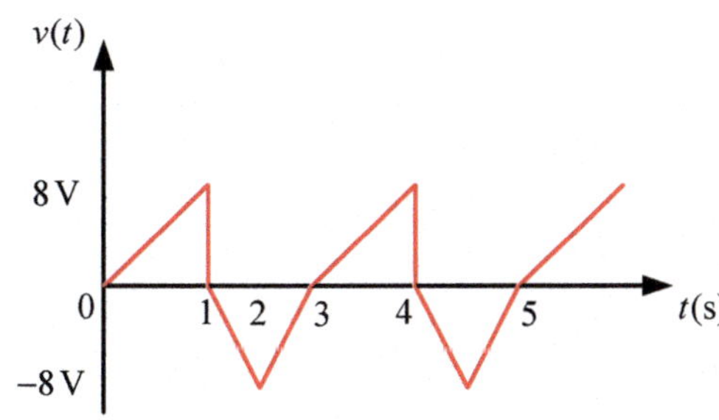

Fig. 6.42 Circuit for Problem 6.11

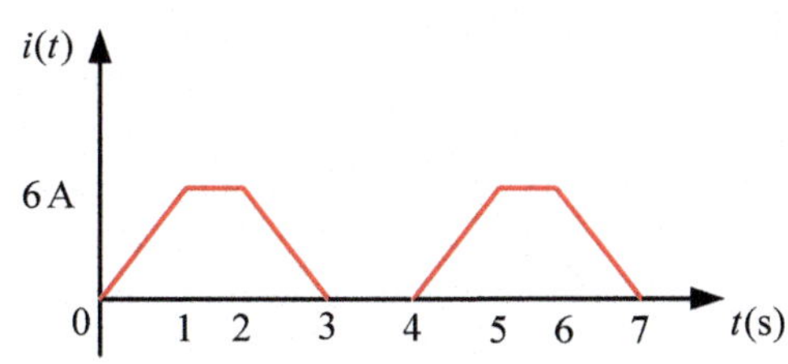

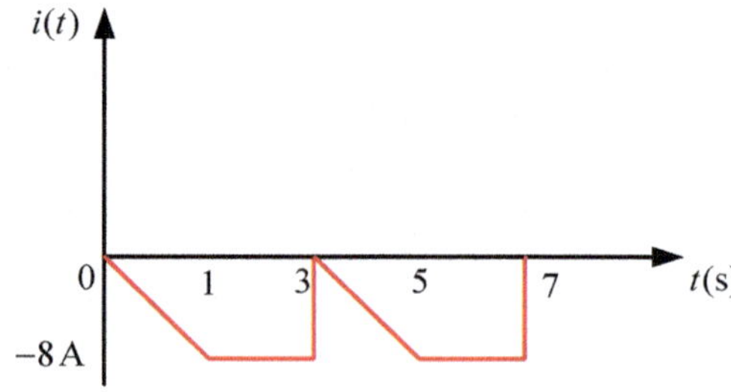

Fig. 6.43 Circuit for Problem 6.12

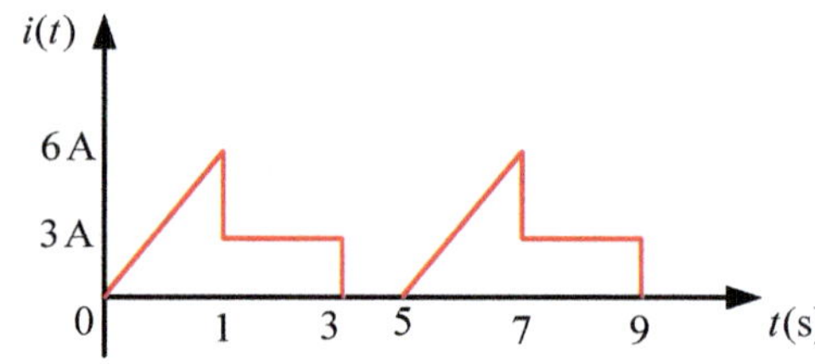

Fig. 6.44 Circuit for Problem 6.13

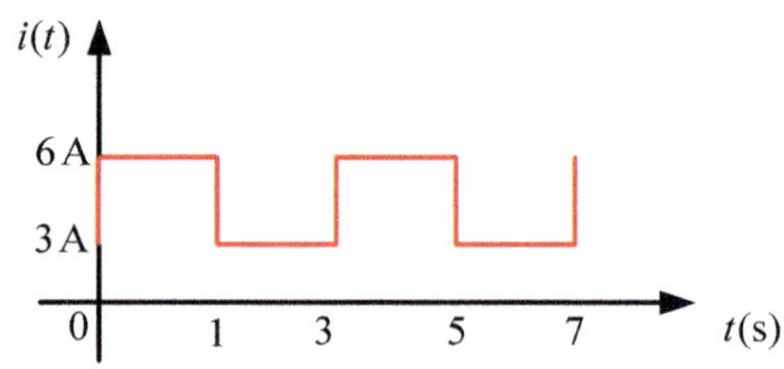

Fig. 6.45 Circuit for Problem 6.14

6.17 The current $i(t) = 12\sin(\omega t + 25°)$ A flows through a 4 Ω resistor. Calculate the voltage in phasor and sinusoidal forms.

6.18 Write down the mathematical expression for the current that flows in a 3 Ω resistor when the voltage across it is given in the following forms:

(i) $v(t) = 10\sin 500t$ V
(ii) $v(t) = 24\sin(\omega t + 15°)$ V
(iii) $v(t) = 10\cos(314t + 65°)$ V

6.19 The voltage across a 3 mH inductor is measured as $v(t) = 8\sin(314t + 12°)$ V. Determine the current in the inductor using phasor and draw the phasor diagram for the voltage and the associated current.

6.20 The voltage across a 2 H inductor is measured as $v(t) = 34\cos(377t + 31°)$ V. Calculate the current in the inductor both in phasor and sinusoidal forms.

6.21 The expression of voltage across a 3 μF capacitor is given as $v(t) = 9\sin(333t + 34°)$ V. Determine the current through the capacitor both in phasor and sinusoidal forms.

6.22 The voltage across a 0.95 μF capacitor is measured as $v(t) = 20\sin(377t - 23°)$ V. Find the current through the capacitor both in phasor and sinusoidal forms. Draw the phasor diagram for the voltage and the associated current.

6.23 A 305 μF capacitor is connected across a 230 V, 50 Hz AC supply. Determine the (i) capacitive reactance, (ii) rms value of the current and (iii) sinusoidal forms for voltage and current (Fig. 6.46).

6.24 An RL series circuit is shown in Fig. 6.47. Calculate the inductive reactance and the source current.

6.25 Figure 6.48 shows an RL series circuit. Find the frequency, source current, voltages across the resistance and the inductor. Consider that the voltage is $v(t) = 23\sin 10^4 t$ V.

6.26 The current in an RL series circuit in Fig. 6.49 is given by $i(t) = 16\sin(377t - 24°)$ A. Calculate the (i) circuit frequency, (ii) total

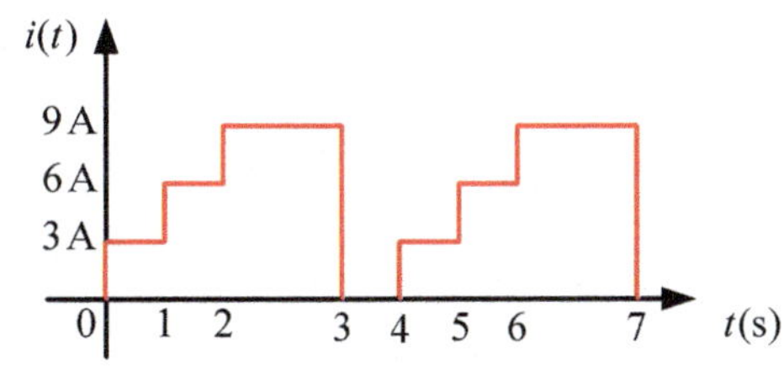

Fig. 6.46 Circuit for Problem 6.15

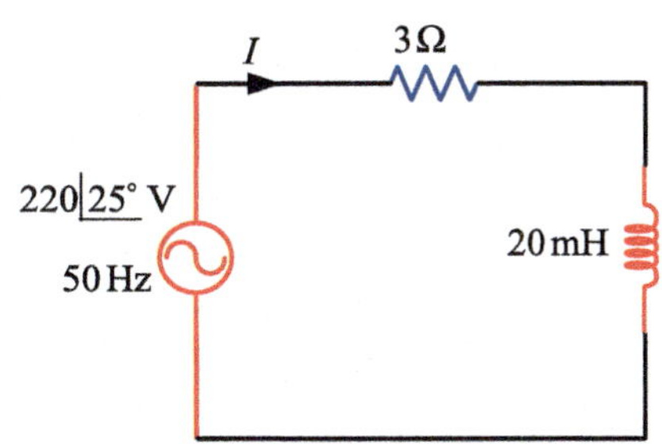

Fig. 6.47 Circuit for Problem 6.24

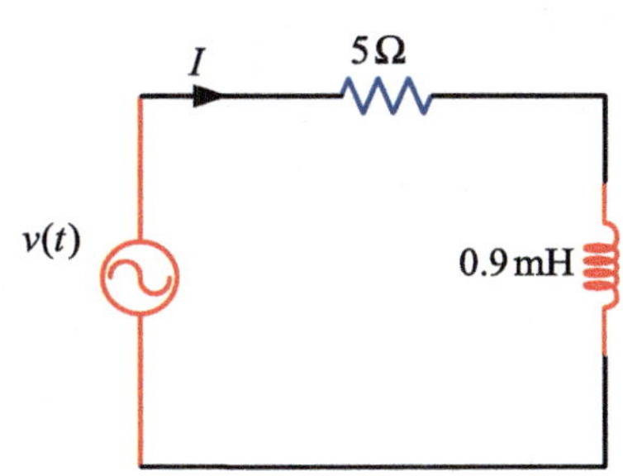

Fig. 6.48 Circuit for Problem 6.25

impedance and (iii) resistance and reactance of the circuit. Consider that the voltage is given by $v(t) = 250\sin 377t\,\mathrm{V}$.

6.27 Figure 6.50 shows an RC series circuit. Find the (i) capacitive reactance, (ii) impedance and (iii) source current.

6.28 An RLC series circuit is shown in Fig. 6.51. Calculate the circuit impedance and source current.

6.29 An RLC series circuit is shown in Fig. 6.52. Calculate the (i) total impedance, (ii) source current and (iii) voltages across the resistor, inductor and capacitor. Consider the voltage is given by $v(t) = 25\sin(377t + 48^\circ)\mathrm{V}$.

6.30 Figure 6.53 shows a series–parallel circuit. Calculate the source current and branch currents.

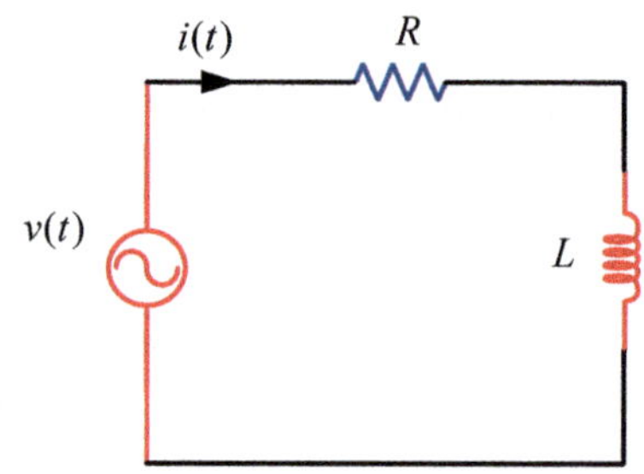

Fig. 6.49 Circuit for Problem 6.26

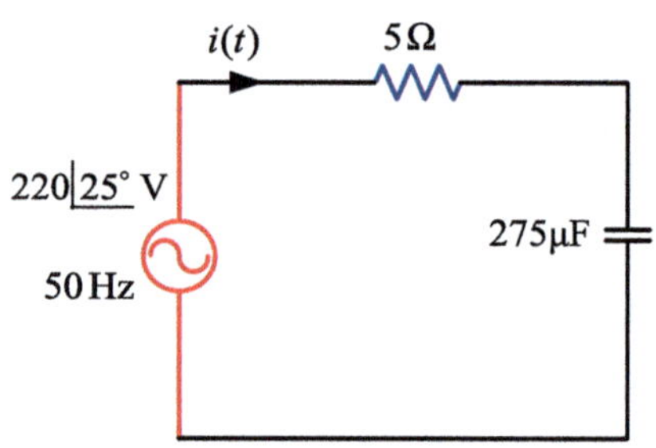

Fig. 6.50 Circuit for Problem 6.27

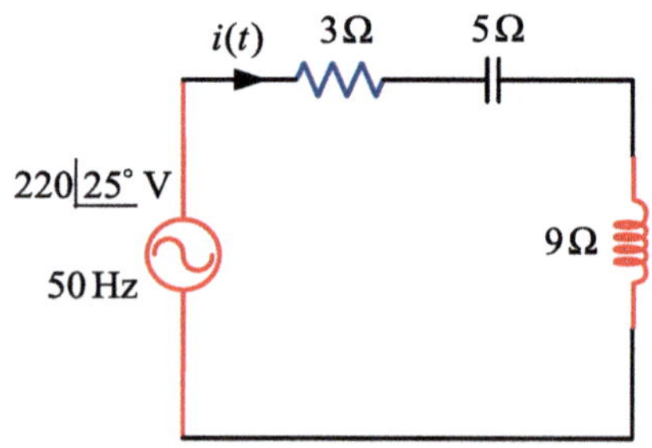

Fig. 6.51 Circuit for Problem 6.28

6.31 From the opening terminals, determine the equivalent impedance of the circuit as shown in Fig. 6.54.

6.32 Determine the equivalent impedance from the opening terminals of the circuit as shown in Fig. 6.55.

6.33 Find the circuit impedance and source current for the circuit shown in Fig. 6.56.

6.34 Calculate the equivalent impedance and source current for the circuit shown in Fig. 6.57.

6.35 Determine the equivalent impedance, admittance and source current in the circuit shown in Fig. 6.58.

Fig. 6.52 Circuit for Problem 6.29

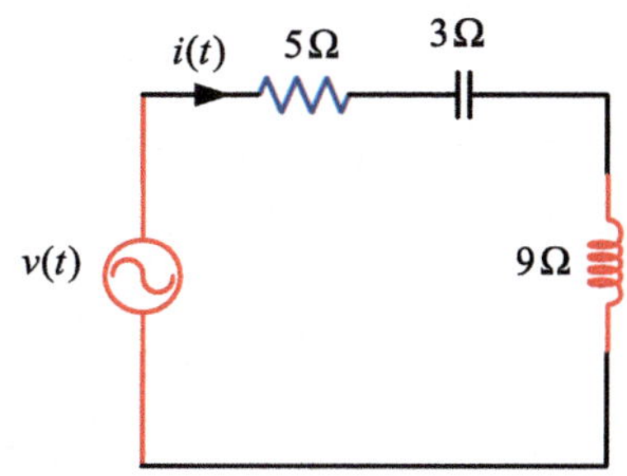

Fig. 6.53 Circuit for Problem 6.30

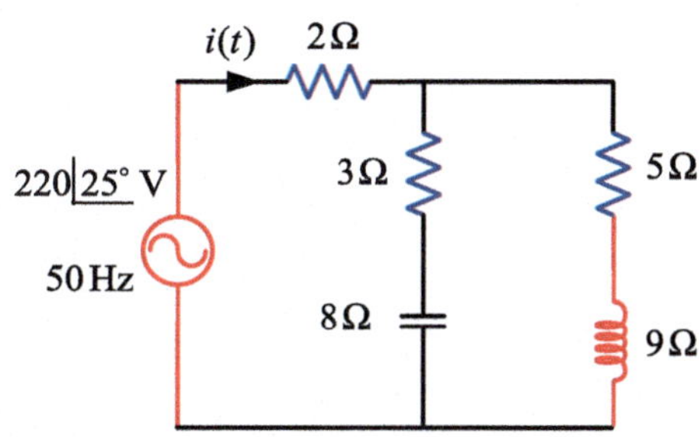

Fig. 6.54 Circuit for Problem 6.31

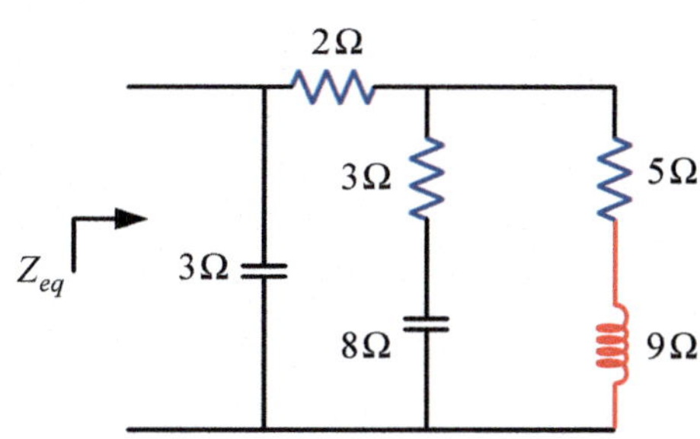

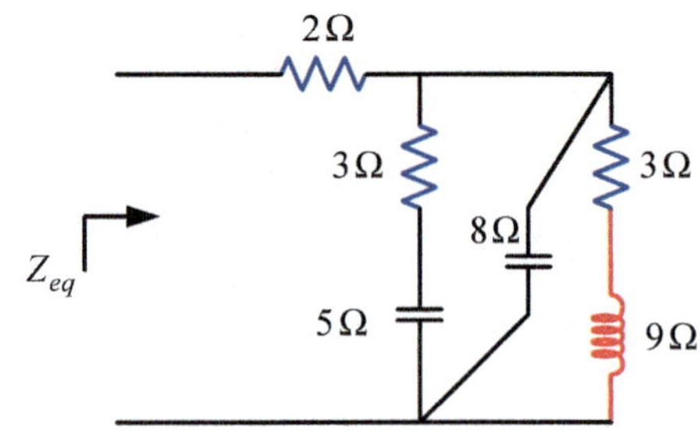

Fig. 6.55 Circuit for Problem 6.32

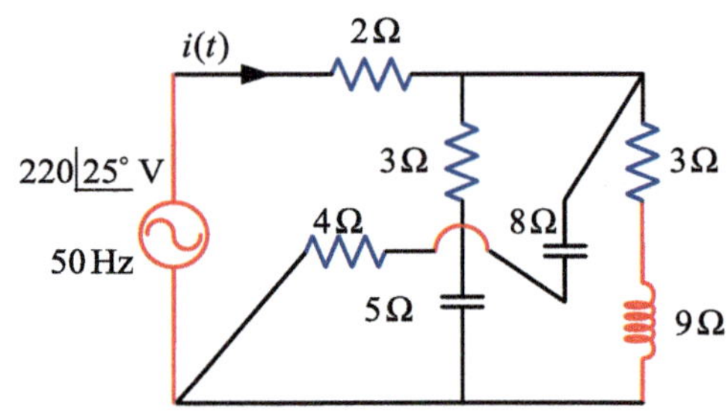

Fig. 6.56 Circuit for Problem 6.33

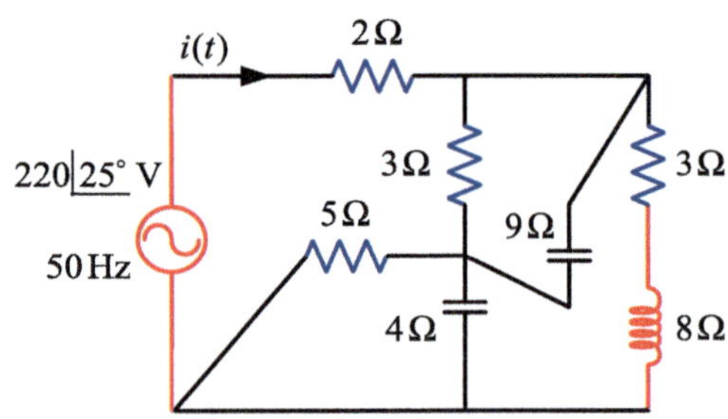

Fig. 6.57 Circuit for Problem 6.34

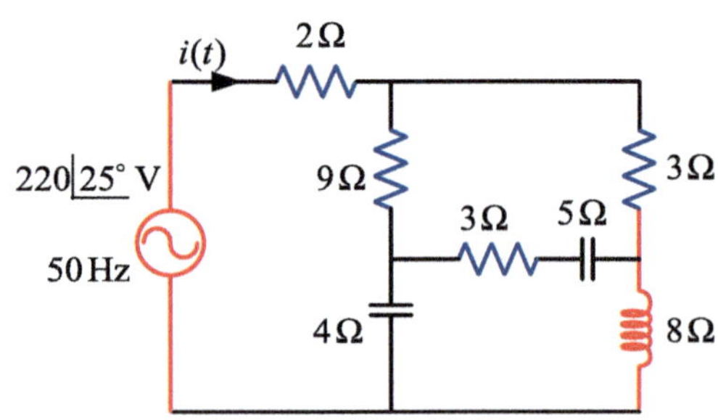

Fig. 6.58 Circuit for Problem 6.35

References

1. H.W. Jackson, D. Temple, B.E. Kelly, *Introduction to Electric Circuits*, 9th edn. (Oxford University Press, 2015)
2. D. Bell, *Fundamentals of Electric Circuits*, 7th edn. (Oxford University Press, 2009)
3. C.K. Alexander, M.N.O. Sadiku, *Fundamentals of Electric Circuits*, 6th edn. (McGraw-Hill Higher Education, 2016)
4. R.L. Boylestad, *Introductory Circuit Analysis*, 3th edn. (Pearson, 2016)
5. J. David Irwin, R. Mark Nelms, *Basic Engineering Circuit Analysis*, 11th edn. (Wiley, USA, 2015)
6. J.W. Nilsson, S.A. Riedel, *Electric Circuits*, 10th edn. (Prentice Hall International Edition, 2015)
7. Md. Abdus Salam, *Basic Electrical Circuits*, 2nd edn. (Shroff Publishers & Distributors Pvt. Ltd, India, 2007)

Chapter 7
AC Circuit Analysis

7.1 Introduction

Analysis of AC circuit plays an important role in designing and testing of power transmission networks, electrical and electronic equipment. In this case, sound knowledge of AC circuit analysis is very important. The AC circuit fundamentals and KVL, KCL and delta–wye conversion in the phasor domain have been discussed in Chap. 6. This chapter presents mesh and nodal analysis on AC circuit along with different network theorems, which have already been introduced in Chap. 4. These theorems include superposition, Thevenin's, Norton's and maximum power transfer theorems.

7.2 Nodal Analysis

In this section, the nodal analysis for an AC circuit with different impedances and current sources has been presented with the aid of the circuit shown in Fig. 7.1. Applying KCL to node 1 (labelled with node voltage V_1) of the circuit in Fig. 7.1 yields [1–3],

$$I_1 = I_x + I_y \tag{7.1}$$

Equation (7.1) can be modified as,

$$I_1 = \frac{V_1}{Z_1} + \frac{V_1 - V_2}{Z_2} \tag{7.2}$$

$$\left(\frac{1}{Z_1} + \frac{1}{Z_2}\right)V_1 - \frac{1}{Z_2}V_2 = I_1 \tag{7.3}$$

Md. A. Salam and Q. Mehbubar Rahman, *Fundamentals of Electrical Circuit Analysis*,
https://doi.org/10.1007/978-981-10-8624-3_7

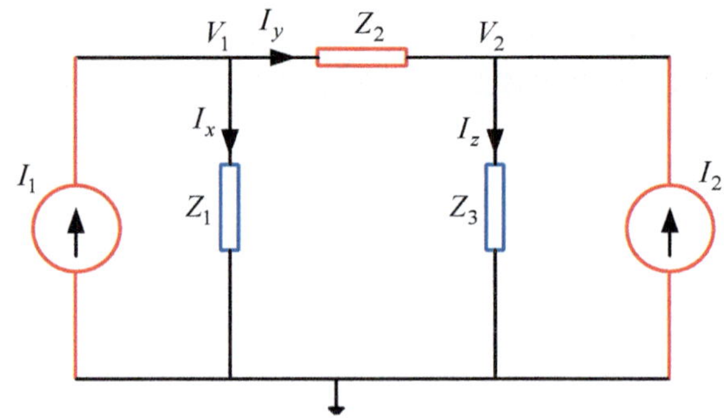

Fig. 7.1 Circuit for nodal analysis

Applying KCL to node 2 (labelled with node voltage V_2) of the circuit in Fig. 7.1 yields,

$$I_z = I_2 + I_y \tag{7.4}$$

$$\frac{V_2}{Z_3} = \frac{V_1 - V_2}{Z_2} + I_2 \tag{7.5}$$

$$-\frac{1}{Z_2} V_1 + \left(\frac{1}{Z_2} + \frac{1}{Z_3} \right) V_2 = I_2 \tag{7.6}$$

The two node voltages V_1 and V_2 can be determined from Eqs. (7.3) and (7.6) if other necessary parameters are given.

Figure 7.2 shows an AC circuit where a voltage source V_s is connected between the two non-reference nodes 2 and 3 (labelled with node voltage V_2 and V_3). Applying KCL to node 1 (labelled with node voltage V_1) of the circuit in Fig. 7.2 yields,

$$I_{s1} = I_x + I_y \tag{7.7}$$

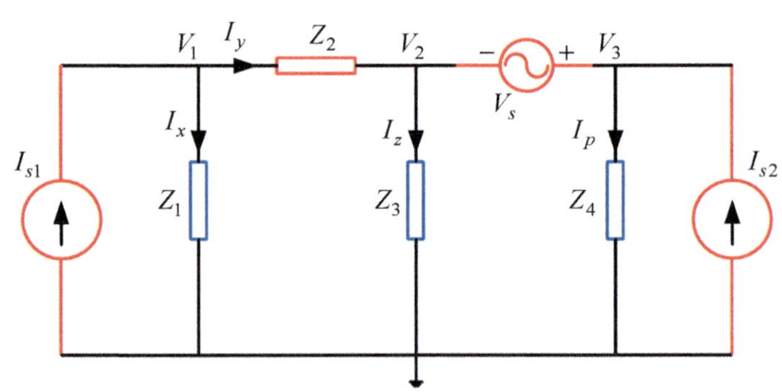

Fig. 7.2 Circuit for nodal analysis with supernode

$$I_{s1} = \frac{V_1}{Z_1} + \frac{V_1 - V_2}{Z_2} \tag{7.8}$$

$$\left(\frac{1}{Z_1} + \frac{1}{Z_2}\right)V_1 - \frac{1}{Z_2}V_2 = I_{s1} \tag{7.9}$$

Applying KCL to the supernode of the circuit in Fig. 7.2 yields,

$$I_y + I_{s2} = I_z + I_p \tag{7.10}$$

$$\frac{V_1 - V_2}{Z_2} + I_{s2} = \frac{V_2}{Z_3} + \frac{V_3}{Z_4} \tag{7.11}$$

$$\frac{1}{Z_2}V_1 - \left(\frac{1}{Z_2} + \frac{1}{Z_3}\right)V_2 - \frac{1}{Z_4}V_3 = -I_{s2} \tag{7.12}$$

Applying KVL to the supernode of the circuit in Fig. 7.2 yields,

$$-V_2 - V_s + V_3 = 0 \tag{7.13}$$

$$0V_1 - V_2 + V_3 = V_s \tag{7.14}$$

The unknown node voltages can be determined from Eqs. (7.9), (7.12) and (7.14) if other necessary parameters are given.

Example 7.1 Determine the non-reference node voltages and the voltage across the inductor with an inductive reactance of 6 Ω for the circuit shown in Fig. 7.3.

Solution:
Applying KVL to node 1 of the circuit in Fig. 7.3 yields,

$$V_1 = 24\angle 15^\circ \text{ V} \tag{7.15}$$

Fig. 7.3 Circuit for Example 7.1

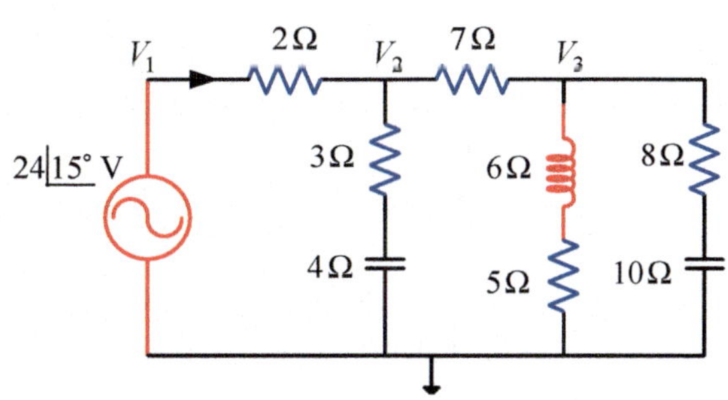

Applying KVL to node 2 of the circuit in Fig. 7.3 yields,

$$\frac{V_1 - V_2}{2} = \frac{V_2}{3 - j4} + \frac{V_2 - V_3}{7} \tag{7.16}$$

Substituting Eq. (7.15) into Eq. (7.16) yields,

$$12\angle 15^\circ + \frac{1}{7} V_3 = \left(0.5 + \frac{1}{3 - j4} + \frac{1}{7}\right) V_2 \tag{7.17}$$

$$12\angle 15^\circ + \frac{1}{7} V_3 = 0.78\angle 11.85^\circ \, V_2 \tag{7.18}$$

$$V_2 = 15.38\angle 3.15^\circ + 0.18\angle -11.85^\circ \, V_3 \tag{7.19}$$

Applying KVL to node 3 of the circuit in Fig. 7.3 yields,

$$\frac{V_2 - V_3}{7} = \frac{V_3}{5 + j6} + \frac{V_3}{8 - j10} \tag{7.20}$$

$$\frac{V_2}{7} = 0.28\angle -7.78^\circ \, V_3 \tag{7.21}$$

Substituting Eq. (7.19) into Eq. (7.21) yields,

$$\frac{15.38\angle 3.15^\circ + 0.18\angle -11.85^\circ V_3}{7} = 0.28\angle -7.78^\circ \, V_3 \tag{7.22}$$

$$1.78\angle -7.37^\circ \, V_3 = 15.38\angle 3.15^\circ \tag{7.23}$$

$$V_3 = \frac{15.38\angle 3.15^\circ}{1.78\angle -7.37^\circ} = 8.64\angle 10.52^\circ \text{ V} \tag{7.24}$$

Substituting Eq. (7.24) into Eq. (7.21) yields,

$$V_2 = 7 \times 0.28\angle -7.78^\circ \times 8.64\angle 10.52^\circ = 16.93\angle 2.74^\circ \text{ V} \tag{7.25}$$

The voltage across the inductor can be calculated as,

$$V_{6\Omega} = \frac{8.64\angle 10.52^\circ}{5 + j6} \times 6\angle 90^\circ = 6.64\angle 50.33^\circ \tag{7.26}$$

Example 7.2 Calculate the non-references node voltages and the voltage across the capacitor with a capacitive reactance of 8 Ω for the circuit shown in Fig. 7.4.

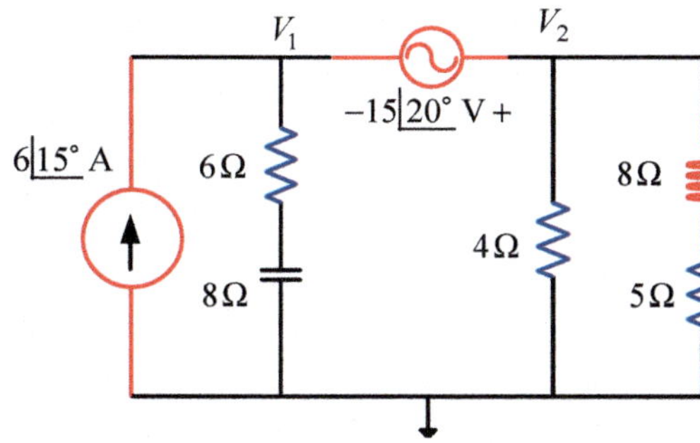

Fig. 7.4 Circuit for Example 7.2

Solution:
Applying KVL to the supernode of the circuit in Fig. 7.4 yields,

$$-V_1 - 15\angle 20^\circ + V_2 = 0 \tag{7.27}$$

$$V_1 = V_2 - 15\angle 20^\circ \tag{7.28}$$

Applying KCL to the supernode of the circuit in Fig. 7.4 yields,

$$\frac{V_2}{4} + \frac{V_2}{5 + j8} + \frac{V_1}{6 - j8} = 6\angle 15^\circ \tag{7.29}$$

$$0.32\angle -16.36^\circ\, V_2 + \frac{V_1}{6 - j8} = 6\angle 15^\circ \tag{7.29}$$

Substituting Eq. (7.28) into Eq. (7.29) yields,

$$0.32\angle -16.36^\circ\, V_2 + \frac{V_2 - 15\angle 20^\circ}{6 - j8} = 6\angle 15^\circ \tag{7.30}$$

$$V_2 = 18.82\angle 27.2^\circ \text{ V} \tag{7.31}$$

Substituting Eq. (7.31) into Eq. (7.28) yields,

$$V_1 = 18.82\angle 27.2^\circ - 15\angle 20^\circ = 4.36\angle 52.72^\circ \text{ V} \tag{7.32}$$

The voltage across the capacitor can be calculated as,

$$V_{-j8\Omega} = \frac{4.36\angle 52.72^\circ}{6 - j8} \times 8\angle -90^\circ = 3.49\angle 15.85^\circ \text{ V} \tag{7.33}$$

Example 7.3 Find the non-reference node voltages and the voltage across the 4 Ω resistor for the circuit shown in Fig. 7.5.

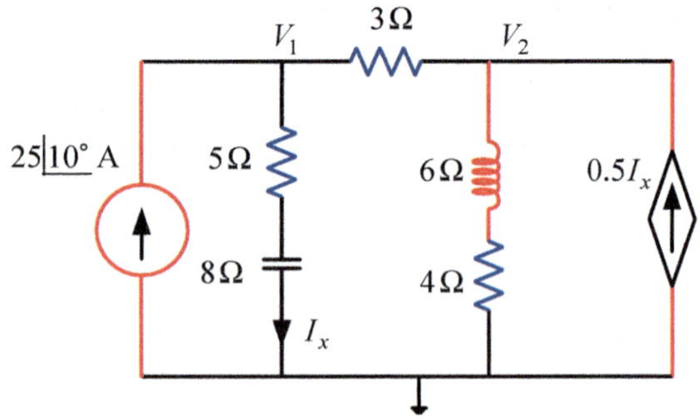

Fig. 7.5 Circuit for Example 7.3

Solution:
Applying KCL to node 1 yields,

$$25\angle 10^\circ = \frac{V_1}{5-j8} + \frac{V_1 - V_2}{3} \tag{7.34}$$

$$V_1 = 62.54\angle -3^\circ + 0.83\angle -13^\circ\, V_2 \tag{7.35}$$

Applying KCL to node 2 yields,

$$\frac{V_1 - V_2}{3} + 0.5I_x = \frac{V_2}{4+j6} \tag{7.36}$$

But, the expression of current is,

$$I_x = \frac{V_1}{5-j8} \tag{7.37}$$

Substituting Eq. (7.37) into Eq. (7.36) yields,

$$\frac{V_1 - V_2}{3} + 0.5\frac{V_1}{5-j8} = \frac{V_2}{4+j6} \tag{7.38}$$

$$\left(\frac{1}{3} + \frac{0.5}{5-j8}\right)V_1 = \left(\frac{1}{3} + \frac{1}{4+j6}\right)V_2 \tag{7.39}$$

$$0.36\angle 7.09^\circ\, V_1 = 0.43\angle -15.71^\circ\, V_2 \tag{7.40}$$

Substituting Eq. (7.35) into Eq. (7.40) yields,

$$0.36\angle 7.09^\circ\,(62.54\angle -3^\circ + 0.83\angle -13^\circ\, V_2) = 0.43\angle -15.71^\circ\, V_2 \tag{7.41}$$

$$V_2 = 90.85\angle 40.36^\circ \text{ V} \tag{7.42}$$

Substituting Eq. (7.42) into Eq. (7.35) yields,

$$V_1 = 62.54\angle{-3^\circ} + 0.83\angle{-13^\circ} \times 90.85\angle{40.36^\circ} = 133.17\angle{13.63^\circ}\ \text{V} \tag{7.43}$$

The voltage drop across the 4 Ω resistor is calculated as,

$$V_{4\Omega} = \frac{90.85\angle{40.36^\circ}}{4 + j6} \times 4 = 50.39\angle{-15.85^\circ}\ \text{V} \tag{7.44}$$

Practice Problem 7.1
Calculate the unknown node voltages and the voltage across the 5 Ω resistor for the circuit shown in Fig. 7.6.

Practice Problem 7.2
Find unknown node voltages and the voltage across the inductor with an inductive reactance of 8 Ω for the circuit shown in Fig. 7.7.

Practice Problem 7.3
Determine the non-references node voltages for the circuit shown in Fig. 7.8.

7.3 Mesh Analysis

In mesh analysis, unknown currents are usually determined from simultaneous equations. In this section, the mesh analysis for an AC circuit with different impedances and current sources has been presented with the aid of the circuit shown in Fig. 7.7. Applying KVL to mesh 1 (labelled by the mesh current I_1) of the circuit in Fig. 7.9 yields [4–6],

$$(Z_1 + Z_2)I_1 - Z_2I_2 - Z_1I_3 = V_1 \tag{7.45}$$

Applying KVL to mesh 2 (labelled by the mesh current I_2) of the circuit in Fig. 7.9 yields,

$$-Z_2I_1 + (Z_2 + Z_3)I_2 - Z_3I_3 = -V_2 \tag{7.46}$$

Fig. 7.6 Circuit for Practice Problem 7.1

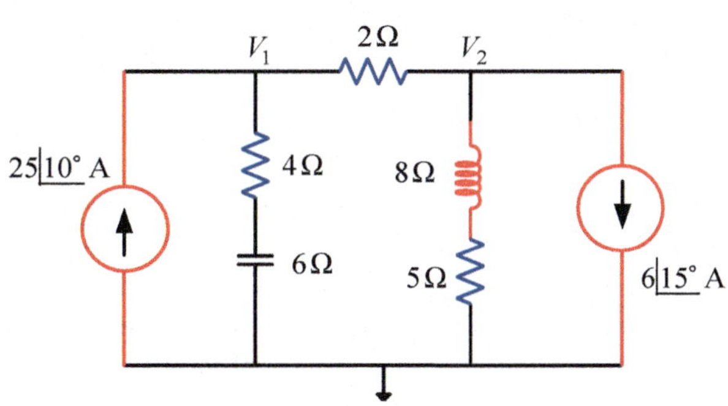

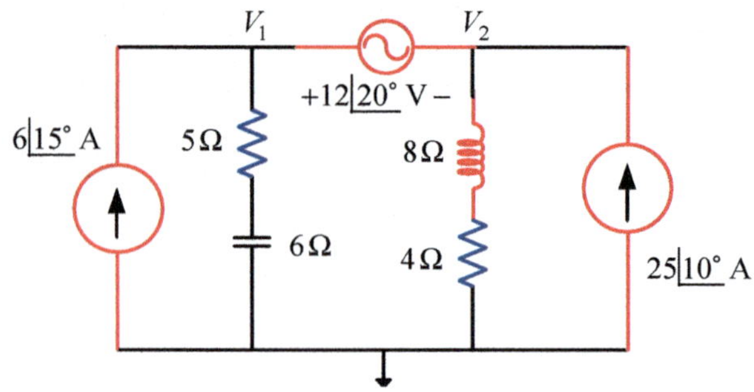

Fig. 7.7 Circuit for Practice Problem 7.2

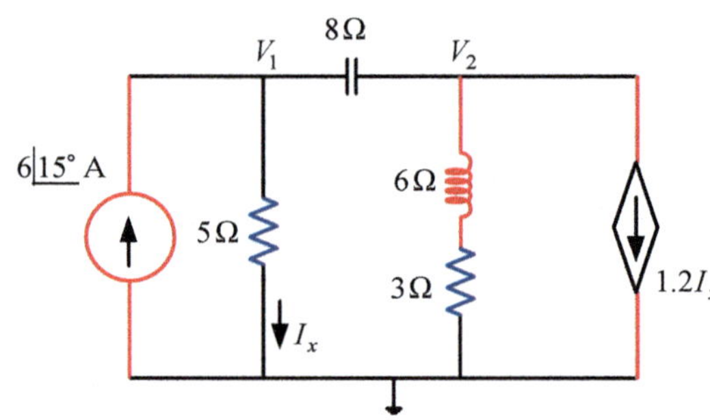

Fig. 7.8 Circuit for Practice Problem 7.3

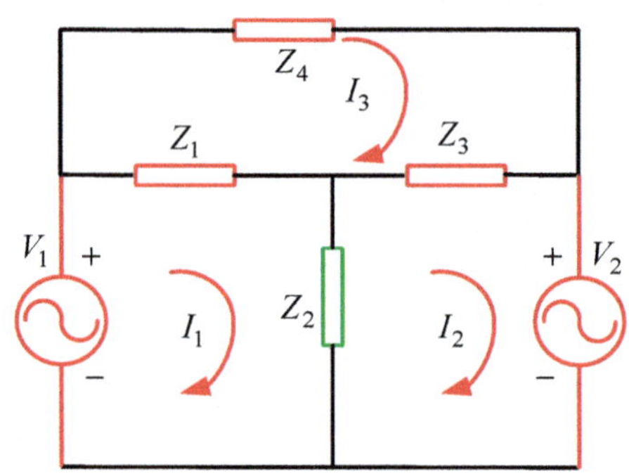

Fig. 7.9 AC circuit for mesh analysis

Applying KVL to mesh 3 (labelled by the mesh current I_3) of the circuit in Fig. 7.9 yields,

$$-Z_1 I_1 - Z_3 I_2 - (Z_1 + Z_3 + Z_4) I_3 = 0 \tag{7.47}$$

The unknown currents can be determined from Eqs. (7.45), (7.46) and (7.47) if other related parameters are known.

Figure 7.10 shows a circuit where the current source I_{s1} is common for both mesh 1 and mesh 2, which represents the supermesh. Applying KVL to the supermesh of the circuit in Fig. 7.10 yields,

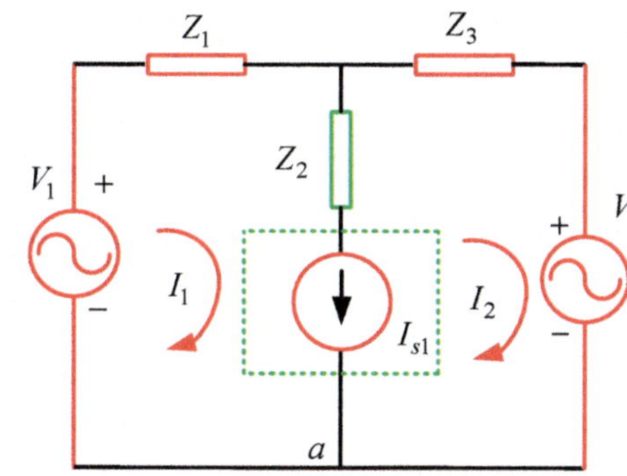

Fig. 7.10 AC circuit with supermesh

$$Z_1I_1 + Z_3I_2 = V_1 - V_2 \tag{7.48}$$

Applying KCL at node *a* of the circuit in Fig. 7.10 yields,

$$I_{s1} + I_2 = I_1 \tag{7.49}$$

The unknown currents can be determined from Eqs. (7.48) and (7.49) if the related parameters are given.

Example 7.4 Determine the mesh currents and the voltage drop across the capacitor with a capacitive reactance of 5 Ω for the circuit shown in Fig. 7.11.

Solution:
Applying KVL to mesh 1 of the circuit in Fig. 7.11 yields,

$$(6 + j8)I_1 - (4 + j8)I_2 = 24\underline{|15^\circ} \tag{7.50}$$

Applying KVL to supermesh of the circuit in Fig. 7.11 yields,

$$-(4 + j8)I_1 + (5 + j8)I_2 + (3 - j5)I_3 = -10 \tag{7.51}$$

Applying KCL to node *a* of the circuit in Fig. 7.11 yields,

$$I_2 = I_3 + 4 \tag{7.52}$$

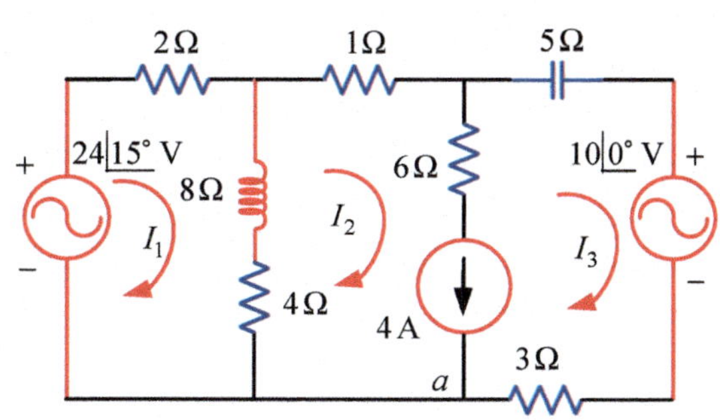

Fig. 7.11 AC circuit for Example 7.4

Substituting Eq. (7.52) into Eq. (7.50) yields,

$$(6+j8)I_1 - (4+j8)(I_3+4) = 24\angle 15^\circ \tag{7.53}$$

$$(6+j8)I_1 = (4+j8)I_3 + (16+j32) + 24\angle 15^\circ \tag{7.54}$$

$$I_1 = \frac{(4+j8)}{(6+j8)}I_3 + \frac{(16+j32)}{(6+j8)} + \frac{24\angle 15^\circ}{(6+j8)} \tag{7.55}$$

Substituting Eq. (7.55) and (7.52) into Eq. (7.51) yields,

$$-(4+j8)\left[\frac{(4+j8)}{(6+j8)}I_3 + \frac{(16+j32)}{(6+j8)} + \frac{24\angle 15^\circ}{(6+j8)}\right] + (5+j8)(I_3+4) + (3-j5)I_3 = -10 \tag{7.56}$$

$$\left[(5+j8) - (4+j8)\frac{(4+j8)}{(6+j8)} + (3-j5)\right]I_3 = (4+j8)\left[\frac{(16+j32)}{(6+j8)} + \frac{24\angle 15^\circ}{(6+j8)}\right] - (20+j32) - 10 \tag{7.57}$$

$$7.42\angle -39.09^\circ I_3 = 8.06\angle 101.69^\circ \tag{7.58}$$

$$I_3 = 1.09\angle 140.78^\circ \text{ A} \tag{7.60}$$

Substituting Eq. (7.60) into Eq. (7.52) yields,

$$I_2 = 4 + 1.09\angle 140.78^\circ = 3.23\angle 12.32^\circ \text{ A} \tag{7.61}$$

Substituting Eq. (7.61) into Eq. (7.50) yields,

$$(6+j8)I_1 - (4+j8)3.23\angle 12.32^\circ = 24\angle 15^\circ \tag{7.62}$$

$$I_1 = 4.57\angle -4.65^\circ \text{ A} \tag{7.63}$$

The voltage drop across the capacitor can be calculated as,

$$V_{-j5\Omega} = 1.09\angle 140.78^\circ \times 5\angle -90^\circ = 5.45\angle 50.78^\circ \text{ V} \tag{7.64}$$

Example 7.5 Calculate the mesh currents and the voltage drop across the capacitor with a capacitive reactance of 5 Ω for the circuit shown in Fig. 7.12.

Solution:
Applying KVL to mesh 1 of the circuit in Fig. 7.12 yields,

$$(6+j8)I_1 - (4+j8)I_2 = 10 \tag{7.65}$$

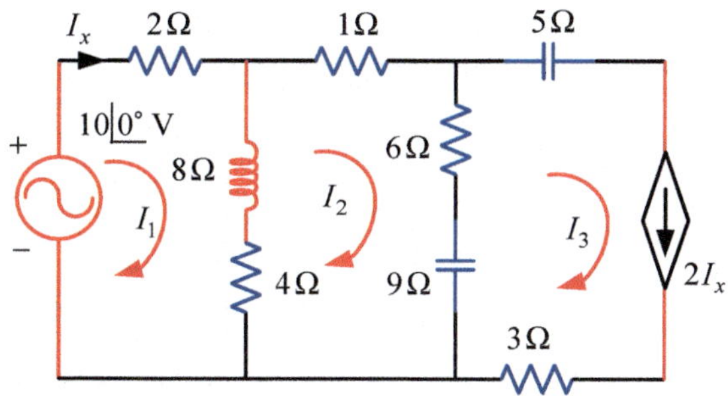

Fig. 7.12 AC circuit for Example 7.5

$$I_2 = \frac{(6+j8)}{(4+j8)}I_1 - \frac{10}{(4+j8)} \tag{7.66}$$

Applying KVL to mesh 2 of the circuit in Fig. 7.12 yields,

$$-(4+j8)I_1 + (11-j1)I_2 - (6-j9)I_3 = 0 \tag{7.67}$$

Applying KVL to mesh 3 of the circuit in Fig. 7.12 yields,

$$I_3 = 2I_x = 2I_1 \tag{7.68}$$

Substituting Eqs. (7.66) and (7.68) into Eq. (7.67) yields,

$$-(4+j8)I_1 + (11-j1)\left[\frac{(6+j8)}{(4+j8)}I_1 - \frac{10}{(4+j8)}\right] - (6-j9)2I_1 = 0 \tag{7.69}$$

$$\left[-(4+j8) + (11-j1)\frac{(6+j8)}{(4+j8)} - (12-j18)\right]I_1 = (11-j1)\frac{10}{(4+j8)} \tag{7.70}$$

$$7.85\underline{|121.46^\circ}\, I_1 = 12.35\underline{|-68.63^\circ} \tag{7.71}$$

$$I_1 = 1.57\underline{|169.91^\circ}\text{ A} \tag{7.72}$$

Substituting Eq. (7.72) into Eqs. (7.66) and (7.68) yields,

$$I_2 = \frac{(6+j8)}{(4+j8)} \times 1.57\underline{|169.91^\circ} - \frac{10}{(4+j8)} = 2.68\underline{|143.08^\circ}\text{ A} \tag{7.73}$$

$$I_3 = 2 \times 1.57\underline{|169.91^\circ} = 3.14\underline{|169.91^\circ}\text{ A} \tag{7.74}$$

The voltage drop across the capacitor can be calculated as,

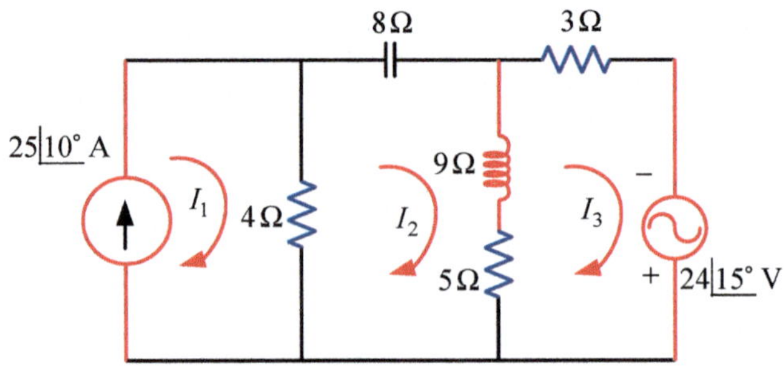

Fig. 7.13 AC circuit for Practice Problem 7.4

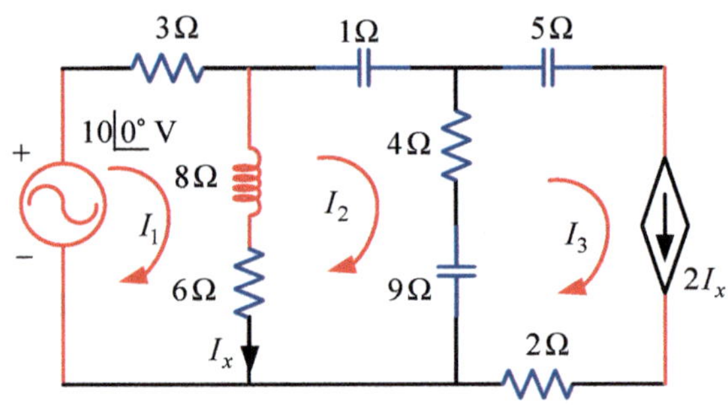

Fig. 7.14 AC circuit for Practice Problem 7.5

$$V_{-j5} = 5\angle -90^\circ \times 3.14\angle 169.91^\circ = 15.7\angle 79.91^\circ \text{ V} \tag{7.75}$$

Practice Problem 7.4
Calculate the mesh currents and the voltage drop across the inductor with an inductive reactance of 9 Ω for the circuit shown in Fig. 7.13.

Practice Problem 7.5
Determine the mesh currents and the voltage drop across the 6 Ω resistor for the circuit shown in Fig. 7.14.

7.4 Superposition Theorem

Here, superposition theorem in an AC circuit has been demonstrated with the aid of the circuit shown in Fig. 7.15, where the current I_x through the circuit component with an impedance of Z_2 needs to be determined.

Consider the voltage source is active and the current source is open circuited. From Fig. 7.16, the total circuit impedance can be determined as,

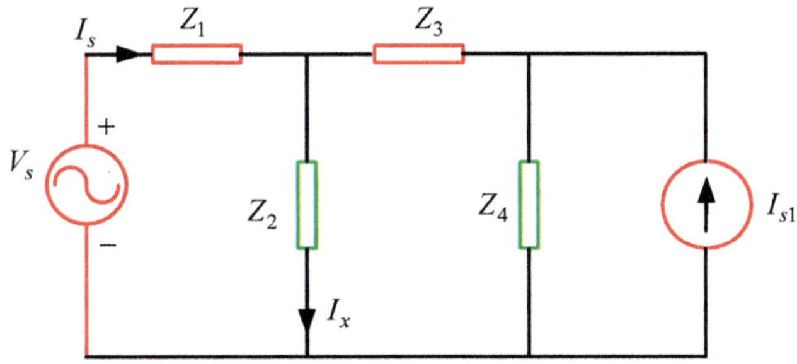

Fig. 7.15 AC circuit for superposition theorem

Fig. 7.16 Circuit with voltage source

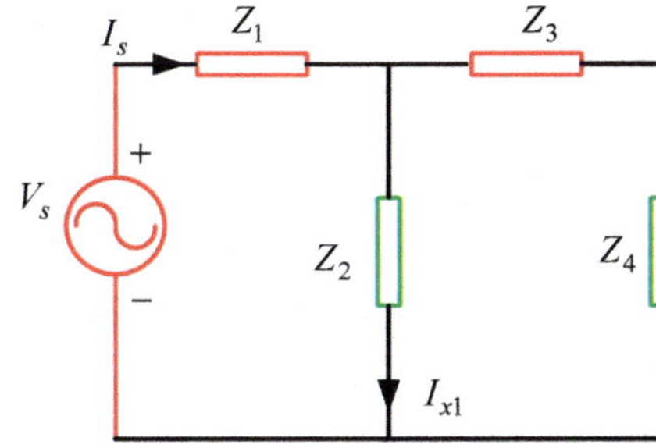

$$Z_t = Z_1 + \frac{Z_2(Z_3 + Z_4)}{Z_2 + Z_3 + Z_4} \tag{7.76}$$

The value of the source current is determined as,

$$I_s = \frac{V_s}{Z_t} \tag{7.77}$$

The current in Z_2 is calculated as,

Fig. 7.17 AC circuit with current source

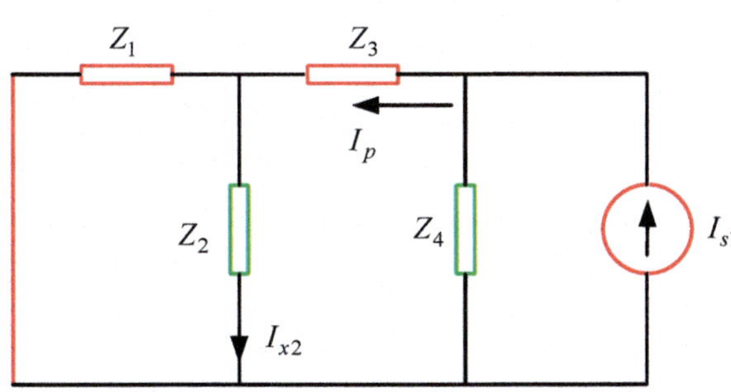

$$I_{x1} = I_s \times \frac{(Z_3 + Z_4)}{Z_2 + Z_3 + Z_4} \tag{7.78}$$

Again, consider the voltage source is short circuited and the current source is active as shown in Fig. 7.17. In this case, the equivalent impedance between Z_1 and Z_2 is,

$$Z_p = \frac{Z_1 Z_2}{Z_1 + Z_2} \tag{7.79}$$

The current I_p is calculated as,

$$I_p = I_{s1} \times \frac{Z_p}{Z_p + Z_4} \tag{7.80}$$

The current I_{x2} is calculated as,

$$I_{x2} = I_p \times \frac{Z_1}{Z_1 + Z_2} \tag{7.81}$$

Finally, the actual current in Z_2 is calculated as,

$$I_x = I_{x1} + I_{x2} \tag{7.82}$$

Example 7.6 Calculate the voltage drop across the 6 Ω resistor for the circuit shown in Fig. 7.18 by superposition theorem.

Solution:
Consider that the first voltage source (10 V) is in active condition and the second one is short circuited as shown in Fig. 7.17. In this case, the equivalent impedance Z_1 between 4Ω, $3 - j4\,\Omega$ and $-j4\,\Omega$ becomes,

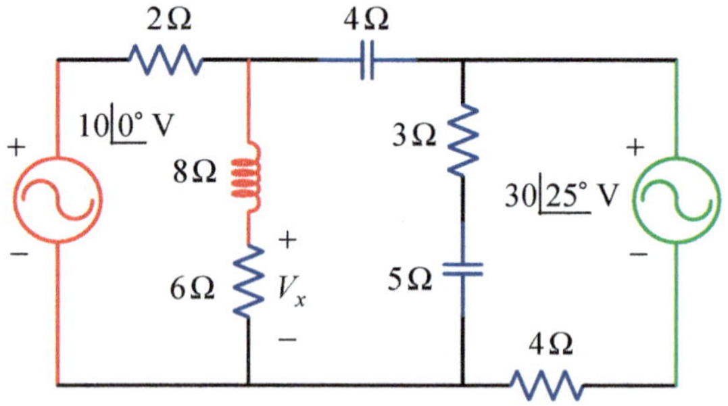

Fig. 7.18 Circuit for Example 7.6

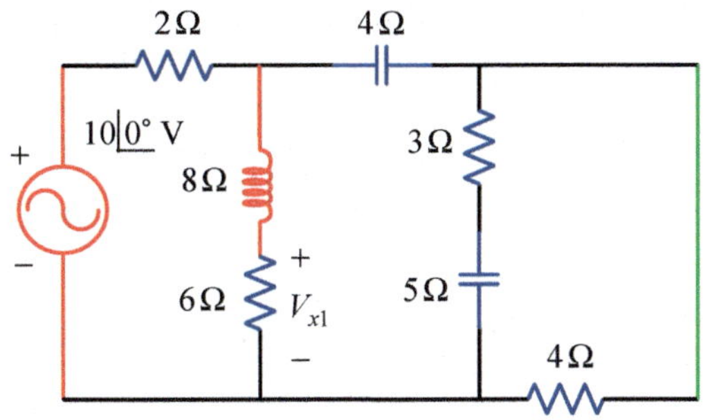

Fig. 7.19 Circuit for Example 7.6 with the first voltage source

$$Z_1 = \frac{4(3 - j5)}{4 + 3 - j5} - j4 = 5.66\angle{-63.92^\circ}\,\Omega \tag{7.83}$$

The overall equivalent circuit impedance is $Z_1 || (6 + j8) + 2$, and this can be calculated as,

$$Z_{\text{eq1}} = 2 + \frac{(6 + j8) \times 5.66\angle 63.92^\circ}{6 + j8 + 5.66\angle{-63.92^\circ}} = 6.31\angle{-29.75^\circ}\,\Omega \tag{7.84}$$

From Fig. 7.19, the source current is calculated as,

$$I_{s1} = \frac{10}{6.31\angle{-29.75^\circ}} = 1.58\angle{-29.75^\circ}\,\text{A} \tag{7.85}$$

The current in the $6\,\Omega$ resistor is,

$$I_{6\Omega} = 1.58\angle{-29.75^\circ} \times \frac{5.66\angle{-63.92^\circ}}{6 + j8 + 5.66\angle{-63.92^\circ}} = 1\angle{-112.63^\circ}\,\text{A} \tag{7.86}$$

The voltage drops across the $6\,\Omega$ resistor due to the first source is,

$$V_{x1} = 1\angle{-112.63^\circ} \times 6 = 6\angle{-112.63^\circ}\,\text{V} \tag{7.87}$$

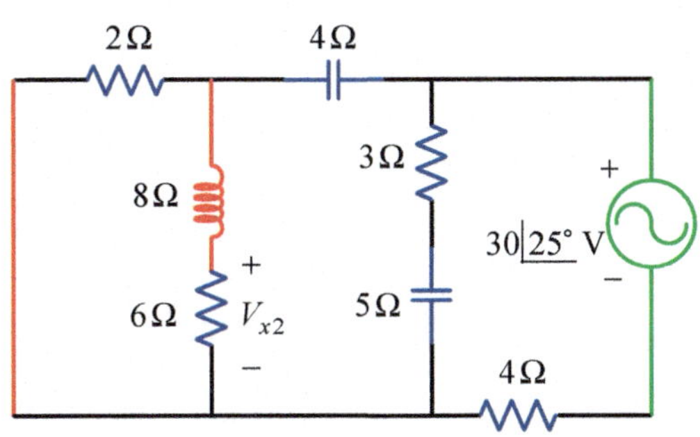

Fig. 7.20 Circuit for Example 7.6 with the second voltage source

Again, consider that the second voltage source (30 V) is active only as shown in the circuit in Fig. 7.20. In this case, the equivalent impedance Z_2 between 2Ω, $(6+j8)\,\Omega$ and $(-j4\,\Omega)$ becomes,

$$Z_2 = \frac{2(6+j8)}{2+6+j8} - j4 = 4.14\angle{-64.98^\circ}\,\Omega \tag{7.88}$$

The overall equivalent circuit impedance $Z_2\|(3-j5)+4$, and this can be calculated as,

$$Z_{\mathrm{eq2}} = 4 + \frac{(3-j5)\times 4.14\angle{-64.98^\circ}}{3-j5+4.14\angle{-64.98^\circ}} = 5.55\angle{-22.78^\circ}\,\Omega \tag{7.89}$$

The source current is,

$$I_{s2} = \frac{30\angle{25^\circ}}{5.55\angle{-22.78^\circ}} = 5.41\angle{47.78^\circ}\,\mathrm{A} \tag{7.90}$$

The branch current is,

$$I_2 = 5.41\angle{47.78^\circ} \times \frac{(3-j5)}{3-j5+4.14\angle{-64.98^\circ}} = 3.17\angle{50.25^\circ}\,\mathrm{A} \tag{7.91}$$

The current in the $6\,\Omega$ resistor is,

$$I_{6\Omega} = 3.17\angle{50.25^\circ} \times \frac{2}{2+6+j8} = 0.56\angle{5.25^\circ}\,\mathrm{A} \tag{7.92}$$

The voltage drops across the $6\,\Omega$ resistor due to the second voltage source is,

$$V_{x2} = 0.56\angle{5.25^\circ} \times 6 = 3.36\angle{5.25^\circ}\,\mathrm{V} \tag{7.93}$$

The actual voltage drops across the $6\,\Omega$ resistor is,

$$V_{6\Omega} = V_{x1} + V_{x2} = 3.36\angle{5.25^\circ} + 6\angle{-112.63^\circ} = 5.33\angle{-78.76^\circ}\,\mathrm{V} \tag{7.94}$$

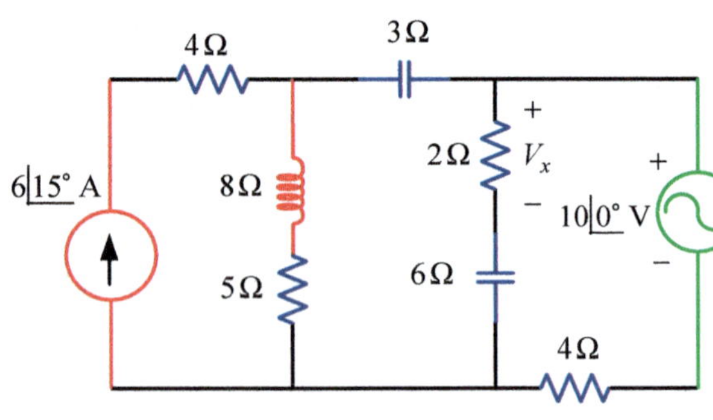

Fig. 7.21 Circuit for Practice Problem 7.6

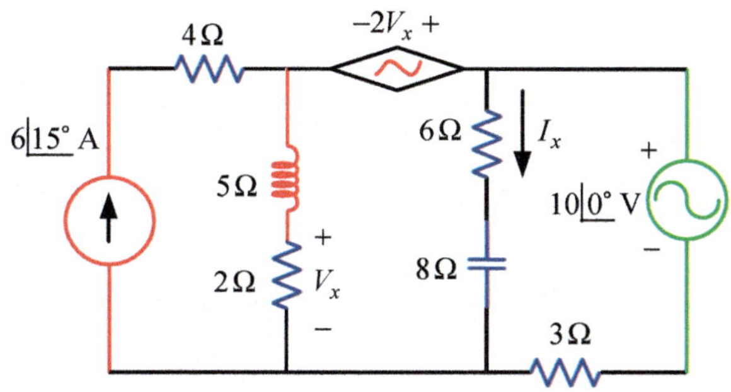

Fig. 7.22 Circuit for Example 7.7

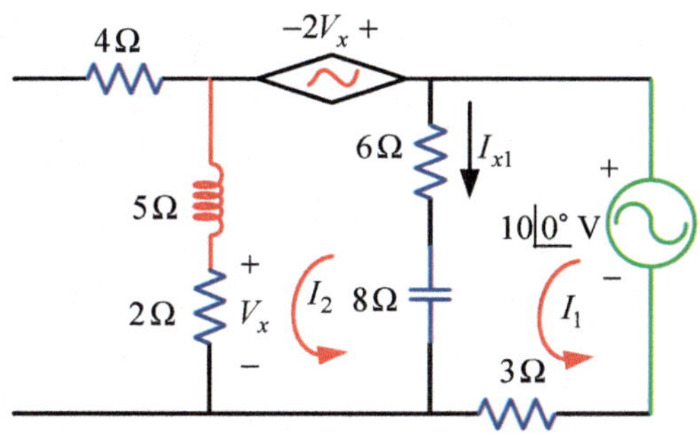

Fig. 7.23 Circuit for Example 7.7 with active voltage source

Practice Problem 7.6

Using superposition theorem, determine the voltage drop across the 2 Ω resistor for the circuit shown in Fig. 7.21.

Example 7.7 Using superposition theorem, calculate the current in the 6 Ω resistor for the circuit shown in Fig. 7.22.

Solution:

Consider that the voltage source is active and the current source is open circuited. Applying KVL to the meshes 1 and 2 of the circuit in Fig. 7.23 yields,

$$(9 - j8)I_1 - (6 - j8)I_2 = 10 \tag{7.95}$$

$$-(6 - j8)I_1 + (8 - j3)I_2 = -2V_x \tag{7.96}$$

However, the voltage drops across the 2 Ω resistor is,

$$V_x = 2I_2 \tag{7.97}$$

Substituting Eq. (7.97) into Eq. (7.96) yields,

$$-(6-j8)I_1+(8-j3)I_2=-4I_2 \tag{7.98}$$

$$(6-j8)I_1=(12-j3)I_2 \tag{7.99}$$

$$I_1=\frac{12-j3}{6-j8}I_2=1.24\angle 39.09^\circ\, I_2 \tag{7.100}$$

Substituting Eq. (7.100) into Eq. (7.95) yields,

$$(9-j8)\times 1.24\angle 39.09^\circ\, I_2-(6-j8)I_2=10 \tag{7.101}$$

$$I_2=\frac{10}{(9-j8)\times 1.24\angle 39.09^\circ-(6-j8)}=0.87\angle -39.45^\circ\ \text{A} \tag{7.102}$$

Substituting Eq. (7.102) into Eq. (7.100) yields,

$$I_1=1.24\angle 39.09^\circ\times 0.87\angle -39.45^\circ=1.08\angle -0.36^\circ\ \text{A} \tag{7.103}$$

In the first case, the current I_{x1} (see Fig. 7.23) is,

$$I_{x1}=I_1-I_2=1.08\angle -0.36^\circ-0.87\angle -39.45^\circ=0.68\angle 53.22^\circ\ \text{A} \tag{7.104}$$

Again, consider that the current source is active and the voltage source is short circuited as shown in Fig. 7.24. Applying KVL to the meshes 1, 2 and 3 of the circuit in Fig. 7.24 yields,

$$I_1=6\angle 15^\circ\ \text{A} \tag{7.105}$$

$$-(2+j5)I_1+(8-j3)I_2-(6-j8)I_3=2V_x \tag{7.106}$$

$$-(6-j8)I_2+(9-j8)I_3=0 \tag{7.107}$$

However, in Fig. 7.24, the voltage drops across the 2 Ω resistor is,

$$V_x=2(I_1-I_2) \tag{7.108}$$

Substituting Eq. (7.108) into Eq. (7.106) yields,

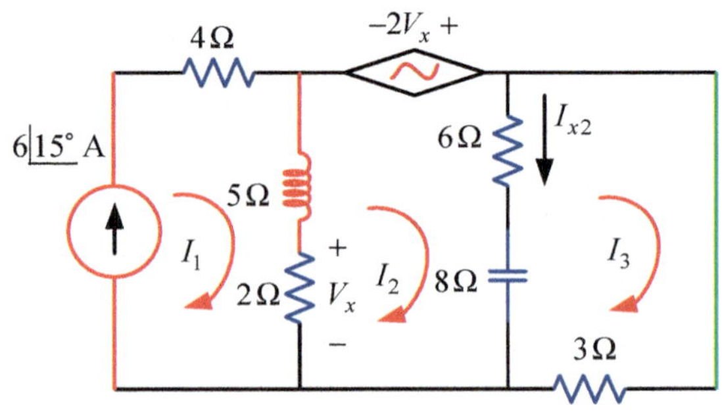

Fig. 7.24 Circuit for Example 7.7 with active current source

$$-(2+j5)I_1 + (8-j3)I_2 - (6-j8)I_3 = 4(I_1 - I_2) \tag{7.109}$$

$$-(6+j5)I_1 + (12-j3)I_2 = (6-j8)I_3 \tag{7.110}$$

Substituting Eq. (7.105) into Eq. (7.110) yields,

$$-(6+j5) \times 6\angle 15^\circ + (12-j3)I_2 = (6-j8)I_3 \tag{7.111}$$

Substituting the value of I_2 from Eq. (7.107) into Eq. (7.111) yields,

$$-(6+j5) \times 6\angle 15^\circ + (12-j3) \times \frac{(9-j8)}{(6-j8)} I_3 = (6-j8)I_3 \tag{7.112}$$

$$11.52\angle 39.58^\circ I_3 = (6+j5) \times 6\angle 15^\circ \tag{7.113}$$

$$I_3 = 4.07\angle 15.25^\circ \text{ A} \tag{7.114}$$

Substituting Eq. (7.114) into Eq. (7.107) yields,

$$-(6-j8)I_2 + (9-j8) \times 4.07\angle 15.23^\circ = 0 \tag{7.115}$$

$$I_2 = 4.9\angle 26.73^\circ \text{ A} \tag{7.116}$$

In the second case, the current I_{x2} (see Fig. 7.24) is,

$$I_{x2} = I_2 - I_3 = 4.9\angle 26.73^\circ - 4.07\angle 15.23^\circ = 1.22\angle 68.4^\circ \text{ A} \tag{7.117}$$

Finally, the current in the 6 Ω resistor is,

$$I_x = I_{x1} + I_{x2} = 0.68\angle 53.22^\circ + 1.22\angle 68.4^\circ = 1.88\angle 62.98^\circ \text{ A} \tag{7.118}$$

Practice Problem 7.7
Using superposition theorem in the circuit shown in Fig. 7.25, determine the current in the 5 Ω resistor.

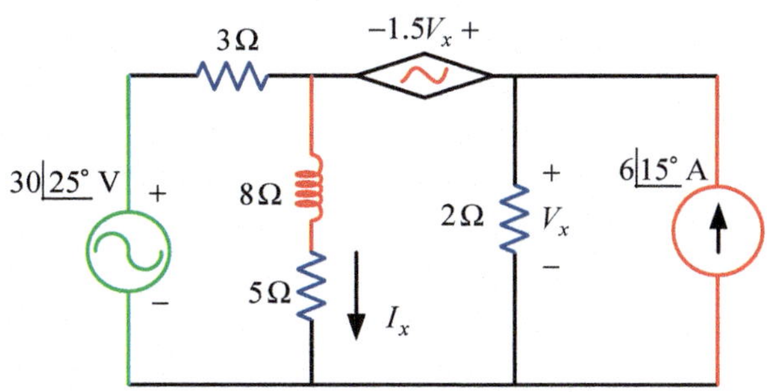

Fig. 7.25 Circuit for Practice Problem 7.7

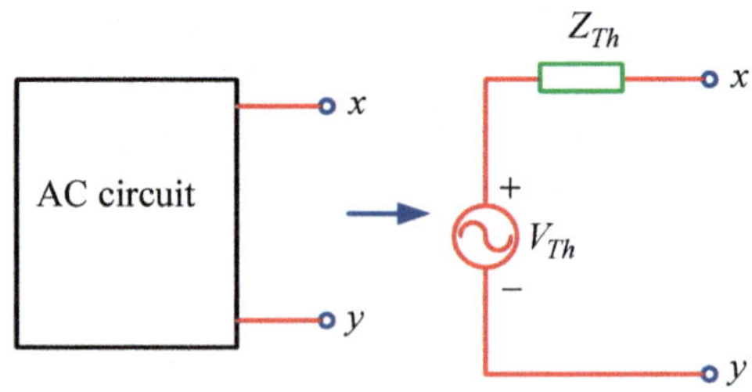

Fig. 7.26 AC circuit for Thevenin's theorem

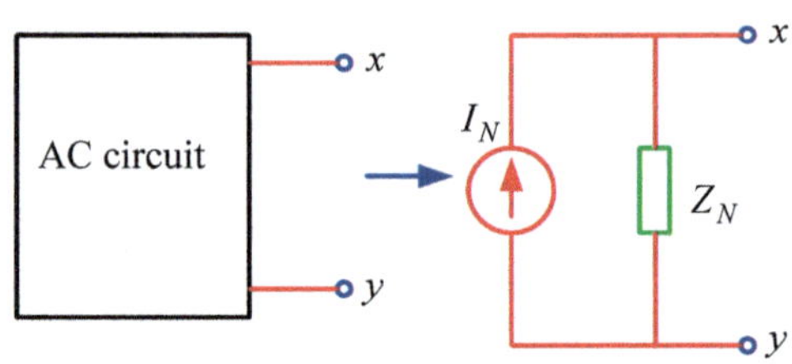

Fig. 7.27 AC circuit for Norton's theorem

7.5 Thevenin's and Norton's Theorems

Thevenin's theorem simplifies any two-terminal AC circuit [6, 7] consisting of sources, and reactive and resistance elements, by replacing it with a Thevenin's equivalent voltage (or open-circuit voltage) source V_{Th} in series with a Thevenin's equivalent impedance Z_{Th} as shown in Fig. 7.26.

In case of Norton's theorem, the whole AC circuit from a specific branch or element is replaced by a Norton's current source in parallel with a Norton impedance as shown in Fig. 7.27.

Example 7.8 Determine the current in the 9 Ω resistor for the circuit shown in Fig. 7.28 using Thevenin's and Norton's theorems.

Solution:
Open the 9 Ω resistor and calculate the Thevenin's impedance from the circuit in Fig. 7.29. This impedance is,

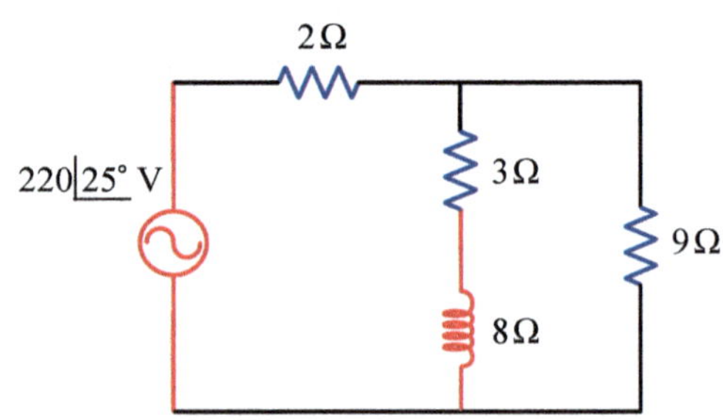

Fig. 7.28 Circuit for Example 7.8

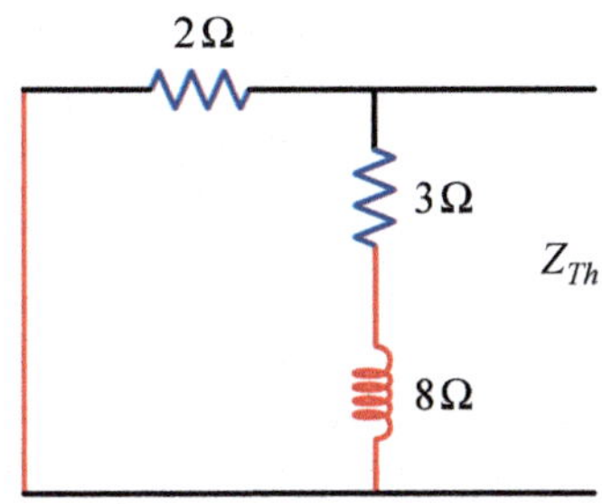

Fig. 7.29 Circuit for calculation Thevenin impedance for Example 7.8

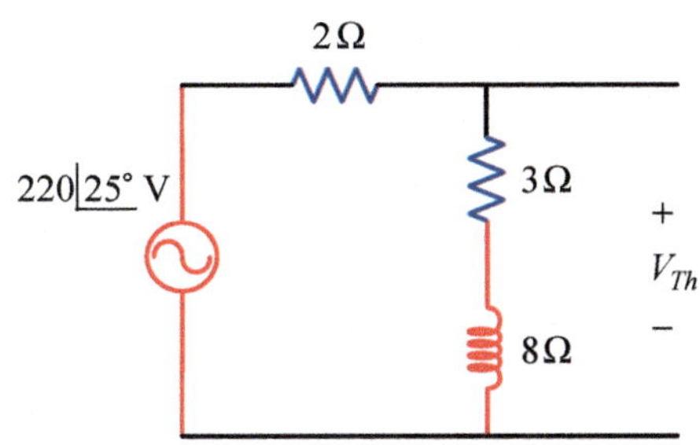

Fig. 7.30 Circuit for calculation Thevenin voltage for Example 7.8

$$Z_{\text{Th}} = \frac{2 \times (3 + j8)}{5 + j8} = 1.81\underline{|11.45}\,\Omega \tag{7.119}$$

Then, calculate the Thevenin or open-circuit voltage from the circuit in Fig. 7.30 as,

$$V_{\text{Th}} = \frac{220\underline{|25^\circ} \times (3 + j8)}{5 + j8} = 199.25\underline{|36.45^\circ}\,\text{V} \tag{7.120}$$

The current through the 9 Ω resistor can be calculated as,

$$I_{9\Omega} = \frac{199.25\underline{|36.45^\circ}}{9 + 1.81\underline{|11.45}} = 18.48\underline{|34.54^\circ}\,\text{A} \tag{7.121}$$

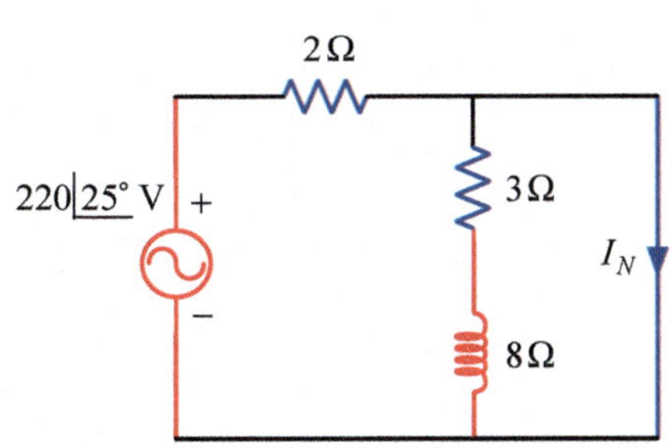

Fig. 7.31 Circuit for calculation Norton current for Example 7.8

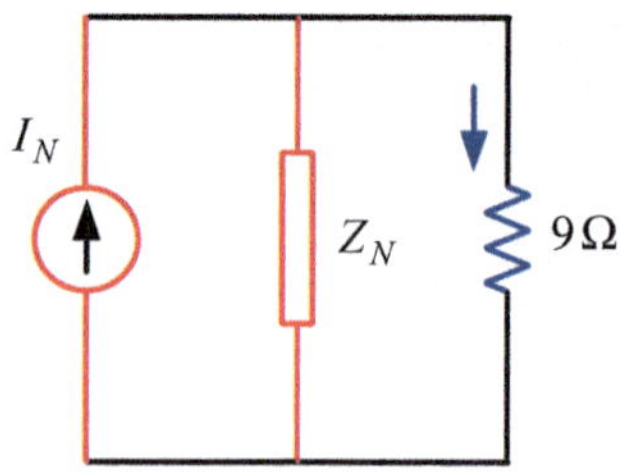

Fig. 7.32 Final circuit for Norton theorem for Example 7.8

Short circuit the open terminals and calculate the Norton's current from the circuit as shown in Fig. 7.31. This current is,

$$I_N = \frac{220\angle 25^\circ}{2} = 110\angle 25^\circ \text{ A} \tag{7.122}$$

The current through the 9 Ω resistor from the circuit in Fig. 7.32 can be calculated as,

$$I_{9\Omega} = \frac{110\angle 25^\circ \times 1.81\angle 11.45}{9 + 1.81\angle 11.45} = 18.47\angle 34.54^\circ \text{ A} \tag{7.123}$$

Practice Problem 7.8
Use Thevenin's and Norton's theorems to calculate the current in the 4 Ω resistor for the circuit shown in Fig. 7.33.

Example 7.9 Use Thevenin's theorem to calculate the current in the 3 Ω resistor for the circuit shown in Fig. 7.34, and compare the results with PSpice simulation.

Fig. 7.33 Circuit for Practice Problem 7.8

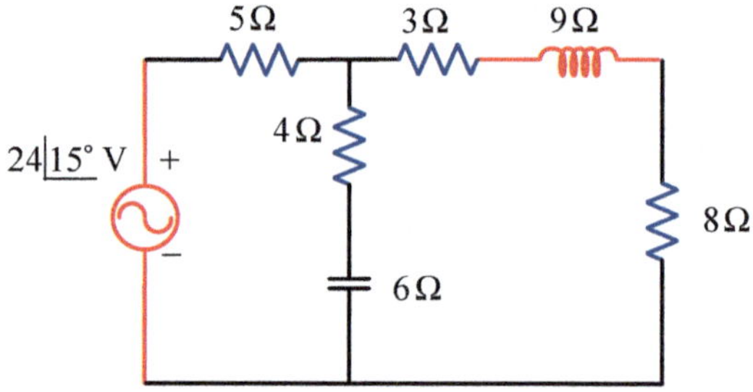

Fig. 7.34 Circuit for Example 7.7

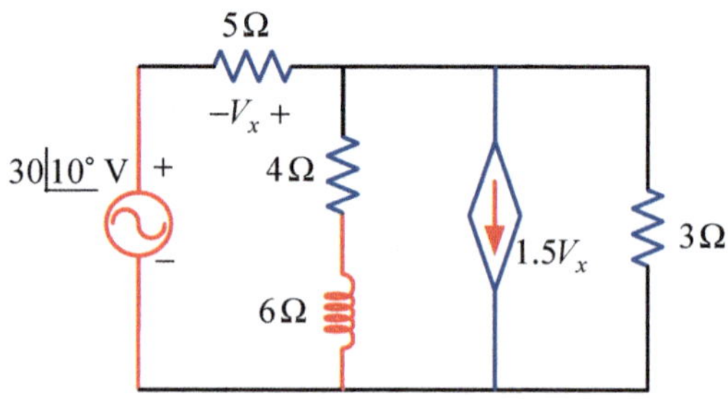

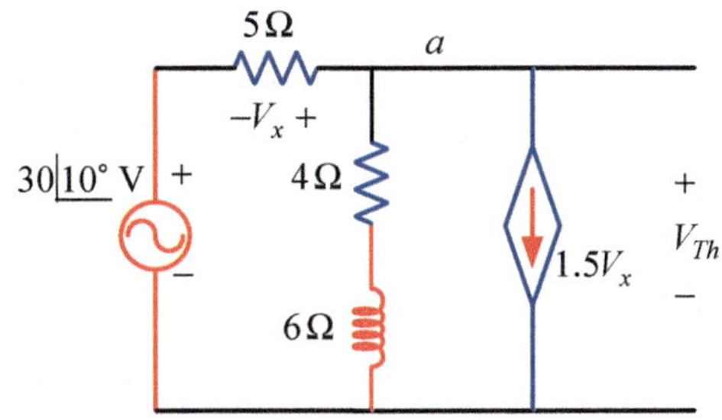

Fig. 7.35 Circuit for Example 7.9 with Thevenin voltage

Solution:

Open the 3 Ω resistor and redraw the circuit as shown in Fig. 7.35. Applying KVL to the circuit in Fig. 7.35 yields,

$$-30\underline{|10^\circ} - V_x + V_{\text{Th}} = 0 \tag{7.124}$$

$$V_{\text{Th}} = 30\underline{|10^\circ} + V_x \tag{7.125}$$

Applying KCL to node a of the circuit in Fig. 7.35 yields,

$$\frac{30\underline{|10^\circ} - V_{\text{Th}}}{5} = 1.5V_x + \frac{V_{\text{Th}}}{4+j6} \tag{7.126}$$

Substituting Eq. (7.125) into Eq. (7.126) yields,

$$\frac{30\underline{|10^\circ} - V_{\text{Th}}}{5} = 1.5(V_{\text{Th}} - 30\underline{|10^\circ}) + \frac{V_{\text{Th}}}{4+j6} \tag{7.127}$$

$$45\underline{|10^\circ} + 6\underline{|10^\circ} = \frac{V_{\text{Th}}}{5} + 1.5V_{\text{Th}} + \frac{V_{\text{Th}}}{4+j6} \tag{7.128}$$

$$V_{\text{Th}} = 28.64\underline{|13.72^\circ}\ \text{V} \tag{7.129}$$

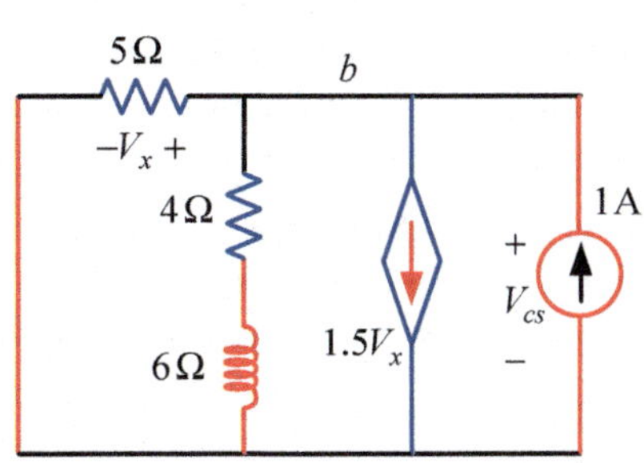

Fig. 7.36 Circuit for Example 7.9 with a current source

To find the Thevenin's impedance, insert a 1 A current source to the open terminals of the circuit and short circuit the independent voltage source as shown in Fig. 7.36.

Applying KVL to the circuit in Fig. 7.36 yields,

$$-V_x + V_{cs} = 0 \tag{7.130}$$

$$V_x = V_{cs} \tag{7.131}$$

Applying KCL to node b of the circuit in Fig. 7.36 yields,

$$1 = \frac{V_{cs}}{4+j6} + \frac{V_{cs}}{5} + 1.5V_x \tag{7.132}$$

Substituting Eq. (7.131) into Eq. (7.132) yields,

$$1 = \left(\frac{1}{4+j6} + \frac{1}{5} + 1.5\right)V_{cs} \tag{7.133}$$

$$V_{cs} = 0.56\underline{|3.72^\circ}\text{ V} \tag{7.134}$$

Thevenin's impedance is then calculated as,

$$Z_{Th} = \frac{V_{cs}}{1} = 0.56\underline{|3.72^\circ}\ \Omega \tag{7.135}$$

The current in the 3 Ω resistor is,

$$I_{3\Omega} = \frac{V_{Th}}{3+Z_{Th}} = \frac{28.64\underline{|13.72^\circ}}{3+0.56\underline{|3.72^\circ}} = 8.05\underline{|13.14^\circ}\text{ A} \tag{7.136}$$

The circuit is simulated with PSpice and the result is shown in Fig. 7.37. Initially, all parameters are placed and connected according to the given circuit, iPrint is set to measure the current. Analysis, setup and AC sweep have been selected. In the AC sweep window, the following simulation parameters have been selected before running the simulation: linear, total points 1, start frequency 50 and end frequency 50. In the iPrint setup, AC, magnitude and phase have been selected before running the simulation.

The PSpice simulation results are found to be the same as the calculated results as shown in Fig. 7.38.

Practice Problem 7.9

Calculate the current in the 3 Ω resistor for the circuit shown in Fig. 7.39 using Thevenin's theorem, and compare the results with PSpice simulation.

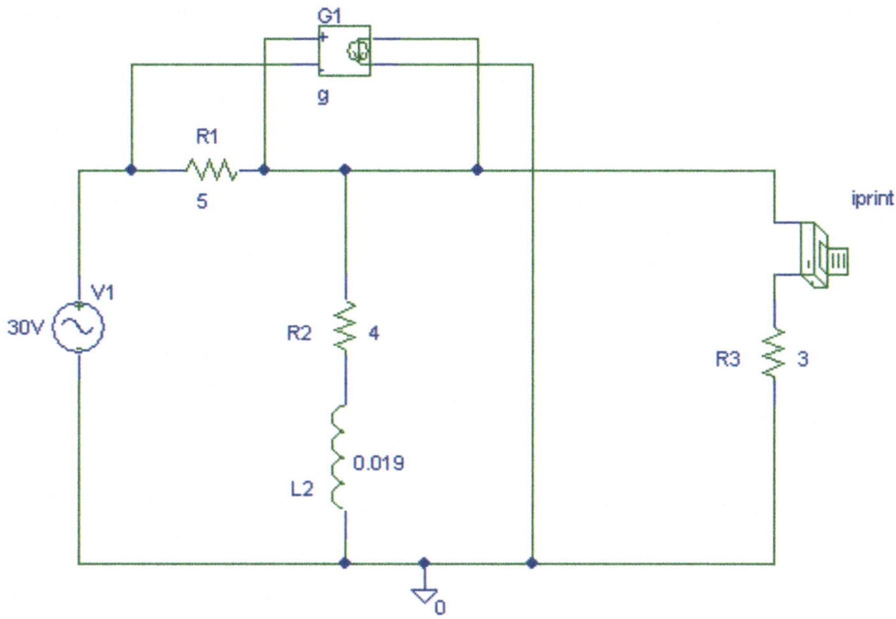

Fig. 7.37 PSpice circuit for Example 7.7

```
FREQ          IM(V_PRINT1)IP(V_PRINT1)IR(V_PRINT1)II(V_PRINT1

 5.000E+01    8.262E+00    1.201E+01    8.081E+00    1.720E+00

          JOB CONCLUDED

          TOTAL JOB TIME              0.00
```

Fig. 7.38 Simulation results for Example 7.7

Fig. 7.39 Circuit for Practice Problem 7.7

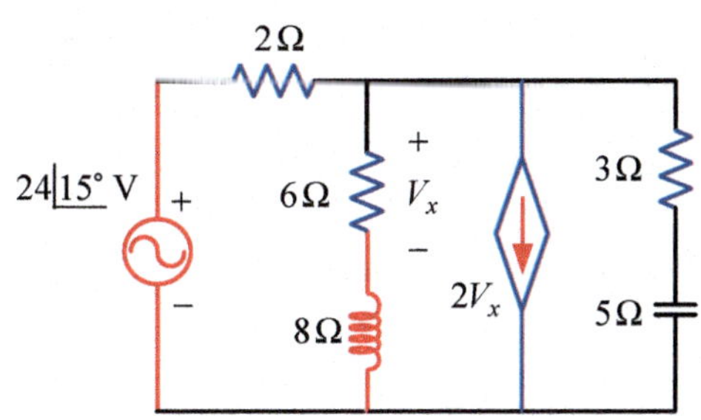

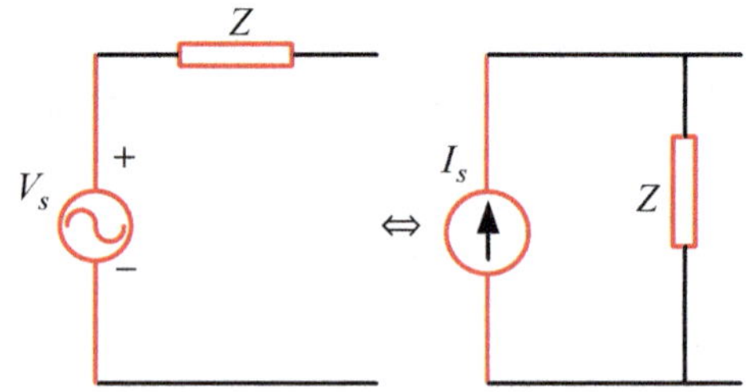

Fig. 7.40 Circuit for source conversion technique

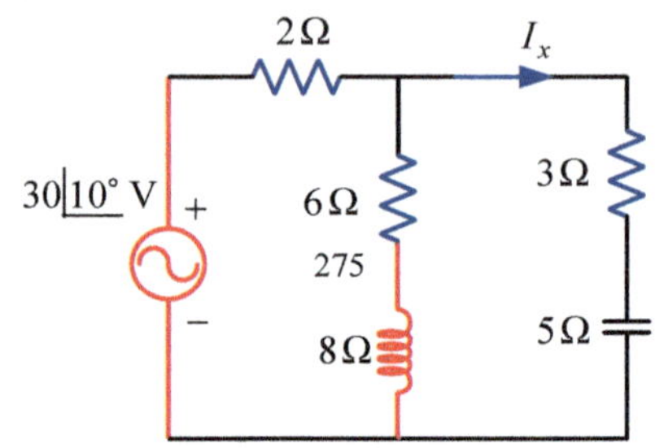

Fig. 7.41 Circuit for Example 7.10

7.6 Source Conversion Technique

The source conversion technique using AC quantities uses the same method as discussed for DC circuits. The circuit with an AC voltage source with an impedance in series can be replaced by a current source in parallel with an impedance as shown in Fig. 7.40 [8, 9].

Example 7.10 Calculate the current I_x in the circuit shown in Fig. 7.41 by source conversion technique, and compare the results with PSpice simulation.
Solution:
Replace the voltage source with a current source as,

$$I_s = \frac{30\underline{|10^\circ}}{2} = 15\underline{|10^\circ}\text{ A} \tag{7.137}$$

The circuit with a current source is shown in Fig. 7.42. In this case, the parallel impedance is,

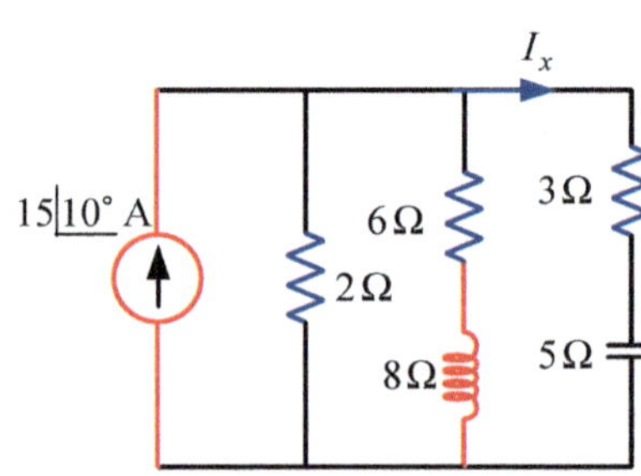

Fig. 7.42 Circuit for Example 7.10 with a current source

$$Z_1 = \frac{2 \times (6 + j8)}{8 + j8} = 1.77\angle 8.13^\circ \text{ A} \tag{7.138}$$

The current can be determined as,

$$I_x = \frac{15\angle 10^\circ \times 1.77\angle 8.13^\circ}{3 - j5 + 1.77\angle 8.13^\circ} = 3.95\angle 63.11^\circ \text{ A} \tag{7.139}$$

PSpice simulation circuit and results are shown in Figs. 7.43 and 7.44, respectively. The results are found to be the same as the calculated results.

Practice Problem 7.10

Determine the current I_x in the circuit shown in Fig. 7.45 by source conversion technique, and compare the results with PSpice simulation.

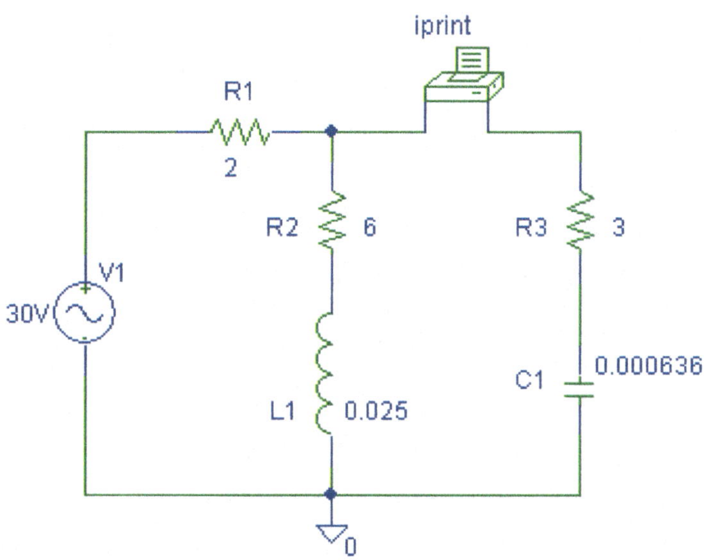

Fig. 7.43 PSpice simulation circuit for Example 7.10

```
FREQ          IM(V_PRINT1)IP(V_PRINT1)

 5.000E+01    3.937E+00    6.321E+01

        JOB CONCLUDED

        TOTAL  JOB TIME            .02
```

Fig. 7.44 PSpice simulation results for Example 7.10

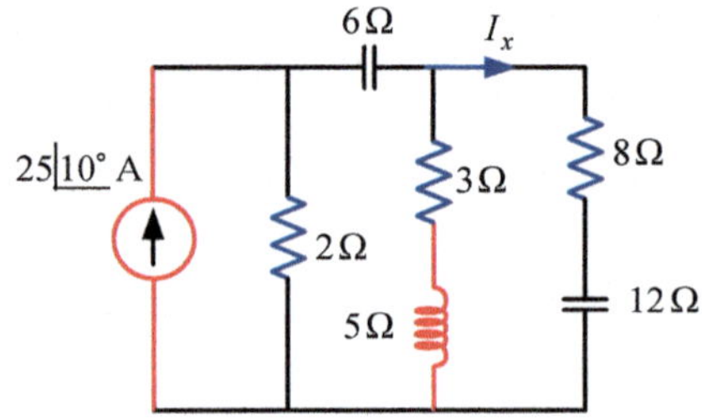

Fig. 7.45 Circuit for Practice Problem 7.10

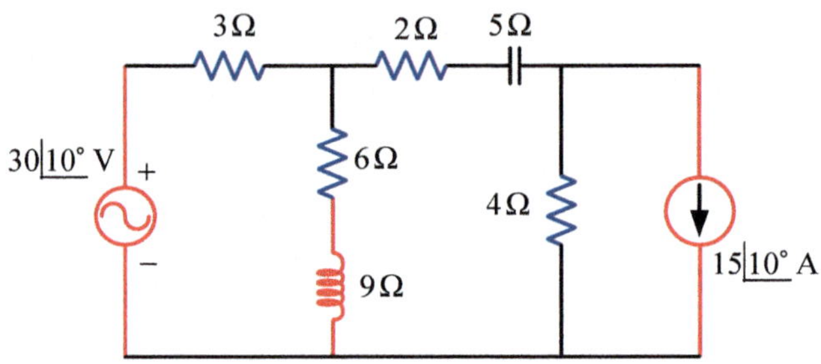

Fig. 7.46 Circuit for Problem 7.1

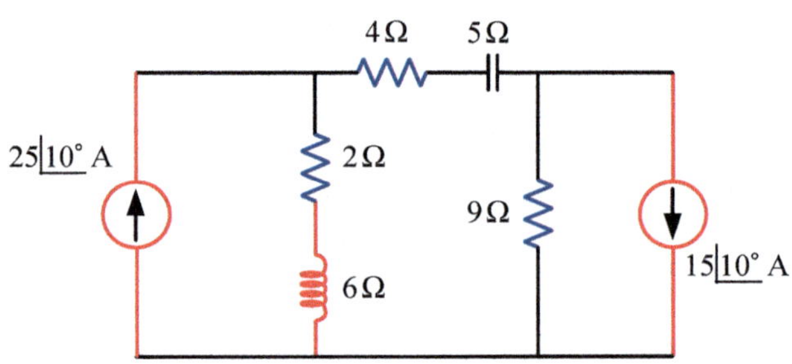

Fig. 7.47 Circuit for Problem 7.2

Exercise Problems

7.1 Find the node voltages, and the current in the 4 Ω resistor for the circuit shown in Fig. 7.46. Verify the result by PSpice simulation.

7.2 Calculate the node voltages, and the current through the 2 Ω resistor for the circuit shown in Fig. 7.47. Compare the result with PSpice simulation.

7.3 Find the voltage drop across the inductor with an inductive reactance of 6 Ω for the circuit shown in Fig. 7.48 by nodal analysis. Verify the result by PSpice simulation.

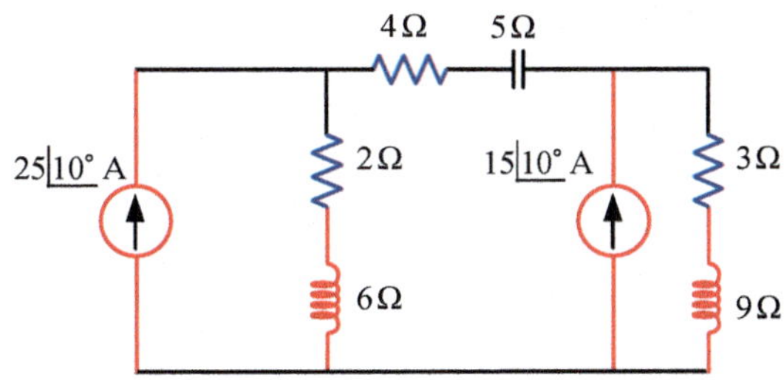

Fig. 7.48 Circuit for Problem 7.3

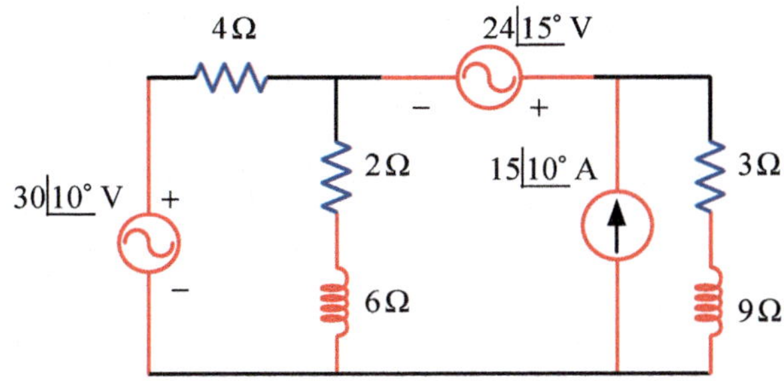

Fig. 7.49 Circuit for Problem 7.4

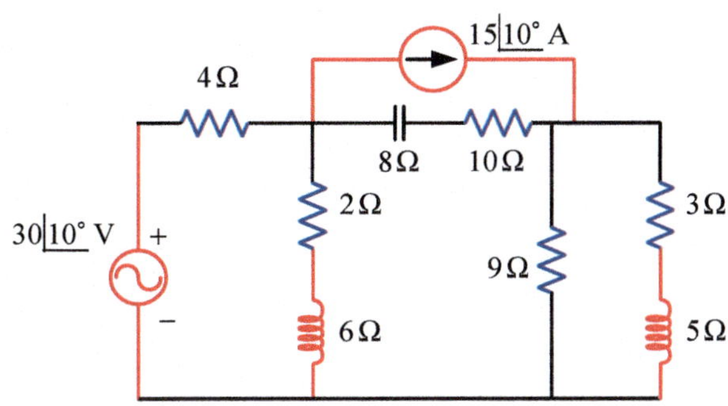

Fig. 7.50 Circuit for Problem 7.5

7.4 Calculate voltage drop across the 2 Ω resistor for the circuit shown in Fig. 7.49 using nodal analysis. Verify the result by PSpice simulation.

7.5 Determine the current through the 9 Ω resistor for the circuit shown in Fig. 7.50 by nodal analysis. Verify the result by PSpice simulation.

7.6 Find the voltage drop across the 3 Ω resistor for the circuit shown in Fig. 7.51 by nodal analysis. Verify the result by PSpice simulation.

7.7 Calculate the voltage drop across the inductor with an inductive reactance of 5 Ω for the circuit shown in Fig. 7.52 by nodal analysis. Compare the result with PSpice simulation.

7.8 Determine the current in the 4 Ω resistor for the circuit shown in Fig. 7.53 by mesh analysis. Verify the result by PSpice simulation.

Fig. 7.51 Circuit for Problem 7.6

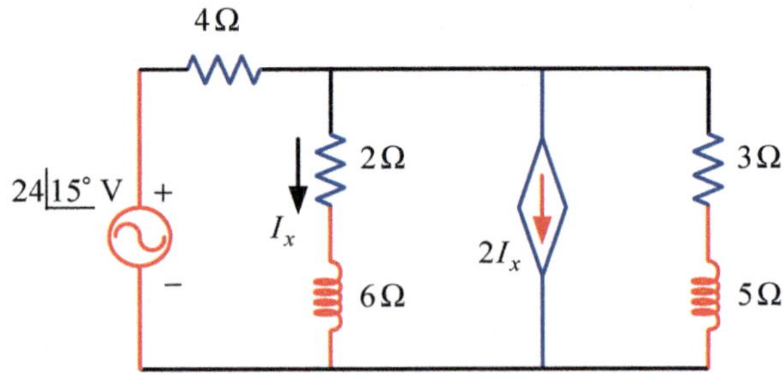

Fig. 7.52 Circuit for Problem 7.7

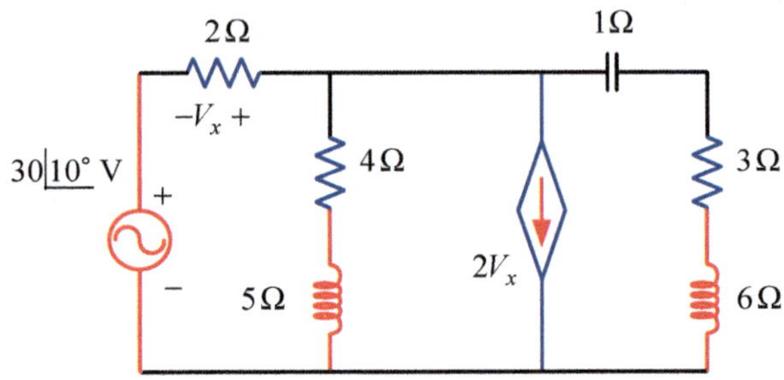

Fig. 7.53 Circuit for Problem 7.8

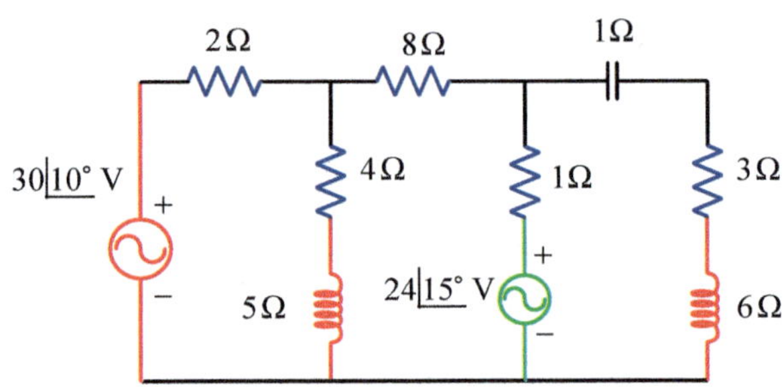

Fig. 7.54 Circuit for Problem 7.9

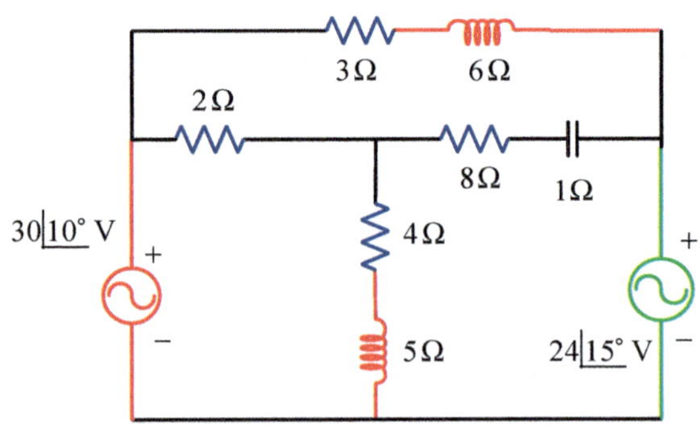

7.9 Calculate the current in the 4 Ω resistor for the circuit shown in Fig. 7.54 by mesh analysis. Verify the result by PSpice simulation.

7.10 Determine the mesh currents in the circuit shown in Fig. 7.55. Verify the result by PSpice simulation.

7.11 Calculate the current in the 4 Ω resistor for the circuit shown in Fig. 7.56 by mesh analysis. Verify the result by PSpice simulation.

7.12 Determine the current in the 2 Ω resistor for the circuit shown in Fig. 7.57 by mesh analysis. Verify the result by PSpice simulation.

7.13 Find current in the 3 Ω resistor for the circuit shown in Fig. 7.58 by mesh analysis and verify the result by PSpice simulation.

7.14 Find the voltage across the 3 Ω resistor for the circuit shown in Fig. 7.59 using mesh analysis. Verify the result by PSpice simulation.

7.15 Determine the current in the 2 Ω resistor for the circuit shown in Fig. 7.60 using mesh analysis and verify the result by PSpice simulation.

7.16 Find the current in the 3 Ω resistor for the circuit shown in Fig. 7.61 using mesh analysis and verify the result by PSpice simulation.

7.17 Determine the current in the 5 Ω resistor for the circuit shown in Fig. 7.62 using superposition theorem. Verify the result by PSpice simulation.

7.18 Find the current in the 3 Ω resistor for the circuit shown in Fig. 7.63 using superposition theorem. Verify the result by PSpice simulation.

7.19 Find the current through the 2 Ω resistor for the circuit shown in Fig. 7.64 using superposition theorem. Verify the result by PSpice simulation.

7.20 Calculate the current I_x in the circuit shown in Fig. 7.65 using superposition theorem. Verify the result by PSpice simulation.

7.21 Using superposition theorem, find the current I_x in the circuit shown in Fig. 7.66. Verify the result by PSpice simulation.

7.22 Using superposition theorem, determine the current I_x in the circuit shown in Fig. 7.67. Verify the result by PSpice simulation.

7.23 Determine the current I_x in the circuit as shown in Fig. 7.68 using superposition theorem. Verify the result by PSpice simulation.

7.24 Calculate the current I_x in the circuit shown in Fig. 7.69 using superposition theorem. Verify the result by PSpice simulation.

7.25 Calculate the current I_s in the circuit shown in Fig. 7.70 using superposition theorem. Verify the result by PSpice simulation.

7.26 Determine the current I_s in the circuit shown in Fig. 7.71 by Thevenin theorem. Verify the result by PSpice simulation.

7.27 Using Thevenin's theorem, calculate the current in the 4 Ω resistor for the circuit shown in Fig. 7.72. Verify the result by PSpice simulation.

7.28 Calculate the current in the 5 Ω resistor for the circuit shown in Fig. 7.73 using Thevenin's theorem. Verify the result by PSpice simulation.

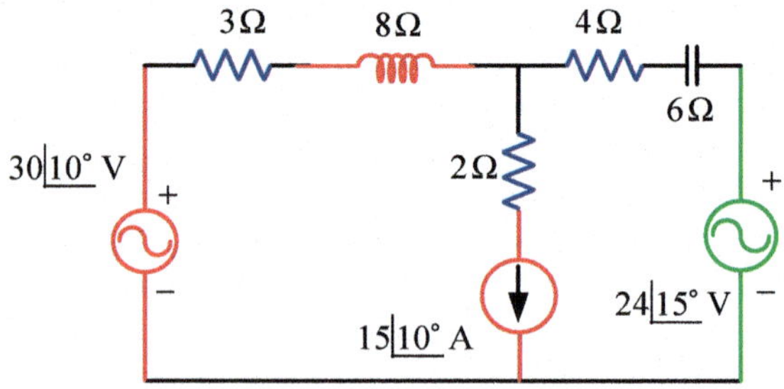

Fig. 7.55 Circuit for Problem 7.10

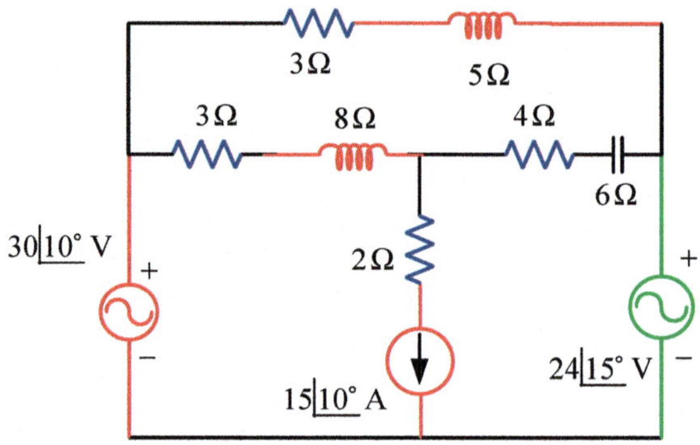

Fig. 7.56 Circuit for Problem 7.11

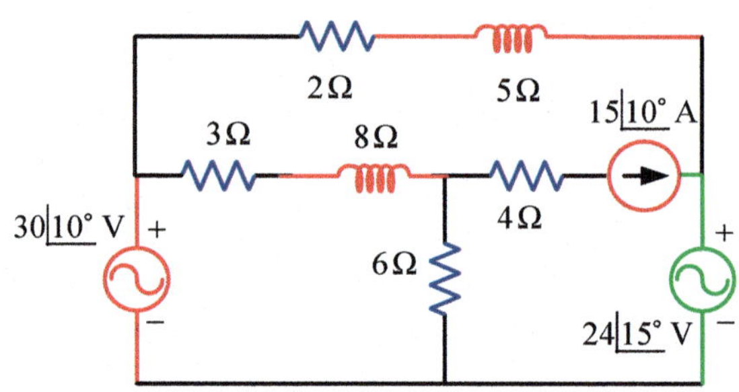

Fig. 7.57 Circuit for Problem 7.12

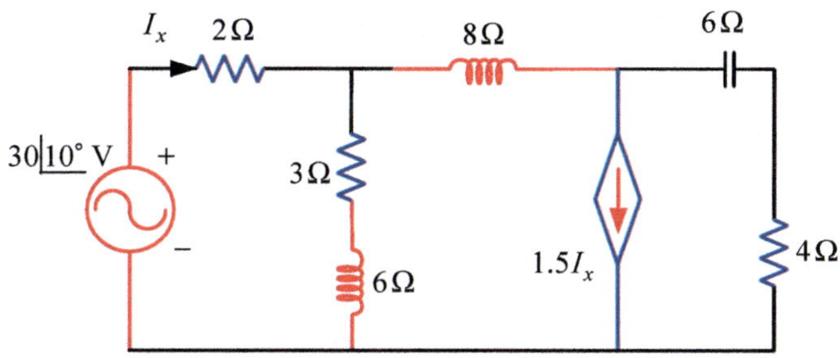

Fig. 7.58 Circuit for Problem 7.13

Fig. 7.59 Circuit for Problem 7.14

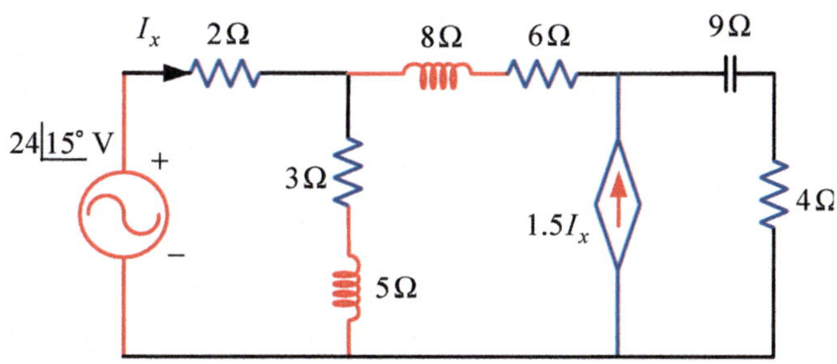

Fig. 7.60 Circuit for Problem 7.15

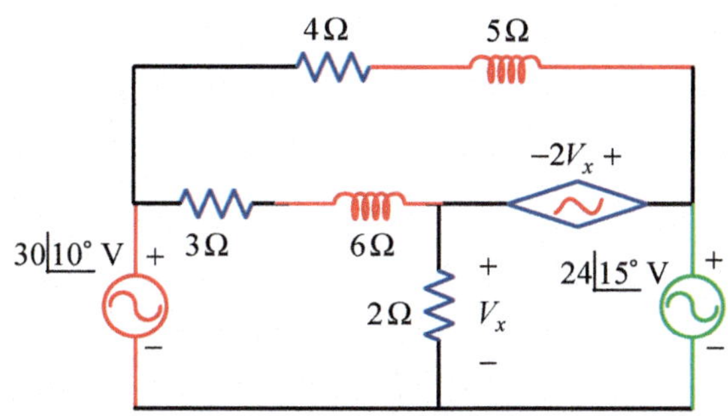

Fig. 7.61 Circuit for Problem 7.16

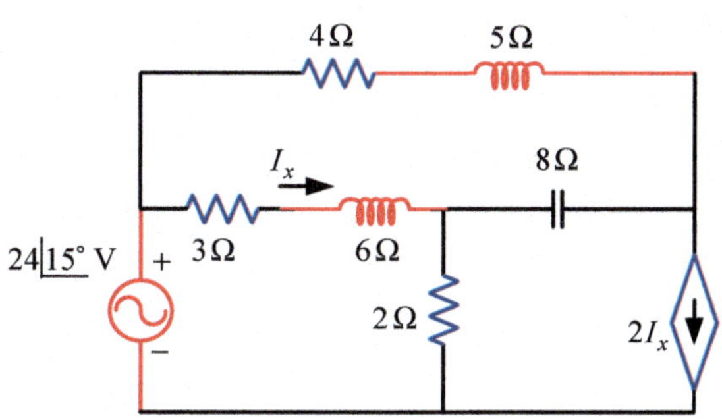

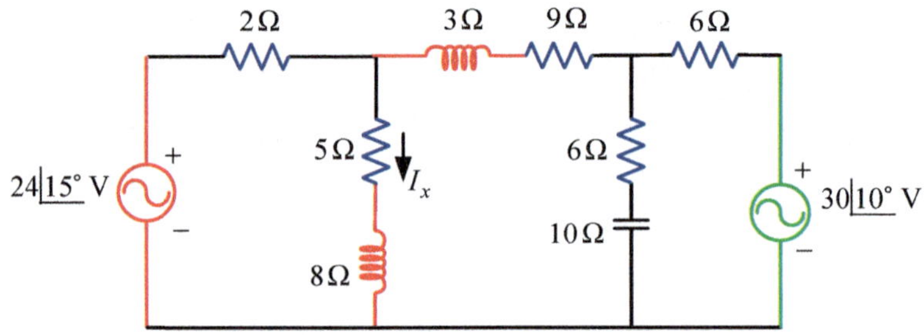

Fig. 7.62 Circuit for Problem 7.17

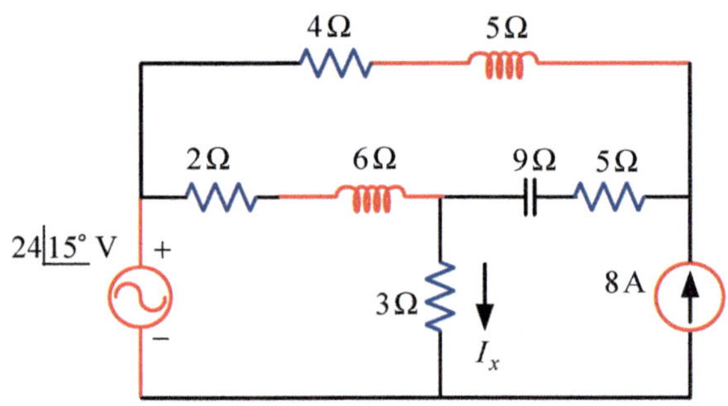

Fig. 7.63 Circuit for Problem 7.18

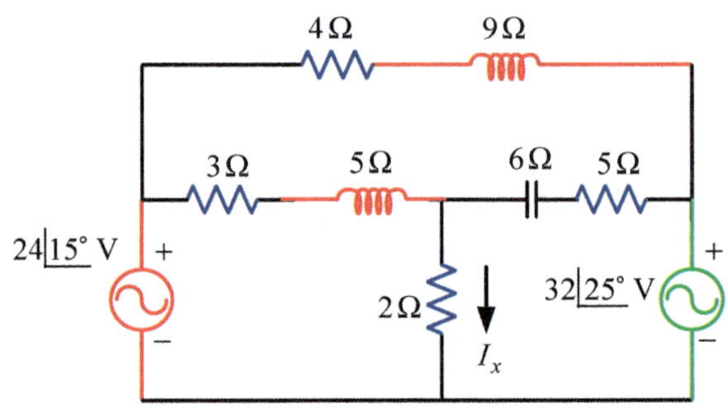

Fig. 7.64 Circuit for Problem 7.19

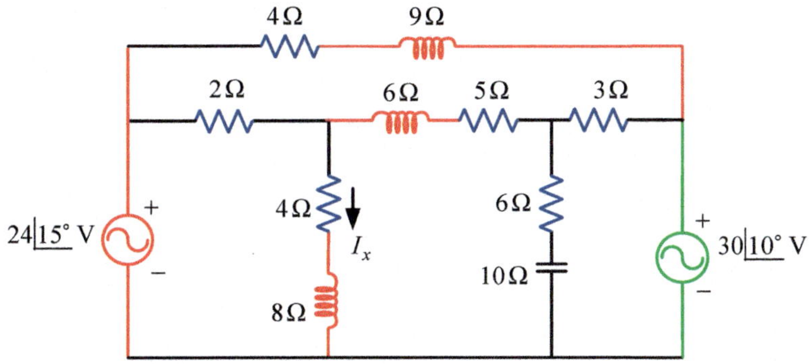

Fig. 7.65 Circuit for Problem 7.20

Fig. 7.66 Circuit for Problem 7.21

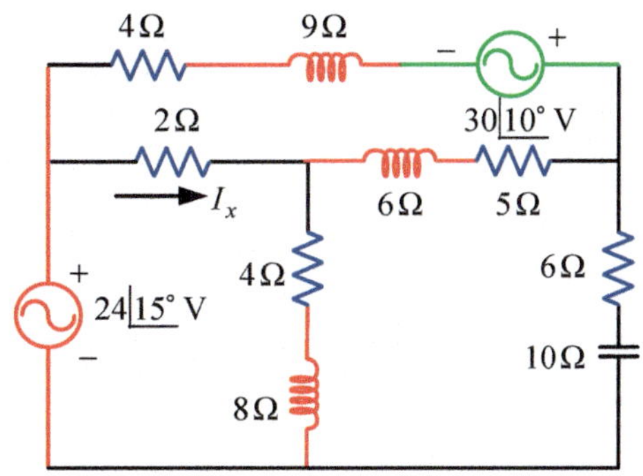

Fig. 7.67 Circuit for Problem 7.22

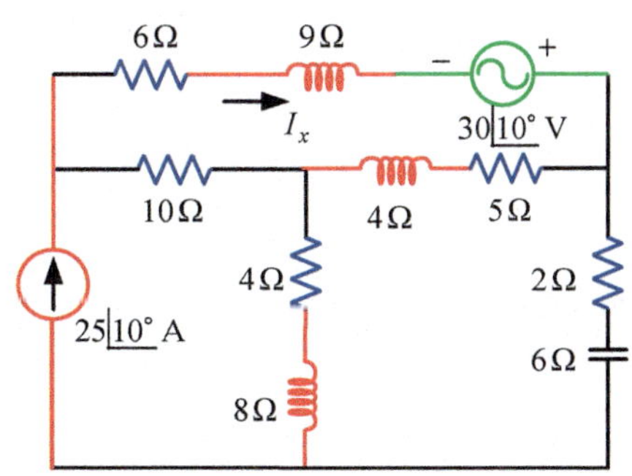

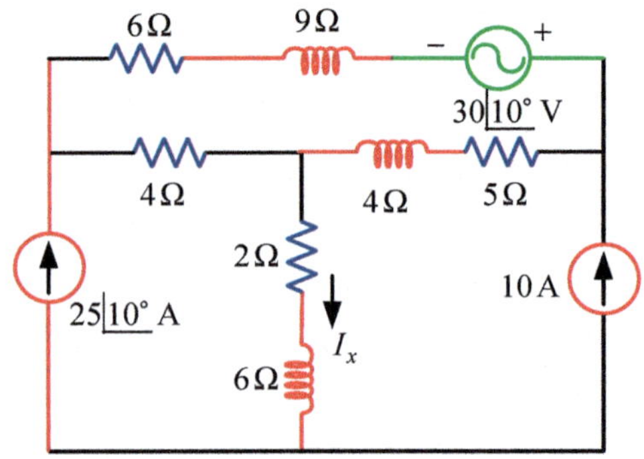

Fig. 7.68 Circuit for Problem 7.23

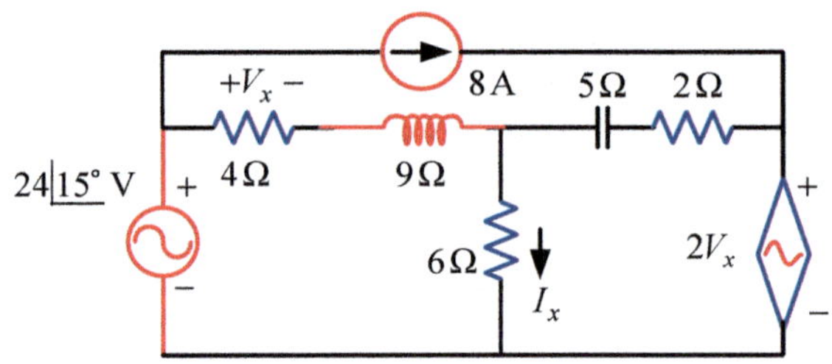

Fig. 7.69 Circuit for Problem 7.24

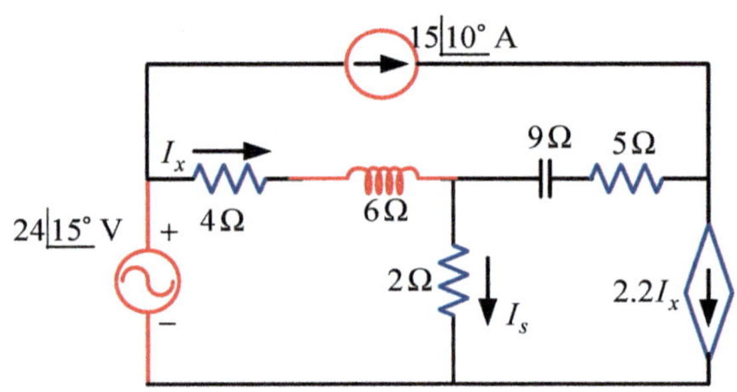

Fig. 7.70 Circuit for Problem 7.25

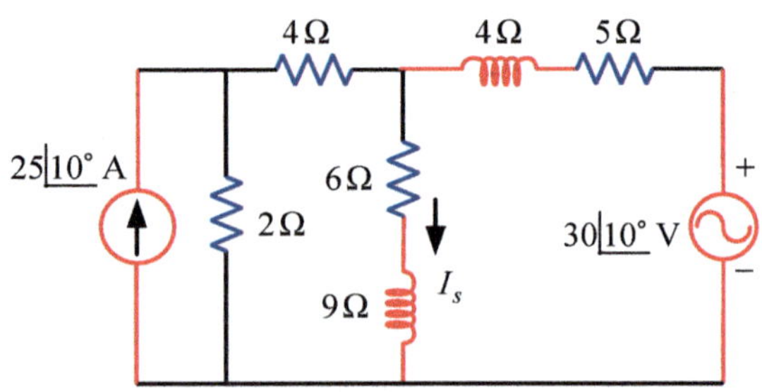

Fig. 7.71 Circuit for Problem 7.26

Fig. 7.72 Circuit for Problem 7.27

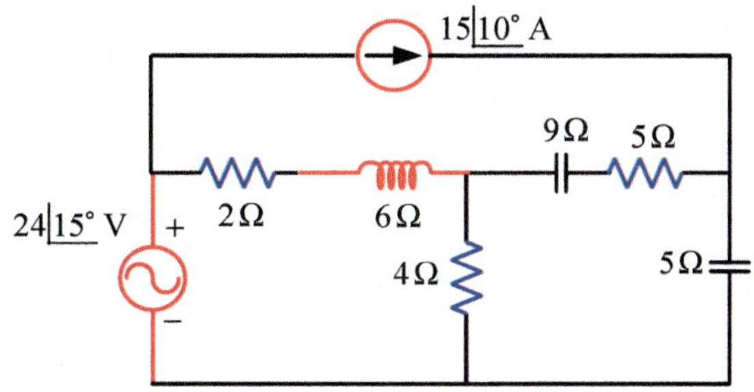

Fig. 7.73 Circuit for Problem 7.28

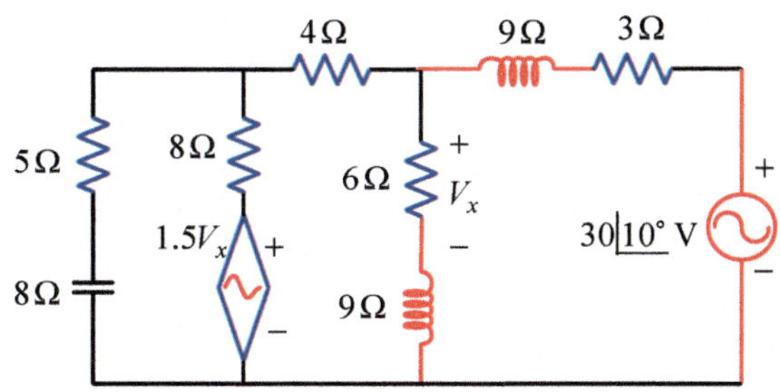

Fig. 7.74 Circuit for Problem 7.27

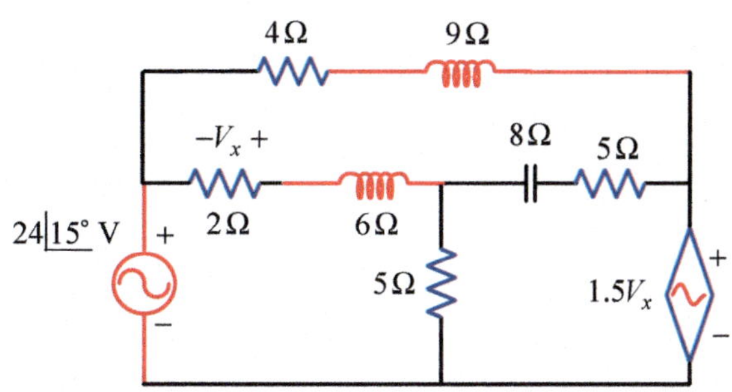

Fig. 7.75 Circuit for Problem 7.30

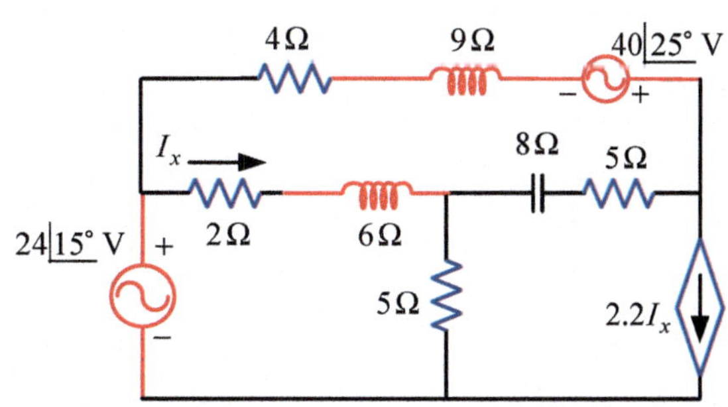

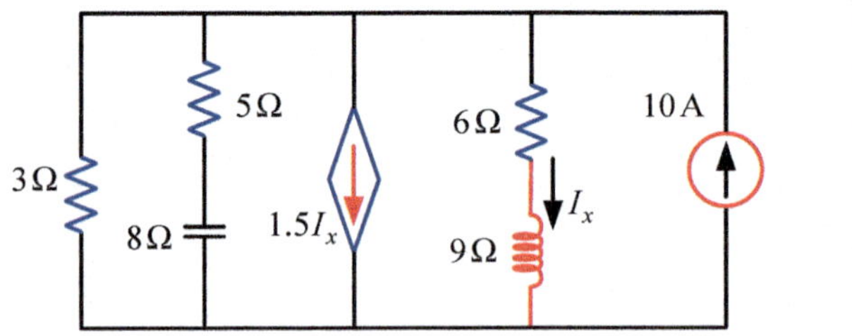

Fig. 7.76 Circuit for Problem 7.31

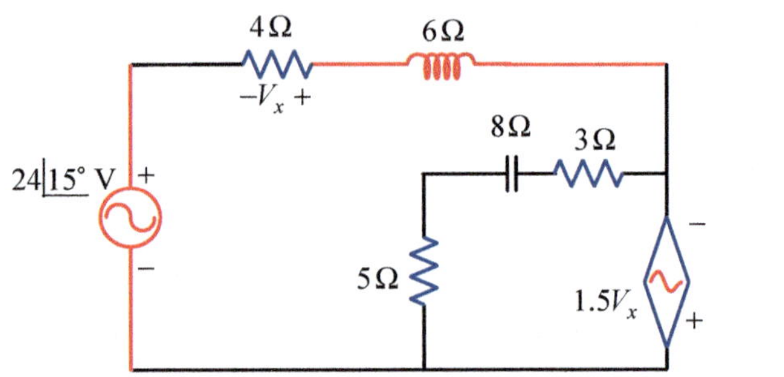

Fig. 7.77 Circuit for Problem 7.32

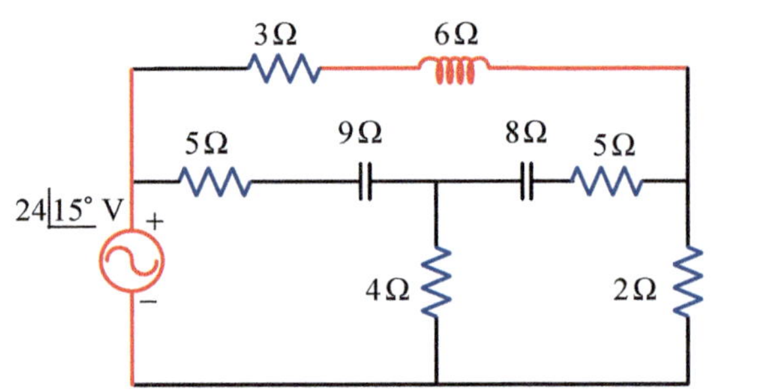

Fig. 7.78 Circuit for Problem 7.33

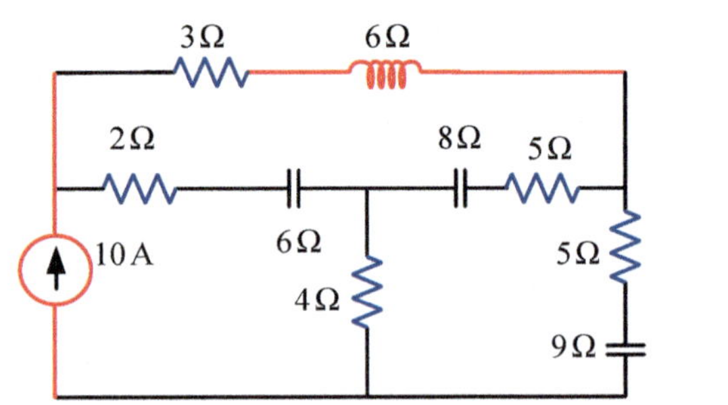

Fig. 7.79 Circuit for Problem 7.34

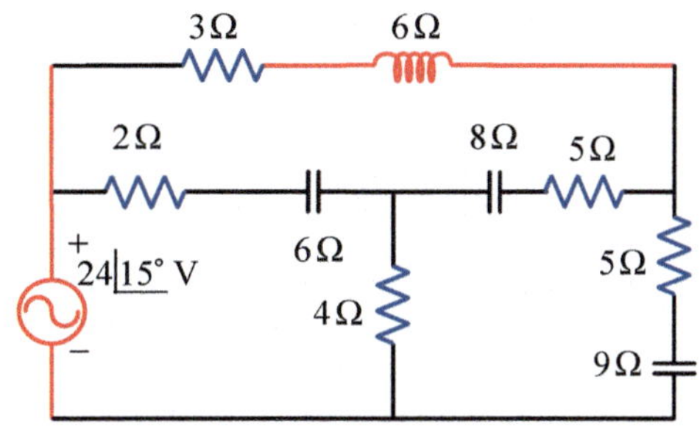

Fig. 7.80 Circuit for Problem 7.35

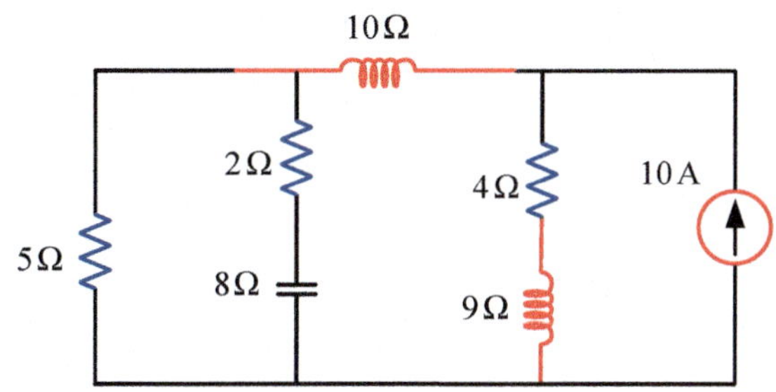

Fig. 7.81 Circuit for Problem 7.36

7.29 Find current in the 5 Ω resistor for the circuit shown in Fig. 7.74 by Thevenin's theorem. Verify the result by PSpice simulation.

7.30 Using Thevenin's theorem, determine the current through the 5 Ω resistor for the circuit shown in Fig. 7.75. Verify the result by PSpice simulation.

7.31 Determine the current in the 3 Ω resistor for the circuit shown in Fig. 7.76 by using Thevenin's theorem. Verify the result by PSpice simulation.

7.32 Calculate the current in the 5 Ω resistor for the circuit shown in Fig. 7.77 by Thevenin's theorem. Verify the result by PSpice simulation.

7.33 Determine the current through the 2 Ω resistor for the circuit shown in Fig. 7.78 by Norton's theorem. Verify the result by PSpice simulation.

7.34 Find the current through the 4 Ω resistor for the circuit shown in Fig. 7.79 by Norton's theorem. Verify the result by PSpice simulation.

7.35 Calculate the current in the inductor with an inductive reactance of 6 Ω for the circuit shown in Fig. 7.80 by Norton's theorem. Verify the result by PSpice simulation.

7.36 Calculate the current in the 5 Ω resistor for the circuit shown in Fig. 7.81 by source conversion technique. Verify the result by PSpice simulation.

References

1. H.W. Jackson, D. Temple, B.E. Kelly, *Introduction to Electric Circuits*, 9th edn.(Oxford University Press, July 2015)
2. D. Bell, *Fundamentals of Electric Circuits*, 7th edn. (Oxford University Press, March 2007)
3. C.K. Alexander, M.N.O. Sadiku, *Fundamentals of Electric Circuits*, 6th edn. (McGraw-Hill Higher Education, January 2016)
4. G. Rizzoni, J. Kearns, *Principles and Applications of Electrical Engineering*, 6th edn. (McGraw-Hill Education, October 2014)
5. R.L. Boylestad, *Introductory Circuit Analysis*, 13th edn. (Pearson, 2016)
6. J.D. Irwin, R.M. Nelms, *Basic Engineering Circuit Analysis*, 11th edn. (John & Wiley Sons Inc., USA, 2015)
7. W. Hayt, J. Kemmerly, S. Durbin, *Engineering Circuit Analysis*, 8th edn. (McGraw-Hill Education, 2012)
8. J.W. Nilsson, S.A. Riedel, *Electric Circuits*, 10th edn. (Prentice Hall International Edition, 2015)
9. M.A. Salam, *Basic Electrical Circuits*, 2nd edn. (Shroff Publishers & Distributors Pvt. Ltd, India, 2007)

Chapter 8
AC Power Analysis

8.1 Introduction

AC generator, transformer and other related high-voltage devices are required to generate, transmit and distribute alternating voltage and current. These pieces of equipment are rated in terms of mega watt (MW) and mega voltage-ampere (MVA). Electrical appliances, such as DVD player, television, microwave oven, light bulb, refrigerator, ceiling fan and many more are rated in terms of watt (W). It requires a careful attention in designing those kinds of equipment to address their power ratings, which in turn address the varying electricity bill. In this case, analysis of AC power plays an important role. This chapter discusses different AC power analytical technique, in terms of instantaneous power, average power, complex power, power factor, maximum power transfer theorem and power factor correction.

8.2 Instantaneous Power

Figure 8.1 shows a circuit with an AC voltage source $v(t)$ with an impedance connected in series. In this case, the resultant current $i(t)$ varies with time. The instantaneous power also varies with time and it is defined as the product of the voltage and the current. Let us consider that the expression of voltage is,

$$v(t) = V_m \sin \omega t \tag{8.1}$$

The expression of the current can be derived as [1, 2],

$$i(t) = \frac{v(t)}{Z\underline{|\theta}} \tag{8.2}$$

Md. A. Salam and Q. Mehbubar Rahman, *Fundamentals of Electrical Circuit Analysis*,
https://doi.org/10.1007/978-981-10-8624-3_8

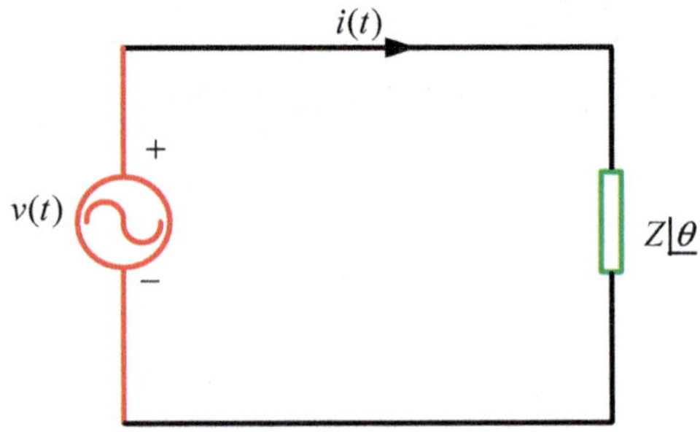

Fig. 8.1 Circuit with an impedance

Substituting Eq. (8.1) into Eq. (8.2) yields,

$$i(t) = \frac{V_m \sin \omega t}{Z\angle\theta} \tag{8.3}$$

$$i(t) = \frac{V_m\angle 0^\circ}{Z\angle\theta} = \frac{V_m}{Z}\angle -\theta \tag{8.4}$$

$$i(t) = I_m\angle\theta = I_m \sin(\omega t - \theta) \tag{8.5}$$

The expression of instantaneous power is,

$$p(t) = v(t) \times i(t) \tag{8.6}$$

Substituting Eqs. (8.1) and (8.5) into Eq. (8.6) yields,

$$p(t) = V_m \sin \omega t \times I_m \sin(\omega t - \theta) \tag{8.7}$$

$$p(t) = \frac{V_m I_m}{2} \times 2 \sin \omega t \, \sin(\omega t - \theta) \tag{8.8}$$

$$p(t) = \frac{V_m I_m}{2}[\cos(\omega t - \omega t + \theta) - \cos(\omega t + \omega t - \theta)] \tag{8.9}$$

$$p(t) = \frac{V_m I_m}{2} \cos\theta - \frac{V_m I_m}{2} \cos(2\omega t - \theta) \tag{8.10}$$

Equation (8.10) provides the expression for the instantaneous power for a series AC circuit.

Example 8.1 The excitation voltage and impedance of a series circuit are given by $v(t) = 15 \sin \omega t$ V and $Z = 5\angle 10^\circ\ \Omega$, respectively. Determine the instantaneous power.

Solution:
The current is,

$$i(t) = \frac{15\angle 0^\circ}{5\angle 10^\circ} = 3\angle -10^\circ \text{ A} \tag{8.11}$$

The instantaneous power is calculated as,

$$p(t) = \frac{15 \times 3}{2} \times 2\sin(\omega t - 10^\circ)\sin\omega t \tag{8.12}$$

$$p(t) = \frac{15 \times 3}{2}[\cos 10^\circ - \cos(2\omega t - 10^\circ)] \tag{8.13}$$

$$p(t) = 22.16 - 22.5\cos(2\omega t - 10^\circ) \text{ W} \tag{8.14}$$

Practice Problem 8.1
The current and impedance of a series circuit are given by $i(t) = 10\sin(\omega t + 15^\circ)$ A and $Z = 2\angle 20^\circ\ \Omega$, respectively. Calculate the instantaneous voltage and power.

8.3 Average Power and Reactive Power

The average power is related to the sinusoidal voltage and current, which are shown in Eqs. (8.1) and (8.5), respectively. The average power for a periodic waveform over one cycle can be derived as [3, 4],

$$P = \frac{1}{T}\int_0^T p(t)\, dt \tag{8.15}$$

Substituting Eq. (8.10) into the Eq. (8.15) yields,

$$P = \frac{V_m I_m}{2T}\int_0^T [\cos\theta - \cos(2\omega t - \theta)]\, dt \tag{8.16}$$

$$P = \frac{V_m I_m}{2T}\cos\theta\, [T] - \frac{V_m I_m}{2T}\int_0^T \cos(2\omega t - \theta)\, dt \tag{8.17}$$

The second term of Eq. (8.17) is a cosine waveform. The average value of any cosine waveform over one cycle is zero. Therefore, from Eq. (8.17), the final expression of the average power can be represented as,

$$P = \frac{V_m I_m}{2}\cos\theta \tag{8.18}$$

Similarly, the expression of the average reactive power can be written as,

$$Q = \frac{V_m I_m}{2} \sin\theta \tag{8.19}$$

The term $\cos\theta$ in Eq. (8.18) is the power factor of the circuit, and it is determined by the phase angle θ of the circuit impedance, where θ is the phase difference between the voltage and current phases, i.e. $\theta = \theta_v - \theta_i$. The average power is often known as the true power or the real power. The units of average power and reactive power are watt (W) and volt-ampere reactive (Var), respectively. The average power from Eq. (8.18) can be represented in terms of rms values of the voltage and current as,

$$P = \frac{V_m}{\sqrt{2}}\frac{I_m}{\sqrt{2}}\cos\theta = \frac{V_m}{\sqrt{2}}\frac{I_m}{\sqrt{2}}\cos(\theta_v - \theta_i) \tag{8.20}$$

Substituting $V_{rms} = \frac{V_m}{\sqrt{2}}$ and $I_{rms} = \frac{I_m}{\sqrt{2}}$ into Eq. (8.20) yields,

$$P = V_{rms} I_{rms} \cos\theta = V_{rms} I_{rms} \cos(\theta_v - \theta_i) \tag{8.21}$$

Similarly, from Eq. (8.19), the reactive power can be represented as,

$$Q = V_{rms} I_{rms} \sin\theta = V_{rms} I_{rms} \sin(\theta_v - \theta_i) \tag{8.22}$$

Due to sufficient magnitude of the reactive power, the current flows back and forth between the source and the network. The reactive power does not dissipate any energy in the load. However, in practice, it produces energy losses in the line. Therefore, extra care needs to be taken in designing a power system network.

For a purely resistive circuit, the voltage $\left(V = V_m \angle \theta_v\right)$ and the current $\left(I = I_m \angle \theta_i\right)$ are in phase. It means that the phase angle between them is zero,

$$\theta = \theta_v - \theta_i = 0 \tag{8.23}$$

Substituting Eq. (8.23) into Eq. (8.18) yields,

$$P_R = \frac{V_m I_m}{2} \cos 0^\circ \tag{8.24}$$

$$P_R = \frac{V_m I_m}{2} \tag{8.25}$$

$$P_R = \frac{V_m I_m}{2} \cos 0^\circ \tag{8.26}$$

$$P_R = \frac{V_m I_m}{2} = \frac{1}{2} I_m^2 R = \frac{V_m^2}{2R} \tag{8.27}$$

The phase difference between the voltage and current due to inductance and capacitance is,

$$\theta = \theta_v - \theta_i = \pm 90^\circ \tag{8.28}$$

Substituting Eq. (8.28) into Eq. (8.18) yields the average power for either an inductance or a capacitance,

$$P_L = P_C = \frac{V_m I_m}{2} \cos 90^\circ = 0 \tag{8.29}$$

The reactive power is usually stored in a circuit and it can be expressed for the inductor and capacitor as,

$$Q_L = I_L^2 X_L = \frac{V_L^2}{X_L} \tag{8.30}$$

$$Q_C = I_C^2 X_C = \frac{V_C^2}{X_C} \tag{8.31}$$

From Eqs. (8.7) and (8.29), it can be concluded that the resistive load absorbs power whereas inductive or capacitive loads do not absorb any power.

Example 8.2 An electrical series circuit with resistance, inductive and capacitive reactance is shown in Fig. 8.2. Calculate the average power supplied by the source and the power absorbed by the resistor.

Solution:
The net impedance is,

$$Z_t = 2 + j5 - j9 = 4.47\underline{|-63.43^\circ}\ \Omega \tag{8.32}$$

The source current in phasor form is,

$$I = \frac{20\underline{|15^\circ}}{4.47\underline{|-63.43^\circ}} = 4.47\underline{|78.43^\circ}\ \text{A} \tag{8.33}$$

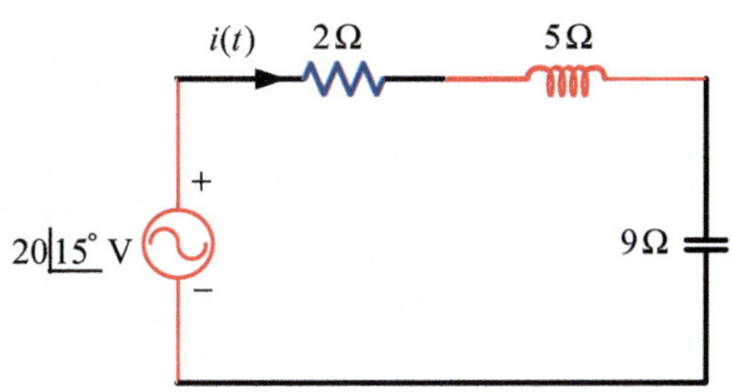

Fig. 8.2 Circuit for Example 8.2

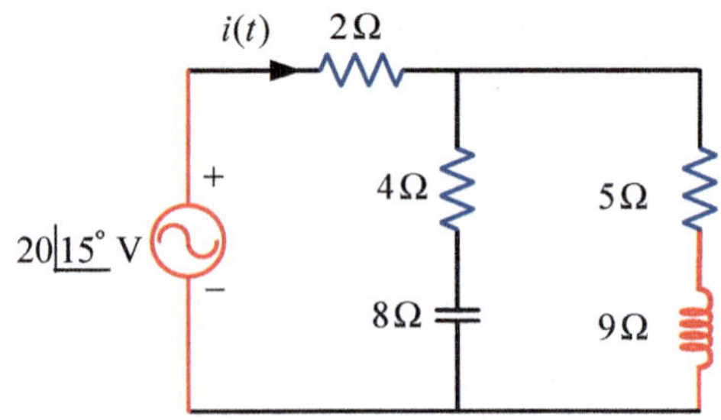

Fig. 8.3 Circuit for Practice Problem 8.2

The average power supplied by the source is,

$$P_s = \frac{1}{2} \times 20 \times 4.47 \times \cos(15^\circ - 78.43^\circ) = 20 \text{ W} \tag{8.34}$$

The average power absorbed by the resistor is,

$$P_R = \frac{1}{2} \times 4.47^2 \times 2 = 20 \text{ W} \tag{8.35}$$

Practice Problem 8.2

A series–parallel circuit with resistance, inductive and capacitive reactance is shown in Fig. 8.3. Determine the average power supplied by the source and the power absorbed by the resistors.

8.4 Apparent Power

The apparent power can be derived from the average power. But, the apparent power is related to the sinusoidal voltage and current. Let us consider that the expressions for sinusoidal voltage and current are [5, 6],

$$v(t) = V_m \sin(\omega t + \theta_v) \tag{8.36}$$

$$i(t) = I_m \sin(\omega t + \theta_i) \tag{8.37}$$

The phasor forms of these voltage and current components are,

$$V = V_m \angle \theta_v \tag{8.38}$$

$$I = I_m \angle \theta_i \tag{8.39}$$

According to Eqs. (8.20) and (8.21), the average power is,

$$P = \frac{V_m I_m}{2} \cos(\theta_v - \theta_i) = V_{\text{rms}} I_{\text{rms}} \cos(\theta_v - \theta_i) \tag{8.40}$$

The apparent power is the product of the rms voltage and rms current. The unit of apparent power is volt-amps (VA) and is denoted by the letter S. The apparent can be expressed as,

$$S = V_{\text{rms}} I_{\text{rms}} = \frac{V_m I_m}{2} \tag{8.41}$$

Substituting Eq. (8.41) into Eq. (8.40) yields,

$$P = S\cos(\theta_v - \theta_i) \tag{8.42}$$

In addition, the apparent power can be determined by the vector sum of the real power (P) and the reactive power (Q). In this case, the expression of reactive power becomes,

$$S = P + jQ \tag{8.43}$$

The power triangle (Power Triangle: A right angle triangle that shows the vector relationship between active power, reactive power and apparent power) with a lagging and leading power factor is shown in Fig. 8.4. The power triangles with an inductance and the capacitance loads will be lagging and leading, respectively.

Example 8.3 An industrial load draws a current of $i(t) = 16\sin(314t + 25^\circ)$ A from an alternating voltage source of $v(t) = 220\sin(314t + 60^\circ)$ V. Determine the apparent power, circuit resistance and inductance.

Solution:
The apparent power is calculated as,

$$S = \frac{V_m I_m}{2} = \frac{220 \times 16}{2} = 1.76 \text{ kVA} \tag{8.44}$$

The circuit impedance is calculated as,

$$Z = \frac{220\angle 60^\circ}{16\angle 25^\circ} = 11.26 + j7.89\ \Omega \tag{8.45}$$

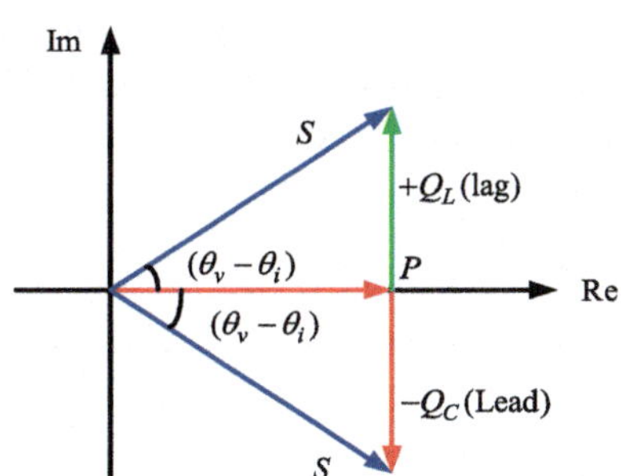

Fig. 8.4 Power triangles

The circuit resistance is,

$$R = 11.26\ \Omega \tag{8.46}$$

The circuit inductance is calculated as,

$$L = \frac{7.89}{314} = 0.025\ \text{H} \tag{8.47}$$

Practice Problem 8.3

An industrial load draws a current of $i(t) = 10\sin(100t + 55^\circ)$ A from an alternating voltage source of $v(t) = 120\sin(100t + 10^\circ)$ V. Calculate the apparent power, circuit resistance and capacitance.

8.5 Complex Power

The complex power is the combination of the real power and the reactive power. The reactive power creates adverse effect on power generation which can be studied by analysing the complex power. Mathematically, the product of half of the phasor voltage and the conjugate of the phasor current is known as complex power. The complex power is represented by the letter S_c and is expressed as,

$$S_c = \frac{1}{2}VI^* \tag{8.48}$$

Substituting Eqs. (8.38) and (8.39) into Eq. (8.48) yields,

$$S_c = \frac{1}{2}V_m\underline{|\theta_v} \times I_m\underline{|-\theta_i} \tag{8.49}$$

$$S_c = V_{\text{rms}}\underline{|\theta_v} \times I_{\text{rms}}\underline{|-\theta_i} \tag{8.50}$$

The complex power in terms of phasor form of rms voltage and current can be written as,

$$S_c = \mathbf{V}_{\text{rms}} \times \mathbf{I}^*_{\text{rms}} \tag{8.51}$$

From Eq. (8.51), the complex power is defined as the product of rms voltage and the conjugate of the rms current.

Equation (8.50) can be rearranged as,

$$S_c = V_{\text{rms}}I_{\text{rms}}\underline{|\theta_v} - \theta_i \tag{8.52}$$

$$S_c = V_{\text{rms}}I_{\text{rms}}\cos(\theta_v - \theta_i) + jV_{\text{rms}}I_{\text{rms}}\sin(\theta_v - \theta_i) \tag{8.53}$$

Consider the circuit as shown in Fig. 8.5 to explain the complex power. The impedance of this circuit is,

$$Z = R + jX \tag{8.54}$$

Equation (8.54) can be represented by the impedance triangle as shown in Fig. 8.6. The rms value of the circuit current is,

$$I_{\text{rms}} = \frac{V_{\text{rms}}}{Z} \tag{8.55}$$

Substituting Eq. (8.55) into Eq. (8.51) yields,

$$S = V_{\text{rms}} \frac{V_{\text{rms}}^*}{Z^*} = \frac{V_{\text{rms}}^2}{Z} \tag{8.56}$$

Equation (8.51) again can be represented as,

$$S_c = I_{\text{rms}} Z I_{\text{rms}}^* = I_{\text{rms}}^2 Z \tag{8.57}$$

Substituting Eq. (8.54) into Eq. (8.57) yields,

$$S_c = I_{\text{rms}}^2 (R + jX) = I_{\text{rms}}^2 R + j I_{\text{rms}}^2 X = P + jQ \tag{8.58}$$

where P and Q are the real and the imaginary parts of the complex power, and in this case, the expressions of P and Q can be written as,

$$P = \text{Re}(S_c) = I_{\text{rms}}^2 R \tag{8.59}$$

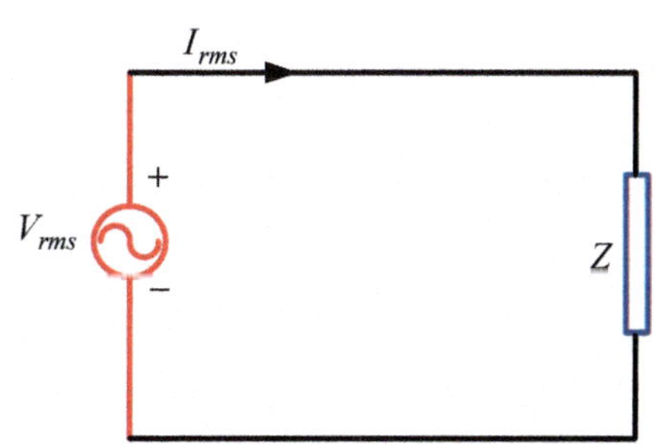

Fig. 8.5 Simple AC circuit

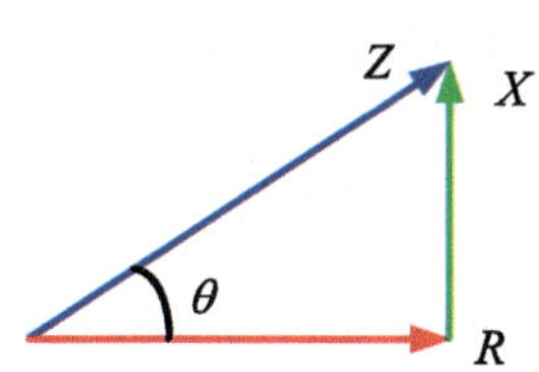

Fig. 8.6 Impedance triangle

$$Q = \mathrm{Im}(S_c) = I_{\mathrm{rms}}^2 X \tag{8.60}$$

The real, reactive and apparent power of Eq. (8.58) is shown in Fig. 8.7. The complex power for a resistive branch can be written as,

$$S_{cR} = P_R + jQ_R = I_{\mathrm{rms}}^2 R \tag{8.61}$$

From Eq. (8.61), the real power and the reactive power for the resistive branch can be expressed as,

$$P_R = I_{\mathrm{rms}}^2 R \tag{8.62}$$

$$Q = 0 \tag{8.63}$$

The complex power for an inductive branch is,

$$S_{cL} = P_L + jQ_L = jI_{\mathrm{rms}}^2 X_L \tag{8.64}$$

From Eq. (8.64), the real power and the reactive power for an inductive branch can be separated as,

$$P_L = 0 \tag{8.65}$$

$$Q_L = I_{\mathrm{rms}}^2 X_L \tag{8.66}$$

The complex power for a capacitive branch is,

$$S_{cC} = P_C + jQ_C = -jI_{\mathrm{rms}}^2 X_C \tag{8.67}$$

From Eq. (8.67), the real power and the reactive power for a capacitive branch can be separated as,

$$P_C = 0 \tag{8.68}$$

$$Q_C = I_{\mathrm{rms}}^2 X_C \tag{8.69}$$

In case of reactive power, the following points are summarized:

$Q = 0$ for resistive load, i.e. unity power factor,
$Q > 0$ for inductive load, i.e. lagging power factor,
$Q < 0$ for capacitive load, i.e. leading power factor.

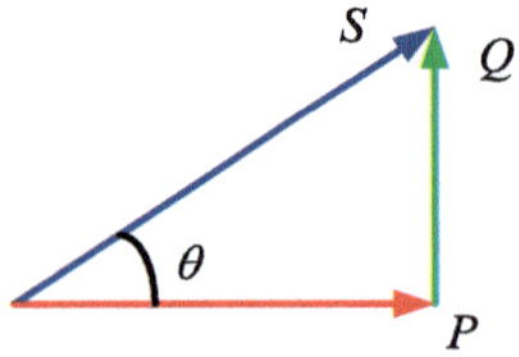

Fig. 8.7 Power triangle

Example 8.4 A series AC circuit is shown in Fig. 8.8. Calculate the source current, apparent, real and reactive powers. The expression of the alternating voltage source is $v(t) = 16\sin(10t + 25°)$ V.

Solution:
The rms value of the source voltage is,

$$V_{\text{rms}} = \sqrt{2} \times 16\angle 25° = 22.63\angle 25° \text{ V} \tag{8.70}$$

The inductive reactance is,

$$X_L = 10 \times 0.9 = 9\ \Omega \tag{8.71}$$

The circuit impedance is,

$$Z = 4 + j9 = 9.85\angle 66.04°\ \Omega \tag{8.72}$$

The source current is,

$$I_{\text{rms}} = \frac{22.63\angle 25°}{9.85\angle 66.04°} = 2.30\angle -41.04° \text{ A} \tag{8.73}$$

The complex power is calculated as,

$$S = V_{\text{rms}} I_{\text{rms}}^* = 22.63\angle 25° \times 2.30\angle 41.04° = 50.02 \text{ W} - j14.38 \text{ Var} \tag{8.74}$$

The apparent power is determined as,

$$S_{ap} = |S| = 52.05 \text{ VA} \tag{8.75}$$

The real power is calculated as,

$$P = \text{Re}(S) = 50.02 \text{ W} \tag{8.76}$$

The reactive power is determined as,

$$Q = \text{Im}(S) = 14.38\text{Var} \tag{8.77}$$

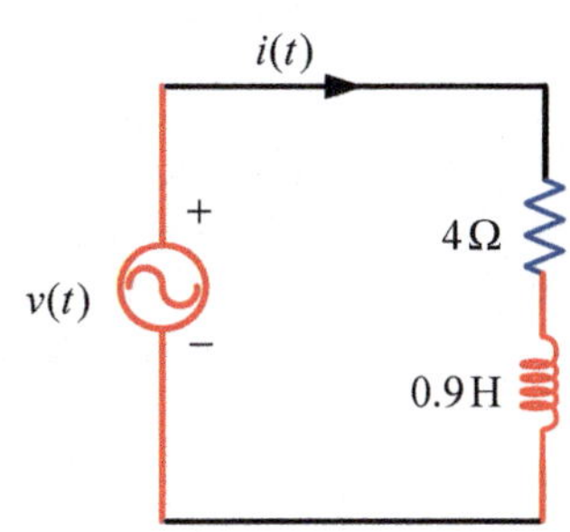

Fig. 8.8 Circuit for Example 8.4

Example 8.5 A 220 V rms delivers a power to a load. The load absorbs an average power of 10 kW at a leading power factor of 0.9. Determine the complex power and the impedance of the load.

Solution:
The power factor is,

$$\cos\theta = 0.9 \tag{8.78}$$

$$\theta = 25.84^\circ \tag{8.79}$$

The reactive component is,

$$\sin\theta = \sin 25.84^\circ = 0.44 \tag{8.80}$$

The magnitude of the complex power is calculated as,

$$|S| = \frac{P}{\cos\theta} = \frac{10}{0.9} = 11.11 \text{ kVA} \tag{8.81}$$

The reactive power is determined as,

$$Q = |S|\sin\theta = 11.11 \times 0.44 = 4.89 \text{ kW} \tag{8.82}$$

The complex power is calculated as,

$$S = P + jQ = 10 \text{ kW} - j4.89 \text{ kVar} \tag{8.83}$$

The rms voltage can be determined as,

$$P = V_{\text{rms}} I_{\text{rms}} \cos\theta = 10000 \tag{8.84}$$

$$I_{\text{rms}} = \frac{10000}{220 \times 0.9} = 50.51 \text{ A} \tag{8.85}$$

The impedance is,

$$|Z| = \frac{|V_{\text{rms}}|}{|I_{\text{rms}}|} = \frac{220}{50.51} = 4.36\ \Omega \tag{8.86}$$

$$Z = 4.36\angle{-25.84^\circ}\ \Omega \tag{8.87}$$

Practice Problem 8.4
A series RC circuit is shown in Fig. 8.9. Find the source current, apparent, real and reactive powers. The expression of alternating voltage source is $v(t) = 10\sin(2t + 12^\circ)$ V.

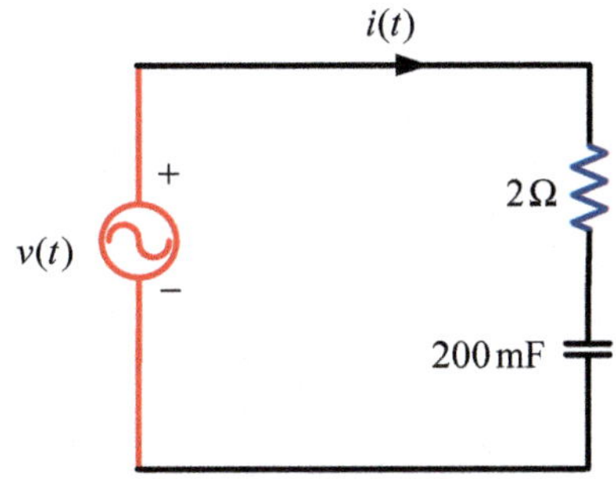

Fig. 8.9 Circuit for Practice Problem 8.4

Practice Problem 8.5
An electrical load absorbs an average power of 12 kW from a source of 230 V rms at a lagging power factor of 0.95. Calculate the complex power and the impedance of the load.

8.6 Complex Power Balance

Two electrical loads are connected in parallel with a voltage source as shown in Fig. 8.10. According to the conservation of energy, the real power delivered by the source will be equal to the total real power absorbed by the loads [7, 8]. Similarly, the complex power delivered by the source will be equal to the total complex power absorbed by the loads.

According to KCL, the rms value of the source current is equal to sum of the rms values of the branch currents I_1 and I_2, i.e.

$$I = I_1 + I_2 \tag{8.88}$$

The total complex power is defined as the product of the rms value of the source voltage and the conjugate of the current supplied by the source and it is expressed as,

$$S = V_{\text{rms}} I^* \tag{8.89}$$

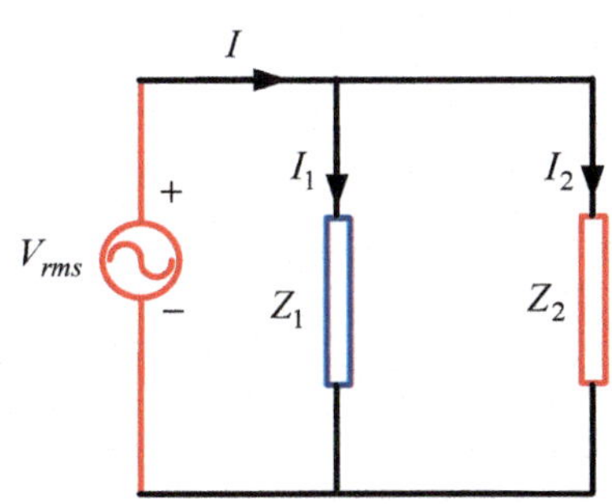

Fig. 8.10 Circuit with parallel impedances

Substituting Eq. (8.88) into Eq. (8.89) yields complex power of the parallel circuit,

$$S_p = V_{\text{rms}}[I_1 + I_2]^* \tag{8.90}$$

$$S_p = V_{\text{rms}}I_1^* + V_{\text{rms}}I_2^* \tag{8.91}$$

$$S_p = S_1 + S_2 \tag{8.92}$$

The electrical loads are again connected in series with a voltage source as shown in Fig. 8.11. According to KVL, the rms value of the source voltage is equal to the sum of the rms values of the load voltages and it is written as,

$$V_{\text{rms}} = V_1 + V_2 \tag{8.93}$$

Substituting Eq. (8.93) into Eq. (8.89) yields the complex power of the series circuit,

$$S_s = (V_1 + V_2)I^* \tag{8.94}$$

$$S_s = V_1 I^* + V_2 I^* \tag{8.95}$$

$$S_s = S_1 + S_2 \tag{8.96}$$

From Eqs. (8.92) and (8.96), it is observed that the total complex power delivered by the source is equal to the sum of the individual complex power, absorbed by the loads.

Example 8.6 A series–parallel circuit is supplied by a source of 60 V rms as shown in Fig. 8.12. Find the complex power for each branch and the total complex power.

Solution:
The circuit impedance is,

$$Z_t = 2 + \frac{(4 + j9)(3 - j8)}{4 + j9 + 3 - j8} = 13.87\underline{|-9.88^\circ}\ \Omega \tag{8.97}$$

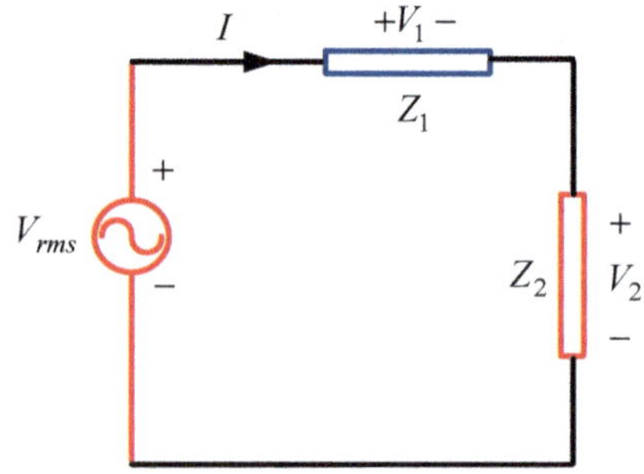

Fig. 8.11 Circuit with series impedances

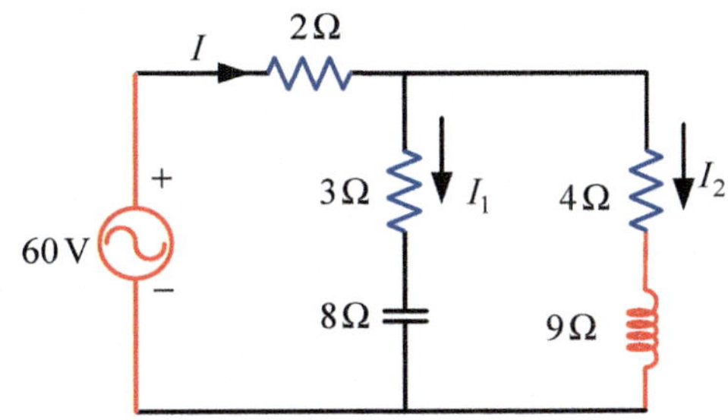

Fig. 8.12 Circuit for Example 8.6

The source current is,

$$I = \frac{60}{13.87\angle{-9.88^\circ}} = 4.33\angle 9.88^\circ \text{ A} \tag{8.98}$$

The branch currents are,

$$I_1 = 4.33\angle 9.88^\circ \times \frac{4 + j9}{7 + j1} = 6.03\angle 67.79^\circ \text{ A} \tag{8.99}$$

$$I_2 = 4.33\angle 9.88^\circ \times \frac{3 - j8}{7 + j1} = 5.23\angle 67.79^\circ \text{ A} \tag{8.100}$$

The voltage across the parallel branches is,

$$V_p = 60 - 2 \times 4.33\angle 9.88^\circ = 51.49\angle{-1.65^\circ} \text{ V} \tag{8.101}$$

The complex power in the branches are,

$$S_1 = 51.49\angle{-1.65^\circ} \times 6.03\angle{-67.79^\circ} = 310.48\angle{-69.44^\circ} \text{ VA} \tag{8.102}$$

$$S_2 = 51.49\angle{-1.65^\circ} \times 5.23\angle 67.69^\circ = 269.29\angle 66.04^\circ \text{ VA} \tag{8.103}$$

The voltage drop across the 2 Ω resistor is,

$$V_{2\Omega} = 2 \times 4.33\angle 9.88^\circ = 8.66\angle 9.88^\circ \text{ V} \tag{8.104}$$

The complex power for 2 Ω resistor is,

$$S_{2\Omega} = 8.66\angle 9.88^\circ \times 4.33\angle{-9.88^\circ} = 37.50 \text{ VA} \tag{8.105}$$

The total complex power is calculated as,

$$S_t = 310.48\angle{-69.44^\circ} + 269.29\angle 66.04^\circ + 37.50 = 259.76\angle{-9.89^\circ} \text{ VA} \tag{8.106}$$

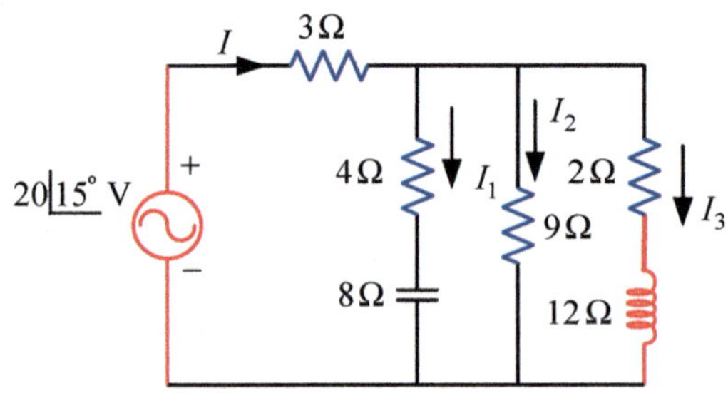

Fig. 8.13 Circuit for Practice Problem 8.6

Alternatively, the total complex power can be calculated as,

$$S = 60 \times 4.33\angle{-9.88^\circ} = 259.8\angle{-9.88^\circ}\ \text{VA} \tag{8.107}$$

Practice Problem 8.6
A series–parallel circuit is supplied by an rms voltage source as shown in Fig. 8.13. Determine the complex power for each branch and the total complex power.

8.7 Power Factor and Reactive Factor

In an AC circuit, power is calculated by multiplying a factor with the rms values of current and voltage. This factor is known as power factor. The power factor is defined as the cosine of the difference in phase angles between the voltage and the current, whereas the reactive factor is defined as the sine of the difference in the phase angles between the voltage and the current. The power factor is also defined as the cosine of the phase angle of the load impedance. A close to unity power factor represents an efficient power transfer from the source to a load, whereas a low power factor identifies inefficient transmission of power. The low power factor usually affects the power generation devices. Mathematically, the power factor is written as [9],

$$\text{pf} = \cos(\theta_v - \theta_i) \tag{8.108}$$

The reactive factor is written as,

$$\text{rf} = \sin(\theta_v - \theta_i) \tag{8.109}$$

From the impedance and power triangles shown in Figs. 8.6 and 8.7, the power factor can be written as,

$$\text{pf} = \cos\theta = \frac{R}{Z} = \frac{\text{kW}}{\text{kVA}} \tag{8.110}$$

The angle θ is positive if the current lags the voltage, and in this case, the power factor is considered as lagging. Whereas the angle θ is negative if the current leads the voltage, and in this case, the power factor is considered as leading. The leading power factor is usually considered for capacitive loads. The industrial loads are inductive in nature, and have a low lagging power factor.

The low power factor has many disadvantages, which are outlined below:

(i) kVA rating of electrical machines is increased,
(ii) larger conductor size is required to transmit or distribute electric power at a constant voltage,
(iii) copper losses are increased, and
(iv) voltage regulation is small.

8.8 Power Factor Correction

In electrical domain, heavy and medium sized industry applications contain inductive loads which draw a lagging current from the source. As a result, the reactive power for these applications are increased. In this scenario, transformer rating and the conductor size need to be increased to carry out the additional reactive power. In order to cancel this reactive component of power, an opposite type of reactance needs to be included in the circuit. Let us consider that a single-phase inductive load is connected across a voltage source as shown in Fig. 8.14 and this load draws a current with a lagging power factor of $\cos\theta_1$.

Figure 8.15 shows a circuit where the capacitor is connected in parallel with the load to improve the power factor. The capacitor will draw a current from the source that leads the source voltage by 90°. The line current is the vector sum of the currents in the inductive load and the capacitor. The current in the inductive load circuit lags the supply voltage by θ_1 and the current in the capacitor leads the voltage by 90° as shown in the vector diagram in Fig. 8.16. The exact value of the capacitor needs to be identified to improve the power factor from $\cos\phi_1$ to $\cos\phi_2$ without changing the real power. A power triangle is drawn using the inductive load

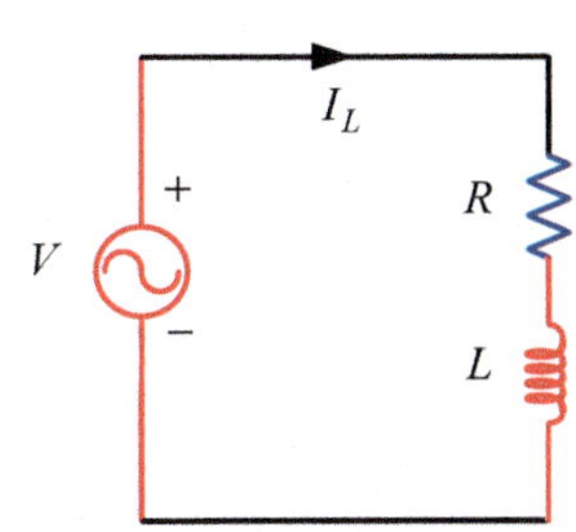

Fig. 8.14 A single-phase inductive load

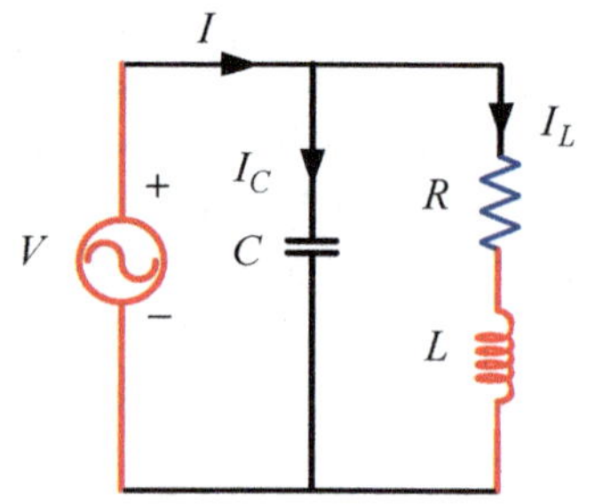

Fig. 8.15 A capacitor across single-phase inductive load

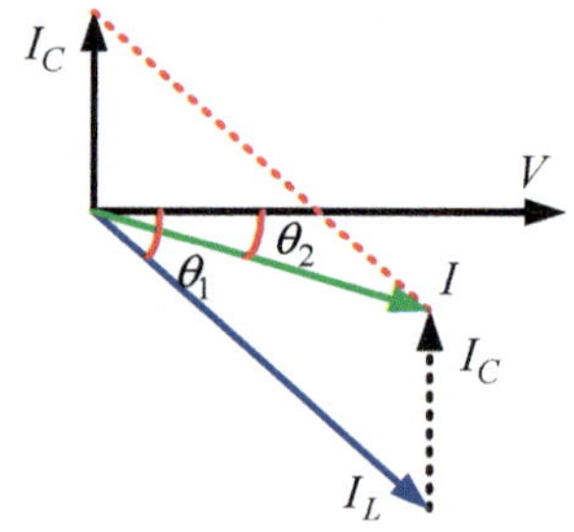

Fig. 8.16 Vector diagram with different currents

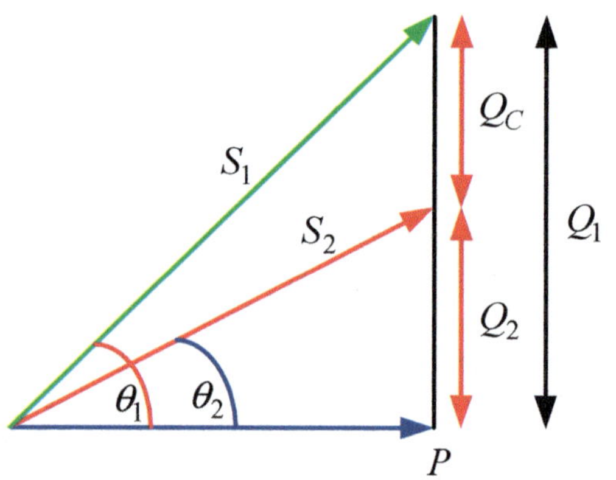

Fig. 8.17 Power triangles for inductive loads and capacitor

and the capacitor as shown in Fig. 8.17. The reactive power of the original inductive load is written as,

$$Q_1 = P \tan \phi_1 \tag{8.111}$$

The expression of new reactive power is written as,

$$Q_2 = P \tan \phi_2 \tag{8.112}$$

The reduction in reactive power due to parallel capacitor is expressed as,

$$Q_C = Q_1 - Q_2 \tag{8.113}$$

Substituting Eqs. (8.111) and (8.112) into the Eq. (8.113) yields,

$$Q_C = P(\tan\phi_1 - \tan\phi_2) \tag{8.114}$$

The reactive power due to the capacitor can be calculated as,

$$Q_C = \frac{V_{\mathrm{rms}}^2}{X_C} = \omega C V_{\mathrm{rms}}^2 \tag{8.115}$$

Substituting Eq. (8.114) into Eq. (8.115) yields the expression for the capacitor as given below:

$$\omega C V_{\mathrm{rms}}^2 = P(\tan\phi_1 - \tan\phi_2) \tag{8.116}$$

$$C = \frac{P(\tan\phi_1 - \tan\phi_2)}{\omega V_{\mathrm{rms}}^2} \tag{8.117}$$

Example 8.7 A load of 6 kVA, 50 Hz, 0.75 lagging power factor is connected across a voltage source of 120 V rms as shown in Fig. 8.18. A capacitor is connected across the load to improve the power factor to 0.95 lagging. Determine the capacitance of the connected capacitor.

Solution:
The initial power factor is,

$$\cos\theta_1 = 0.75 \tag{8.118}$$

$$\theta_1 = 41.41^\circ \tag{8.119}$$

The highest power factor is,

$$\cos\theta_2 = 0.95 \tag{8.120}$$

$$\theta_2 - 18.19^\circ \tag{8.121}$$

Fig. 8.18 Circuit for Example 8.7

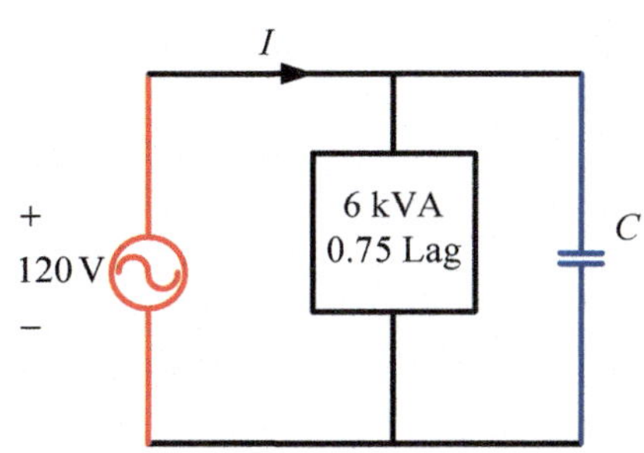

The real power is,

$$P = 6 \times 0.75 = 4.5\ \text{kW} \tag{8.122}$$

The value of the parallel capacitance can be calculated as,

$$C = \frac{4.5 \times 1000(\tan 41.41^\circ - \tan 18.19^\circ)}{2\pi \times 50 \times 120^2} = 0.55\ \text{mF} \tag{8.123}$$

Example 8.8 A load of 10 kVA, 50 Hz, 0.8 lagging power factor is connected across the voltage source shown in Fig. 8.19. A capacitor is connected across the load to improve the power factor to 0.90 lagging. Calculate the value of the capacitance and line loss with and without the capacitor.

Solution:
The initial power factor is,

$$\cos\theta_1 = 0.8 \tag{8.124}$$

$$\theta_1 = 36.87^\circ \tag{8.125}$$

The final power factor is,

$$\cos\theta_2 = 0.9 \tag{8.126}$$

$$\theta_2 = 25.84^\circ \tag{8.127}$$

The power of the load is calculated as,

$$P = 10 \times 0.8 = 8\ \text{kW} \tag{8.128}$$

The capacitor is calculated as,

$$C = \frac{8 \times 1000(\tan 36.87^\circ - \tan 25.84^\circ)}{2\pi \times 50 \times 120^2} = 0.47\ \text{mF} \tag{8.129}$$

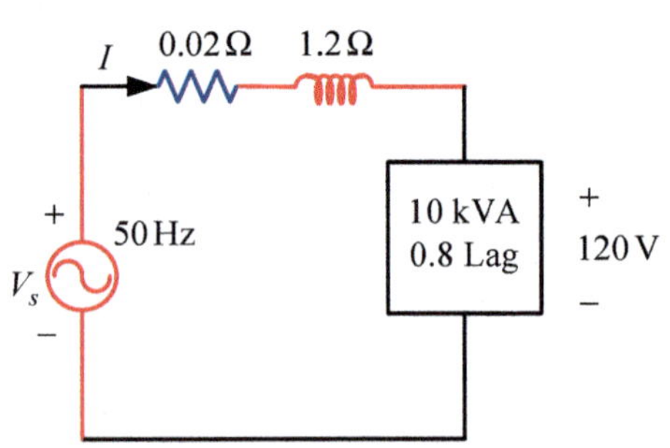

Fig. 8.19 Circuit for Example 8.8

The line current before adding capacitor is,

$$I_1 = \frac{8000}{0.8 \times 120} = 83.33 \text{ A} \tag{8.130}$$

The power loss in the line before adding capacitor is,

$$P_1 = 83.33^2 \times 0.02 = 138.88 \text{ W} \tag{8.131}$$

The apparent power with a power factor of 0.9 lagging is,

$$S = \frac{8000}{0.9} = 8888.89 \text{ VA} \tag{8.132}$$

The line current after adding capacitor is,

$$I_2 = \frac{8888.89}{120} = 74.07 \text{ A} \tag{8.133}$$

The power loss in the line after adding capacitor is,

$$P_2 = 74.07^2 \times 0.02 = 109.73 \text{ W} \tag{8.134}$$

Practice Problem 8.7
A load of 0.85 lagging power factor is connected across a voltage source of 220 V rms as shown in Fig. 8.20. A 0.56 mF capacitor is connected across the load to improve the power factor to 0.95 lagging. Find the value of the load.

Practice Problem 8.8
Two loads are connected with a source through a line as shown in Fig. 8.21. Determine the value of the voltage source.

8.9 Maximum Power Transfer

The primary goal of any power utility company is to transfer power from the power station to the consumer terminals with maximum efficiency. In this case, maximum power transfer theorem can play an important role [9]. Consider an AC source which is connected to a variable AC load as shown in Fig. 8.22.

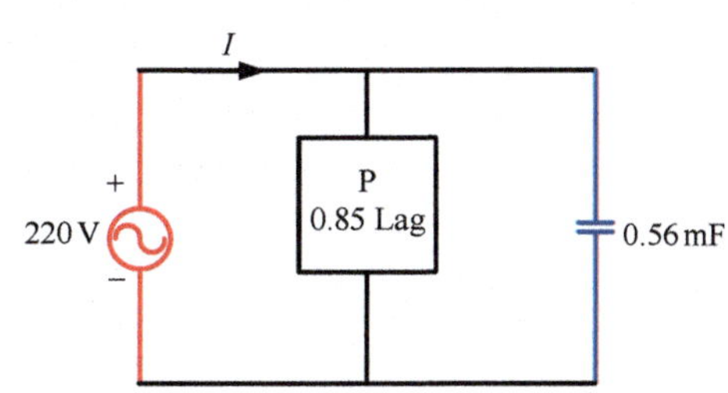

Fig. 8.20 Circuit for Practice Problem Example 8.7

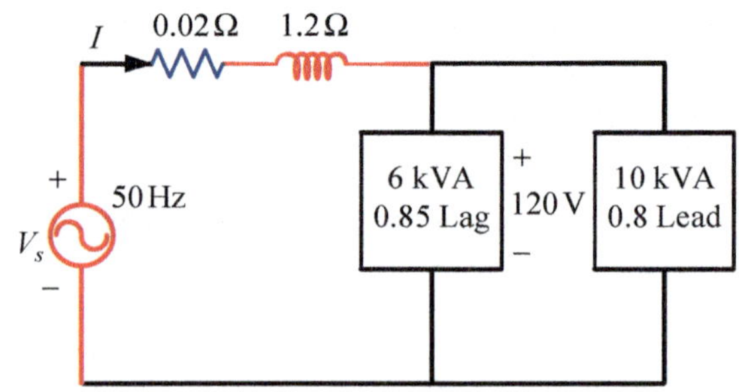

Fig. 8.21 Circuit for Practice Problem Example 8.8

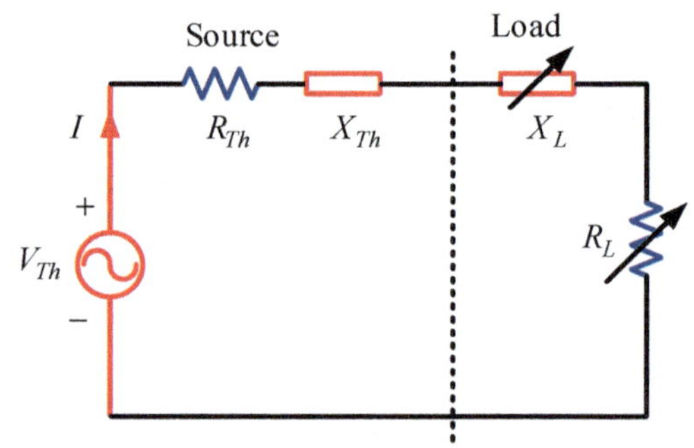

Fig. 8.22 Circuit for maximum power transfer theorem

In a rectangular form, Thevenin's and load impedances can be expressed as,

$$Z_{\text{Th}} = R_{\text{Th}} + jX_{\text{Th}} \tag{8.135}$$

$$Z_L = R_L + jX_L \tag{8.136}$$

The expression of the load current is,

$$I = \frac{V_{\text{Th}}}{Z_{\text{Th}} + Z_L} \tag{8.137}$$

Substituting Eqs. (8.135) and (8.136) into Eq. (8.137) yields,

$$I = \frac{V_{\text{Th}}}{(R_{\text{Th}} + R_L) + j(X_{\text{Th}} + X_L)} \tag{8.138}$$

The expression of average power supplied to the load is,

$$P_L = \frac{1}{2}|I|^2 R_L \tag{8.139}$$

Substituting Eq. (8.138) into Eq. (8.139) yields,

$$P_L = \frac{|V_{\text{Th}}|^2 R_L}{(R_{\text{Th}} + R_L)^2 + (X_{\text{Th}} + X_L)^2} \tag{8.140}$$

The variable load resistance and load reactance need to be varied to transfer maximum power. In this case, the following conditions must be satisfied:

$$\frac{\partial P}{\partial X_L} = 0 \tag{8.141}$$

$$\frac{\partial P}{\partial R_L} = 0 \tag{8.142}$$

Substituting Eq. (8.140) into Eq. (8.141) yields,

$$\frac{\partial P}{\partial X_L} = -\frac{1}{2}\frac{|V_{\text{Th}}|^2 R_L 2(X_{\text{Th}} + X_L)}{\left[(R_{\text{Th}} + R_L)^2 + (X_{\text{Th}} + X_L)^2\right]^2} = 0 \tag{8.143}$$

$$X_{\text{Th}} + X_L = 0 \tag{8.144}$$

$$X_L = -X_{\text{Th}} \tag{8.145}$$

Substituting Eq. (8.140) into Eq. (8.142) yields,

$$\frac{\partial P}{\partial R_L} = \frac{|V_{\text{Th}}|^2 \left\{(R_{\text{Th}} + R_L)^2 + (X_{\text{Th}} + X_L)^2 - R_L 2(R_{\text{Th}} + R_L)\right\}}{\left[(R_{\text{Th}} + R_L)^2 + (X_{\text{Th}} + X_L)^2\right]^2} = 0 \tag{8.146}$$

$$(R_{\text{Th}} + R_L)^2 + (X_{\text{Th}} + X_L)^2 - R_L 2(R_{\text{Th}} + R_L) = 0 \tag{8.147}$$

$$R_{\text{Th}}^2 - R_L^2 + (X_{\text{Th}} + X_L)^2 = 0 \tag{8.148}$$

$$R_L^2 = R_{\text{Th}}^2 + (X_{\text{Th}} + X_L)^2 = R_{\text{Th}}^2 + 0 \tag{8.149}$$

$$R_L = R_{\text{Th}} \tag{8.150}$$

The expression of load impedance can be modified as,

$$Z_L = R_L + jX_L = R_{\text{Th}} - jX_{\text{Th}} = Z_{\text{Th}}^* \tag{8.151}$$

Substituting Eq. (8.150) into Eq. (8.140) yields the expression of maximum power,

$$P_{\max} = \frac{|V_{\text{Th}}|^2}{8R_{\text{Th}}} \tag{8.152}$$

If the load is purely resistive, then Eq. (8.151) can be modified as,

$$R_L = \sqrt{R_{\text{Th}}^2 + X_{\text{Th}}^2} = |Z_{\text{Th}}| \tag{8.153}$$

Example 8.9 A load absorbs maximum power from the circuit shown in Fig. 8.23. Determine the load impedance and the maximum power.

Solution:
After opening the load impedance, calculate the value of the Thevenin impedance as,

$$Z_{\text{Th}} = 3 + \frac{(2+j8)(4-j9)}{6-j1} = 15.59 + j4.43 = 16.21\underline{|15.87^\circ}\ \Omega \tag{8.154}$$

The source current after opening the load impedance is,

$$I = \frac{60}{2+j8+4-j9} = 9.86\underline{|9.46^\circ}\ \text{A} \tag{8.155}$$

The open circuit or Thevenin's voltage is calculated as,

$$V_{\text{Th}} = 9.86\underline{|9.46^\circ} \times (4-j9) = 97.15\underline{|-56.58^\circ}\ \text{V} \tag{8.156}$$

The load impedance is calculated as,

$$Z_L = Z_{\text{Th}}^* = 15.59 - j4.43 = 16.21\underline{|-15.87^\circ}\ \Omega \tag{8.157}$$

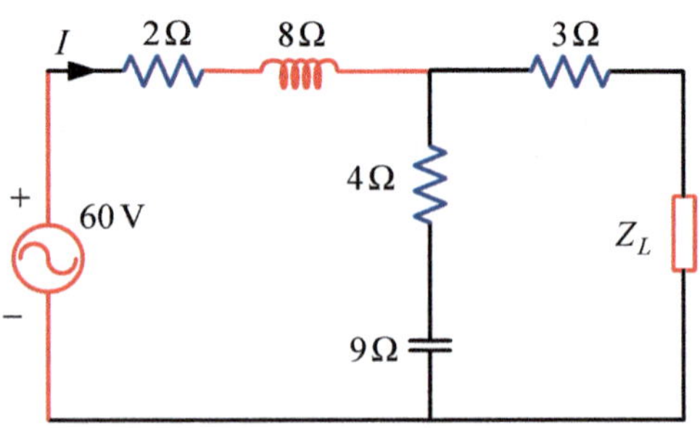

Fig. 8.23 Circuit for Example 8.9

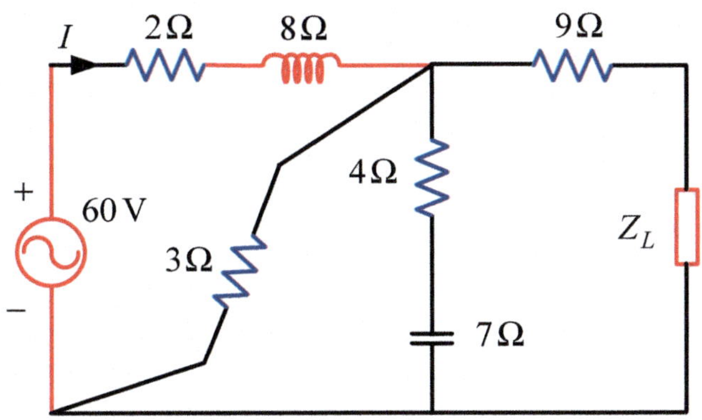

Fig. 8.24 Circuit for Practice Problem 8.9

The maximum power is determined as,

$$P_{\max} = \frac{|V_{\mathrm{Th}}|^2}{8R_{\mathrm{Th}}} = \frac{97.15^2}{8 \times 15.59} = 75.67\ \mathrm{W} \tag{8.158}$$

Practice Problem 8.9
A load absorbs maximum power from the circuit shown in Fig. 8.24. Determine the load impedance and the maximum power.

Exercise Problems

8.1 The excitation voltage and the impedance of a series circuit are given by $v(t) = 8\sin 10t$ V and $Z = 5\underline{|10^\circ}\ \Omega$, respectively. Calculate the instantaneous power.

8.2 The excitation current and the impedance of a series circuit are given by $i(t) = 4\sin(100t - 20^\circ)$A and $Z = 5\underline{|10^\circ}\ \Omega$, respectively. Determine the instantaneous power.

8.3 Calculate the average power supplied by the source and the power absorbed by the resistors as shown in Fig. 8.25.

Fig. 8.25 Circuit for Problem 8.3

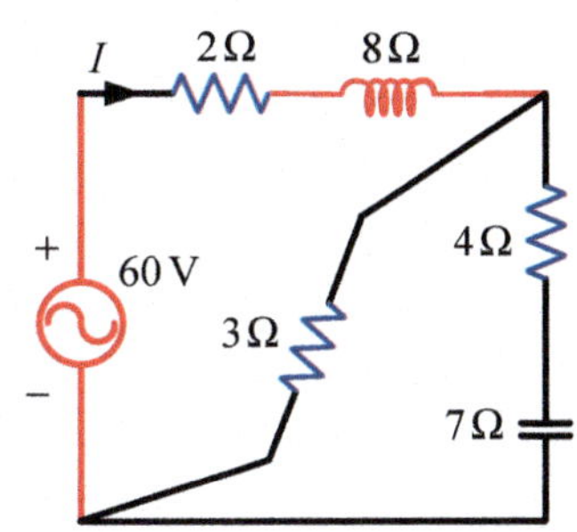

8.4 Determine the average power supplied by the source and the power absorbed by the resistors shown in Fig. 8.26.

8.5 Calculate the average power supplied by the source and the power absorbed by the resistors as shown in Fig. 8.27.

8.6 Determine the average power supplied by the source and the power absorbed by the resistors shown in Fig. 8.28.

8.7 Find the total average power absorbed by all the resistors in the circuit shown in Fig. 8.29.

8.8 An industrial load is connected across an alternating voltage source of $v(t) = 230\sin(314t + 20°)$ V that draws a current of $i(t) = 15\sin(314t + 45°)$ A. Determine the apparent power, circuit resistance and capacitance.

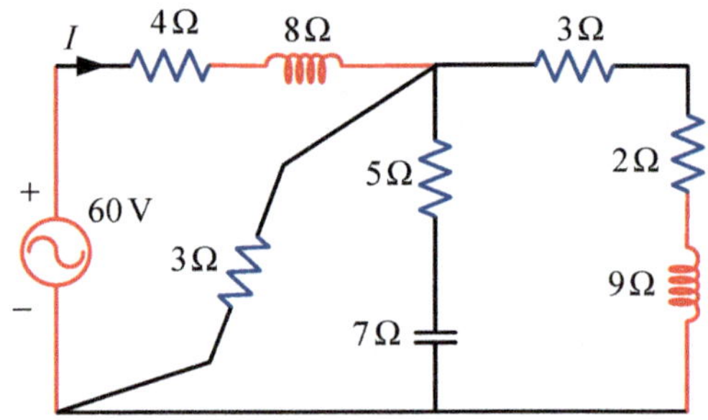

Fig. 8.26 Circuit for Problem 8.4

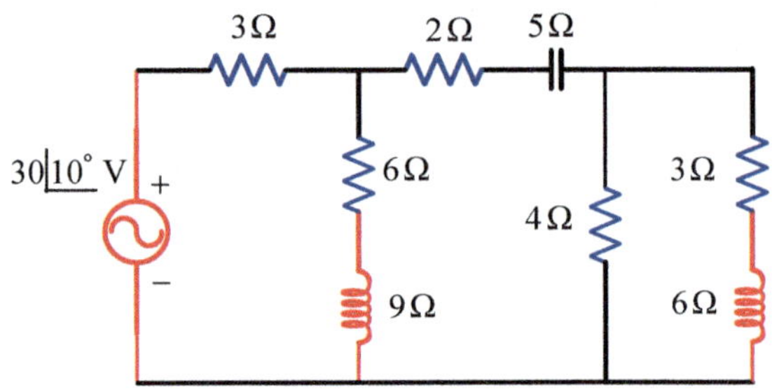

Fig. 8.27 Circuit for Problem 8.5

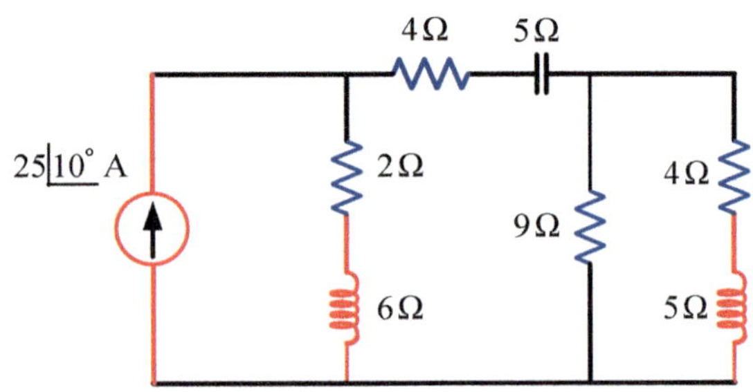

Fig. 8.28 Circuit for Problem 8.6

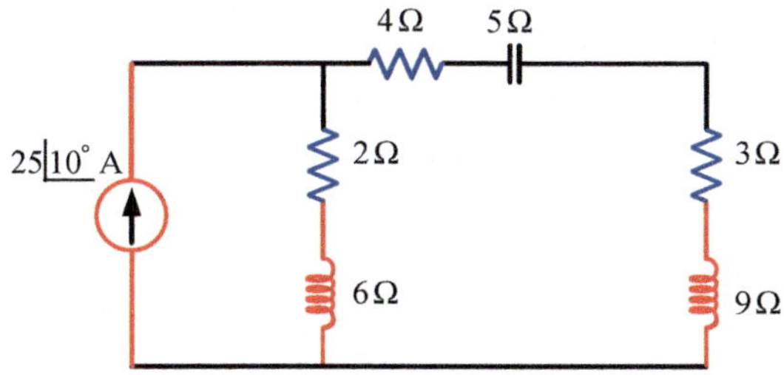

Fig. 8.29 Circuit for Problem 8.7

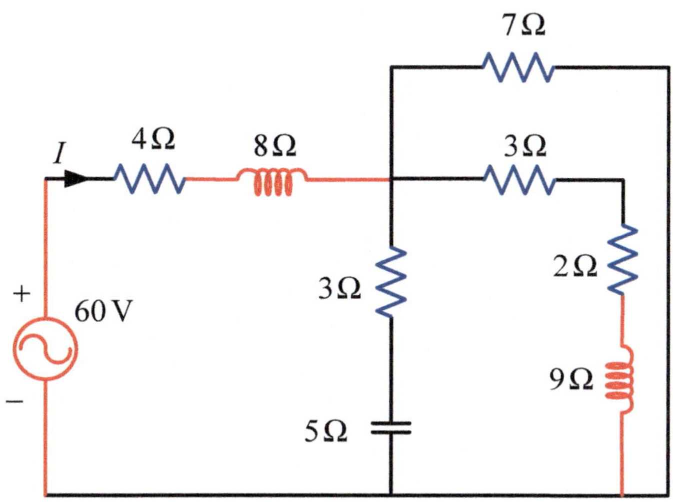

Fig. 8.30 Circuit for Problem 8.11

8.9 The rms values of voltage and current are given by $V = 20\angle{-15^\circ}$ V, and $I = 3\angle{25^\circ}$ A. Calculate the complex power, real power and reactive power.

8.10 The rms values of voltage are given by $V = 34\angle{25^\circ}$ V and the impedance is $Z = 6\angle{-15^\circ}\ \Omega$. Determine the complex power, real power and reactive power.

8.11 A series–parallel circuit is supplied by an rms source of 60 V as shown in Fig. 8.30. Find the complex power for each branch and the total complex power.

8.12 Calculate the total complex power and complex power of all branches of the circuit shown in Fig. 8.31.

8.13 A 10 kVA, 50 Hz, 0.6 lagging power factor load is connected across an rms voltage source of 220 V as shown in Fig. 8.32. A capacitor is connected across the load to improve the power factor to 0.85 lagging. Find the capacitance of the connected capacitor.

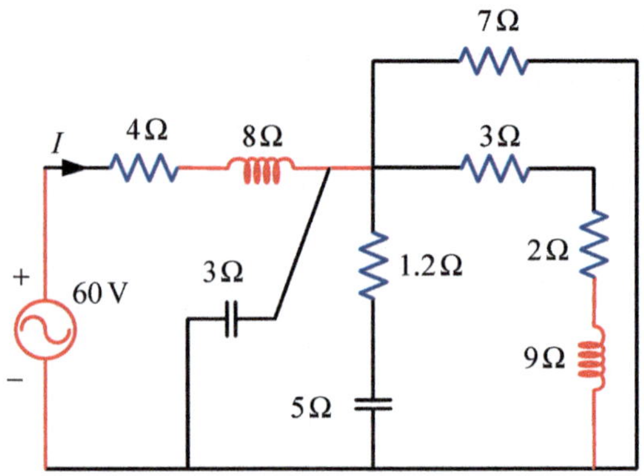

Fig. 8.31 Circuit for Problem 8.12

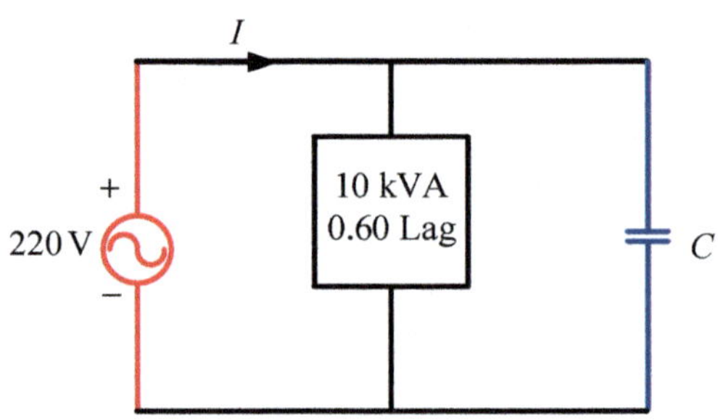

Fig. 8.32 Circuit for Problem 8.13

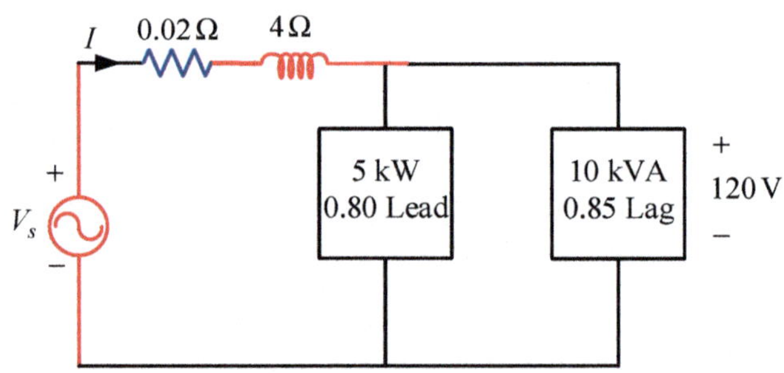

Fig. 8.33 Circuit for Problem 8.14

8.14 Two loads with different power factor are connected with the source through a transmission line as shown in Fig. 8.33. Determine the source current and the source voltage.

8.15 A voltage source delivers power to the three loads shown in Fig. 8.34. Find the source current and the source voltage.

8.16 A load absorbs maximum power from the circuit shown in Fig. 8.35. Determine the load impedance and the maximum power.

8.17 A load absorbs maximum power from the circuit shown in Fig. 8.36. Calculate the load impedance and the maximum power.

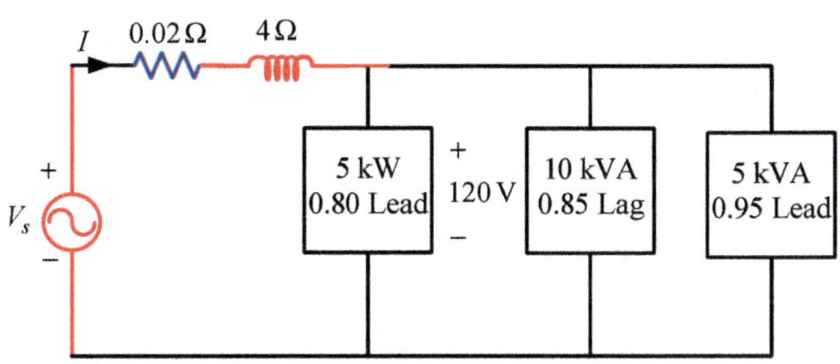

Fig. 8.34 Circuit for Problem 8.15

Fig. 8.35 Circuit for Problem 8.16

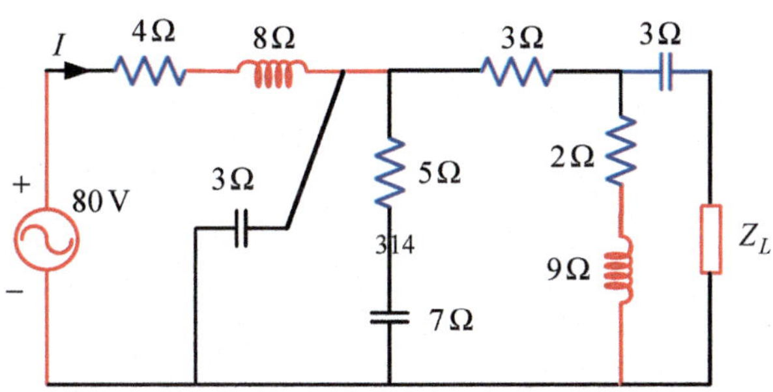

Fig. 8.36 Circuit for Problem 8.17

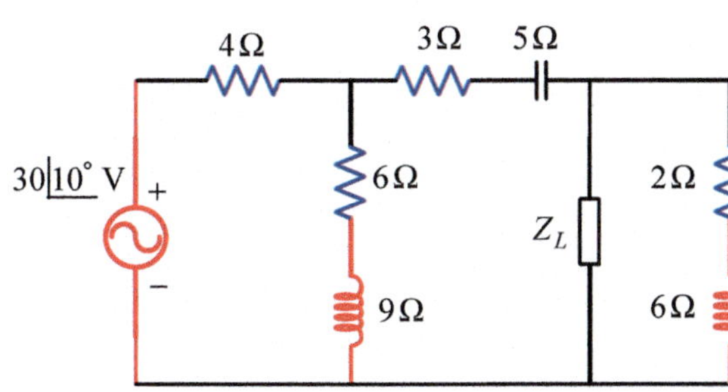

References

1. C.K. Alexander, M.N.O. Sadiku, *Fundamentals of Electric Circuits*, 6th edn. (McGraw-Hill Higher Education, Jan 2016)
2. R.L. Boylestad, *Introductory Circuit Analysis*, 13th edn. (Pearson, 2016)
3. J.W. Nilsson, S.A. Riedel, *Electric Circuits*, 10th edn. (Prentice Hall International Edition, 2015)
4. H.W. Jackson, Dale Temple, B.E. Kelly, *Introduction to Electric Circuits*, 9th edn. (Oxford University Press, July 2015)
5. D. Bell, *Fundamentals of Electric Circuits*, 7th edn. (Oxford University Press, March 2007)
6. G. Rizzoni, J. Kearns, *Principles and Applications of Electrical Engineering*, 6th edn. (McGraw-Hill Education, Oct 2014)
7. J. David Irwin, R. Mark Nelms, *Basic Engineering Circuit Analysis*, 11th edn. (Wiley, USA, 2015)
8. W. Hayt, J. Kemmerly, S. Durbin, *Engineering Circuit Analysis*, 8th edn. (McGraw-Hill Education, 2012)
9. M.A. Salam, *Basic Electrical Circuits*, 2nd edn. (Shroff Publishers & Distributors Pvt. Ltd, India, 2007)

Chapter 9
Three-Phase Circuits

9.1 Introduction

Single-phase circuit with its detail analysis has been presented in the previous chapter. Three-phase circuits are most common for generation, transmission and distribution of electric power. Over single-phase circuits, three-phase circuits or systems have many advantages such as better voltage regulation and require only three-fourth weight of the copper to transmit the same amount of power over a fixed distance at a given voltage, more economical for transmitting power at a constant power loss. Three-phase machines are smaller, simpler in construction and have better operating characteristics. In this chapter, three-phase voltage generation, voltage and current relationship, different types of power for wye connection and delta connection have been discussed.

9.2 Three-Phase Voltage Generation

Figure 9.1 shows a two-pole three-phase AC generator for three-phase voltage generation. Coils aa', bb' and cc' represent the whole coils into a three-phase system as shown in Fig. 9.1a. The rotor of the AC machine is energized by the dc source which creates the magnetic field. This rotor is attached to the turbine through a soft coupling and this turbine rotates the rotor. According to Faraday's law of electromagnetic induction, three-phase voltages V_{an}, V_{bn} and V_{cn} will be been generated across the generator terminals. The magnitude of these voltages is constant and is displaced from each other by 120 electrical degrees as shown in Fig. 9.1b.

The waveforms of the generated voltages are shown in Fig. 9.2. The expression of the generated voltages can be represented as [1, 2],

Md. A. Salam and Q. Mehbubar Rahman, *Fundamentals of Electrical Circuit Analysis*,
https://doi.org/10.1007/978-981-10-8624-3_9

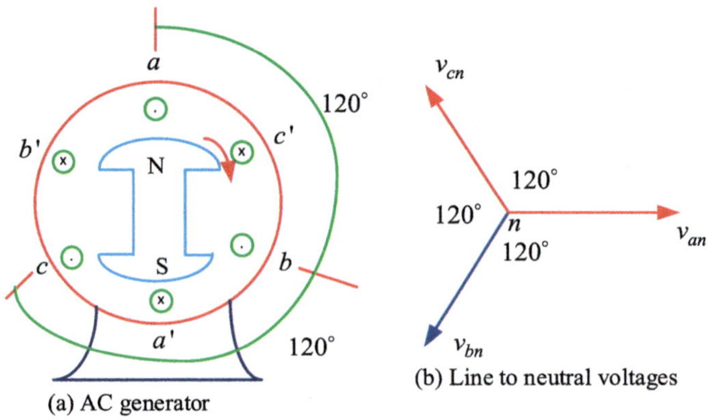

Fig. 9.1 Schematic of three-phase AC generator and generation voltages

Fig. 9.2 Three-phase voltage waveforms

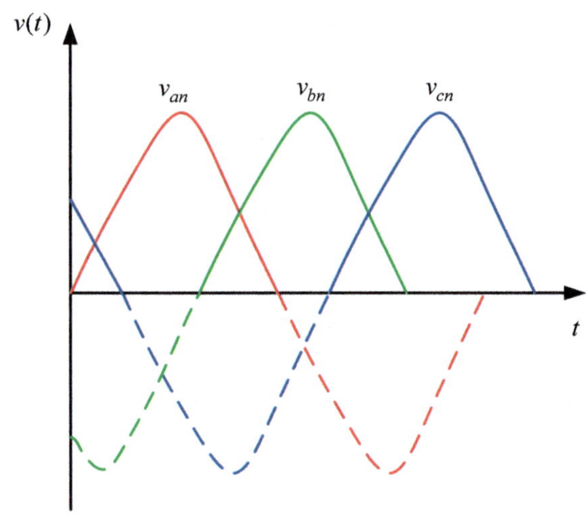

$$v_{an} = V_{an} \sin \omega t \tag{9.1}$$

$$v_{bn} = V_{bn} \sin(\omega t - 120^\circ) \tag{9.2}$$

$$v_{cn} = V_{cn} \sin(\omega t - 240^\circ) \tag{9.3}$$

where V_{an}, V_{bn} and V_{cn} are the magnitudes of the line to neutral or phase voltages. These voltages are constant in magnitude and it can be expressed as,

$$|V_{an}| = |V_{bn}| = |V_{cn}| = V_p \tag{9.4}$$

The phasor of the generated voltages can be written as,

$$V_{\mathrm{an}} = V_{\mathrm{P}}\angle 0^{\circ} \tag{9.5}$$

$$V_{\mathrm{bn}} = V_{\mathrm{P}}\angle -120^{\circ} \tag{9.6}$$

$$V_{\mathrm{cn}} = V_{\mathrm{P}}\angle -240^{\circ} = V_{\mathrm{P}}\angle 120^{\circ} \tag{9.7}$$

The sum of phasor voltages and the sum of sinusoidal voltages are zero and these can be expressed as,

$$V = V_{\mathrm{an}} + V_{\mathrm{bn}} + V_{\mathrm{cn}} = V_{\mathrm{P}}\angle 0^{\circ} + V_{\mathrm{P}}\angle -120^{\circ} + V_{\mathrm{P}}\angle -240^{\circ} = 0 \tag{9.8}$$

$$v = V_{\mathrm{an}} + v_{\mathrm{bn}} + v_{\mathrm{cn}} = V_{\mathrm{an}} \sin \omega t + V_{\mathrm{bn}} \sin(\omega t - 120^{\circ}) + V_{\mathrm{cn}} \sin(\omega t - 240^{\circ}) = 0 \tag{9.9}$$

9.3 Phase Sequence

The phase sequence is very important for the interconnection of three-phase transformer, motor and other high voltage equipment. The three-phase systems are numbered either by the numbers 1, 2 and 3 or by the letters *a*, *b* and *c*. Sometimes, these are labelled by the colours red, yellow and blue or *RYB* in short. The generator is said to have a positive phase sequence when the generated voltages reach to their maximum or peak values in the sequential order of *abc*. Whereas the generator is said to have a negative phase sequence when the generated voltages reach to their maximum or peak values in the sequential order of *acb*. Figure 9.3 shows the positive and negative phase sequences.

Here, the voltage V_{an} is considered to be the reference voltage while the direction of rotation is considered to be anticlockwise. In the positive phase sequence, the crossing sequence of voltage rotation is identified by $V_{\mathrm{an}} - V_{\mathrm{bn}} - V_{\mathrm{cn}}$, whereas for negative phase sequence, it is identified as $V_{\mathrm{an}} - V_{\mathrm{cn}} - V_{\mathrm{bn}}$.

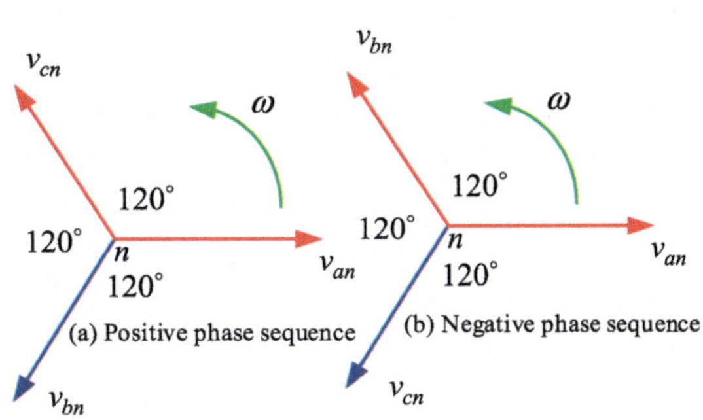

Fig. 9.3 Phase sequence identification

9.4 Wye Connection

Three-phase transformer, AC generator and induction motor are connected either in wye or in delta connection. In wye connection, one terminal of each coil is connected together to form a common or neutral point and other terminals are connected to the three-phase supply. The voltage between any line and neutral is known as the phase voltage and the voltage between any two lines is called the line voltage.

The line voltage and phase voltage are usually represented by V_L and V_P, respectively. The important points of this connection are the line voltage is equal to $\sqrt{3}$ times the phase voltage, the line current is equal to the phase current and the current (I_n) in the neutral wire is equal to the phasor sum of the three-line currents. For a balanced three-phase load, the neutral current is zero, i.e. $I_n = 0$. The wye-connected generator and load are shown in Fig. 9.4.

9.5 Analysis for Wye Connection

A three-phase wye-connected AC generator shown in Fig. 9.5 is considered for analysis. Here, V_{an}, V_{bn} and V_{cn} are the phase voltages and V_{ab}, V_{bc} and V_{ca} are the line voltages, respectively. Applying KVL to the circuit to find the line voltage between lines a and b yields,

$$V_{an} - V_{bn} - V_{ab} = 0 \tag{9.10}$$

$$V_{ab} = V_{an} - V_{bn} \tag{9.11}$$

Substituting Eqs. (9.5) and (9.6) into Eq. (9.11) yields,

$$V_{ab} = V_P\angle 0^\circ V_P\angle -120^\circ \tag{9.12}$$

$$V_{ab} = \sqrt{3}\, V_P\angle 30^\circ \tag{9.13}$$

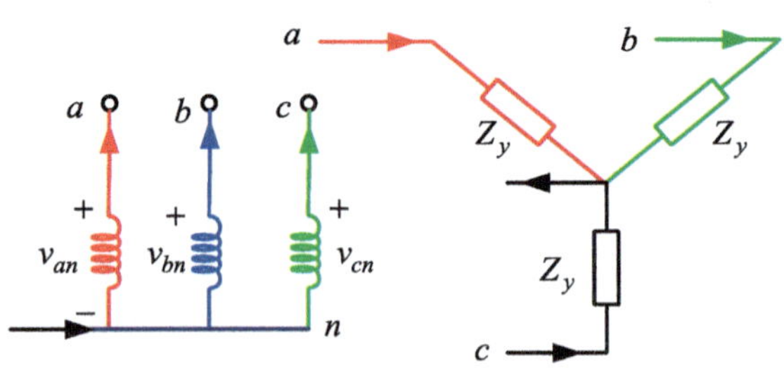

Fig. 9.4 Wye connection generator and load

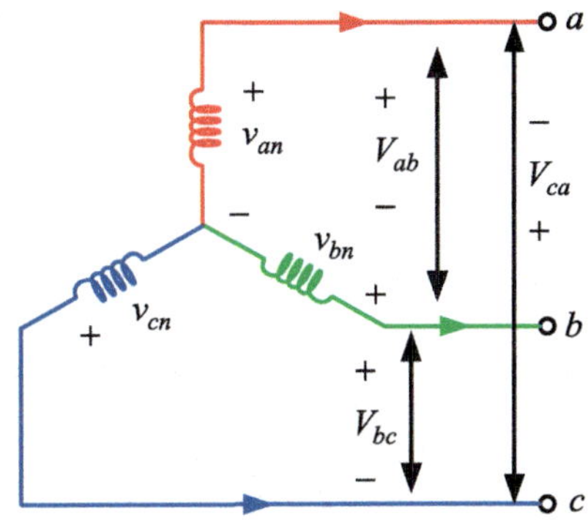

Fig. 9.5 Wye connection generator

Applying KVL between lines b and c yields the expression of line voltage as,

$$V_{\mathrm{bc}} = V_{\mathrm{bn}} - V_{\mathrm{cn}} \tag{9.14}$$

Substituting Eqs. (9.6) and (9.7) into Eq. (9.14) yields,

$$V_{\mathrm{bc}} = V_{\mathrm{P}}\angle{-120^\circ} - V_{\mathrm{P}}\angle{-240^\circ} = \sqrt{3}\,V_{\mathrm{P}}\angle{-90^\circ} \tag{9.15}$$

Applying KVL between lines c and a yields the expression of the line voltage as,

$$V_{\mathrm{ca}} = V_{\mathrm{cn}} - V_{\mathrm{an}} \tag{9.16}$$

Substituting Eqs. (9.5) and (9.7) into Eq. (9.16) yields,

$$V_{\mathrm{ca}} = V_{\mathrm{P}}\angle{-240^\circ} - V_{\mathrm{P}}\angle{0^\circ} = \sqrt{3}\,V_{\mathrm{P}}\angle{150^\circ} \tag{9.17}$$

Line voltages with angles are drawn as shown in Fig. 9.6. From Eqs. (9.13), (9.15) and (9.17), it is seen that the magnitude of the line voltage is equal to $\sqrt{3}$ times the magnitude of the phase voltage. The general relationship between the line voltage and the phase voltage can be written as,

$$V_{\mathrm{L}} = \sqrt{3}\,V_{\mathrm{P}} \tag{9.18}$$

From Fig. 9.5, it is also observed that the phase current is equal to the line current and it is written as,

$$I_{\mathrm{L}} = I_{\mathrm{P}} \tag{9.19}$$

Alternative approach: A vector diagram with phase voltages is drawn using the lines a and c as shown in Fig. 9.7. A perpendicular line is drawn from point A, which divides the line BD equally. From the triangle ABC, the following expression relation can be written:

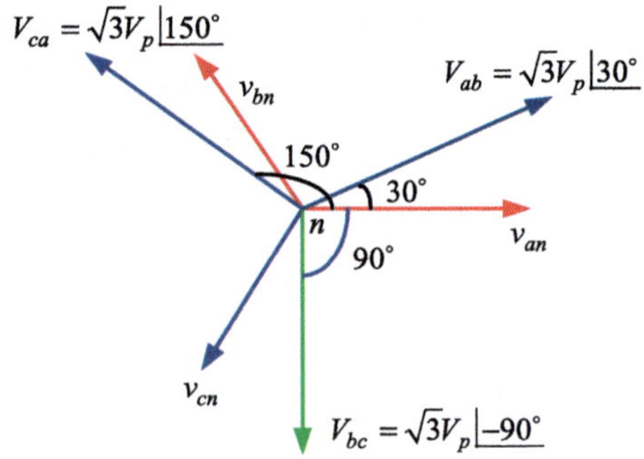

Fig. 9.6 Line voltages with angles

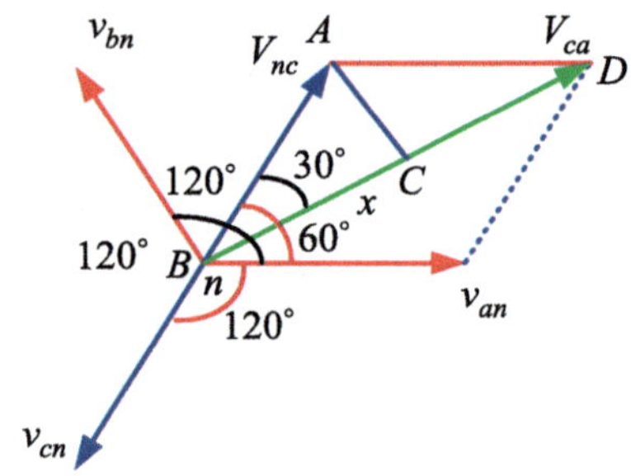

Fig. 9.7 A resultant vector with two lines

$$\cos 30° = \frac{\mathrm{BC}}{\mathrm{AB}} \tag{9.20}$$

$$\frac{\sqrt{3}}{2} = \frac{x}{|V_{\mathrm{nc}}|} \tag{9.21}$$

According to Fig. 9.7, the following expression can be written:

$$\mathrm{BD} = 2\mathrm{AC} \tag{9.22}$$

$$V_{\mathrm{ca}} = 2x \tag{9.23}$$

Substituting Eq. (9.21) into Eq. (9.23) yields,

$$V_{\mathrm{ca}} = 2 \times \frac{\sqrt{3}}{2}|V_{\mathrm{nc}}| \tag{9.24}$$

$$V_{\mathrm{ca}} = \sqrt{3}V_{\mathrm{nc}} \tag{9.25}$$

$$V_{\mathrm{L}} = \sqrt{3}V_{\mathrm{p}}$$

Example 9.1 The phase voltage is given by $V_{an} = 230\angle 10^\circ$ V. For abc phase sequence, determine V_{bn} and V_{cn}.

Solution:
The phase voltage for line *a* is,

$$V_{an} = 230\angle 10^\circ \text{ V} \tag{9.26}$$

The phase voltage for line *b* is,

$$V_{bn} = 230\angle 10^\circ - 120^\circ = 230\angle -110^\circ \text{ V} \tag{9.27}$$

The phase voltage for line *c* is,

$$V_{cn} = 230\angle 10^\circ - 240^\circ = 230\angle -230^\circ \text{ V} \tag{9.28}$$

Practice Problem 9.1
The phase voltage is given by $V_{bn} = 200\angle 10^\circ$ V. For abc phase sequence, calculate V_{an} and V_{cn}.

Example 9.2 A wye-connected generator generates a voltage of 180 V rms as shown in Fig. 9.8. For abc phase sequence, write down the phase and line voltages.

Solution:
The phase voltages are,

$$V_{an} = 180\angle 0^\circ \text{ V} \tag{9.29}$$

$$V_{bn} = 180\angle -120^\circ \text{ V} \tag{9.30}$$

$$V_{cn} = 180\angle -240^\circ \text{ V} \tag{9.31}$$

The line voltages are calculated as,

$$V_{ab} = \sqrt{3} \times 180\angle 30^\circ = 311.77\angle 30^\circ \text{ V} \tag{9.32}$$

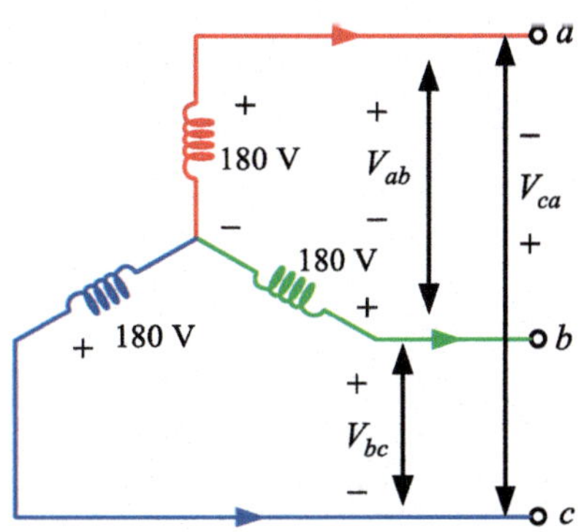

Fig. 9.8 Circuit for Example 9.2

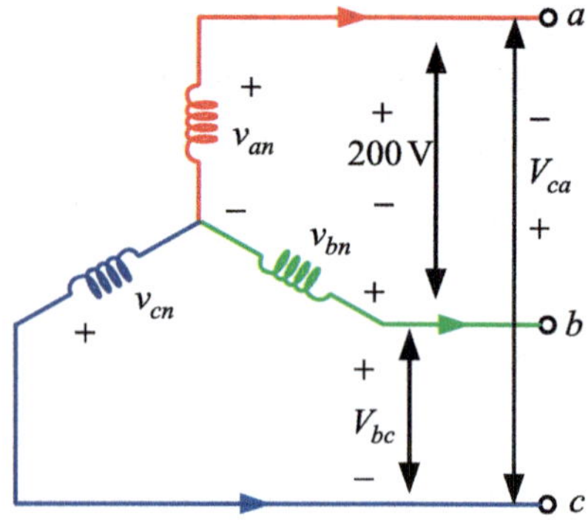

Fig. 9.9 Circuit for Practice Problem 9.2

$$V_{\mathrm{bc}}=\sqrt{3}\times 180\angle -120^{\circ}+30^{\circ}=311.77\angle -90^{\circ}\ \mathrm{V}$$

$$V_{\mathrm{ca}}=\sqrt{3}\times 180\angle -240^{\circ}+30^{\circ}=311.77\angle -210^{\circ}\ \mathrm{V} \tag{9.34}$$

Practice Problem 9.2
A wye-connected generator generates the line-to-line voltage of 200 V rms as shown in Fig. 9.9. For abc phase sequence, write down the phase voltages.

9.6 Wye–Wye Connection

Consider that a balanced three-phase wye-connected generator delivers a power to a wye-connected load as shown in Fig. 9.10. Transmission lines are used to connect the generator and load. The line currents can be determined as [3, 4],

$$I_{\mathrm{aA}}=\frac{V_{\mathrm{p}}\angle 0^{\circ}}{Z_{\mathrm{line}}+Z_y}=\frac{V_{\mathrm{p}}}{Z_t}\angle -\theta_z \tag{9.35}$$

$$I_{\mathrm{bB}}=\frac{V_{\mathrm{p}}\angle -120^{\circ}}{Z_{\mathrm{line}}+Z_y}=\frac{V_{\mathrm{p}}}{Z_t}\angle -120^{\circ}-\theta_z \tag{9.36}$$

$$I_{\mathrm{cC}}=\frac{V_{\mathrm{p}}\angle -240^{\circ}}{Z_{\mathrm{line}}+Z_y}=\frac{V_{\mathrm{p}}}{Z_t}\angle -240^{\circ}-\theta_z \tag{9.37}$$

The current in the neutral line is expressed as,

$$I_n=I_{\mathrm{aA}}+I_{\mathrm{bB}}+I_{\mathrm{cC}} \tag{9.38}$$

Example 9.3 A balanced wye-connected generator delivers power to the balanced wye-connected load as shown in Fig. 9.11. The line voltage of the generator is 200 V rms. Determine the line currents for abc phase sequence.

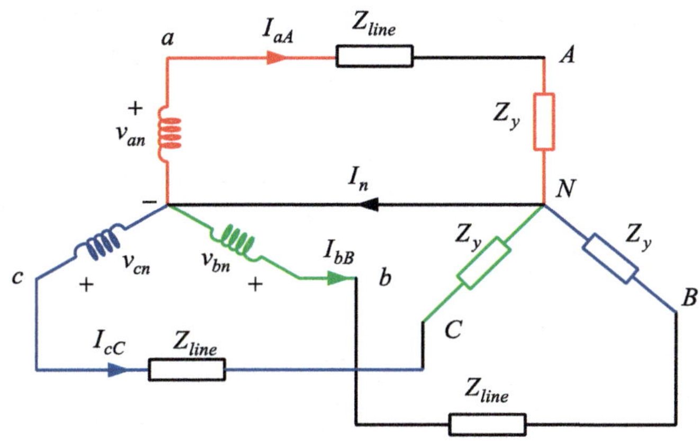

Fig. 9.10 A balanced wye–wye circuit

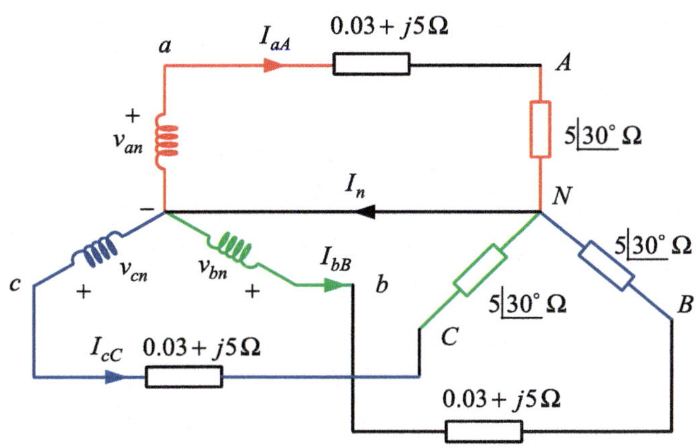

Fig. 9.11 Circuit for Example 9.3

Solution:
For abc phase sequence, the phase voltages are,

$$V_{an} = \frac{V_L}{\sqrt{3}}\angle -30^\circ = \frac{200}{\sqrt{3}}\angle -30^\circ = 115.47\angle -30^\circ \text{ V} \tag{9.39}$$

$$V_{bn} = 115.47\angle -150^\circ \text{ V} \tag{9.40}$$

$$V_{cn} = 115.47\angle 90^\circ \text{ V} \tag{9.41}$$

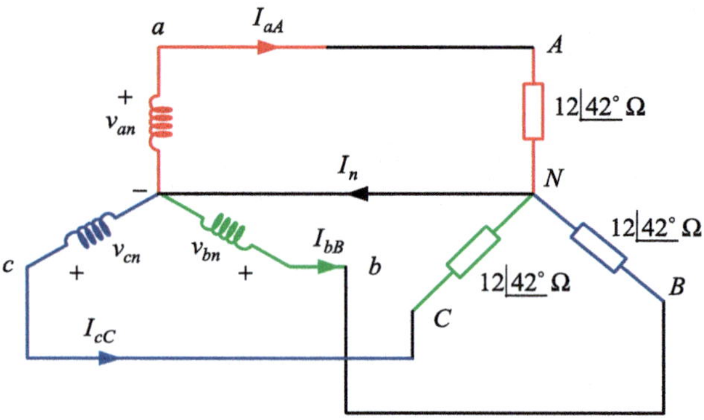

Fig. 9.12 Circuit for Practice Problem 9.3

The total impedance is,

$$Z_t = 0.03 + j5 + 5\angle 30° = 8.68\angle 59.83°\ \Omega \tag{9.42}$$

The line currents are calculated as,

$$I_{\mathrm{aA}} = \frac{V_{\mathrm{an}}}{Z_t} = \frac{115.47\angle -30°}{8.68\angle 59.83°} = 13.30\angle -89.83°\ \mathrm{A} \tag{9.43}$$

$$I_{\mathrm{bB}} = \frac{V_{\mathrm{bn}}}{Z_t} = \frac{115.47\angle -150°}{8.68\angle 59.83°} = 13.30\angle 150.17°\ \mathrm{A} \tag{9.44}$$

$$I_{\mathrm{bB}} = \frac{V_{\mathrm{bn}}}{Z_t} = \frac{115.47\angle 90°}{8.68\angle 59.83°} = 13.30\angle 30.17°\ \mathrm{A} \tag{9.45}$$

Practice Problem 9.3
A balanced wye-connected load is energized by a balanced wye-connected generator as shown in Fig. 9.12. The line voltage of the generator is 400 V rms. Calculate the line currents for abc phase sequence.

9.7 Delta Connection

The coils in a delta-connected circuit are arranged in such a way that a looking structure is formed [5]. The delta connection is formed by connecting point a_2 of a_1a_2 coil to the point b_1 of b_1b_2 coil, point b_2 of b_1b_2 coil to point c_1 of c_1c_2 coil, and point c_2 of c_1c_2 coil to point a_1 of a_1a_2 coil. In this connection, the phase voltage is equal to the line voltage, and line current is equal to $\sqrt{3}$ times the phase current. Figure 9.13 shows delta-connected generator and load.

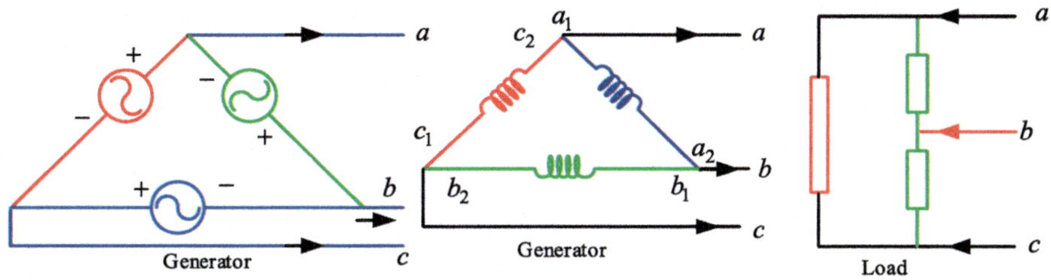

Fig. 9.13 Delta-connected generator and load

9.8 Analysis for Delta Connection

A three-phase delta-connected load is shown in Fig. 9.14. In this connection, $I_{\text{ab}}, I_{\text{bc}}$ and I_{ca} are the phase currents and I_a, I_b and I_c are the line currents. For abc phase sequence, the phase currents can be written as,

$$I_{\text{ab}} = I_{\text{P}}\angle 0^\circ \tag{9.46}$$

$$I_{\text{bc}} = I_{\text{P}}\angle -120^\circ \tag{9.47}$$

$$I_{\text{ca}} = I_{\text{P}}\angle -240^\circ \tag{9.48}$$

Applying KCL at node a of the circuit in Fig. 9.14 yields,

$$I_a = I_{\text{ab}} - I_{\text{ca}} \tag{9.49}$$

Substituting Eqs. (9.46) and (9.48) into Eq. (9.49) yields,

$$I_a = I_{\text{P}}\angle 0^\circ - I_{\text{P}}\angle -120^\circ = \sqrt{3}\, I_{\text{P}}\angle -30^\circ \tag{9.50}$$

Applying KCL at node b of the circuit in Fig. 9.14 yields,

$$I_b = I_{\text{bc}} - I_{\text{ab}} \tag{9.51}$$

Fig. 9.14 Delta-connected load

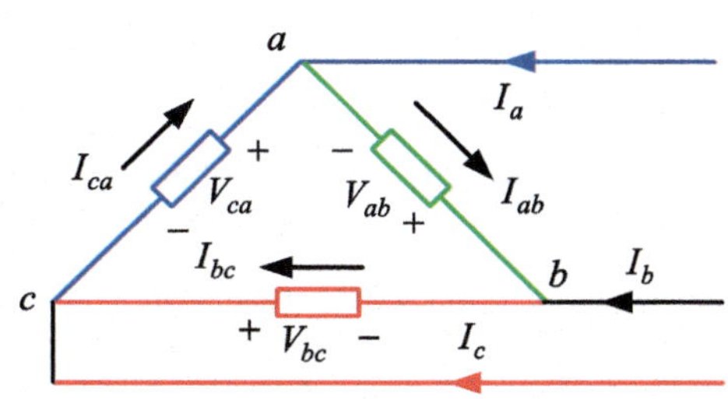

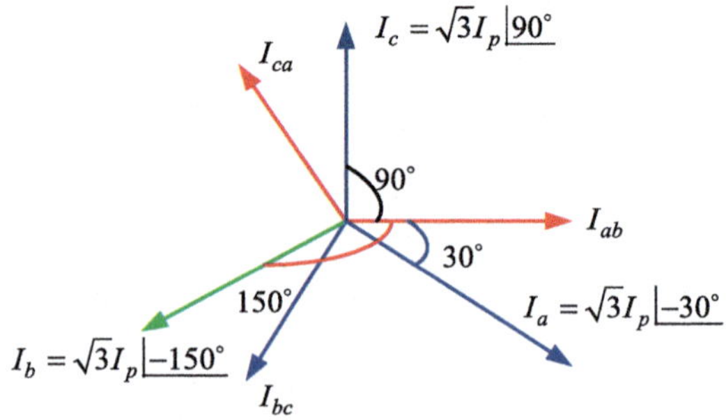

Fig. 9.15 Phasor diagram using line and phase currents

Substituting Eqs. (9.46) and (9.47) into Eq. (9.51) yields,

$$I_b = I_P\angle -120^\circ - I_P\angle 0^\circ = \sqrt{3}\, I_P\angle -150^\circ \tag{9.52}$$

Applying KCL at node c of the circuit in Fig. 9.14 yields,

$$I_c = I_{ca} - I_{bc} \tag{9.53}$$

Substituting Eqs. (9.47) and (9.48) into Eq. (9.53) yields,

$$I_c = I_P\angle -240^\circ - I_P\angle -120^\circ = \sqrt{3}\, I_P\angle 90^\circ \tag{9.54}$$

From Eqs. (9.50), (9.52) and (9.54), it is found that the magnitude of the line current is equal to $\sqrt{3}$ times the phase current. The general relationship between the line current and the phase current is,

$$I_L = \sqrt{3}\, I_P \tag{9.55}$$

Consider the phasor diagram, as shown in Fig. 9.15, where the phase current I_{ab} is arbitrarily chosen as reference. According to Fig. 9.14, it is observed that the phase voltage is equal to the line voltage, i.e.

$$V_L = V_P \tag{9.56}$$

The line and phase currents with their phase angles are drawn as shown in Fig. 9.15.

9.9 Wye–Delta Connection Without Line Impedance

A balanced three-phase wye-connected generator is connected to a balanced three-phase delta-connected load as shown in Fig. 9.16. The load consists of resistance and inductance, and per phase load impedance is considered $Z_\Delta = Z\angle\theta$. According to

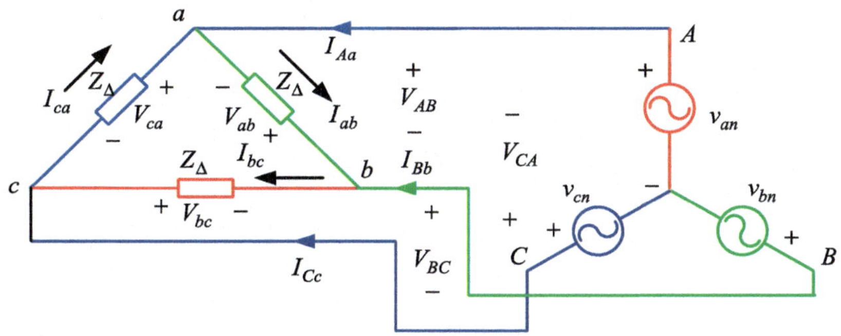

Fig. 9.16 A wye–delta connection without line impedance

the circuit in Fig. 9.16, the line voltages are equal to the load phase voltages and these can be written as [6, 7],

$$V_{ab} = V_{AB} = \sqrt{3}V_p\angle 30^\circ \tag{9.57}$$

$$V_{bc} = V_{BC} = \sqrt{3}V_p\angle -90^\circ \tag{9.58}$$

$$V_{ca} = V_{CA} = \sqrt{3}V_p\angle 150^\circ \tag{9.59}$$

The expressions of the phase currents can be determined as,

$$I_{ab} = \frac{V_{ab}}{Z\angle\theta} = \frac{\sqrt{3}V_p}{Z}\angle 30^\circ - \theta \tag{9.60}$$

$$I_{bc} = \frac{V_{bc}}{Z\angle\theta} = \frac{\sqrt{3}V_p}{Z}\angle -90^\circ - \theta \tag{9.61}$$

$$I_{ca} = \frac{V_{ca}}{Z\angle\theta} = \frac{\sqrt{3}V_p}{Z}\angle 150^\circ - \theta \tag{9.62}$$

Example 9.4 A balanced three-phase wye–delta system is shown in Fig. 9.17. Per phase load impedance and generator voltage are $Z_\Delta = 5\angle 20^\circ\,\Omega$ and 240 V rms, respectively. For ABC phase sequence, calculate the load phase currents and line currents.

Solution:
The line voltages are,

$$V_{AB} = \sqrt{3} \times 240\angle 30^\circ = 415.69\angle 30^\circ\,\text{V} \tag{9.63}$$

$$V_{BC} = \sqrt{3} \times 240\angle -90^\circ = 415.69\angle -90^\circ\,\text{V} \tag{9.64}$$

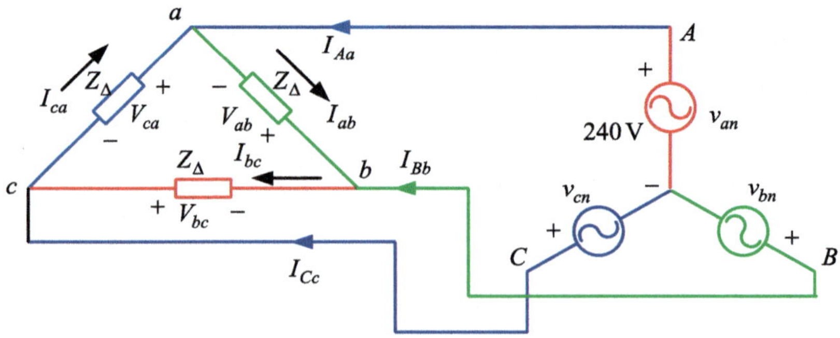

Fig. 9.17 Circuit for Example 9.4

$$V_{CA} = \sqrt{3} \times 240\angle 150^\circ = 415.69\angle 150^\circ \text{ V} \tag{9.65}$$

The phase currents in the load can be determined as,

$$I_{ab} = \frac{V_{AB}}{Z} = \frac{415.69\angle 30^\circ}{25\angle 20^\circ} = 16.63\angle 10^\circ \text{ A} \tag{9.66}$$

$$I_{bc} = \frac{V_{BC}}{Z} = \frac{415.69\angle -90^\circ}{25\angle 20^\circ} = 16.63\angle -110^\circ \text{ A} \tag{9.67}$$

$$I_{ca} = \frac{V_{CA}}{Z} = \frac{415.69\angle 150^\circ}{25\angle 20^\circ} = 16.63\angle 130^\circ \text{ A} \tag{9.68}$$

The line currents can be calculated by applying KCL at nodes *a*, *b* and *c* of the circuit in Fig. 9.17 as,

$$I_{Aa} = I_{ab} - I_{ca} = 16.63\angle 10^\circ - 16.63\angle 130^\circ = 28.80\angle -20^\circ \text{ A} \tag{9.69}$$

$$I_{Bb} = I_{bc} - I_{ab} = 16.63\angle -110^\circ - 16.63\angle 10^\circ = 28.80\angle -140^\circ \text{ A} \tag{9.70}$$

$$I_{Cc} = I_{ca} - I_{bc} = 16.63\angle 130^\circ - 16.63\angle -110^\circ = 28.80\angle 100^\circ \text{ A} \tag{9.71}$$

Practice Problem 9.4

A balanced three-phase wye-connected generator is connected to an unbalanced delta-connected load as shown in Fig. 9.18. Per phase load impedance are $Z_{ab} = 42\angle 25^\circ\ \Omega$, $Z_{bc} = 35\angle -35^\circ\ \Omega$ and $Z_{ca} = 55\angle 45^\circ\ \Omega$. For ABC phase sequence and $V_{AN} = 220\angle 10^\circ$ V, find the load phase currents and line currents.

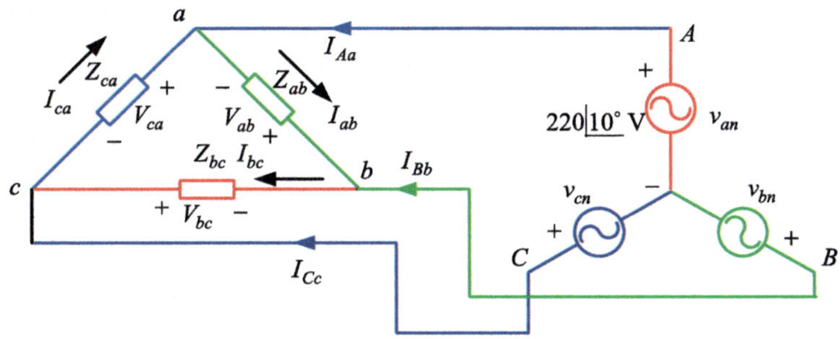

Fig. 9.18 Circuit for Practice Problem 9.4

9.10 Wye–Delta Connection with Line Impedance

A balanced three-phase wye-connected generator is connected to a balanced three-phase delta-connected load through line impedances as shown in Fig. 9.19.

The balanced delta-connected load is converted to a balanced wye-connected load by using the following expression:

$$Z_y = \frac{Z_\Delta}{3} \tag{9.72}$$

From Fig. 9.19, a single-phase circuit can be drawn as shown in Fig. 9.20. For ABC phase sequence, the following expressions can be written:

$$I_{\text{Aa}} = \frac{v_{\text{an}}}{Z_t} = \frac{V_{\text{p}}}{Z_L + Z_y}\underline{|0°} \tag{9.73}$$

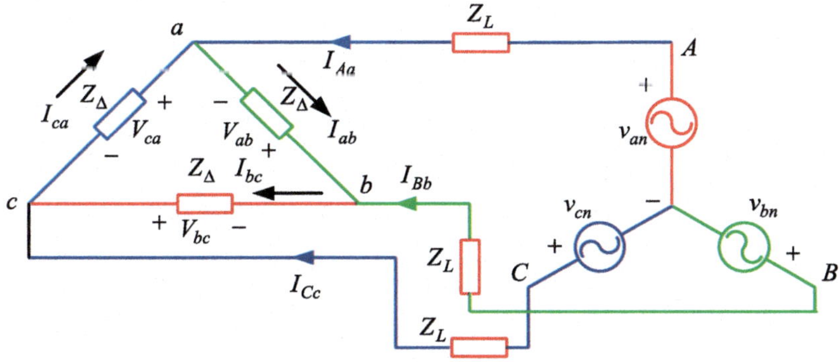

Fig. 9.19 Wye–delta circuit with line impedance

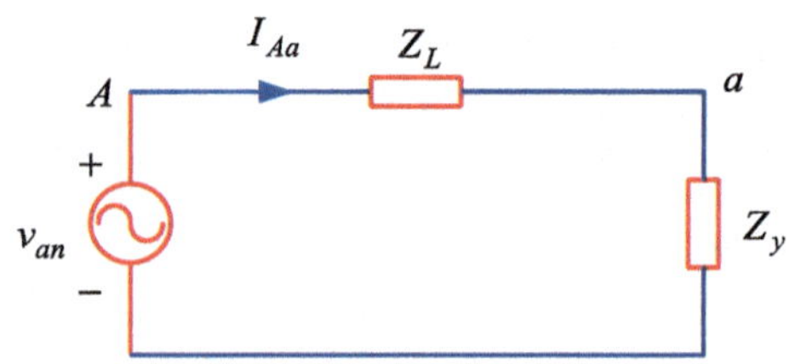

Fig. 9.20 Single-phase circuit for wye-connected load

$$I_{\mathrm{Bb}} = \frac{v_{\mathrm{bn}}}{Z_t} = \frac{V_{\mathrm{p}}}{Z_L + Z_y}\angle{-120^\circ} \tag{9.74}$$

$$I_{\mathrm{Cc}} = \frac{v_{\mathrm{cn}}}{Z_t} = \frac{V_{\mathrm{p}}}{Z_L + Z_y}\angle{-240^\circ} \tag{9.75}$$

Example 9.5 A balanced three-phase wye-connected generator delivers power to a balanced three-phase delta-connected load through a line as shown in Fig. 9.21. Line impedance, per phase load impedance and generator voltage are $Z_L = 2\angle{15^\circ}\,\Omega$, $Z_\Delta = 12\angle{30^\circ}\,\Omega$ and 230 V rms, respectively. For ABC phase sequence, calculate the line currents and phase currents.

Solution:
Converting delta-connected load to a wye-connected load yields,

$$Z_y = \frac{Z_\Delta}{3} = \frac{12\angle{30^\circ}}{3} = 4\angle{30^\circ}\,\Omega \tag{9.76}$$

The line currents can be calculated as,

$$I_{\mathrm{Aa}} = \frac{v_{\mathrm{an}}}{Z_y + Z_L} = \frac{230\angle{0^\circ}}{2\angle{15^\circ} + 4\angle{30^\circ}} = 38.63\angle{-25.01^\circ}\ \mathrm{A} \tag{9.77}$$

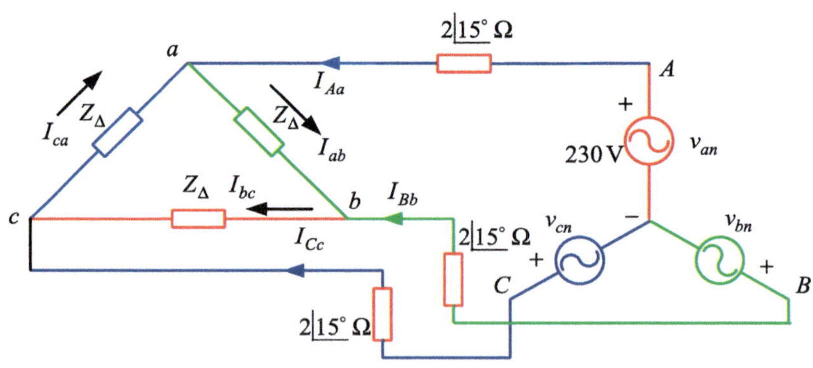

Fig. 9.21 Wye–delta circuit for Example 9.5

$$I_{\text{Bb}} = \frac{v_{\text{bn}}}{Z_y + Z_L} = \frac{230\angle -120^\circ}{2\angle 15^\circ + 4\angle 30^\circ} = 38.63\angle -145.01^\circ \text{ A} \quad (9.78)$$

$$I_{\text{Cc}} = \frac{v_{\text{cn}}}{Z_y + Z_L} = \frac{230\angle 120^\circ}{2\angle 15^\circ + 4\angle 30^\circ} = 38.63\angle 94.99^\circ \text{ A} \quad (9.79)$$

The line voltage is calculated as,

$$V_{\text{AB}} = \sqrt{3} v_{\text{an}}\angle 30^\circ = \sqrt{3} \times 230\angle 30^\circ = 398.37\angle 30^\circ \text{ V} \quad (9.80)$$

Applying KVL between lines A and B of the circuit in Fig. 9.21 yields,

$$V_{\text{ab}} = V_{\text{AB}} - I_{\text{Aa}} Z_L + I_{\text{Bb}} Z_L \quad (9.81)$$

$$V_{\text{ab}} = 398.37\angle 30^\circ + 2\angle 15^\circ (38.63\angle -145.01^\circ - 38.63\angle -25.01^\circ) = 267.60\angle 35^\circ \text{ V} \quad (9.82)$$

The phase current is calculated as,

$$I_{\text{ab}} = \frac{V_{\text{ab}}}{Z_\Delta} = \frac{267.60\angle 35^\circ}{12\angle 30^\circ} = 22.3\angle 5^\circ \text{ A} \quad (9.83)$$

Other phase currents can be written as,

$$I_{\text{bc}} = I_{\text{ab}}\angle -120^\circ = 22.3\angle 5^\circ - 120^\circ = 22.33\angle -115^\circ \text{ A} \quad (9.84)$$

$$I_{\text{ca}} = I_{\text{ab}}\angle 120^\circ = 22.3\angle 5^\circ + 120^\circ = 22.33\angle 125^\circ \text{ A} \quad (9.85)$$

Practice Problem 9.5

A balanced wye–delta system is shown in Fig. 9.22. Line impedance, per phase load impedance and generator voltage are $Z_L = 3\angle 34^\circ\,\Omega$, $Z_\Delta = 30\angle -25^\circ\,\Omega$ and $220\angle 10^\circ$ V rms, respectively. For ABC phase sequence, determine the line currents.

9.11 Delta–Wye System Without Line Impedance

A balanced three-phase delta-connected generator delivers power to the balanced three-phase wye-connected load as shown in Fig. 9.23. Consider the load is inductive and the per phase load impedance is $Z_y = Z\angle\theta$. According to the circuit in Fig. 9.23, the per phase load voltages are equal to the line voltages and these are,

$$V_{\text{ab}} = V_{\text{AB}} = V_{\text{p}}\angle 0^\circ \quad (9.86)$$

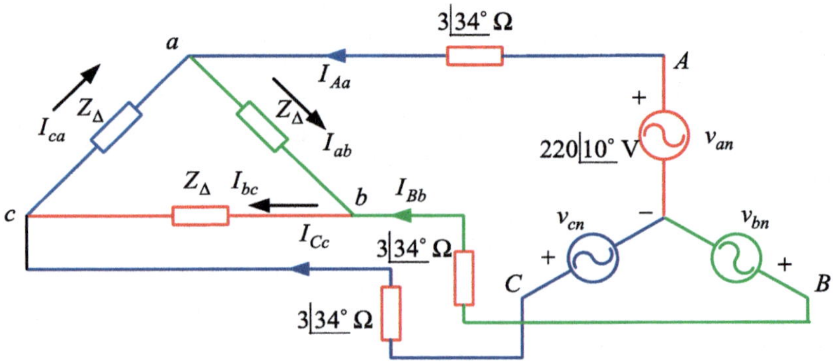

Fig. 9.22 Wye–delta circuit for Practice Problem 9.5

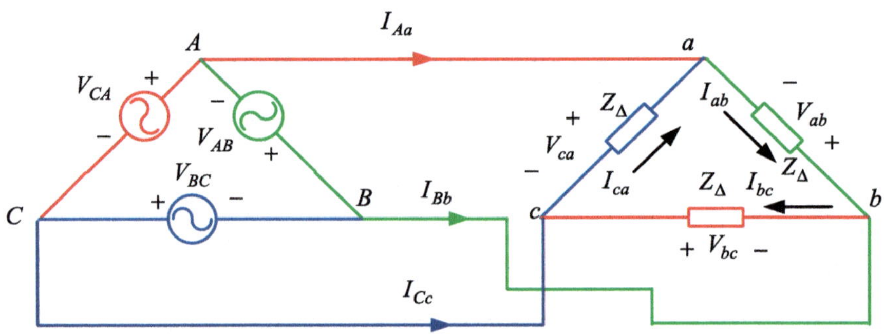

Fig. 9.23 Delta–delta system without line impedance

$$V_{\mathrm{bc}} = V_{\mathrm{BC}} = V_{\mathrm{p}}\angle -120^\circ \tag{9.87}$$

$$V_{\mathrm{ca}} = V_{\mathrm{CA}} = V_{\mathrm{p}}\angle 120^\circ \tag{9.88}$$

The phase currents can be calculated as,

$$I_{\mathrm{ab}} = \frac{V_{\mathrm{ab}}}{Z\angle\theta} = \frac{V_{\mathrm{p}}}{Z}\angle -\theta \tag{9.89}$$

$$I_{\mathrm{bc}} = \frac{V_{\mathrm{bc}}}{Z\angle\theta} = \frac{V_{\mathrm{p}}}{Z}\angle -120^\circ - \theta \tag{9.90}$$

$$I_{\mathrm{ca}} = \frac{V_{\mathrm{ca}}}{Z\angle\theta} = \frac{V_{\mathrm{p}}}{Z}\angle 120^\circ - \theta \tag{9.91}$$

Applying KCL to the nodes a, b and c yields the line currents,

$$I_{\mathrm{Aa}} = I_{\mathrm{ab}} - I_{\mathrm{ca}} \tag{9.92}$$

$$I_{\mathrm{Bb}} = I_{\mathrm{bc}} - I_{\mathrm{ab}} \tag{9.93}$$

$$I_{\mathrm{Cc}} = I_{\mathrm{ca}} - I_{\mathrm{bc}} \tag{9.94}$$

Example 9.6 A balanced three-phase delta-connected generator delivers power to a balanced three-phase delta-connected load as shown in Fig. 9.24. Per phase load impedance and generator voltage are $16\underline{|35^\circ}\ \Omega$ and $230\underline{|15^\circ}$ V rms, respectively. For ABC phase sequence, find the phase currents in the load and line currents.

Solution:
The magnitude of the phase currents in the load are calculated as,

$$I_{\mathrm{ab}} = \frac{V_{\mathrm{ab}}}{Z_\Delta} = \frac{230\underline{|15^\circ}}{16\underline{|35^\circ}} = 14.38\underline{|-20^\circ}\ \mathrm{A} \tag{9.95}$$

$$I_{\mathrm{bc}} = \frac{V_{\mathrm{bc}}}{Z_\Delta} = \frac{230\underline{|15^\circ - 120^\circ}}{16\underline{|35^\circ}} = 14.38\underline{|-140^\circ}\ \mathrm{A} \tag{9.96}$$

$$I_{\mathrm{ca}} = \frac{V_{\mathrm{ca}}}{Z_\Delta} = \frac{230\underline{|15^\circ + 120^\circ}}{16\underline{|35^\circ}} = 14.38\underline{|100^\circ}\ \mathrm{A} \tag{9.97}$$

The line currents can be determined by applying KCL at nodes *a*, *b* and *c* as,

$$I_{\mathrm{Aa}} = I_{\mathrm{ab}} - I_{\mathrm{ca}} = 14.38\underline{|-20^\circ} - 14.38\underline{|100^\circ} = 24.91\underline{|-50^\circ}\ \mathrm{A} \tag{9.98}$$

$$I_{\mathrm{Bb}} = I_{\mathrm{bc}} - I_{\mathrm{ab}} = 14.38\underline{|-140^\circ} - 14.38\underline{|-20^\circ} = 24.91\underline{|-170^\circ}\ \mathrm{A} \tag{9.99}$$

$$I_{\mathrm{Cc}} = I_{\mathrm{ca}} - I_{\mathrm{bc}} = 14.38\underline{|100^\circ} - 14.38\underline{|-140^\circ} = 24.91\underline{|70^\circ}\ \mathrm{A} \tag{9.100}$$

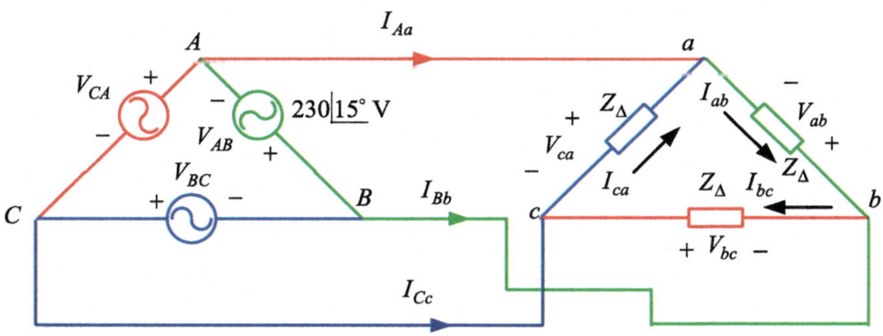

Fig. 9.24 Delta–delta system for Example 9.6

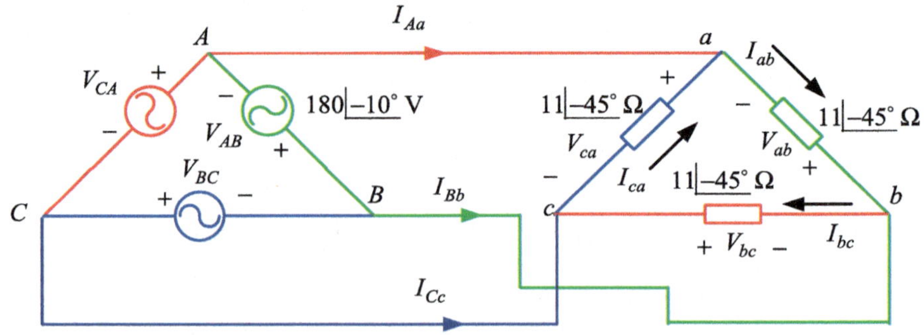

Fig. 9.25 Delta–delta system for Practice Problem 9.6

Practice Problem 9.6
Figure 9.25 shows a balanced three-phase delta-connected generator that delivers power to the balanced three-phase delta-connected load. Per phase load impedance and generator voltage are $11\underline{|-45°}\,\Omega$ and $180\underline{|-10°}$ V rms, respectively. For *ABC* phase sequence, find the phase currents in the load and line currents.

9.12 Unbalanced Systems

A three-phase delta-connected or wye-connected load is considered unbalanced when their impedances are different from each other. Generally, a known three-phase balanced system supplies power to these types of loads. There are four ways to make unbalanced systems. These are wye–wye, wye–delta, delta–delta and delta–wye unbalanced systems. For analysis, consider a wye–wye unbalanced system as shown in Fig. 9.26. The line currents can be determined as,

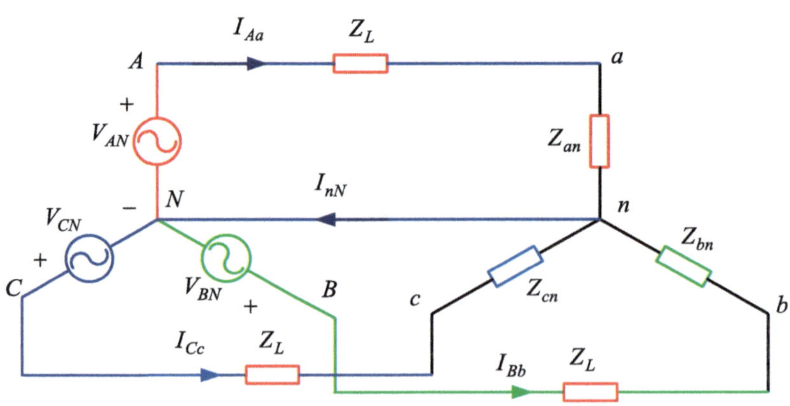

Fig. 9.26 Wye–wye system with unbalanced load

$$I_{\text{Aa}} = \frac{V_{\text{AN}}}{Z_L + Z_{\text{an}}} \tag{9.101}$$

$$I_{\text{Bb}} = \frac{V_{\text{BN}}}{Z_L + Z_{\text{bn}}} \tag{9.102}$$

$$I_{\text{Cc}} = \frac{V_{\text{CN}}}{Z_L + Z_{\text{cn}}} \tag{9.103}$$

Example 9.7 An unbalanced three-phase wye-connected load is connected to a balanced three-phase wye-connected generator as shown in Fig. 9.27. The generator voltage is $220\angle 10^\circ$ V rms. For ABC phase sequence, find the line currents.

Solution:
The line currents can be determined as,

$$I_{\text{Aa}} = \frac{220\angle 10^\circ}{2\angle 15^\circ + 2\angle 15^\circ} = 55\angle -5^\circ \text{ A} \tag{9.104}$$

$$I_{\text{Bb}} = \frac{220\angle -110^\circ}{2\angle 15^\circ + 4\angle -35^\circ} = 39.98\angle -91.16^\circ \text{ A} \tag{9.105}$$

$$I_{\text{Cc}} = \frac{220\angle 130^\circ}{2\angle 15^\circ + 5\angle -45^\circ} = 35.23\angle 158.90^\circ \text{ A} \tag{9.106}$$

Practice Problem 9.7

An unbalanced three-phase delta-connected load is connected with a balanced three-phase wye-connected generator as shown in Fig. 9.28. The generator voltage is $180\angle -10^\circ$ V rms. For ABC phase sequence, find the line currents.

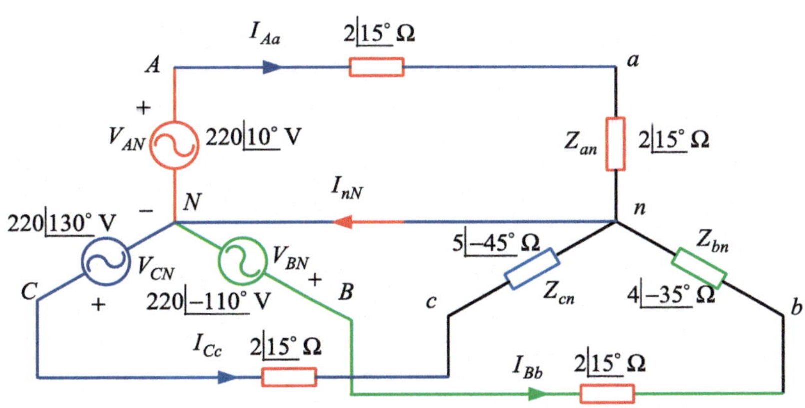

Fig. 9.27 Wye–wye system for Example 9.7

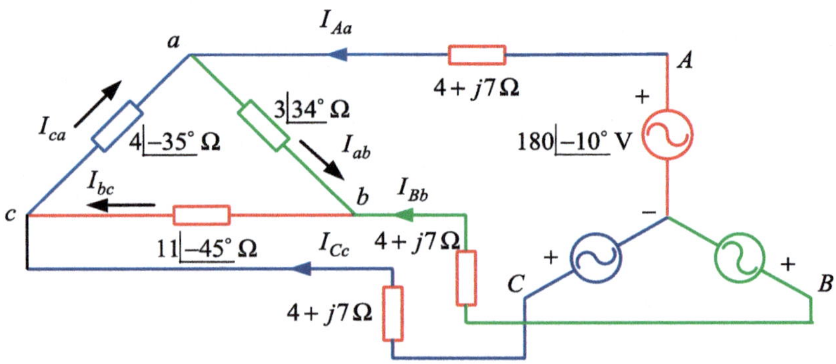

Fig. 9.28 Wye–delta system for Practice Problem 9.7

9.13 Three-Phase Power Analysis

Consider a balanced three-phase wye-connected generator that delivers power to the balanced three-phase wye-connected load as shown in Fig. 9.29. The total power of the three-phase system is calculated by considering the instantaneous voltages and currents.

The instantaneous voltages are,

$$v_{\text{AN}} = V_m \sin \omega t \tag{9.107}$$

$$v_{\text{BN}} = V_m \sin(\omega t - 120^\circ) \tag{9.108}$$

$$v_{\text{CN}} = V_m \sin(\omega t - 240^\circ) \tag{9.109}$$

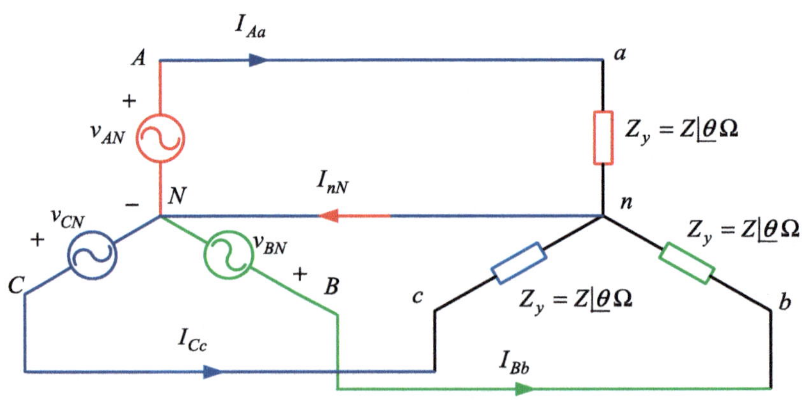

Fig. 9.29 Wye–wye system for power calculation

The phase currents of the three-phase wye-connected load can be expressed as,

$$i_{\mathrm{Aa}} = \frac{v_{\mathrm{AN}}}{Z_y} = \frac{V_m \sin \omega t}{Z\underline{|\theta}} = I_m \sin(\omega t - \theta) \tag{9.110}$$

$$i_{\mathrm{Bb}} = \frac{v_{\mathrm{BN}}}{Z_y} = \frac{V_m \sin(\omega t - 120^\circ)}{Z\underline{|\theta}} = I_m \sin(\omega t - \theta - 120^\circ) \tag{9.111}$$

$$i_{\mathrm{Cc}} = \frac{v_{\mathrm{CN}}}{Z_y} = \frac{V_m \sin(\omega t - 240^\circ)}{Z\underline{|\theta}} = I_m \sin(\omega t - \theta - 240^\circ) \tag{9.112}$$

The instantaneous power for phase a can be expressed as [8, 9],

$$p_a(t) = \frac{1}{T}\int_0^T v_{\mathrm{AN}} i_{\mathrm{Aa}} \mathrm{d}t \tag{9.113}$$

Substituting Eqs. (9.110) and (9.111) into Eq. (9.113) yields,

$$p_a(t) = \frac{V_m I_m}{T}\int_0^T \sin \omega t \times \sin(\omega t - \theta)\mathrm{d}t \tag{9.114}$$

$$p_a(t) = \frac{V_m I_m}{2T}\int_0^T 2\sin \omega t \times \sin(\omega t - \theta)\mathrm{d}t \tag{9.115}$$

$$p_a(t) = \frac{V_m I_m}{2T}\int_0^T [\cos\theta - \cos(2\omega t - \theta)]\mathrm{d}t \tag{9.116}$$

$$p_a(t) = \frac{V_m I_m}{2T} \times \cos\theta \times T - 0 \tag{9.117}$$

$$p_a(t) = \frac{V_m I_m}{\sqrt{2} \times \sqrt{2}} \cos\theta = V_{\mathrm{P}} I_{\mathrm{P}} \cos\theta \tag{9.118}$$

where V_{p} and I_p are the rms values of phase voltage and phase current.

Similarly, the expressions of the instantaneous power for the phase b and phase c can be written as,

$$p_b(t) = V_{\mathrm{P}} I_{\mathrm{P}} \cos\theta \tag{9.119}$$

$$p_c(t) = V_{\mathrm{P}} I_{\mathrm{P}} \cos\theta \tag{9.120}$$

Therefore, the average three-phase power P can be calculated as,

$$P = p_a(t) + p_b(t) + p_c(t) \tag{9.121}$$

Substituting Eqs. (9.118), (9.119) and (9.120) into Eq. (9.121) yields,

$$P_t = 3V_{\mathrm{P}}I_{\mathrm{P}}\cos\theta \tag{9.122}$$

Similarly, the expression of three-phase reactive power can be expressed as,

$$Q_t = 3V_{\mathrm{p}}I_p\sin\theta \tag{9.123}$$

Therefore, the per phase average (P_{pp}) and reactive (Q_{pp}) power can be written as,

$$P_{\mathrm{pp}} = V_{\mathrm{P}}I_{\mathrm{P}}\cos\theta \tag{9.124}$$

$$Q_{\mathrm{pp}} = V_{\mathrm{P}}I_{\mathrm{P}}\sin\theta \tag{9.125}$$

The complex power per phase S_{pp} is represented as,

$$S_{\mathrm{pp}} = P_{\mathrm{pp}} + jQ_{\mathrm{pp}} \tag{9.126}$$

Substituting Eqs. (9.124) and (9.125) into Eq. (9.126) yields,

$$S_{\mathrm{pp}} = V_{\mathrm{P}}I_{\mathrm{P}}\cos\theta + jV_{\mathrm{P}}I_{\mathrm{P}}\sin\theta \tag{9.127}$$

Equation (9.127) can be expressed as,

$$S_{\mathrm{pp}} = V_{\mathrm{P}}I_{\mathrm{P}}^{*} \tag{9.128}$$

From Eq. (9.128), it is seen that the per phase complex power is equal to the product of the voltage per phase and the conjugate of the phase current.

Y-connection: Substituting Eqs. (9.18) and (9.19) into Eq. (9.122) yields,

$$P_{\mathrm{tY}} = 3 \times \frac{V_{\mathrm{L}}}{\sqrt{3}} \times I_{\mathrm{L}} \times \cos\theta \tag{9.129}$$

$$P_{\mathrm{tY}} = \sqrt{3}\,V_{\mathrm{L}}I_{\mathrm{L}}\cos\theta \tag{9.130}$$

Substituting Eqs. (9.18) and (9.19) into Eq. (9.123) yields,

$$Q_{\mathrm{tY}} = 3 \times \frac{V_{\mathrm{L}}}{\sqrt{3}} \times I_{\mathrm{L}} \times \sin\theta \tag{9.131}$$

$$Q_{\mathrm{tY}} = \sqrt{3}\,V_{\mathrm{L}}I_{\mathrm{L}}\sin\theta \tag{9.132}$$

Delta connection: Again, substituting Eqs. (9.55) and (9.56) into Eq. (9.122) yields,

$$P_{t\Delta} = 3 \times \frac{V_L}{\sqrt{3}} \times I_L \times \cos\theta \tag{9.133}$$

$$P_{t\Delta} = \sqrt{3}\, V_L I_L \cos\theta \tag{9.134}$$

Substituting Eqs. (9.55) and (9.56) into Eq. (9.123) yields,

$$Q_{t\Delta} = 3 \times \frac{V_L}{\sqrt{3}} \times I_L \times \sin\theta \tag{9.135}$$

$$Q_{t\Delta} = \sqrt{3}\, V_L I_L \sin\theta \tag{9.136}$$

In general, the total real and reactive power can be expressed as,

$$P_t = \sqrt{3}\, V_L I_L \cos\theta \tag{9.137}$$

$$Q_t = \sqrt{3}\, V_L I_L \sin\theta \tag{9.138}$$

The total complex power can be written as,

$$S_t = P_t + jQ_t \tag{9.139}$$

Substituting Eqs. (9.137) and (9.138) into Eq. (9.139) yields,

$$S_t = \sqrt{3}\, V_L I_L \cos\theta + j\sqrt{3}\, V_L I_L \sin\theta \tag{9.140}$$

$$S_t = \sqrt{3}\, V_L I_L \underline{|\theta} \tag{9.141}$$

Three-phase system uses less amount of copper wire than the single-phase system for the same line voltage and same power factor to transmit the same amount of power over a fixed distance. From Fig. 9.30, the real power for a single-phase two-wire system is,

$$P_{1\phi 2w} = V_L I_{1\phi L} \cos\theta \tag{9.142}$$

From Fig. 9.30, the real power for a three-phase three wire system is,

$$P_{3\phi 3w} = \sqrt{3} V_L I_{3\phi L} \cos\theta \tag{9.143}$$

Equations (9.142) and (9.143) will be equal for transmitting or delivering same amount of power over a fixed distance. It can be expressed as,

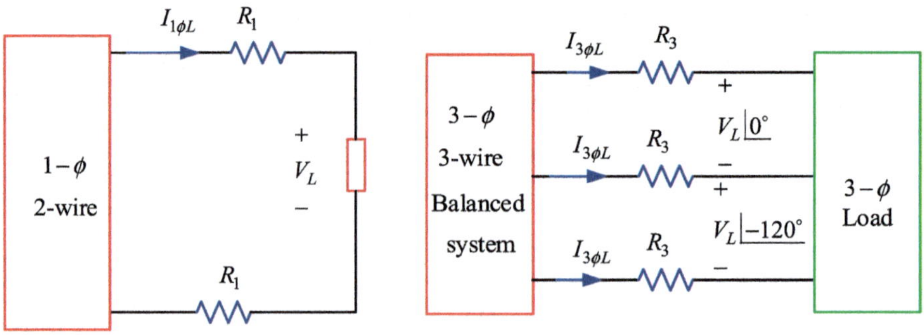

Fig. 9.30 Single-phase and three-phase with loads

$$V_L I_{1\phi L} \cos\theta = \sqrt{3} V_L I_{3\phi L} \cos\theta \quad (9.144)$$

$$I_{1\phi L} = \sqrt{3} I_{3\phi L} \quad (9.145)$$

The power loss in the single-phase wire is,

$$P_{1\phi 2w\text{loss}} = 2I_{1\phi L}^2 R_1 \quad (9.146)$$

$$P_{3\phi 3w\text{loss}} = 3I_{3\phi L}^2 R_3 \quad (9.147)$$

From Eqs. (9.146) and (9.147), the ratio of power loss of a single-phase system to a three-phase system can be derived as,

$$\frac{P_{1\phi 2w\text{loss}}}{P_{3\phi 3w\text{loss}}} = \frac{2I_{1\phi L}^2 R_1}{3I_{3\phi L}^2 R_3} \quad (9.148)$$

For equal losses ($P_{1\phi 2w\text{loss}} = P_{3\phi 3w\text{loss}}$), Eq. (9.148) can be modified as,

$$1 = \frac{2I_{1\phi L}^2 R_1}{3I_{3\phi L}^2 R_3} \quad (9.149)$$

$$\frac{3I_{3\phi L}^2}{2I_{1\phi L}^2} = \frac{R_1}{R_3} \quad (9.150)$$

Substituting Eq. (9.145) into Eq. (9.150) yields,

$$\frac{3I_{3\phi L}^2}{2 \times 3I_{3\phi L}^2} = \frac{R_1}{R_3} \quad (9.151)$$

$$\frac{R_1}{R_3} = \frac{1}{2} \tag{9.152}$$

The following ratio can be written:

$$\frac{\text{Copper for } 3\phi \text{ system}}{\text{Copper for } 1\phi \text{ system}} = \frac{\text{number of wires in } 3\phi \text{ system}}{\text{number of wires in } 1\phi \text{ system}} \times \frac{R_1}{R_3} \tag{9.153}$$

Substituting Eq. (9.152) and the number of wires for both systems in Eq. (9.153) yields,

$$\frac{\text{Copper for } 3\phi \text{ system}}{\text{Copper for } 1\phi \text{ system}} = \frac{3}{2} \times \frac{1}{2} \tag{9.154}$$

$$\text{Copper for } 3\phi \text{ system} = \frac{3}{4} \times \text{Copper for } 1\phi \text{ system} \tag{9.155}$$

From Eq. (9.155), it is seen that the copper required for three-phase system is equal to the three-fourths of the copper required for a single-phase system.

Example 9.8 A balanced three-phase wye–wye system is shown in Fig. 9.31. For ABC phase sequence, calculate the line current, power supplied to each phase, power absorbed by each phase and the total complex power supplied by the source.

Solution:
The line currents are calculated as,

$$I_{\text{Aa}} = \frac{230\angle 15^\circ}{2 + j8} = 27.89\angle -60.96^\circ \text{ A} \tag{9.156}$$

$$I_{\text{Bb}} = I_{\text{Aa}}\angle -120^\circ = 27.89\angle -180.96^\circ \text{ A} \tag{9.157}$$

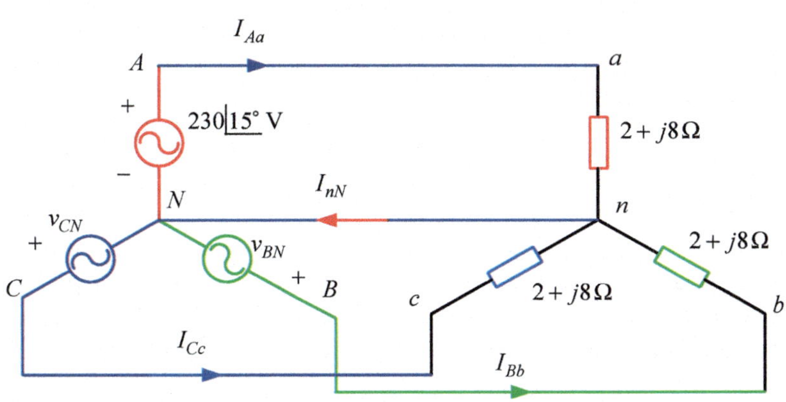

Fig. 9.31 Circuit for Example 9.8

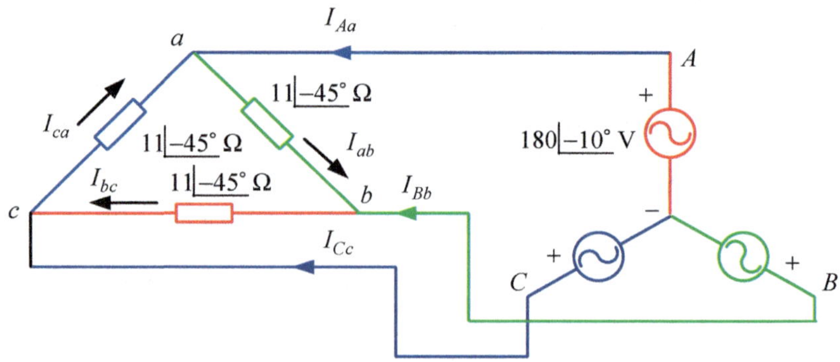

Fig. 9.32 Circuit for Practice Problem 9.8

$$I_{\text{Cc}} = I_{\text{Aa}}\angle{+120^\circ} = 27.89\angle{59.04^\circ}\ \text{A} \tag{9.158}$$

Power supplied to each phase is,

$$\begin{aligned} P_A &= V_{\text{p}} I_p \cos\theta = V_{\text{AN}} I_{\text{Aa}} \cos(\theta_v - \theta_i) = 230 \times 27.89 \times \cos(15 + 60.96) \\ &= 1556.20\ \text{W} \end{aligned} \tag{9.159}$$

Per phase power absorbed by the load is,

$$P_{L1\phi} = 27.89^2 \times 2 = 1555.70\ \text{W} \tag{9.160}$$

Total complex power supplied by the source is,

$$S_t = 3V_{\text{An}} I_{\text{Aa}}^* = 3 \times 230\angle{15^\circ} \times 27.89\angle{60.96^\circ} = 4668.60 + j18669.21\ \text{VA} \tag{9.161}$$

Practice Problem 9.8
A balanced three-phase wye–delta system is shown in Fig. 9.32. For ABC phase sequence, calculate the line current, power supplied to each phase, power absorbed by each phase and the total complex power supplied by the source.

9.14 Parallel Delta–Wye Load

Consider a wye-connected generator delivers power to the delta–wye connected parallel loads as shown in Fig. 9.33. In this case, the delta-connected load is converted to a wye-connected load, and connected in parallel to an existing wye-connected load as shown in Fig. 9.34. The delta-connected load is converted to a wye-connected load as,

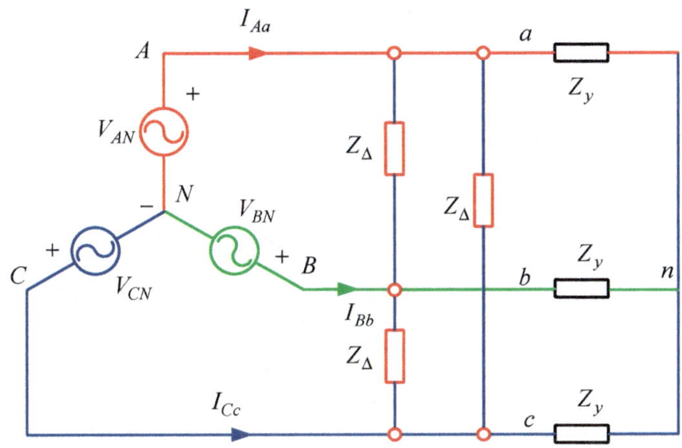

Fig. 9.33 Delta–wye parallel loads

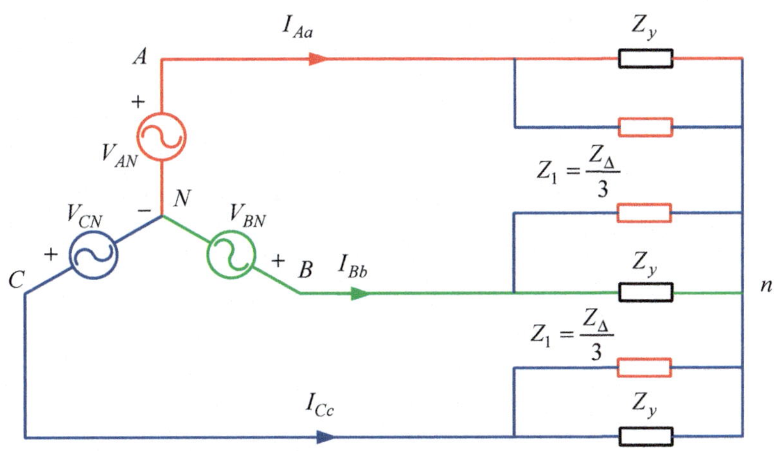

Fig. 9.34 Delta load converted to wye load

$$Z_1 = Z_{\Delta-y} = \frac{Z_\Delta}{3} \tag{9.162}$$

The parallel load impedance is,

$$Z_{\text{PL}} = \frac{Z_1 Z_y}{Z_1 + Z_y} \tag{9.163}$$

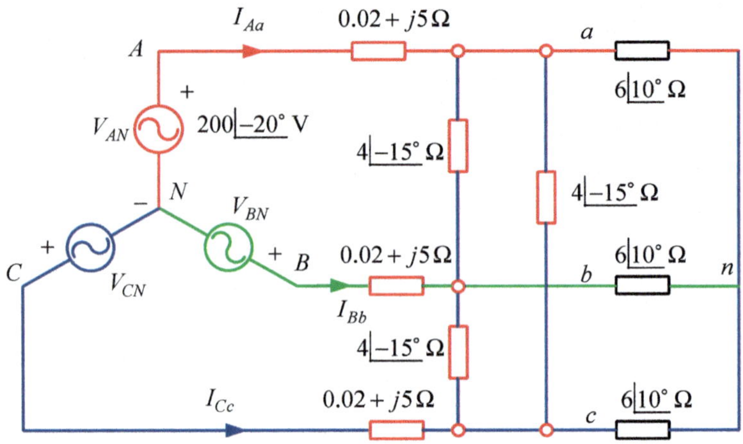

Fig. 9.35 Balanced wye–delta load for Example 9.9

The line current is calculated as,

$$I_{\mathrm{Aa}} = \frac{V_{\mathrm{AN}}}{Z_{\mathrm{PL}}} \tag{9.164}$$

Example 9.9 A balanced three-phase wye-connected generator delivers power to a balanced three-phase delta–wye connected load as shown in Fig. 9.35. Determine the power per phase and the total power delivered by the source.

Solution:
Delta load is converted to wye load as,

$$Z_1 = \frac{4\angle -15^\circ}{3} = 1.33\angle -15^\circ\,\Omega \tag{9.165}$$

The value of the parallel loads is,

$$Z_2 = \frac{1.33\angle -15^\circ \times 6\angle 10^\circ}{1.33\angle -15^\circ + 6\angle 10^\circ} = 1.10\angle -10.54^\circ\,\Omega \tag{9.166}$$

The line current is calculated as,

$$I_{\mathrm{Aa}} = \frac{200\angle -20^\circ}{0.02 + j5 + 1.10\angle -10.54^\circ} = 40.62\angle -97.07^\circ\ \mathrm{A} \tag{9.167}$$

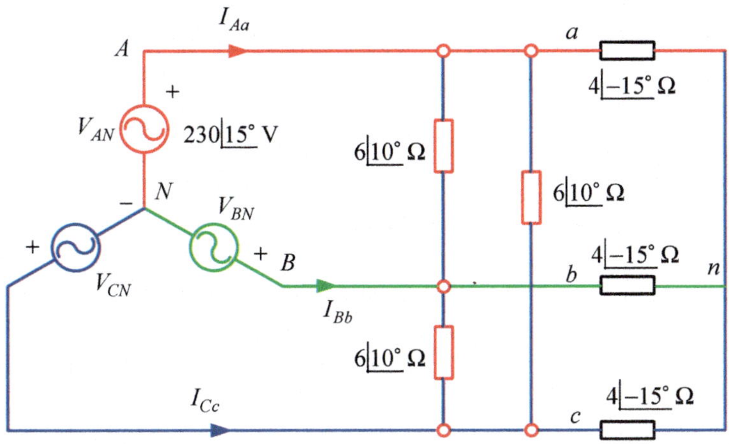

Fig. 9.36 Circuit for Practice Problem 9.9

Power delivers by the first line is calculated as,

$$P_A = V_{\mathrm{AN}} I_{\mathrm{Aa}} \cos(\theta_v - \theta_i) = 200 \times 40.62 \cos(-20^\circ + 97.07^\circ) = 1817.83\,\mathrm{W} \tag{9.168}$$

Total power delivered by the source is,

$$P_t = 3P_A = 3 \times 1817.83 = 5453.49\,\mathrm{W} \tag{9.169}$$

Practice Problem 9.9
A balanced three-phase wye connected generator delivers power to a delta–wye connected load as shown in Fig. 9.36. Calculate the power per phase and the total power delivered by the source.

9.15 Measurement of Three-Phase Power

Wattmeter is used to measure the electrical power in any circuit. The wattmeter has two coils, namely, current coil and voltage coil, which are connected in series and in parallel with the circuit, respectively. Two single-phase wattmeters can be used to measure three-phase power instead of a three-phase wattmeter [10]. The connection diagram of two-wattmeter method to measure the three-phase power is shown in Fig. 9.37.

The total power measured by the two-wattmeter method is given by,

$$P_t = P_1 + P_2 \tag{9.170}$$

Consider V_{an}, V_{bn} and V_{cn} are the rms values of the phase voltages across the wye-connected load and I_a, I_b and I_c are the rms values of the line currents. Again,

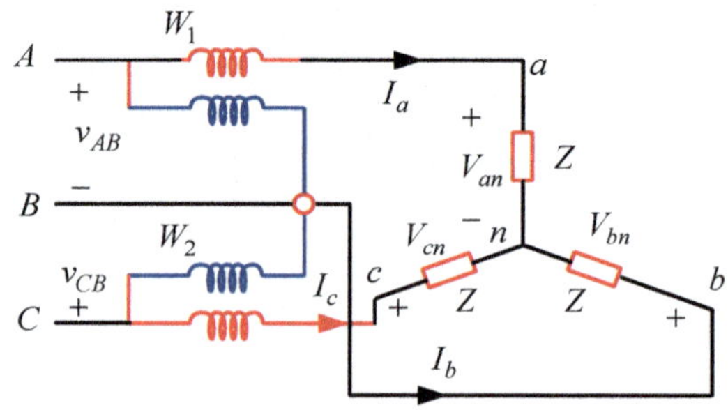

Fig. 9.37 Two-wattmeter with wye-connected load

consider that the currents lag the corresponding phase voltages by an angle θ. The average power recorded by the first wattmeter W_1 is expressed as,

$$P_1 = V_{\mathrm{AB}} I_a \cos\theta_{\mathrm{AB}} \tag{9.171}$$

The average power recorded by the second wattmeter W_2 is expressed as,

$$P_2 = V_{\mathrm{CB}} I_c \cos\theta_{\mathrm{CB}} \tag{9.172}$$

According to the circuit, the following expressions can be drawn:

$$V_{\mathrm{AB}} = V_{\mathrm{ab}} \tag{9.173}$$

$$\cos\theta_{\mathrm{AB}} = \cos\theta_{\mathrm{ab}} \tag{9.174}$$

Equation (9.171) can be modified as,

$$P_1 = V_{\mathrm{ab}} I_a \cos\theta_{\mathrm{ab}} \tag{9.175}$$

Similarly, Eq. (9.172) can be modified as,

$$P_2 = V_{\mathrm{cb}} I_c \cos\theta_{\mathrm{cb}} \tag{9.176}$$

where θ_{ab} and θ_{cb} are the phase angles between the voltage and the current of the respective phases. The magnitude of those phase angles can be found with the help of a phasor diagram, which is shown in Fig. 9.38. The phase angle between the current I_a and the voltage V_{ab} is,

$$\theta_{\mathrm{ab}} = 30^\circ + \theta \tag{9.177}$$

The phase angle between the current I_c and the voltage V_{cb} is,

$$\theta_{\mathrm{bc}} = 30^\circ - \theta \tag{9.178}$$

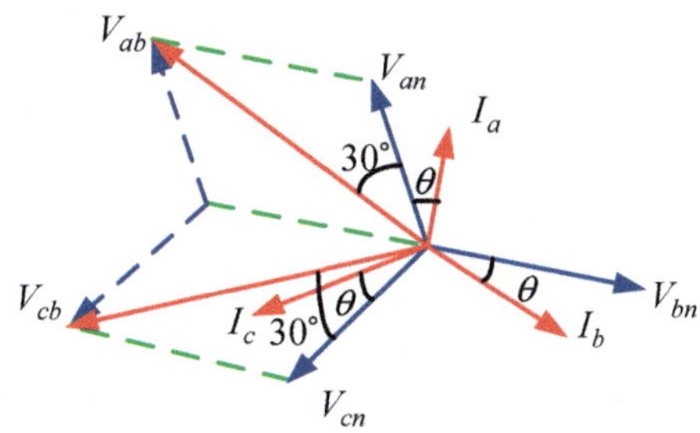

Fig. 9.38 Phasor diagram with two lines

For a balanced wye-connected load, the following expression can be written:

$$I_a = I_b = I_c = I_{\mathrm{L}} \tag{9.179}$$

$$V_{\mathrm{ab}} = V_{\mathrm{bc}} = V_{\mathrm{ca}} = V_{\mathrm{L}} \tag{9.180}$$

Substituting Eqs. (9.178), (9.179) and (9.180) into Eq. (9.175) yields,

$$P_1 = V_{\mathrm{L}} I_{\mathrm{L}} \cos(30^\circ + \theta) \tag{9.181}$$

Substituting Eqs. (9.177), (9.179) and (9.180) into Eq. (9.176) yields,

$$P_2 = V_{\mathrm{L}} I_{\mathrm{L}} \cos(30^\circ - \theta) \tag{9.182}$$

Substituting Eqs. (9.181) and (9.182) into Eq. (9.170) yields,

$$P_t = V_{\mathrm{L}} I_{\mathrm{L}} \cos(30^\circ + \theta) + V_{\mathrm{L}} I_{\mathrm{L}} \cos(30^\circ - \theta) \tag{9.183}$$

$$P_t = 2 V_{\mathrm{L}} I_{\mathrm{L}} \cos 30^\circ \cos\theta \tag{9.184}$$

$$P_1 + P_2 = \sqrt{3} V_{\mathrm{L}} I_{\mathrm{L}} \cos\theta \tag{9.185}$$

Similarly, the difference of the two wattmeter readings is calculated as,

$$P_2 - P_1 = V_{\mathrm{L}} I_{\mathrm{L}} \cos(30^\circ - \theta) - V_{\mathrm{L}} I_{\mathrm{L}} \cos(30^\circ + \theta) \tag{9.186}$$

$$P_2 - P_1 = 2 V_{\mathrm{L}} I_{\mathrm{L}} \sin 30^\circ \sin\theta \tag{9.187}$$

$$P_2 - P_1 = V_{\mathrm{L}} I_{\mathrm{L}} \sin\theta \tag{9.188}$$

Substituting Eq. (9.138) into Eq. (9.188) yields,

$$Q_t = \sqrt{3}(P_2 - P_1) \tag{9.189}$$

$$P_2 - P_1 = 0.58 Q_t \tag{9.190}$$

From Eq. (9.190), it is seen that the difference in wattmeter readings is equal to 0.58 times the total reactive power.

Dividing Eq. (9.188) by Eq. (9.185) yields,

$$\frac{1}{\sqrt{3}}\tan\theta = \frac{P_2 - P_1}{P_2 + P_1} \tag{9.191}$$

$$\tan\theta = \frac{Q_t}{P_t} = \sqrt{3}\frac{P_2 - P_1}{P_2 + P_1} \tag{9.192}$$

Therefore, the power factor angle θ and the power factor can be determined by using Eq. (9.192).

Example 9.10 A balanced three-phase wye-connected load receives supply from a source as shown in Fig. 9.39. Calculate the wattmeter readings, and total power absorbed by the load.

Solution:
For positive phase sequence, the line currents can be determined as,

$$I_a = \frac{110}{15\angle 10^\circ} = 7.33\angle 10^\circ \text{ A} \tag{9.193}$$

$$I_b = \frac{110\angle -120^\circ}{15\angle 10^\circ} = 7.33\angle -130^\circ \text{ A} \tag{9.194}$$

$$I_c = \frac{110\angle 120^\circ}{15\angle 10^\circ} = 7.33\angle 110^\circ \text{ A} \tag{9.195}$$

Wattmeter readings can be calculated as,

$$P_1 = \text{Re}(V_{\text{an}} I_a^*) = V_{\text{an}} I_a \cos(\theta_v - \theta_i) = 110 \times 7.33\cos(0^\circ - 10^\circ) = 794.05 \text{ W} \tag{9.196}$$

$$P_2 = \text{Re}(V_{\text{cn}} I_c^*) = V_{\text{cn}} I_c \cos(\theta_v - \theta_i) = 110 \times 7.33\cos(120^\circ - 110^\circ) = 794.05 \text{ W} \tag{9.197}$$

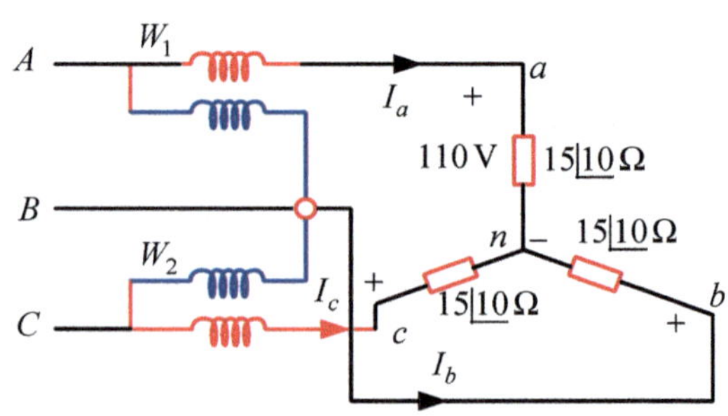

Fig. 9.39 Circuit for Example 9.10

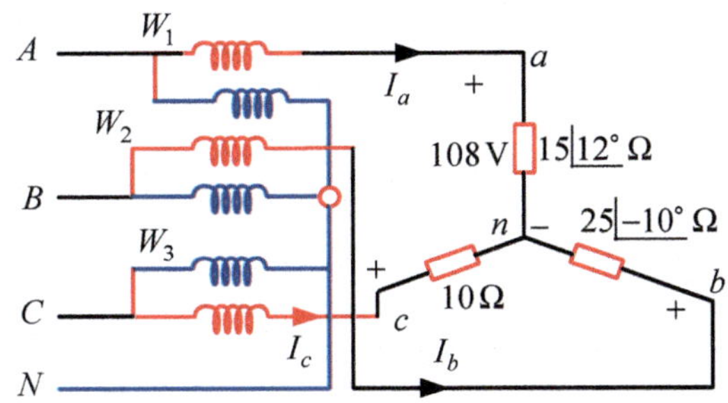

Fig. 9.40 Circuit for Practice Problem 9.10

Total power absorbed by the load is,

$$P_t = 2P_1 = 1588.10\,\text{W} \tag{9.198}$$

Practice Problem 9.10

An unbalanced three-phase wye-connected load receives supply from a balanced source as shown in Fig. 9.40. Calculate the wattmeter readings, total power and total reactive power absorbed by the load.

Exercise Problems

9.1 The line voltage of a three-phase wye-connected generator is found to be 440 V. For abc phase sequence, calculate the phase voltages.

9.2 The phase voltage of a three-phase wye-connected generator is given by $V_{\text{an}} = 100\angle{-10^\circ}\,\text{V}$. Determine the voltages V_{bn} and V_{cn} for abc phase sequence.

9.3 For abc phase sequence, calculate the line currents for a balanced three-phase wye–wye system as shown in Fig. 9.41.

9.4 A balanced three-phase wye-connected source delivers power to a balanced three-phase wye-connected load as shown in Fig. 9.42. For abc phase sequence, calculate the line currents.

9.5 A balanced three-phase wye–wye system is shown in Fig. 9.43. The rms value of the voltage at the load terminals for the phase a is $215\angle 20^\circ\,\text{V}$. For abc phase sequence, calculate the source phase voltages.

9.6 A balanced three-phase wye-connected source delivers power to a balanced three-phase wye-connected load as shown in Fig. 9.44. For positive phase sequence, find the line voltages at the source terminals.

9.7 A balanced three-phase wye–wye system is shown in Fig. 9.45. For positive phase sequence, determine the load impedance .

9.8 A balanced three-phase wye–wye system is shown in Fig. 9.46. For a positive phase sequence, calculate the line impedance.

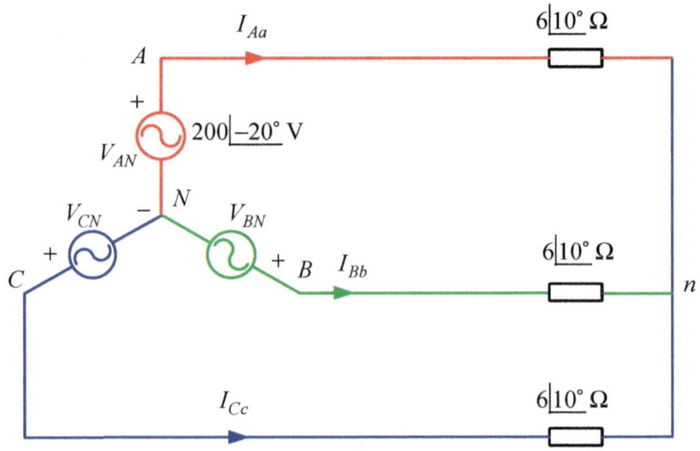

Fig. 9.41 Circuit for Problem 9.3

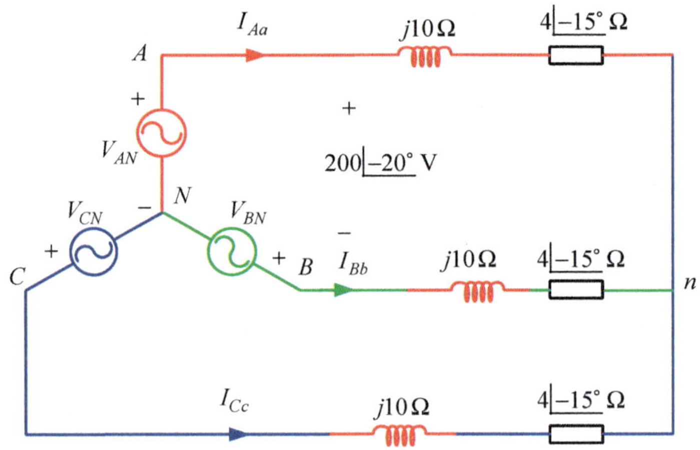

Fig. 9.42 Circuit for Problem 9.4

9.9 A balanced three-phase wye–wye system is shown in Fig. 9.47. The total power loss in the line is 600 W and the load power factor is 0.85 lagging. Determine the load impedance.

9.10 A balanced three-phase wye-connected source delivers power to a balanced three-phase wye-connected load as shown in Fig. 9.48. The total power absorbed by the load is 1200 W. Calculate the total power loss in the lines.

9.11 A balanced three-phase wye-connected source delivers power to a balanced three-phase delta-connected load as shown in Fig. 9.49. For positive phase sequence, calculate the phase and line currents.

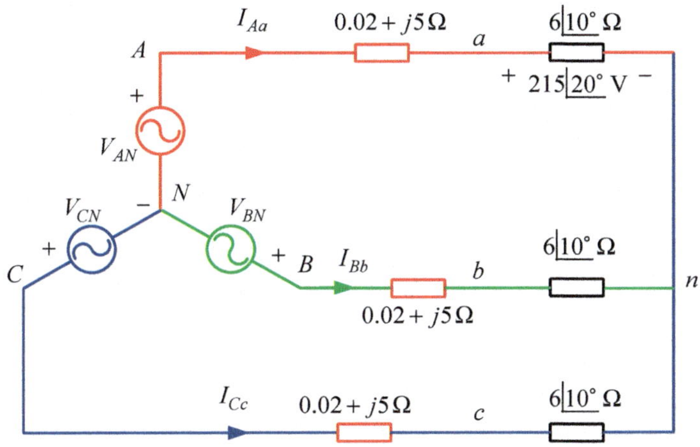

Fig. 9.43 Circuit for Problem 9.5

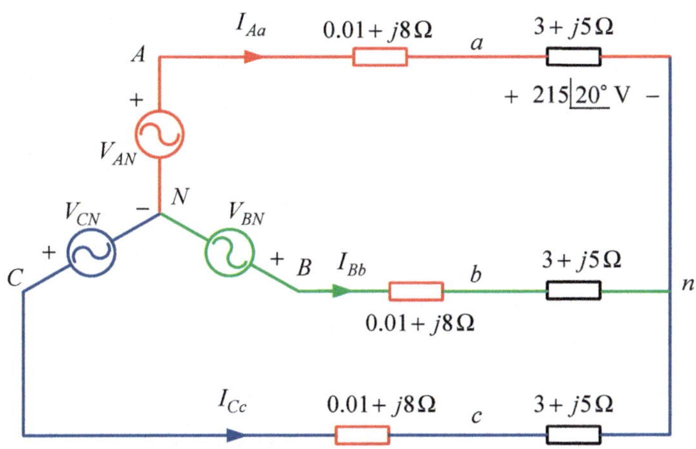

Fig. 9.44 Circuit for Problem 9.6

9.12 A balanced three-phase delta–wye system is shown in Fig. 9.50. For positive phase sequence, calculate the line currents and the phase voltage across the load terminals.

9.13 A balanced three-phase wye-connected source delivers power to a balanced three-phase delta-connected load as shown in Fig. 9.51. Determine the phase and line currents for positive phase sequence.

9.14 A balanced three-phase delta-connected source delivers power to a balanced three-phase delta-connected load as shown in Fig. 9.52. For positive phase sequence, calculate the phase and line currents.

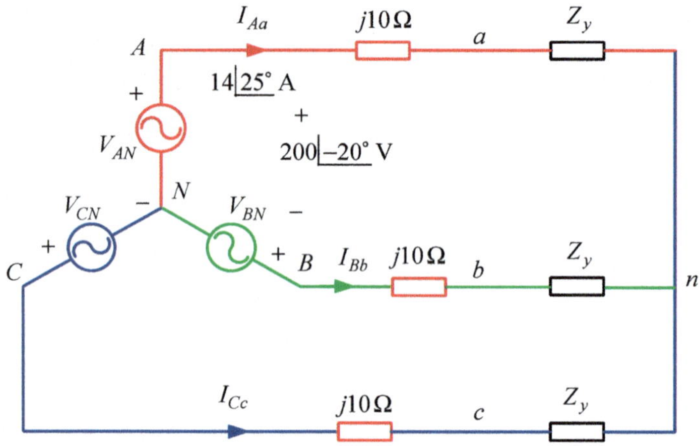

Fig. 9.45 Circuit for Problem 9.7

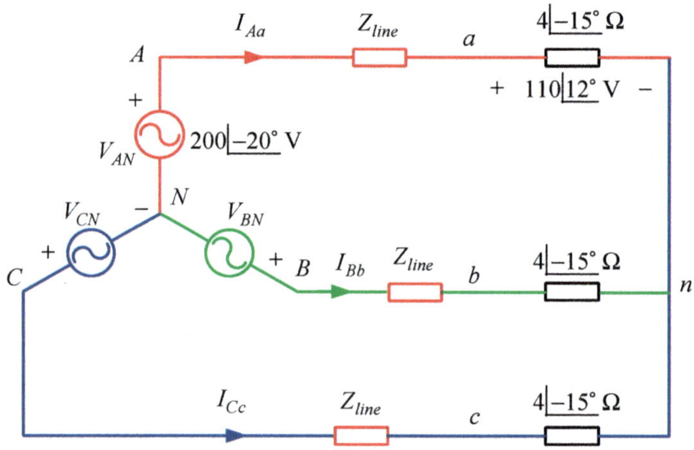

Fig. 9.46 Circuit for Problem 9.8

9.15 An unbalanced three-phase wye-connected load is given supply by a balanced three-phase source as shown in Fig. 9.53. The line to neutral voltage is 140 V. For positive phase sequence, calculate the line currents.

9.16 An unbalanced three-phase delta-connected load is given supply by a balanced three-phase wye-connected source as shown in Fig. 9.54. The line-to-line voltage is 220 V. For negative phase sequence, determine the line currents.

9.17 A balanced delta-connected source delivers power to an unbalanced delta-connected load as shown in Fig. 9.55. Calculate the total real power absorbed by the load.

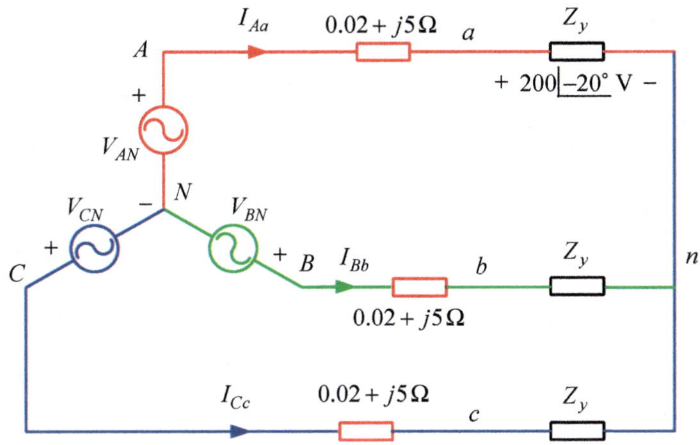

Fig. 9.47 Circuit for Problem 9.9

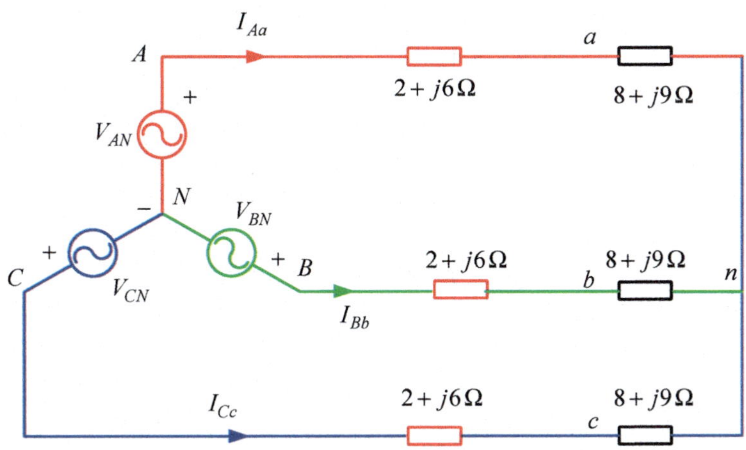

Fig. 9.48 Circuit for Problem 9.10

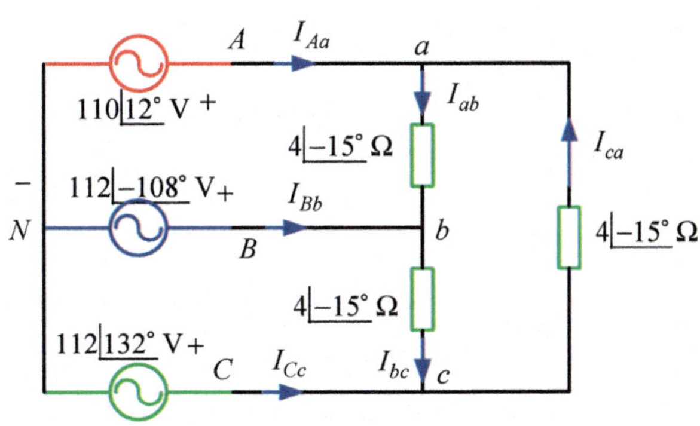

Fig. 9.49 Circuit for Problem 9.11

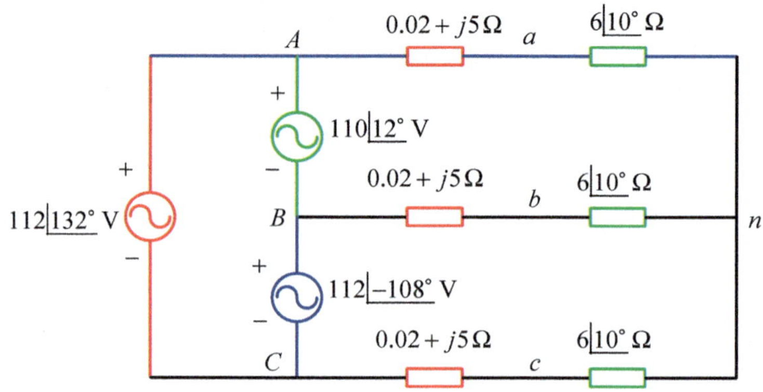

Fig. 9.50 Circuit for Problem 9.12

Fig. 9.51 Circuit for Problem 9.13

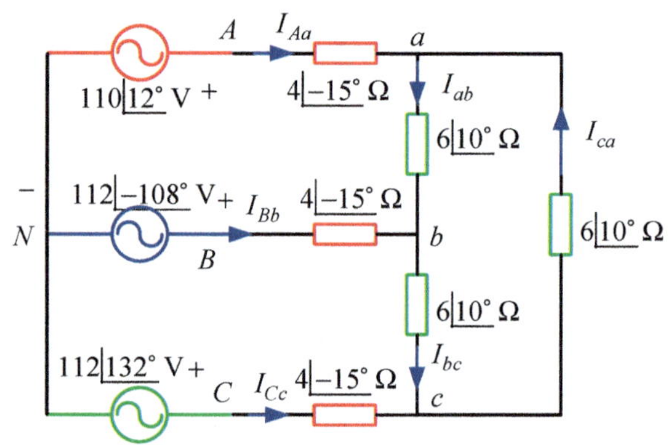

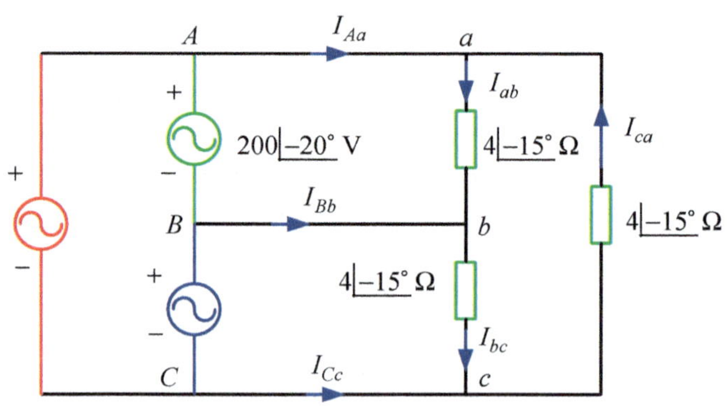

Fig. 9.52 Circuit for Problem 9.14

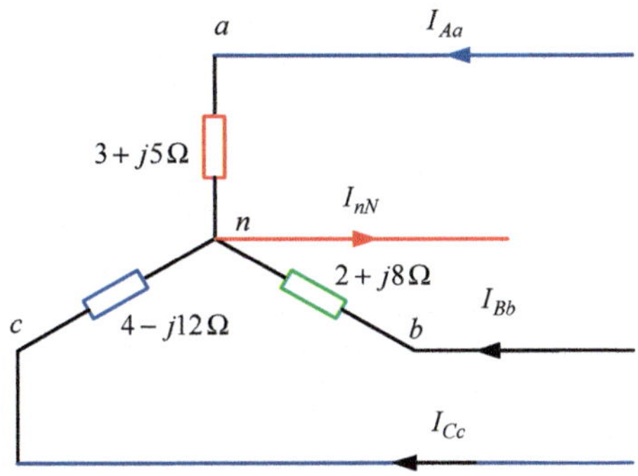

Fig. 9.53 Circuit for Problem 9.15

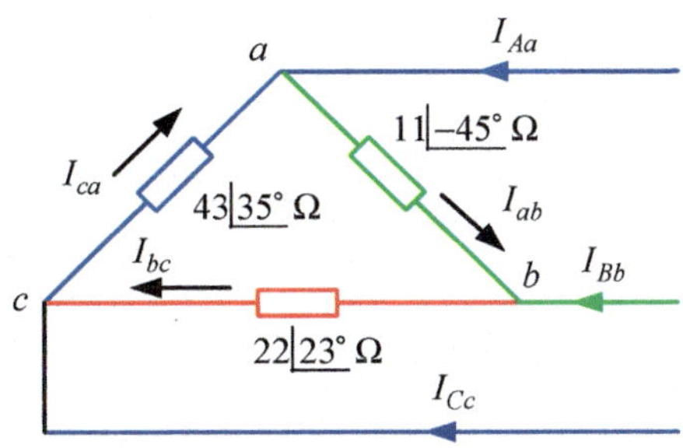

Fig. 9.54 Circuit for Problem 9.16

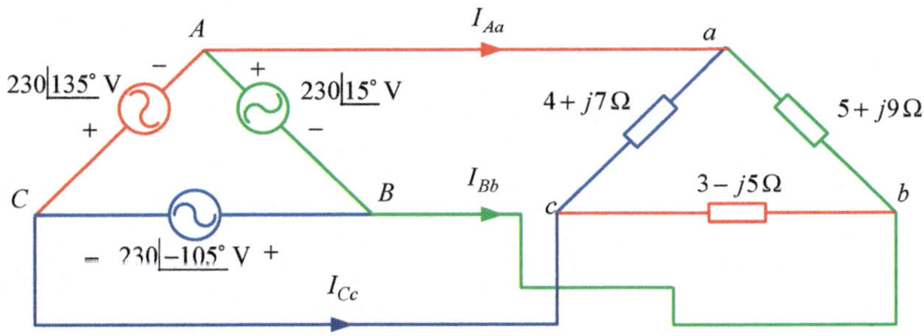

Fig. 9.55 Circuit for Problem 9.17

9.18 A balanced wye–wye system is shown in Fig. 9.56. Find the total real power absorbed by the load.

9.19 A balanced three-phase delta–wye system is shown in Fig. 9.57. Calculate the total power loss in the line, and average power absorbed by the load.

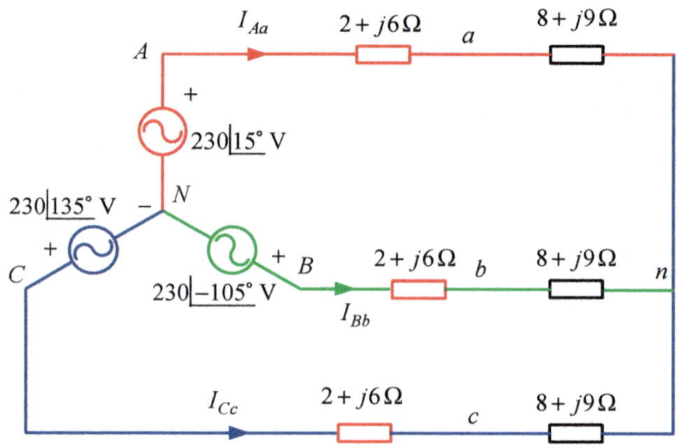

Fig. 9.56 Circuit for Problem 9.18

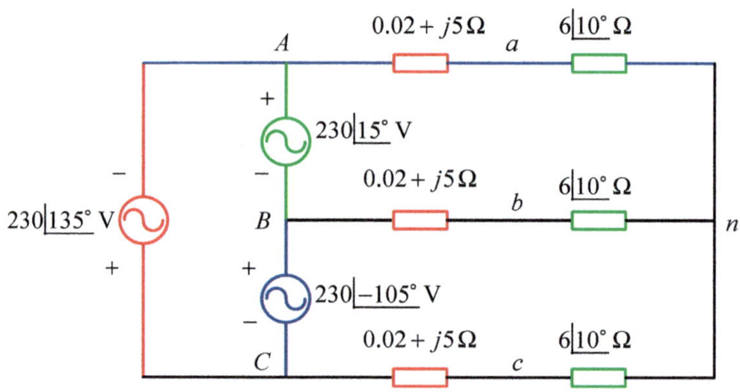

Fig. 9.57 Circuit for Problem 9.19

9.20 A balanced three-phase wye–delta system is shown in Fig. 9.58. Determine the line currents and the complex power absorbed by the loads.

9.21 A balanced three-phase source with a line voltage of 11 kV supplies power to a 1.5MVA, 0.95 lagging power factor load through three-phase transmission line with an impedance of $0.2 + j4\ \Omega$. Calculate the line current, and the total power loss in the line.

9.22 A balanced three-phase source delivers power to the balanced three-phase loads of 2 kVA at 0.85 power factor lagging, 3 kW at 0.85 leading power factor and 5 kVA at 0.9 power factor lagging. Determine the line current if the line-to-line voltage of the source is 440 V rms.

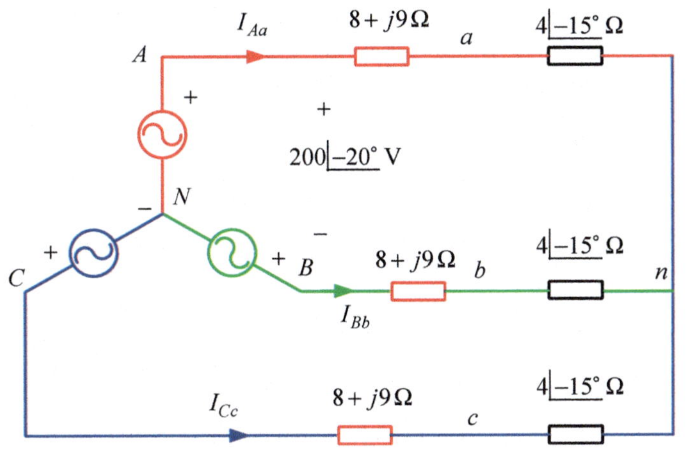

Fig. 9.58 Circuit for Problem 9.20

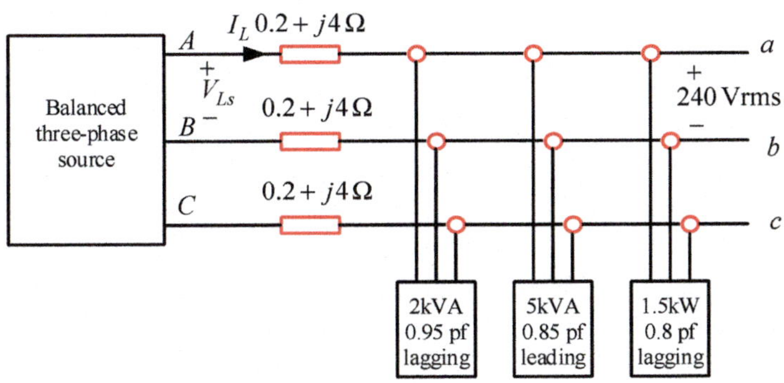

Fig. 9.59 Circuit for Problem 9.23

9.23 A balanced three-phase source delivers power to three different loads as shown in Fig. 9.59. Calculate the line current, complex power supplied by the source, source line voltage and power factor.

9.24 Three loads are given supply by a balanced three-phase source. The line voltage at the load terminals is 300 Vrms. The complex power supplied by the source is 20 kVA at 0.9 power factor lagging. Load 1: 5 kW at 0.75 lagging power factor, Load 2: 8 kVA at 0.85 power factor leading. Calculate the kVA rating of Load 3 and its power factor with a negligible line impedance. In addition, calculate the line current.

9.25 A three-phase wye-connected motor draws 5A current from a three-phase source whose line voltage is 415 V. The two-wattmeter method is used to measure the power and the readings are found to be $P_1 = 200\,\mathrm{W}$ and $P_2 = 1000\,\mathrm{W}$. Determine the power factor and phase impedance of the motor.

9.26 An unbalanced three-phase wye-connected load takes supply from a balanced supply as shown in Fig. 9.60. Calculate the wattmeter readings, and total power absorbed by the load.

9.27 Two-wattmeter method is connected to line a and line b of a three-phase wye-connected load to measure the total power as shown in Fig. 9.61. The line voltage of the source is 200 V and the wattmeter readings are found to be $P_1 = -400\,\mathrm{W}$ and $P_2 = 900\,\mathrm{W}$. Determine the per phase average power, reactive power, power factor and phase impedance of the load.

9.28 A balanced three-phase delta-connected load is supplied by a source as shown in Fig. 9.62. Two-wattmeter is connected to the lines. Calculate the wattmeter readings of the circuit.

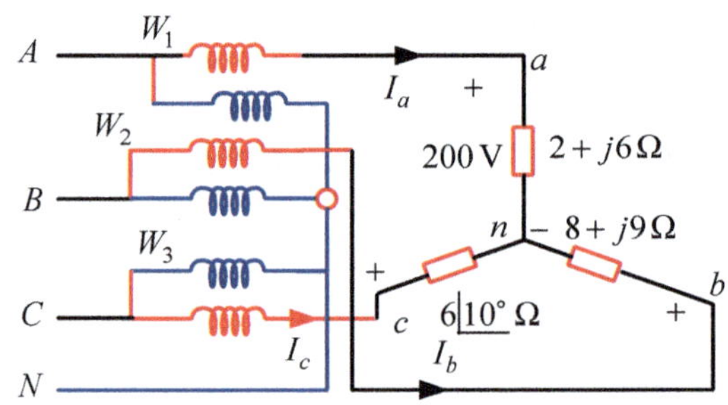

Fig. 9.60 Circuit for Problem 9.26

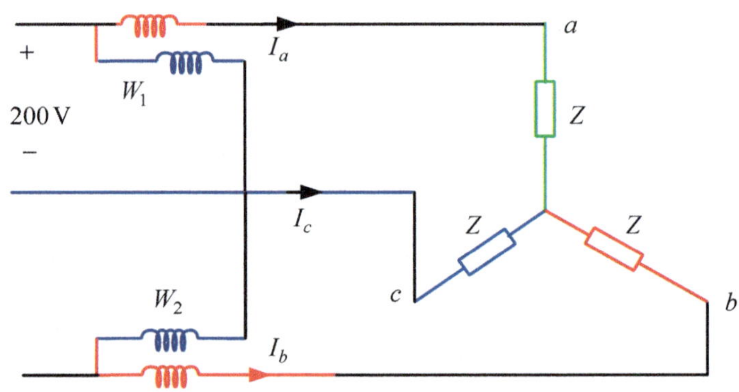

Fig. 9.61 Circuit for Problem 9.27

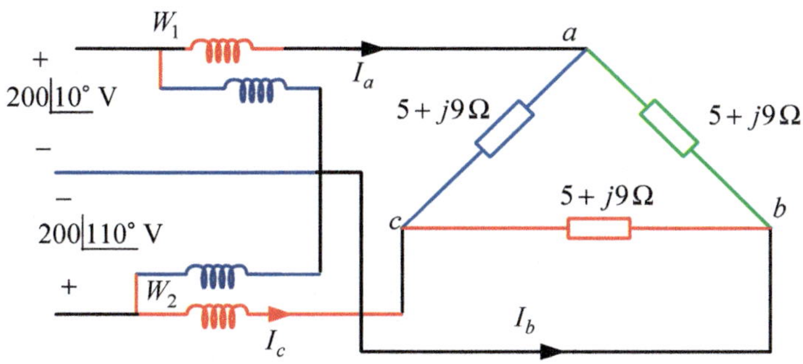

Fig. 9.62 Circuit for Problem 9.28

References

1. A. Hambley, *Electrical Engineering: Principles & Applications*, 7th edn. (Pearson, 2018)
2. R.L. Boylestad, *Introductory Circuit Analysis*, 13th edn. (Pearson, 2016)
3. J.W. Nilsson, S.A. Riedel, *Electric Circuits*, 10th edn. (Prentice Hall International Edition, 2015)
4. H.W. Jackson, D. Temple, B.E. Kelly, *Introduction to Electric Circuits*, 9th edn. (Oxford University Press, July 2015)
5. C.K. Alexander, M.N.O. Sadiku, *Fundamentals of Electric Circuits*, 6th edn. (McGraw-Hill Higher Education, January 2016)
6. J.D. Irwin, R.M. Nelms, *Basic Engineering Circuit Analysis*, 11th edn. (John & Wiley Sons Inc., USA, 2015)
7. D. Bell, *Fundamentals of Electric Circuits*, 7th edn. (Oxford University Press, March 2007)
8. G. Rizzoni, J. Kearns, *Principles and Applications of Electrical Engineering*, 6th edn. (McGraw-Hill Education, October 2014)
9. W. Hayt, J. Kemmerly, S. Durbin, *Engineering Circuit Analysis*, 8th edn. (McGraw-Hill Education, 2012)
10. M.A. Salam, *Basic Electrical Circuits*, 2nd edn. (Shroff Publishers & Distributors Pvt. Ltd, India, 2007)

Chapter 10
Frequency Response

10.1 Introduction

Resistance, inductance, capacitance, voltage and current are important parameters in an electrical circuit. Amplifier and filter circuits use these parameters to produce the gain and alter the circuit characteristics with different frequencies. When dealing with AC circuits, the source frequency is generally held constant either at 50 or at 60 Hz. If this frequency is varied, by keeping a constant magnitude of the source of a circuit, the frequency response of this circuit can be obtained. The frequency response of a circuit provides information on how the output behaviour of the circuit, in terms of magnitude and/or phase, changes with the change in the operating frequency. Generally, frequency response of a circuit is obtained by plotting its frequency-dependent gain (also known as transfer function) versus frequency. In this chapter, attention is given to the topics on (i) variation of resistance, inductance and capacitance with frequency, (ii) transfer function, (iii) filter circuits and (iv) resonance, bandwidth and Q-factor for series and parallel circuits.

10.2 Frequency Response with Resistance, Inductance and Capacitance

The impedance and angular frequency have already been discussed in an AC circuit. The impedance due to resistance is represented as [1, 2],

$$Z_R = R + j0 \tag{10.1}$$

$$\underline{|Z_R} = 0^\circ \tag{10.2}$$

Md. A. Salam and Q. Mehbubar Rahman, *Fundamentals of Electrical Circuit Analysis*,
https://doi.org/10.1007/978-981-10-8624-3_10

From Eqs. (10.1) and (10.2), it is seen that the magnitude and the phase are not dependent on frequency. The magnitude and phase due to resistance are shown in Fig. 10.1.

The impedance due to inductance is,

$$Z_L = 0 + j\omega L \tag{10.3}$$

The phase of the inductance is,

$$\underline{|Z_L} = 90^\circ \tag{10.4}$$

The inductor acts as a short circuit ($Z_L = 0$) at low frequency (DC) and an open circuit ($Z_L = \infty$) at high frequency. From Eqs. (10.3) and (10.4), it is seen that the magnitude of the impedance due to inductance is directly proportional to the frequency, while its phase remains constant. The frequency response of an inductor is shown in Fig. 10.2. The impedance due to capacitor is written as,

$$Z_C = 0 - j\frac{1}{\omega C} \tag{10.5}$$

The phase of the capacitance is,

$$\underline{|Z_C} = -90^\circ \tag{10.6}$$

The capacitor acts as an open circuit ($Z_C = \infty$) at DC and at high frequencies, the capacitor acts as a short circuit ($Z_C = 0$). From Eqs. (10.5) and (10.6), it is seen that the magnitude of Z_C varies inversely with frequency and its phase is independent of frequency. The frequency response of a capacitor is shown in Fig. 10.3.

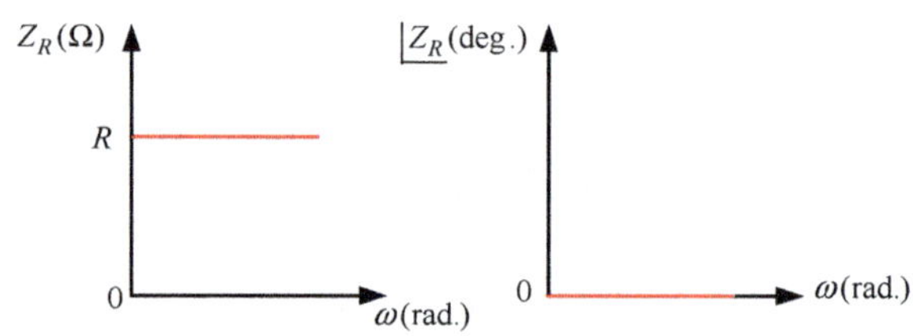

Fig. 10.1 Magnitude and phase response for resistance

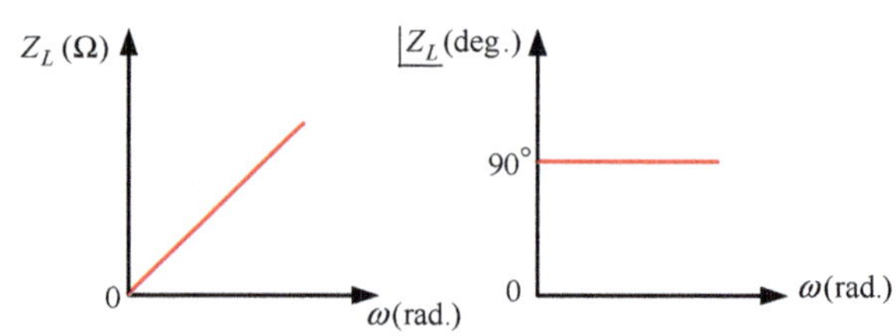

Fig. 10.2 Magnitude and phase response for inductance

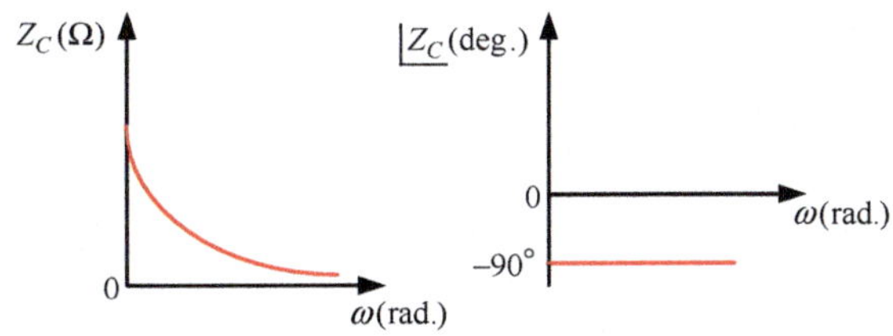

Fig. 10.3 Magnitude and phase response for capacitance

10.3 Transfer Function

The transfer function is a useful analytical tool for evaluating the frequency response of a circuit. It is defined as the ratio of output parameter to the input parameter of a circuit. The transfer function, represented by the letter H, can be written in different ways in frequency domain. In terms of voltage and current gains, the transfer functions can be written as [3, 4],

$$H(\omega) = \frac{V_0(\omega)}{V_i(\omega)} \tag{10.7}$$

$$H(\omega) = \frac{I_0(\omega)}{I_i(\omega)} \tag{10.8}$$

The transfer functions due to impedance and admittance of a circuit can be written as,

$$H(\omega) = \frac{V_0(\omega)}{I_i(\omega)} \tag{10.9}$$

$$H(\omega) = \frac{I_0(\omega)}{V_i(\omega)} \tag{10.10}$$

Poles and zeros of a transfer function are the frequencies for which the value of the denominator and numerator of a transfer function becomes zero, respectively. Poles and zeros are usually used in control engineering for system stability analysis. The transfer function due to poles and zeros are written as,

$$H(\omega) = \frac{N(\omega)}{D(\omega)} \tag{10.11}$$

Here, the roots of numerator $N(\omega) = 0$ are the zeros and the roots of denominator $D(\omega) = 0$ are the poles.

Example 10.1 A series–parallel circuit is shown in Fig. 10.4. Calculate the current gain $\frac{I_0(\omega)}{I_i(\omega)}$ and its zeros and poles.

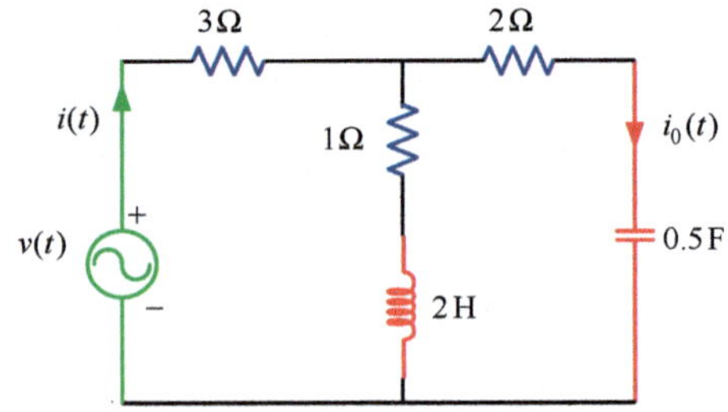

Fig. 10.4 Circuit for Example 10.1

Solution:
The output current can be determined as,

$$I_0(\omega) = I_i(\omega) \times \frac{(1 + j2\omega)}{1 + j2\omega + 2 + \frac{1}{j0.5\omega}} \tag{10.12}$$

$$\frac{I_0(\omega)}{I_i(\omega)} = \frac{j0.5\omega(1 + j2\omega)}{(1 + j2\omega + 2)j0.5\omega + 1} \tag{10.13}$$

Considering $s = j\omega$, Eq. (10.13) can be modified as,

$$\frac{I_0(\omega)}{I_i(\omega)} = \frac{0.5s(1 + 2s)}{(1 + 2s + 2)0.5s + 1} \tag{10.14}$$

$$\frac{I_0(\omega)}{I_i(\omega)} = \frac{s(s + 0.5)}{s^2 + 1.5s + 1} \tag{10.15}$$

Setting the numerator of Eq. (10.15) equal to zero yields,

$$s(s + 0.5) = 0 \tag{10.16}$$

$$z_1 = 0,\ z_2 = -0.5 \tag{10.17}$$

Setting the denominator of Eq. (10.15) equal to zero yields,

$$s^2 + 1.5s + 1 = 0 \tag{10.18}$$

$$s_{1,2} = \frac{-1.5 \pm \sqrt{1.5^2 - 4}}{2} \tag{10.19}$$

$$p_1 = -0.75 + j0.66,\ p_2 = -0.75 - j0.66 \tag{10.20}$$

Practice Problem 10.1

Figure 10.5 shows a series–parallel circuit. Determine the current gain $\frac{I_0(\omega)}{I_i(\omega)}$ and its zeros and poles.

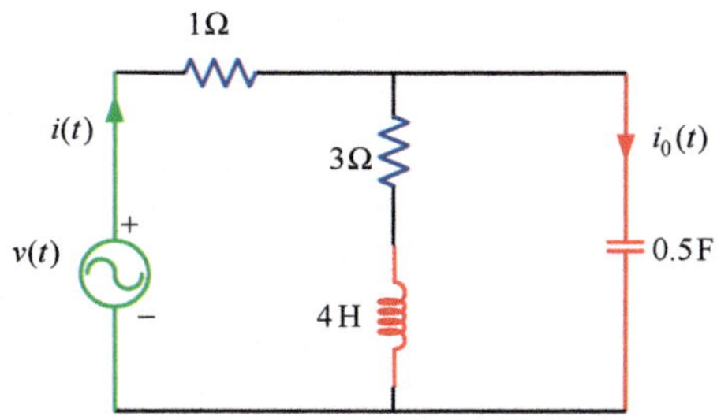

Fig. 10.5 Circuit for Practice Problem 10.1

10.4 Decibel

Decibel is abbreviated as dB. It is a relative unit of measurement commonly used in communications for providing a reference for input and output powers. This unit is named in honour of the Scottish-American scientist Alexander Graham Bell (1847–1922) who invented telephone system. The unit dB is equal to ten times log of the ratio of the calculated power (P_2) to the reference power (P_1), and it is represented as [5, 6],

$$G_{dB} = 10\log_{10}\frac{P_2}{P_1} \tag{10.21}$$

The input and output parameters of a network are shown in Fig. 10.6, where V_1, I_1, R_1 and P_1 represent the input voltage, current, resistance and power, respectively, while V_2, I_2, R_2 and P_2 represent the corresponding output quantities, respectively.

The input and output powers of this network are,

$$P_1 = \frac{V_1^2}{R_1} \tag{10.22}$$

$$P_2 = \frac{V_2^2}{R_2} \tag{10.23}$$

Substituting Eqs. (10.22) and (10.23) into Eq. (10.21) yields,

$$G_{dB} = 10\log_{10}\frac{\frac{V_2^2}{R_2}}{\frac{V_1^2}{R_1}} \tag{10.24}$$

$$G_{dB} = 10\log_{10}\left(\frac{V_2^2}{V_1^2}\times\frac{R_1}{R_2}\right) \tag{10.25}$$

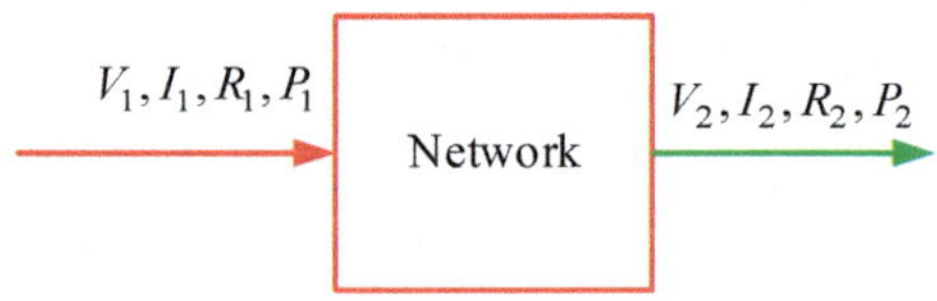

Fig. 10.6 Sample network

$$G_{dB} = 10\log_{10}\left(\frac{V_2^2}{V_1^2}\right) + 10\log_{10}\left(\frac{R_1}{R_2}\right) \tag{10.26}$$

$$G_{dB} = 20\log_{10}\frac{V_2}{V_1} + 10\log_{10}\frac{R_1}{R_2} \tag{10.27}$$

If the input and output resistance of a network are the same, then Eq. (10.27) is modified as,

$$G_{dB} = 20\log_{10}\frac{V_2}{V_1} \tag{10.28}$$

Again, substituting $P_1 = I_1^2 R_1$ and $P_2 = I_2^2 R_2$ into Eq. (10.21) yields,

$$G_{dB} = 10\log_{10}\left(\frac{I_2^2 R_2}{I_1^2 R_1}\right) \tag{10.29}$$

For the input and output resistance of a network, Eq. (10.29) can be modified as,

$$G_{dB} = 20\log_{10}\frac{I_2}{I_1} \tag{10.30}$$

10.5 Low Pass Filter

Filter circuit is usually designed to pass the signal with a specific frequency and reject the signal with other frequencies. In a low pass filter, the signal is allowed to pass with a low frequency ranging from DC to a certain cut-off frequency, beyond which it is stopped. A low pass filter circuit consists of resistor and capacitor and the expected output of this filter is measured across the capacitor as shown in Fig. 10.7. The current in this circuit is [7, 8],

$$i(t) = \frac{v_i(t)}{R + \frac{1}{j\omega C}} \tag{10.31}$$

The phasor form of the current is,

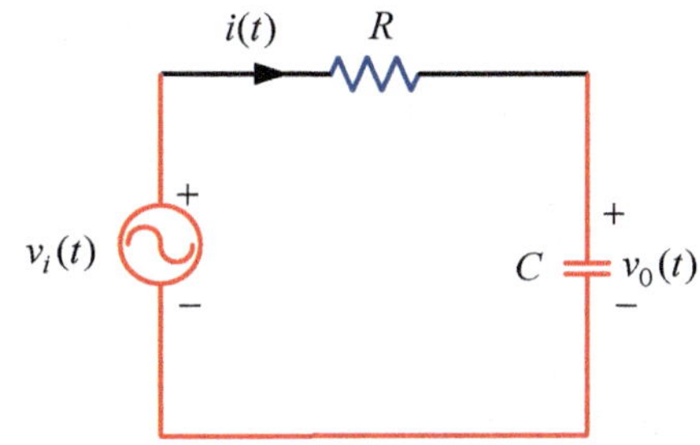

Fig. 10.7 Circuit for low pass filter

$$I_i = \frac{V_i}{R + \frac{1}{j\omega C}} \tag{10.32}$$

The magnitude of the current is,

$$|I_i| = \frac{V_i}{\sqrt{\left\{R^2 + \left(\frac{1}{\omega C}\right)^2\right\}}} \tag{10.33}$$

The output voltage in phasor form can be expressed as,

$$V_o = \frac{1}{j\omega C} I_i \tag{10.34}$$

Substituting Eq. (10.32) into Eq. (10.34) yields,

$$V_o = \frac{1}{j\omega C} \times \frac{V_i}{R + \frac{1}{j\omega C}} \tag{10.35}$$

$$V_o = \frac{V_i}{j\omega CR + 1} \tag{10.36}$$

From Eq. (10.36), the transfer function can be expressed as,

$$H(\omega) = \frac{V_o}{V_i} = \frac{1}{1 + j\omega CR} \tag{10.37}$$

The magnitude and the phase angle of the transfer function can be written as,

$$|H(\omega)| = \frac{1}{\sqrt{1 + (\omega CR)^2}} \tag{10.38}$$

$$\angle H(\omega) = -\tan^{-1}(\omega CR) \tag{10.39}$$

From Eqs. (10.38) and (10.39), the following approximations can be made:

$$\text{At}\,\omega \to 0, |H(\omega)| = 1, \underline{\left|H(\omega)\right.} = 0^\circ \tag{10.40}$$

$$\text{At}\,\omega \to \infty, |H(\omega)| = 0, \underline{\left|H(\omega)\right.} = -90^\circ \tag{10.41}$$

The corner frequency, also known as cut-off frequency or break frequency or roll-off frequency or half-power frequency, is the frequency at which $|H(\omega)|$ is equal to $\frac{1}{\sqrt{2}}$ (shown later) of its peak value, and it is found to be,

$$\omega_c = \frac{1}{RC} \tag{10.42}$$

Substituting Eq. (10.42) into Eq. (10.38) yields,

$$|H(\omega)| = \frac{1}{\sqrt{1 + \left(\frac{\omega}{\omega_c}\right)^2}} \tag{10.43}$$

Again, substituting Eq. (10.42) into Eq. (10.39) yields,

$$\theta = \underline{\left|H(\omega)\right.} = -\tan^{-1}\left(\frac{\omega}{\omega_c}\right) \tag{10.44}$$

Equation (10.43) can be rearranged as,

$$|V_0| = \frac{|V_i|}{\sqrt{1 + \left(\frac{\omega}{\omega_c}\right)^2}} = \frac{|V_i|}{\sqrt{1 + \left(\frac{f}{f_c}\right)^2}} \tag{10.45}$$

At $\omega = \omega_c$, Eqs. (10.43) and (10.44) can be modified as,

$$|H(\omega)| = \frac{1}{\sqrt{1+1}} = \frac{1}{\sqrt{2}} \tag{10.46}$$

$$\theta = \underline{\left|H(\omega)\right.} = -45^\circ \tag{10.47}$$

With respect to frequency, Figs. 10.8 and 10.9 show the magnitude of the transfer function and the phase, respectively. It is seen that the low frequency components of the signal are less affected. Therefore, the low frequency components of the signal are almost passed without change in magnitude or phase. However, this filter rejects the high frequency components of the signal.

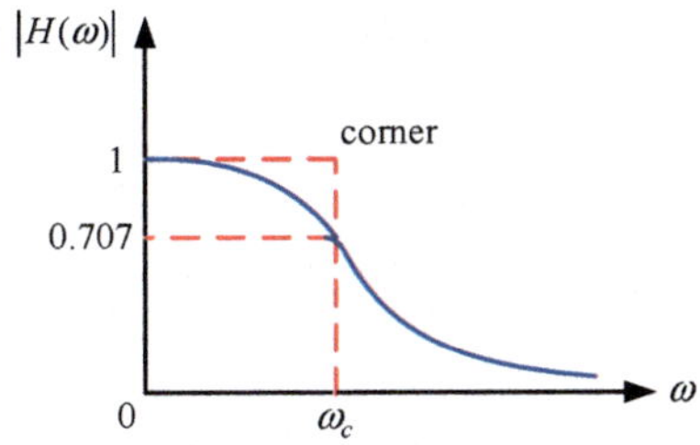

Fig. 10.8 Transfer function versus frequency for low pass filter

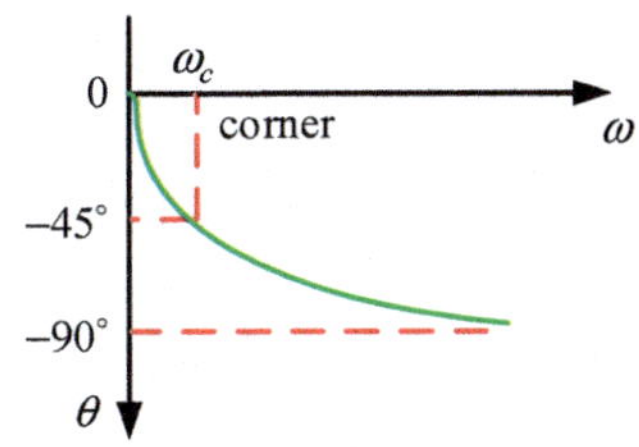

Fig. 10.9 Phase versus frequency for low pass filter

The power dissipation in the resistance is,

$$P_R = |I|^2 R \tag{10.48}$$

Substituting Eq. (10.32) into Eq. (10.48) yields,

$$P_R = \frac{V_i^2 R}{R^2 + \left(\frac{1}{\omega C}\right)^2} \tag{10.49}$$

At $\omega = 0$, the expression of the power dissipation from Eq. (10.49) can be derived as,

$$P_R = \frac{V_i^2 R}{R^2} = \frac{V_i^2}{R} \tag{10.50}$$

At $\omega = \infty$, the expression of the power dissipation from Eq. (10.49) can be written as,

$$P_R = 0 \tag{10.51}$$

At the corner frequency, the expression of the power can be derived from Eq. (10.49) as,

$$P_R = \frac{V_i^2 R}{R^2 + \left(\frac{\omega_c R}{\omega}\right)^2} \tag{10.52}$$

Substituting $\omega = \omega_c$ into Eq. (10.52) yields,

$$P_R = \frac{V_i^2 R}{R^2 + R^2} = \frac{V_i^2}{2R} = \frac{1}{2} P_m, \tag{10.53}$$

where the expression of maximum power is,

$$P_m = \frac{V_i^2}{R} \tag{10.54}$$

Example 10.2 Figure 10.10 shows a low pass filter. Calculate the cut-off frequency, and the magnitude of the output voltage if the frequency of the input voltage is 400 Hz.
Solution:
The cut-off frequency is,

$$f_c = \frac{1}{2\pi RC} = \frac{1}{2\pi \times 10 \times 10^3 \times 0.03 \times 10^{-6}} = 530.52\,\text{Hz} \tag{10.55}$$

The output voltage is,

$$|V_0| = \frac{|V_i|}{\sqrt{\left(\frac{f}{f_c}\right)^2 + 1}} = \frac{120}{\sqrt{\left(\frac{400}{530.52}\right)^2 + 1}} = 96\,\text{V} \tag{10.56}$$

Practice Problem 10.2
A low pass filter is shown in Fig. 10.11. Determine the cut-off frequency, and the value of the resistance by considering 500 Hz as the frequency of the input voltage.

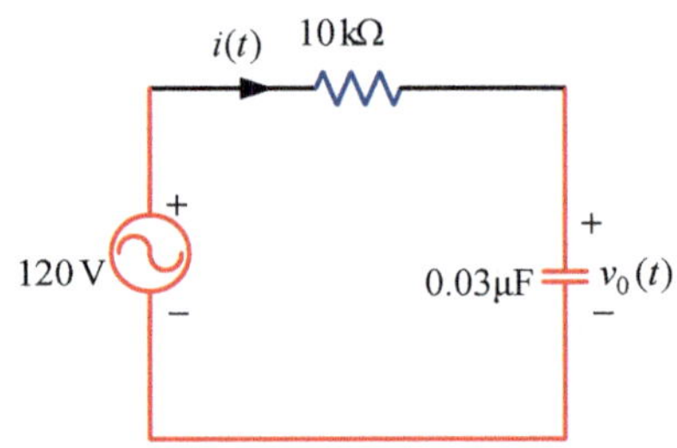

Fig. 10.10 Circuit for Example 10.2

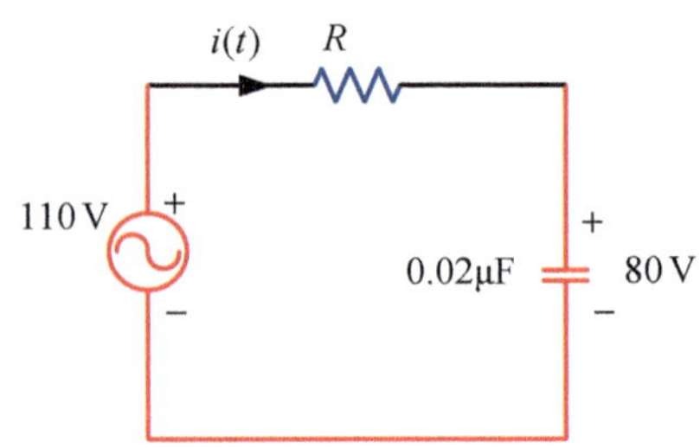

Fig. 10.11 Circuit for Practice Problem 10.2

10.6 High Pass Filter

A high pass filter allows high frequency signal to flow through it, and rejects low frequency signal. The circuit diagram is the same as the low pass filter except for the fact that in this case, the expected output is measured across the resistor as shown in Fig. 10.12. In this circuit, the expression of the phasor form of the current is given by [9],

$$I = \frac{V_i}{R + \frac{1}{j\omega C}} \tag{10.57}$$

The output voltage across the resistor is,

$$V_o = RI \tag{10.58}$$

Substituting Eq. (10.57) into Eq. (10.58) yields,

$$V_o = R \times \frac{V_i}{R + \frac{1}{j\omega C}} \tag{10.59}$$

$$\frac{V_o}{V_i} = \frac{j\omega RC}{1 + j\omega RC} \tag{10.60}$$

The transfer function for the high pass filter is,

$$H(\omega) = \frac{V_o}{V_i} = \frac{j\omega CR}{1 + j\omega CR} \tag{10.61}$$

The magnitude of the transfer function is,

$$|H(\omega)| = \frac{\omega CR}{\sqrt{1 + (\omega CR)^2}} \tag{10.62}$$

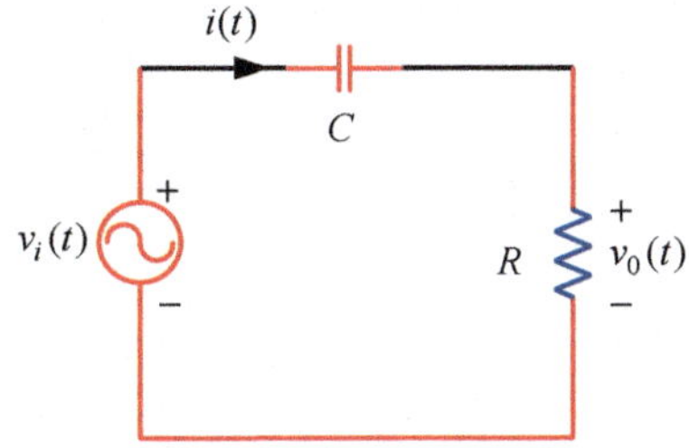

Fig. 10.12 Circuit for high pass filter

The phase of the transfer function for this filter is equal to the phase of the numerator (which is 90°) minus the phase of the denominator. From Eq. (10.61), the phase of the transfer function can be written as,

$$\theta = \angle H(\omega) = 90^\circ - \tan^{-1}(\omega CR) \tag{10.63}$$

Again, consider that the cut-off frequency is,

$$\omega_1 = \frac{1}{RC} \tag{10.64}$$

Substituting Eq. (10.64) into Eq. (10.62) yields,

$$|H(\omega)| = \frac{\frac{\omega}{\omega_1}}{\sqrt{1 + \left(\frac{\omega}{\omega_1}\right)^2}} \tag{10.65}$$

$$|H(\omega)| = \frac{1}{\frac{\omega_1}{\omega}} \frac{1}{\sqrt{1 + \left(\frac{\omega}{\omega_1}\right)^2}} \tag{10.66}$$

$$|H(\omega)| = \frac{1}{\sqrt{\left(\frac{\omega_1}{\omega}\right)^2 + 1}} \tag{10.67}$$

$$|H(f)| = \frac{1}{\sqrt{\left(\frac{f_1}{f}\right)^2 + 1}} \tag{10.68}$$

In terms of the voltage, Eq. (10.68) can be written as,

$$|V_0| = \frac{1}{\sqrt{\left(\frac{f_1}{f}\right)^2 + 1}} |V_i| \tag{10.69}$$

From Eqs. (10.62), (10.63) and (10.68), the following approximations can be drawn:

$$\text{At}\,\omega = 0, H(\omega) = 0, \theta = \angle H(\omega) = 90^\circ \tag{10.70}$$

$$\text{At}\,\omega \to \infty, H(\omega) = 1, \theta = \angle H(\omega) = 0^\circ \tag{10.71}$$

$$\text{At}\,\omega_1 = \frac{1}{RC}, f = f_1,\, H(\omega) = \frac{1}{\sqrt{2}},\, \theta = \underline{|H(\omega)} = 90^\circ - 45^\circ = 45^\circ \tag{10.72}$$

With respect to frequency, the magnitude of the transfer function and its phase are shown in Figs. 10.13 and 10.14, respectively.

Example 10.3 Figure 10.15 shows a high pass filter. Find the cut-off frequency, the magnitude of the output voltage at 15 kHz, and the phase angle of the output voltage at 30 kHz.

Solution:
The cut-off frequency is,

$$f_1 = \frac{1}{2\pi RC} = \frac{1}{2\pi \times 100 \times 0.03 \times 10^{-6}} = 53.05\,\text{kHz} \tag{10.73}$$

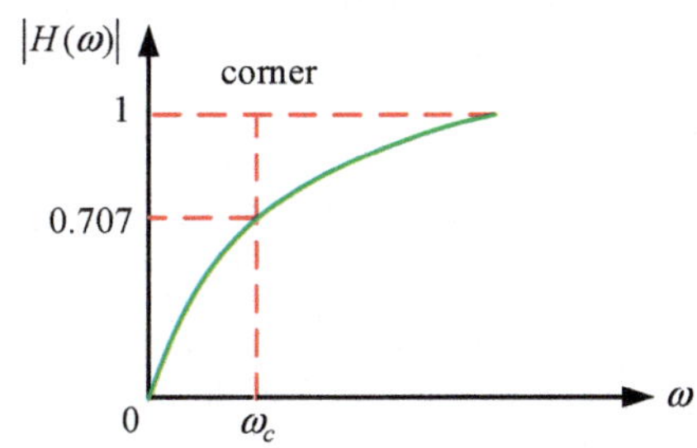

Fig. 10.13 Transfer function versus frequency for high pass filter

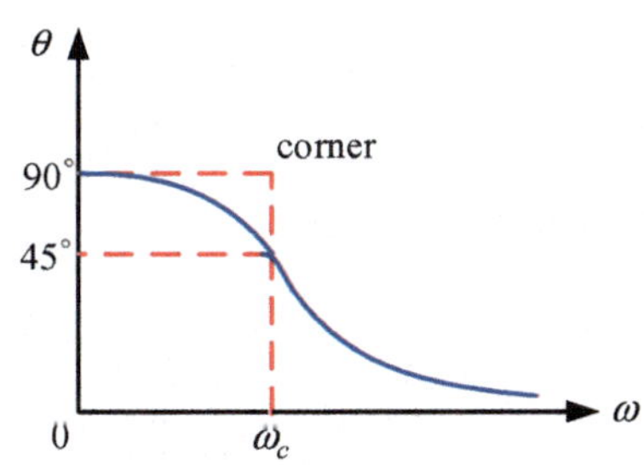

Fig. 10.14 Phase angle versus frequency for high pass filter

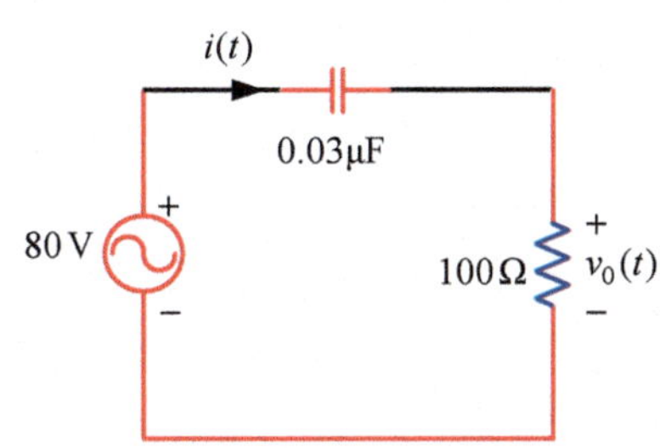

Fig. 10.15 Circuit for Example 10.3

The output voltage is calculated as,

$$|V_O| = \frac{|V_i|}{\sqrt{\left(\frac{f_1}{f}\right)^2 + 1}} = \frac{80}{\sqrt{\left(\frac{53.05}{15}\right)^2 + 1}} = 21.80\,\text{V} \tag{10.74}$$

At 30 kHz, the phase angle of the output voltage is determined as,

$$\theta = \underline{|H(\omega)} = 90^\circ - \tan^{-1}(2\pi \times 30 \times 10^3 \times 100 \times 0.03 \times 10^{-6}) = 60.53^\circ \tag{10.75}$$

Practice Problem 10.3
Figure 10.16 shows a high pass filter. Calculate the cut-off frequency, the magnitude of the output voltage at 30 kHz, and the phase angle of the output voltage at 50 kHz.

10.7 Series Resonance

Resonance is a valuable property of an AC circuit, employed in many applications that include tuning of a radio and TV receiver, voltage amplifier (series resonance), etc. Resonance occurs in a circuit when it contains inductance and capacitance. Figure 10.17 shows a *RLC* series circuit. The inductive reactance and the capacitive reactance varies with the frequency. If the frequency increases, inductive reactance increases and capacitive reactance decreases. At a certain frequency, known as the

Fig. 10.16 Circuit for Practice Problem 10.3

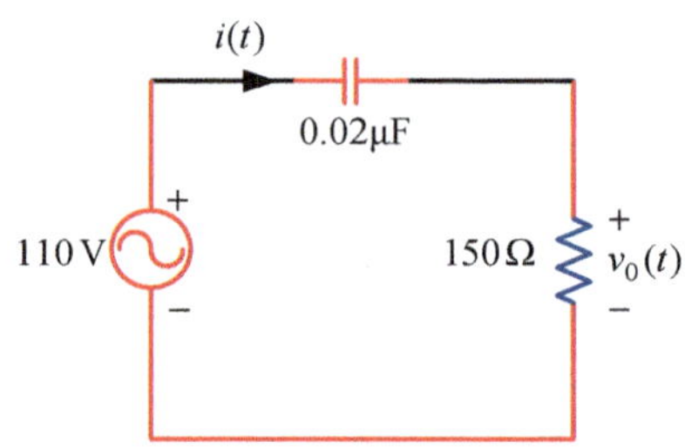

Fig. 10.17 Circuit for series resonance

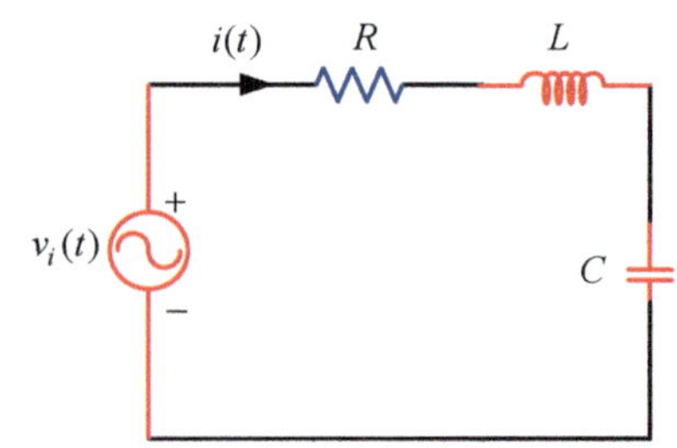

resonant frequency, and represented by f_0, the circuit power factor becomes unity. In this condition, the net reactance in the circuit becomes zero. Therefore, at resonance, the circuit impedance equates to the circuit resistance. In other words, at the resonant frequency f_0, the imaginary part of the total impedance becomes zero, i.e. the inductive reactance becomes equal to the capacitive reactance.

The impedance of a *RLC* series circuit in Fig. 10.17 is given by,

$$Z = R + j(X_L - X_C) \tag{10.76}$$

The resonant frequency can be evaluated by setting the imaginary part of Eq. (10.76) equal to zero as shown below:

$$X_L - X_C = 0 \tag{10.77}$$

$$2\pi f_0 L = \frac{1}{2\pi f_0 C} \tag{10.78}$$

$$f_0 = \frac{1}{2\pi\sqrt{LC}} \tag{10.79}$$

$$\omega_0 = \frac{1}{\sqrt{LC}} \tag{10.80}$$

10.8 Quality Factor for Series Resonance

A circuit with an inductor and capacitor always contains resistor, and the quality factor of a circuit is associated to the voltage drops across the resistor, inductor and capacitor. The quality factor, also known as Q-factor, is defined as the ratio of voltage drop across the inductor or the capacitor to the voltage drop across the resistor. Alternatively, quality factor is defined as the ratio of the reactive power associated to the inductor or the capacitor to the average power associated to the resistor at resonance. Quality factor is used to identify the sharpness of the resonance in a resonant circuit. According to the definition, the quality factor (Q_s), in terms of voltage drop, is expressed as,

$$Q_s = \frac{V_L}{V_R} \tag{10.81}$$

In terms of power, it can be expressed as,

$$Q_{sL} = \frac{I^2 X_L}{I^2 R} = \frac{X_L}{R} \tag{10.82}$$

$$Q_{sC} = \frac{I^2 X_C}{I^2 R} = \frac{X_C}{R} \tag{10.83}$$

Since, $X_L = \omega_0 L = 2\pi f_0 L$, Eq. (10.82) can be rewritten as,

$$Q_{sL} = \frac{2\pi f_0 L}{R} \tag{10.84}$$

Substituting Eq. (10.79) into Eq. (10.84) yields,

$$Q_{sL} = \frac{2\pi L}{R} \times \frac{1}{2\pi\sqrt{LC}} \tag{10.85}$$

$$Q_{sL} = \frac{1}{R}\sqrt{\frac{L}{C}} \tag{10.86}$$

Equation (10.86) can also be derived from Eq. (10.83) as,

$$Q_{sC} = \frac{1}{2\pi f_0 CR} \tag{10.87}$$

Again, substituting Eq. (10.79) into Eq. (10.87) yields,

$$Q_{sC} = \frac{1}{2\pi CR} \times \frac{1}{\frac{1}{2\pi\sqrt{LC}}} \tag{10.88}$$

$$Q_{sC} = \frac{1}{2\pi CR} \times 2\pi\sqrt{LC} \tag{10.89}$$

$$Q_{sC} = \frac{1}{R}\sqrt{\frac{L}{C}} \tag{10.90}$$

In general, the quality factor for a series resonant circuit is,

$$Q_s = \frac{1}{R}\sqrt{\frac{L}{C}} \tag{10.91}$$

10.9 Bandwidth for Series Resonance

The range of frequency determined by the two frequencies at which the source current of the circuit is equal to $\frac{1}{\sqrt{2}}$ times the maximum current is known as the bandwidth of the circuit. The phasor form of the source current in the circuit in Fig. 10.17 can be written as,

$$I = \frac{V_m}{\sqrt{R^2 + (\omega L - \frac{1}{\omega C})^2}} \tag{10.92}$$

The average power in this circuit is,

$$P(\omega) = \frac{1}{2} I^2 R \tag{10.93}$$

The current and power dissipation at resonance frequency are,

$$I_0 = \frac{V_m}{R} \tag{10.94}$$

$$P(\omega_0) = \frac{1}{2} \frac{V_m^2}{R} \tag{10.95}$$

The current versus frequency is shown in Fig. 10.18. The expression of the source current at the half-power frequencies is,

$$I = \frac{1}{\sqrt{2}} I_0 \tag{10.96}$$

Substituting Eq. (10.94) into Eq. (10.96) yields,

$$I = \frac{1}{\sqrt{2}} \frac{V_m}{R} \tag{10.97}$$

Substituting Eq. (10.98) into Eq. (10.93) yields the expression of the dissipated power at the half-power frequencies,

$$P(\omega_1) = P(\omega_2) = \frac{1}{2} \left(\frac{1}{\sqrt{2}} \frac{V_m}{R} \right)^2 R = \frac{V_m^2}{4R} \tag{10.98}$$

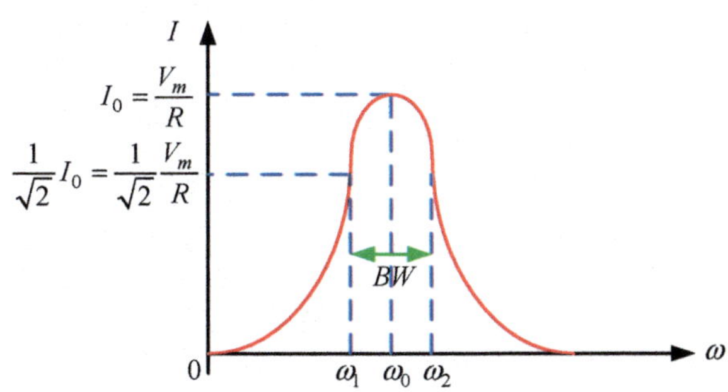

Fig. 10.18 Current versus frequency for series resonance

Substituting Eq. (10.97) into Eq. (10.92) yields,

$$\frac{1}{\sqrt{2}}\frac{V_m}{R} = \frac{V_m}{\sqrt{R^2 + (\omega L - \frac{1}{\omega C})^2}} \tag{10.99}$$

$$\sqrt{R^2 + \left(\omega L - \frac{1}{\omega C}^2\right)} = \sqrt{2}R \tag{10.100}$$

$$R^2 = \left(\omega L - \frac{1}{\omega C}\right)^2 \tag{10.101}$$

$$\omega_{1,2} = \pm\frac{R}{2L} \pm \sqrt{\frac{R^2}{4L^2} + \frac{1}{LC}} \tag{10.102}$$

The half-power frequencies are,

$$\omega_1 = -\frac{R}{2L} + \sqrt{\frac{R^2}{4L^2} + \frac{1}{LC}} \tag{10.103}$$

$$\omega_2 = \frac{R}{2L} + \sqrt{\frac{R^2}{4L^2} + \frac{1}{LC}} \tag{10.104}$$

Since $\frac{1}{LC} > \frac{R^2}{4L^2}$, the term $\frac{R^2}{4L^2}$ can be neglected from Eqs. (10.103) and (10.104), and that yields,

$$\omega_1 = -\frac{R}{2L} + \frac{1}{\sqrt{LC}} \tag{10.105}$$

$$\omega_2 = \frac{R}{2L} + \frac{1}{\sqrt{LC}} \tag{10.106}$$

Substituting Eq. (10.80) into Eqs. (10.105) and (10.106) yields,

$$\omega_1 = -\frac{R}{2L} + \omega_0 \tag{10.107}$$

$$\omega_2 = \frac{R}{2L} + \omega_0 \tag{10.108}$$

The lower band and upper band frequencies can be written as,

$$f_1 = f_0 - \frac{R}{4\pi L} \tag{10.109}$$

$$f_2 = f_0 + \frac{R}{4\pi L} \tag{10.110}$$

The difference between the half-power frequencies is known as bandwidth and it can be expressed as,

$$BW = f_2 - f_1 \tag{10.111}$$

Substituting Eqs. (10.109) and (10.110) into Eq. (10.111) yields,

$$BW = f_0 + \frac{R}{4\pi L} - f_0 + \frac{R}{4\pi L} \tag{10.112}$$

$$BW = \frac{R}{2\pi L} \tag{10.113}$$

$$R = (2\pi L)BW \tag{10.114}$$

Substituting Eq. (10.114) into the Eq. (10.82) yields,

$$Q_s = \frac{X_L}{(2\pi L)BW} \tag{10.115}$$

$$(2\pi L)BW = \frac{2\pi f_0 L}{Q_s} \tag{10.116}$$

$$BW = \frac{f_0}{Q_s} \tag{10.117}$$

From Eq. (10.117), it can be concluded that the bandwidth of a series resonance is directly proportional to the resonant frequency and inversely proportional to the quality factor.

Again, multiplying Eqs. (10.107) and (10.108) yields,

$$\omega_1\omega_2 = \left(\sqrt{\frac{R^2}{4L^2} + \frac{1}{LC}}\right)^2 - \left(\frac{R}{2L}\right)^2 \tag{10.118}$$

$$\omega_1\omega_2 = \frac{R^2}{4L^2} + \frac{1}{LC} - \frac{R^2}{4L^2} = \frac{1}{LC} \tag{10.119}$$

Substituting Eq. (10.80) into the Eq. (10.119) yields,

$$\omega_0 = \sqrt{\omega_1\omega_2} \tag{10.120}$$

From Eq. (10.120), it is seen that resonant angular frequency is the geometric mean of the half-power angular frequencies.

Again, the bandwidth is expressed as,

$$BW = \omega_2 - \omega_1 \tag{10.121}$$

Substituting Eqs. (10.107) and (10.108) into Eq. (10.122) yields,

$$BW = \frac{R}{2L} + \sqrt{\frac{R^2}{4L^2} + \frac{1}{LC}} + \frac{R}{2L} - \sqrt{\frac{R^2}{4L^2} + \frac{1}{LC}} \tag{10.122}$$

$$BW = \frac{R}{L} \tag{10.123}$$

From Eq. (10.123), it is seen that the bandwidth is directly proportional to the resistance and inversely proportional to the inductance.

Substituting Eq. (10.123) into Eq. (10.84) yields the general quality factor as,

$$Q_s = \frac{2\pi f_0}{BW} = \frac{\omega_0}{BW} \tag{10.124}$$

Similar to the finding in Eq. (10.117), it can be concluded from Eq. (10.124) that the quality factor is directly proportional to the resonant angular frequency and inversely proportional to the system bandwidth.

Example 10.4 Figure 10.19 shows a series *RLC* circuit. Calculate the resonant frequency, quality factor, bandwidth and half-power frequencies.

Solution:
The resonant frequency is,

$$f_0 = \frac{1}{2\pi\sqrt{600 \times 10^{-3} \times 80 \times 10^{-12}}} = 22.97\,\text{kHz} \tag{10.125}$$

The quality factor is calculated as,

$$Q_s = \frac{1}{R}\sqrt{\frac{L}{C}} = \frac{1}{540}\sqrt{\frac{600 \times 10^{-3}}{80 \times 10^{-12}}} = 160.38 \tag{10.126}$$

The bandwidth is calculated as,

$$BW = \frac{f_0}{Q_s} = \frac{22.97}{160.38} = 167.8\,\text{kHz/s} \tag{10.127}$$

Fig. 10.19 Circuit for Example 10.4

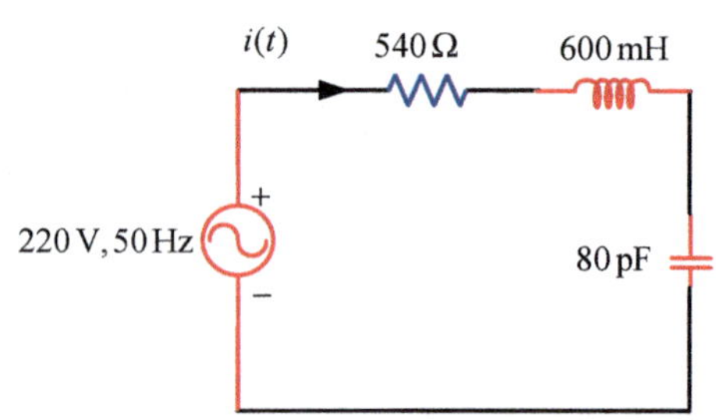

The half-power frequencies are determined as,

$$f_1 = 22970 - \frac{540}{4\pi \times 600 \times 10^{-3}} = 22.90\,\text{kHz} \tag{10.128}$$

$$f_2 = 22970 + \frac{540}{4\pi \times 600 \times 10^{-3}} = 23.04\,\text{kHz} \tag{10.129}$$

Practice Problem 10.4
A series *RLC* circuit is shown in Fig. 10.20. Determine the resonant frequency, quality factor, bandwidth and half-power frequencies.

10.10 Parallel Resonance

A parallel *RLC* circuit is shown in Fig. 10.21. In this circuit, resistor, inductor and capacitor are connected in parallel with the voltage source. The total admittance of this circuit can be written as,

$$Y = \frac{1}{R} + j\omega C - j\frac{1}{\omega L} \tag{10.130}$$

$$Y = \frac{1}{R} + j\left(\omega C - \frac{1}{\omega L}\right) \tag{10.131}$$

$$Z = \frac{1}{\frac{1}{R} + j\left(\omega C - \frac{1}{\omega L}\right)} \tag{10.132}$$

Fig. 10.20 Circuit for Practice Problem 10.4

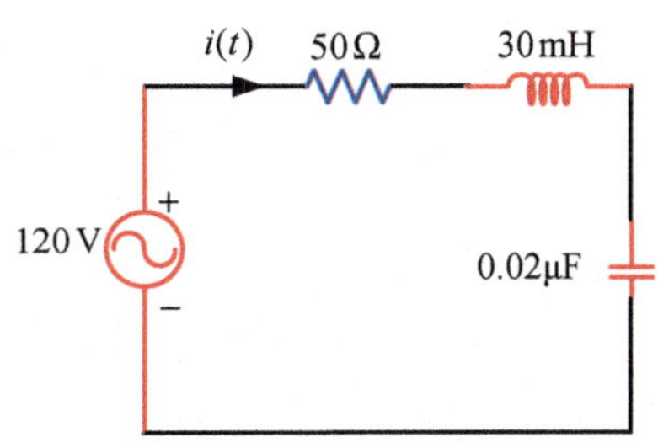

Fig. 10.21 *RLC* parallel circuit

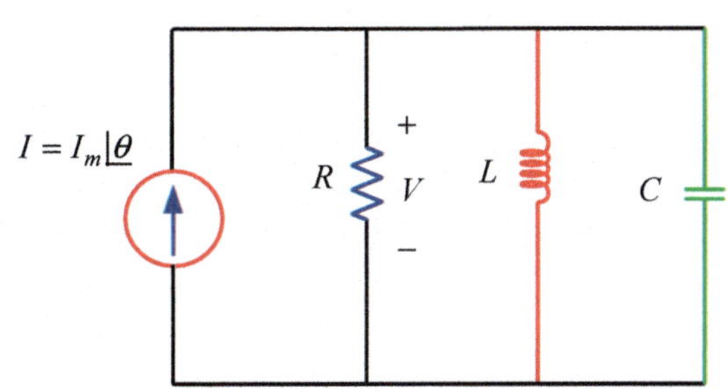

The magnitude of the impedance is,

$$|Z| = \frac{1}{\sqrt{\frac{1}{R^2} + \left(\omega C - \frac{1}{\omega L}\right)^2}} \tag{10.133}$$

Setting the imaginary part of Eq. (10.131) equal to zero yields,

$$\omega_0 C - \frac{1}{\omega_0 L} = 0 \tag{10.134}$$

$$\omega_0 = \frac{1}{\sqrt{LC}} \tag{10.135}$$

At half-power frequencies, the expression of voltage is,

$$V(\omega_1, \omega_2) = \frac{1}{\sqrt{2}} I_m R \tag{10.136}$$

$$\frac{V(\omega_1, \omega_2)}{I_m} = \frac{1}{\sqrt{2}} R \tag{10.137}$$

$$|Z| = \frac{1}{\sqrt{2}} R \tag{10.138}$$

Substituting Eq. (10.133) into Eq. (10.138) yields,

$$\frac{R}{\sqrt{2}} = \frac{1}{\sqrt{\frac{1}{R^2} + \left(\omega C - \frac{1}{\omega L}\right)^2}} \tag{10.139}$$

$$\frac{1}{R^2} + \left(\omega C - \frac{1}{\omega L}\right)^2 = \frac{2}{R^2} \tag{10.140}$$

$$\omega C - \frac{1}{\omega L} = \pm \frac{1}{R} \tag{10.141}$$

$$\omega^2 - \frac{1}{LC} = \pm \frac{1}{RC} \omega \tag{10.142}$$

$$\omega^2 \mp \frac{1}{RC} \omega - \frac{1}{LC} = 0 \tag{10.143}$$

$$\omega_{1,2} = \frac{\pm \frac{1}{RC} \pm \sqrt{\left(\frac{1}{RC}\right)^2 + \frac{4}{LC}}}{2} \tag{10.144}$$

$$\omega_{1,2} = \pm\frac{1}{2RC} + \sqrt{\left(\frac{1}{2RC}\right)^2 + \frac{1}{LC}} \tag{10.145}$$

$$\omega_1 = -\frac{1}{2RC} + \sqrt{\left(\frac{1}{2RC}\right)^2 + \frac{1}{LC}} \tag{10.146}$$

$$\omega_2 = \frac{1}{2RC} + \sqrt{\left(\frac{1}{2RC}\right)^2 + \frac{1}{LC}} \tag{10.147}$$

Substituting Eqs. (10.146) and (10.147) into Eq. (10.121) yields the expression of the bandwidth as,

$$BW = \frac{1}{2RC} + \sqrt{\left(\frac{1}{2RC}\right)^2 + \frac{1}{LC}} + \frac{1}{2RC} - \sqrt{\left(\frac{1}{2RC}\right)^2 + \frac{1}{LC}} \tag{10.148}$$

$$BW = \frac{1}{RC} \tag{10.149}$$

From Eq. (10.149), it is seen that the bandwidth is inversely proportional to the product of resistance and capacitance.

According to Eq. (10.124), the expression of quality factor for parallel resonance can be written as,

$$Q_p = \frac{\omega_0}{BW} \tag{10.150}$$

Substituting Eqs. (10.135) and (10.149) into Eq. (10.150) yields,

$$Q_p = \frac{RC}{\sqrt{LC}} \tag{10.151}$$

$$Q_p = R\sqrt{\frac{C}{L}} = \frac{1}{Q_s} \tag{10.152}$$

From Eq. (10.152), it is seen that the quality factor in parallel resonance is inversely proportional to the quality factor in series resonance.

Substituting Eq. (10.149) into Eqs. (10.146) and (10.147) yields,

$$\omega_1 = -\frac{BW}{2} + \sqrt{\left(\frac{BW}{2}\right)^2 + \omega_0^2} \tag{10.153}$$

$$\omega_2 = \frac{BW}{2} + \sqrt{\left(\frac{BW}{2}\right)^2 + \omega_0^2} \tag{10.154}$$

Substituting Eq. (10.150) into Eqs. (10.153) and (10.154) yields,

$$\omega_1 = -\frac{BW}{2} + \sqrt{\omega_0^2 \frac{1}{(2Q_p)^2} + \omega_0^2} \tag{10.155}$$

$$\omega_2 = \frac{BW}{2} + \sqrt{\omega_0^2 \frac{1}{(2Q_p)^2} + \omega_0^2} \tag{10.156}$$

For high quality factor, Eqs. (10.155) and (10.156) can be modified as,

$$\omega_1 = \omega_0 - \frac{BW}{2} \tag{10.157}$$

$$\omega_2 = \omega_0 + \frac{BW}{2} \tag{10.158}$$

A parallel resonance can also be produced with different arrangement of resistance, inductance and capacitance. Parallel resonance circuit with a useful combination of such parameters is shown in Fig. 10.22.

The total admittance of this circuit is,

$$Y = \frac{1}{R + jX_L} + \frac{1}{-jX_C} \tag{10.159}$$

$$Y = \frac{R - jX_L}{R^2 + X_L^2} - \frac{1}{jX_C} \tag{10.160}$$

$$Y = \frac{R}{R^2 + X_L^2} + j\left(\frac{1}{X_C} - \frac{X_L}{R^2 + X_L^2}\right) \tag{10.161}$$

Setting the imaginary part of Eq. (10.161) equal to zero yields,

$$\frac{1}{X_C} - \frac{X_L}{R^2 + X_L^2} = 0 \tag{10.162}$$

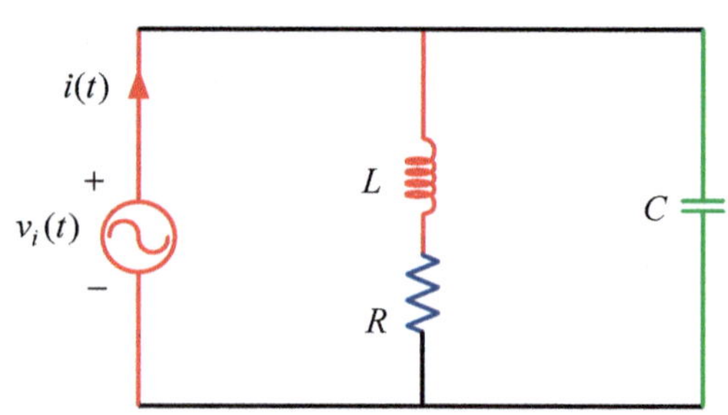

Fig. 10.22 A useful *RLC* parallel circuit

$$X_L X_C = R^2 + X_L^2 = Z_L^2 \tag{10.163}$$

Substituting $X_L = \omega L$ and $X_C = \frac{1}{\omega C}$ into Eq. (10.163) yields,

$$\frac{\omega L}{\omega C} = \frac{L}{C} = R^2 + X_L^2 = Z_L^2 \tag{10.164}$$

$$\frac{L}{C} = R^2 + (2\pi f_0 L)^2 \tag{10.165}$$

$$(2\pi f_0 L)^2 = \frac{L}{C} - R^2 \tag{10.166}$$

$$f_0 = \frac{1}{2\pi}\sqrt{\frac{1}{LC} - \frac{R^2}{L^2}} \tag{10.167}$$

The impedance at resonance condition can be derived as,

$$Y = \frac{1}{Z_r} = \frac{R}{R^2 + X_L^2} + 0 \tag{10.168}$$

$$\frac{1}{Z_r} = \frac{R}{Z_L^2} \tag{10.169}$$

Substituting Eq. (10.164) into Eq. (10.169) yields,

$$\frac{1}{Z_r} = \frac{R}{\frac{L}{C}} \tag{10.170}$$

$$Z_r = \frac{L}{RC} \tag{10.171}$$

From Eq. (10.171), it is seen that the impedance at resonance condition is equal to the ratio of inductance to the product resistance and capacitance.

Example 10.5 A parallel *RLC* circuit is shown in Fig. 10.23. At parallel resonance condition determine the resonant frequency, quality factor and resonant impedance.

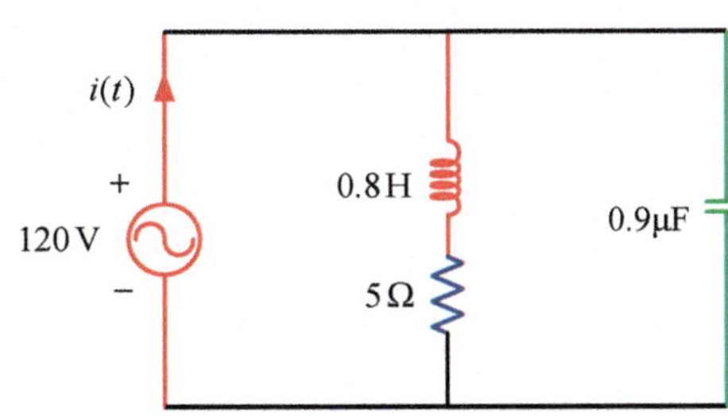

Fig. 10.23 Circuit for Example 10.5

Solution:
The resonant frequency is calculated as,

$$f_0 = \frac{1}{2\pi}\sqrt{\frac{1}{0.8 \times 0.9 \times 10^{-6}} - \frac{5^2}{0.8^2}} = 187.56\,\text{Hz} \tag{10.172}$$

The quality factor is determined as,

$$Q_p = \frac{2\pi \times 187.56 \times 0.8}{5} = 188.56 \tag{10.173}$$

Resonance impedance is calculated as,

$$Z_r = \frac{0.8}{5 \times 0.9 \times 10^{-6}} = 177.78\,\text{k}\,\Omega \tag{10.174}$$

Practice Problem 10.5
A parallel *RLC* circuit is shown in Fig. 10.24. At parallel resonance condition, determine the resonant frequency, quality factor, bandwidth and power dissipation at resonance condition.

10.11 Bode Plot

In 1930 and 1940, at Bell Laboratories, an American engineer by the name Hendrik Wade Bode (1905–1982) introduced the semi-log plots of the magnitude (in dBs) and the phase angle (in radians) of the transfer functions against the frequencies for avoiding linear scale for the frequency axis. Bode plots are used to interpret how the input affects the output both in magnitude and phase over frequency. Nowadays, these plots are widely used for displaying and communicating the necessary information regarding frequency response. Bode plots are also used for analysing power system stability, designing lead compensator and control system.

Consider the following transfer function to understand the Bode plot:

$$H(\omega) = \frac{k(s+z_1)}{s(s+p_1)}, \tag{10.175}$$

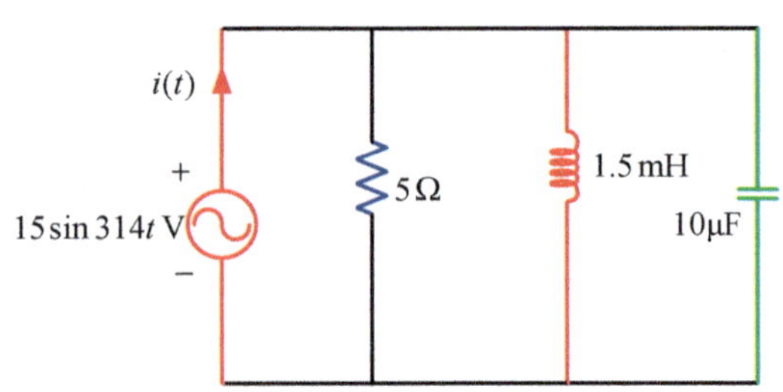

Fig. 10.24 Circuit for Practice Problem 10.5

where

z represents zero,
p represents pole.

Rewrite Eq. (10.175) in the following way:

$$H(\omega) = \frac{kz_1\left(\frac{s}{z_1}+1\right)}{sp_1\left(\frac{s}{p_1}+1\right)} \tag{10.176}$$

Substituting $s = j\omega$ into Eq. (10.170) provides,

$$H(\omega) = \frac{kz_1\left(\frac{j\omega}{z_1}+1\right)}{j\omega p_1\left(\frac{j\omega}{p_1}+1\right)} \tag{10.177}$$

Now, taking $20\log_{10}$ of Eq. (10.177) yields,

$$20\log_{10}(H(\omega)) = 20\log_{10}\left(\frac{kz_1\left(\frac{j\omega}{z_1}+1\right)}{j\omega p_1\left(\frac{j\omega}{p_1}+1\right)}\right) \tag{10.178}$$

$$\begin{aligned}20\log_{10}(H(\omega)) = {} & 20\log_{10}|k| + 20\log_{10}|z_1| + 20\log_{10}\left|\left(\frac{j\omega}{z_1}+1\right)\right| - 20\log_{10}|(j\omega)| \\ & - 20\log_{10}|(p_1)| - 20\log_{10}\left|\left(\frac{j\omega}{p_1}+1\right)\right|\end{aligned} \tag{10.179}$$

Initially, individual Bode plots are drawn by considering the effect of constants, poles or zeros at the origin, and not at the origin. Then, all separate Bode plots are merged together to get the complete Bode plots. The constant term of Eq. (10.179) creates a straight line with a magnitude of $20\log_{10}(k) = k_1$ dB and its phase provides 0° as shown in Figs. 10.25 and 10.26, respectively. In these plots, both the magnitude and phase do not change with frequency.

A zero at the origin occurs when the numerator is multiplied by the term s or j. Each occurrence creates a positively sloped line that passing through 1 with a rise 20 dB per decade (*one decade corresponds to a ratio of* 10 *between two frequencies*), as shown in Fig. 10.27. In case of zero at the origin, the phase is 90° as shown in Fig. 10.28. The pole at the origin occurs when the denominator is multiplied by s or j. For the pole at the origin, the magnitude is $-20\log_{10}(\omega)$ and the phase is −90°. Therefore, in this case, a negatively sloped line passes through 1 with a drop of 20 dB over a decade. Here, the phase remains constant with frequency. The Bode plots for pole at the origin are shown in Figs. 10.29 and 10.30, respectively.

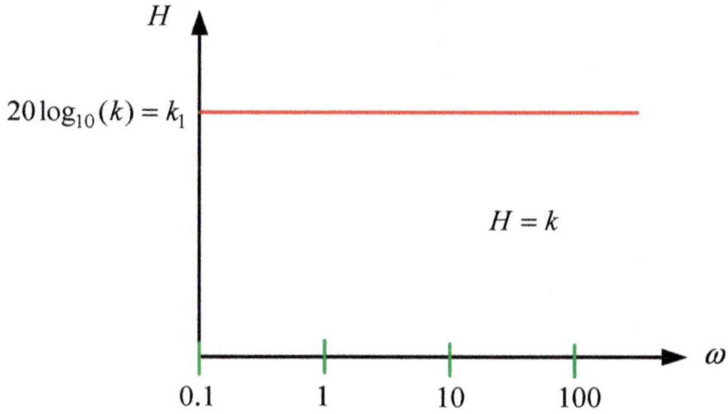

Fig. 10.25 Magnitude versus frequency for constant term

Fig. 10.26 Phase versus frequency for constant term

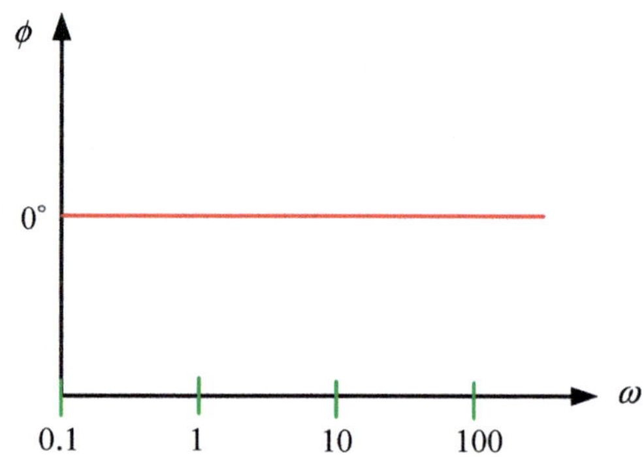

Fig. 10.27 Magnitude versus frequency for zero at the origin

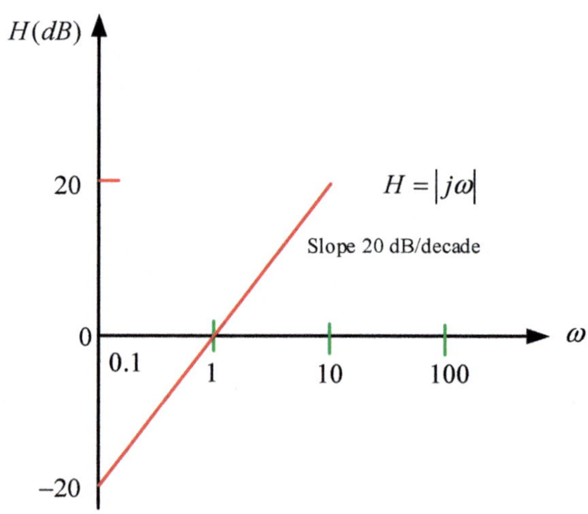

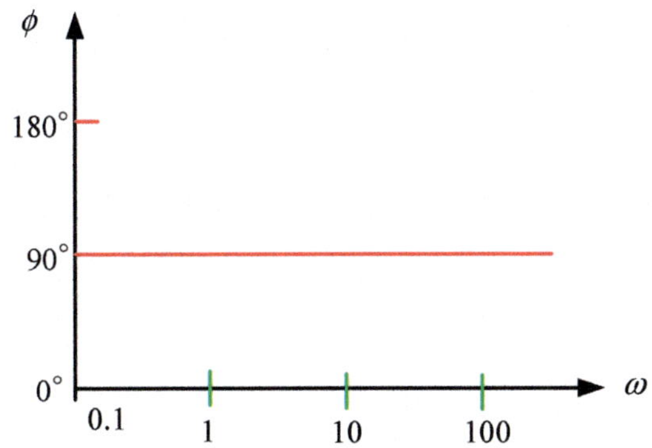

Fig. 10.28 Phase versus frequency for zero at the origin

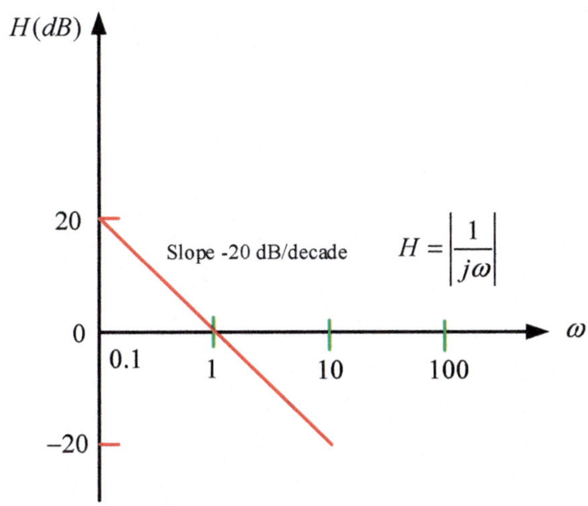

Fig. 10.29 Magnitude versus frequency for pole at the origin

Fig. 10.30 Phase versus frequency for pole at the origin

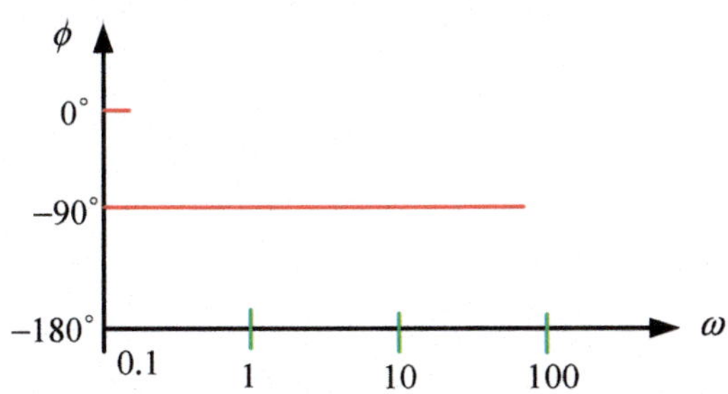

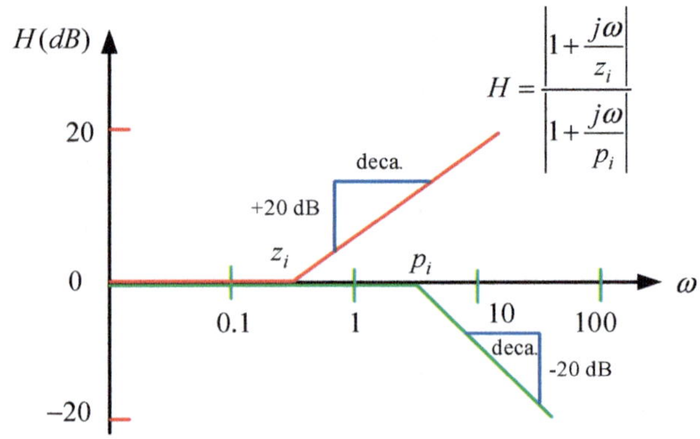

Fig. 10.31 Bode plots of zero and pole not at the origin

The zeros and poles not at the origin are generally represented by the terms $\left(\frac{j\omega}{z_i}+1\right)$ and $\left(\frac{j\omega}{p_i}+1\right)$, respectively. The terms z_i and p_i represent the critical frequency or break frequency of those two expressions. If $\omega \to 0$, then the magnitudes are,

$$H = 20\log_{10}\left|1+\frac{j\omega}{z_i}\right| = 0 \tag{10.180}$$

$$H = 20\log_{10}\left|1+\frac{j\omega}{p_i}\right| = 0 \tag{10.181}$$

Therefore, the terms in Eqs. (10.180) and (10.181) do not contribute to the overall plots below their critical frequency. However, they represent a ramp function of 20 dB per decade above the critical frequency as shown in Fig. 10.31.

From the term $\left(\frac{j\omega}{z_i}+1\right)$, the phase for this zero can be written as,

$$\phi = \tan^{-1}\left(\frac{\omega}{z_i}\right) \tag{10.182}$$

From Eq. (10.182), the following phase angles can be determined:

$$\text{At}\,\omega = 0, \phi = 0^\circ \tag{10.183}$$

$$\text{At}\,\omega = z_i, \phi = 45^\circ \tag{10.184}$$

$$\text{At}\,\omega \to \infty, \phi = 90^\circ \tag{10.185}$$

As an approximation, let $\phi = 0^\circ$ for $\omega \leq 0.1z_i$, $\phi = 45^\circ$ for $\omega = z_i$ and $\phi = 90^\circ$ for $\omega \geq 10z_i$. The phase plot for the zero not at the origin is drawn as shown in Fig. 10.32.

Fig. 10.32 Phase plot of zero not at the origin

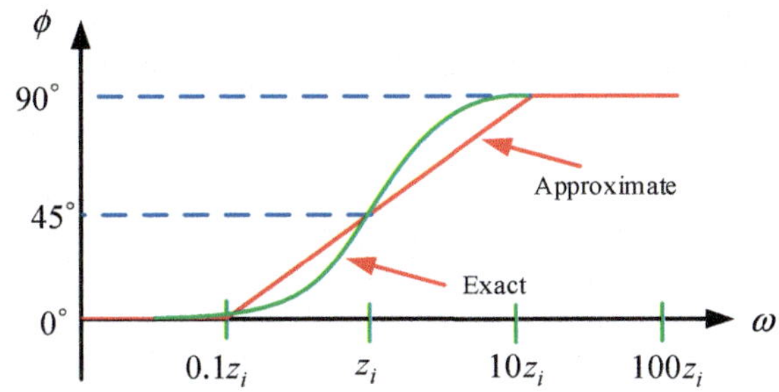

From the term $\left(\frac{j\omega}{p_i}+1\right)$, the phase for the pole can be written as,

$$\phi = -\tan^{-1}\left(\frac{\omega}{p_i}\right) \tag{10.186}$$

The following phase angles can be determined:

$$\text{At}\, \omega = 0, \phi = 0^\circ \tag{10.187}$$

$$\text{At}\, \omega = p_i, \phi = -45^\circ \tag{10.188}$$

$$\text{At}\, \omega \to \infty, \phi = -90^\circ \tag{10.189}$$

As an approximation, let $\phi = 0^\circ$ for $\omega \leq 0.1p_i$, $\phi = 45^\circ$ for $\omega = p_i$ and $\phi = 90^\circ$ for $\omega \geq 10p_i$. The phase plot for the pole not at the origin is drawn as shown in Fig. 10.33.

Example 10.6 A transfer function is given by $H(s) = \frac{72s}{(s+3)(s+6)}$. Sketch the Bode plots.

Fig. 10.33 Phase plot of pole not at the origin

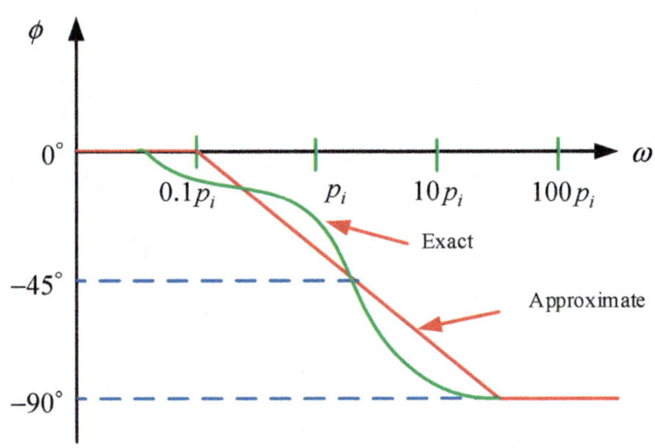

Solution:
Substituting $s = j\omega$ in the transfer function yields,

$$H(\omega) = \frac{72(j\omega)}{(j\omega+3)(j\omega+6)} \tag{10.190}$$

$$H(\omega) = \frac{72(j\omega)}{3\left(\frac{j\omega}{3}+1\right) \times 6\left(\frac{j\omega}{6}+1\right)} \tag{10.191}$$

$$H(\omega) = \frac{4(j\omega)}{\left(\frac{j\omega}{3}+1\right)\left(\frac{j\omega}{6}+1\right)} \tag{10.192}$$

The expressions of the magnitude and phase are,

$$H_{\mathrm{dB}} = 20\log_{10}(4) + 20\log_{10}(j\omega) - 20\log_{10}\left|\frac{j\omega}{3}+1\right| - 20\log_{10}\left|\frac{j\omega}{6}+1\right| \tag{10.193}$$

$$\phi = 90^\circ - \tan^{-1}\left(\frac{\omega}{3}\right) - \tan^{-1}\left(\frac{\omega}{6}\right) \tag{10.194}$$

The magnitude and phase plots are shown in Figs. 10.34 and 10.35, respectively.

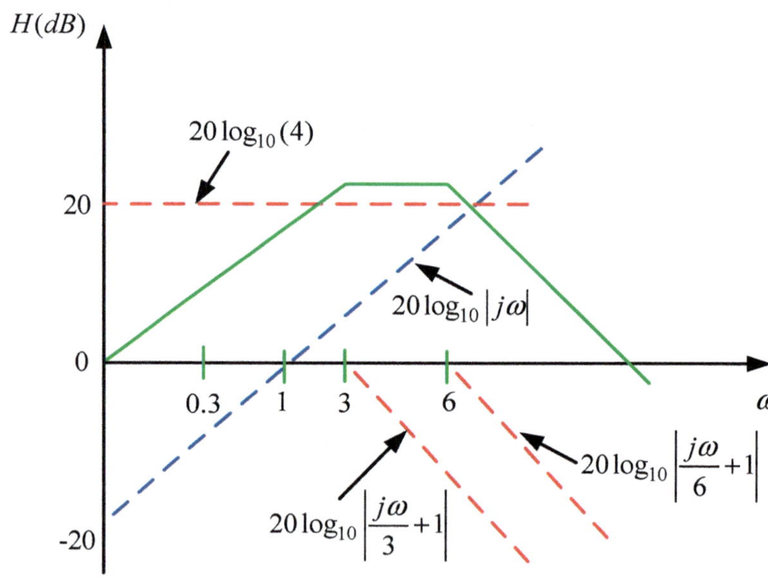

Fig. 10.34 Magnitude plots for Example 10.6

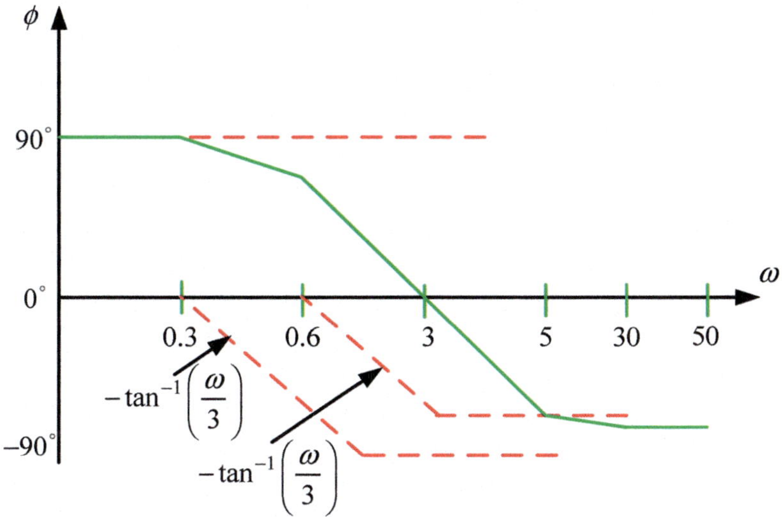

Fig. 10.35 Phase plot for Example 10.6

Practice Problem 10.6
Draw the Bode plots for the transfer function $H(s) = \frac{9}{(s+4)}$.

10.12 Bode Plots for Second Order

Consider a *RLC* series circuit for describing second-order Bode plot. In this case, the expected output is measured across the capacitor. The voltage across the capacitor is (Fig. 10.36),

$$V_C = \frac{i}{j\omega C} \tag{10.195}$$

Fig. 10.36 *RLC* series circuit for second-order Bode plots

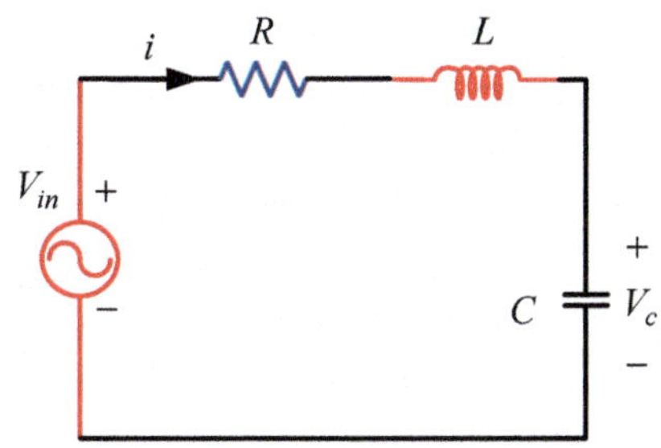

The expression of the input voltage is,

$$V_{in} = i\left(R + j\omega L + \frac{1}{j\omega C}\right) \tag{10.196}$$

The transfer function is defined as,

$$H(j\omega) = \frac{V_C}{V_{in}} \tag{10.197}$$

Substituting Eqs. (10.195) and (10.196) into Eq. (10.197) yields,

$$H(j\omega) = \frac{i\frac{1}{j\omega C}}{i\left(R + j\omega L + \frac{1}{j\omega C}\right)} \tag{10.198}$$

$$H(j\omega) = \frac{1}{(j\omega)^2 LC + j\omega RC + 1} \tag{10.199}$$

$$H(j\omega) = \frac{\frac{1}{LC}}{(j\omega)^2 + j\omega\frac{R}{L} + \frac{1}{LC}} \tag{10.200}$$

The resonant frequency is,

$$\omega_0 = \frac{1}{\sqrt{LC}} \tag{10.201}$$

Substituting Eq. (10.201) into Eq. (10.200) yields,

$$H(j\omega) = \frac{\omega_0^2}{-\omega^2 + j\omega\frac{R}{L} + \omega_0^2} \tag{10.202}$$

The damping ratio is defined as,

$$\xi = \frac{R}{2}\sqrt{\frac{C}{L}}, \tag{10.203}$$

where the value of the damping ratio is $0 < \xi < 1$.
Multiplying Eqs. (10.201) and (10.203) yields,

$$\xi\omega_0 = \frac{1}{\sqrt{LC}}\frac{R}{2}\sqrt{\frac{C}{L}} \tag{10.204}$$

$$2\xi\omega_0 = \frac{R}{L} \tag{10.205}$$

Substituting Eq. (10.205) into Eq. (10.202) yields,

$$H(j\omega) = \frac{\omega_0^2}{-\omega^2 + 2\xi\omega_0 j\omega + \omega_0^2} \tag{10.206}$$

$$H(j\omega) = \frac{1}{-\frac{\omega^2}{\omega_0^2} + \frac{2\xi\omega_0 j\omega}{\omega_0^2} + 1} \tag{10.207}$$

$$H(j\omega) = \frac{1}{1 - \frac{\omega^2}{\omega_0^2} + j\frac{2\xi\omega}{\omega_0}} \tag{10.208}$$

Considering $u = \frac{\omega}{\omega_0}$, Eq. (10.208) modified as,

$$H(j\omega) = \frac{1}{1 - u^2 + j2\xi u} \tag{10.209}$$

$$H(j\omega) = \frac{(1 - u^2) - j2\xi u}{(1 - u^2)^2 + (2\xi u)^2} \tag{10.210}$$

The real and imaginary parts of the transfer function are,

$$\text{Re}[H(j\omega)] = \frac{(1 - u^2)}{(1 - u^2)^2 + (2\xi u)^2} \tag{10.211}$$

$$\text{Im}[H(j\omega)] = \frac{-2\xi u}{(1 - u^2)^2 + (2\xi u)^2} \tag{10.212}$$

For the condition, $u \ll 1$, or $\frac{\omega}{\omega_0} \ll 1$, or $\omega \ll \omega_0$, the following expressions can be written:

$$\text{Re}[H(j\omega)] = \frac{(1 - 0)}{(1 - 0)^2 + (\text{very small})^2} = 1 \tag{10.213}$$

$$\text{Im}[H(j\omega)] = \frac{-\text{very small}}{(1 - 0)^2 + (\text{very small})^2} = \text{very small} \approx 0 \tag{10.214}$$

In this condition, the magnitude and phase are,

$$|H(j\omega)| = \sqrt{(\text{Re})^2 + (\text{Im})^2} \tag{10.215}$$

$$\phi = \underline{\left|H(j(\omega)\right.} = \tan^{-1}\left(\frac{0}{1}\right) = 0^\circ \tag{10.216}$$

Substituting Eqs. (10.213) and (10.214) into Eq. (10.215) yields,

$$|H(j\omega)| = \sqrt{(1)^2 + (0)^2} = 1 \tag{10.217}$$

The magnitude in Decibel is,

$$M(\mathrm{dB})_{\omega \ll \omega_0} = 20\log_{10}(1) = 0 \tag{10.218}$$

When $u \gg 1$ or $\frac{\omega}{\omega_0} \gg 1$ or $\omega \gg \omega_0$, then Eqs. (10.211) and (10.212) yields,

$$\mathrm{Re}[H(j\omega)] = \frac{(-u^2)}{(-u^2)^2 + (\text{very small})^2} = \frac{(-u^2)}{(-u^2)^2} = -u^{-2} = -\left(\frac{\omega}{\omega_0}\right)^{-2} \tag{10.219}$$

$$\mathrm{Im}[H(j\omega)] = \frac{-2\xi u}{(-u^2)^2 + (\text{very small})^2} = -2\xi u^{-3} \tag{10.220}$$

Then, the magnitude of the transfer function is,

$$|H(j\omega)| = \sqrt{(-u^{-2})^2 + (-2\xi u^{-3})^2} = \sqrt{u^{-4} + (\text{very small})^2} = \sqrt{u^{-4}} = u^{-2} \tag{10.221}$$

The magnitude in Decibel is,

$$M(\mathrm{dB})_{\omega \gg \omega_0} = 20\log_{10}(u^{-2}) = -40\log_{10}(u) = -40\log_{10}\left(\frac{\omega}{\omega_0}\right) \tag{10.222}$$

From Eq. (10.222), the following magnitudes are calculated:

$$\text{At}\,\omega = \omega_0, M_{\omega_0} = -40\log_{10}(1) = 0\,\mathrm{dB} \tag{10.223}$$

$$\text{At}\,\omega = 10\omega_0, M_{10\omega_0} = -40\log_{10}(10) = -40\,\mathrm{dB} \tag{10.224}$$

$$\text{At}\,\omega = 100\omega_0, M_{10\omega_0} = -40\log_{10}(10^2) = -80\,\mathrm{dB} \tag{10.225}$$

The phase can be written as,

$$\phi = \underline{|H(j(\omega)} = \tan^{-1}\left(\frac{-2\xi u^{-3}}{-u^{-2}}\right) \tag{10.226}$$

$$\phi = \underline{|H(j(\omega)} = \tan^{-1}(2\xi u^{-1}) \tag{10.227}$$

$$\phi = \underline{|H(j(\omega)} = \tan^{-1}\left(\frac{2\xi}{\frac{\omega}{\omega_0}}\right) \tag{10.228}$$

From Eq. (10.228), it is observed that when the expression $u = \frac{\omega}{\omega_0}$ is very high then $\phi = 0°$ and for a lower value, the phase $\phi = -180°$.

From Eq. (10.209), the magnitude of the transfer function can be written as,

$$H(j\omega) = \frac{1}{\sqrt{(1-u^2)^2 + (2\xi u)^2}} \tag{10.229}$$

Differentiating Eq. (10.229) with respect to u and setting it equal to zero yields,

$$\frac{dH(j\omega)}{du} = \frac{d}{du}\left[1 - 2u^2 + u^4 + 4\xi^2 u^2\right]^{\frac{-1}{2}} = 0 \tag{10.230}$$

$$-4u + 4u^3 + 8\xi^2 u = 0 \tag{10.231}$$

$$4u(u^2 + 2\xi^2 - 1) = 0 \tag{10.232}$$

$$u^2 + 2\xi^2 - 1 = 0 \tag{10.233}$$

$$u = \sqrt{1 - 2\xi^2} \tag{10.234}$$

$$\frac{\omega_r}{\omega_0} = \sqrt{1 - 2\xi^2} \tag{10.235}$$

$$\omega_r = \omega_0\sqrt{1 - 2\xi^2} \tag{10.236}$$

Substituting Eq. (10.234) into Eq. (10.229) yields the resonant peak,

$$M_r = \frac{1}{\sqrt{(1 - 1 + 2\xi^2)^2 + 4\xi^2(1 - 2\xi^2)}} \tag{10.237}$$

$$M_r = \frac{1}{\sqrt{4\xi^4 + 4\xi^2 - 8\xi^4}} \tag{10.238}$$

$$M_r = \frac{1}{2\xi\sqrt{1 - \xi^2}} \tag{10.239}$$

When $u = 1$ or $\frac{\omega}{\omega_0} = 1$ or $\omega = \omega_0$, then Eqs. (10.211) and (10.212) yields,

$$\text{Re}[H(j\omega)] = \frac{(1 - 1^2)}{(1 - 1^2)^2 + (2\xi)^2} = 0 \tag{10.240}$$

$$\text{Im}[H(j\omega)] = \frac{-2\xi}{(1 - 1)^2 + (2\xi)^2} = \frac{-1}{2\xi} \tag{10.241}$$

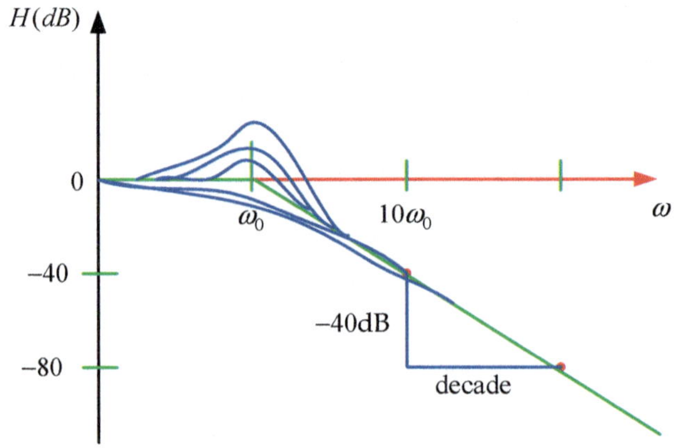

Fig. 10.37 Magnitude Bode plots for second order

Then, the magnitude in Decibels is,

$$M(\mathrm{dB})_{\omega=\omega_0} = 20\log_{10}(2\xi^{-1}) = -20\log_{10}(2\xi) \tag{10.242}$$

From Eq. (10.242), the following magnitudes are calculated:

$$\text{At } \xi = 0.5, M_{\xi=0.5} = -20\log_{10}(1) = 0\,\mathrm{dB} \tag{10.243}$$

$$\text{At } \xi > 0.5, M_{\xi>0.5} = -20\log_{10}(2\times 0.9) = -5.1\,\mathrm{dB} \tag{10.244}$$

$$\text{At } \xi < 0.5, M_{\xi<0.5} = -20\log_{10}(2\times 0.1) = 13.9\,\mathrm{dB} \tag{10.245}$$

The value of the phase is,

$$\phi = \underline{|H(j(\omega)} = \tan^{-1}\left(\frac{\frac{-1}{2\xi}}{0}\right) = \tan^{-1}(\infty) = -90^\circ \tag{10.246}$$

The Bode plots that include magnitude, peak and phase are drawn as shown in Figs. 10.37 and 10.38, respectively.

Exercise Problems

10.1 Figure 10.39 shows a series–parallel circuit. Determine the current gain $\frac{I_0(\omega)}{I_i(\omega)}$ and its zeros and poles.

10.2 A series–parallel circuit is shown in Fig. 10.40. Calculate the current gain $\frac{I_0(\omega)}{I_i(\omega)}$ and its zeros and poles.

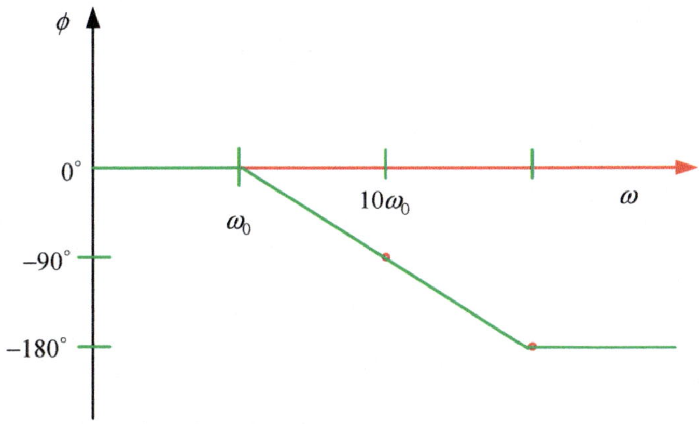

Fig. 10.38 Phase Bode plots for second order

Fig. 10.39 Circuit for Problem 10.1

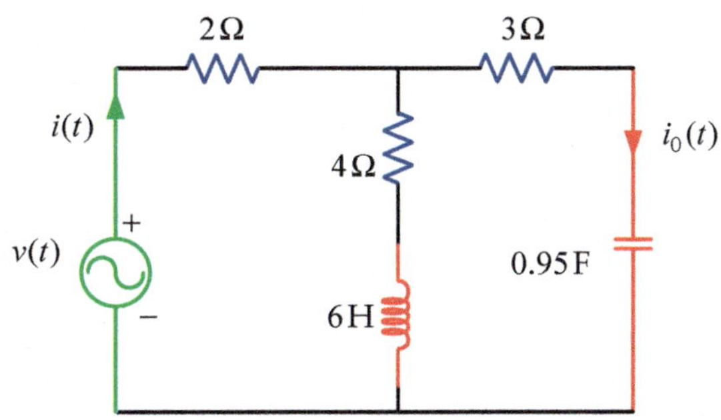

Fig. 10.40 Circuit for Problem 10.2

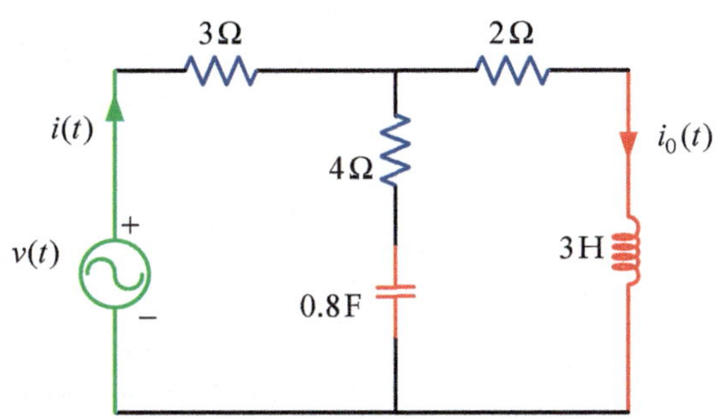

10.3 A series RL low pass filter circuit is shown in Fig. 10.41. The cut-off frequency of this filter is 1 kHz. Find the inductance and the magnitude of the transfer function at 60 kHz frequency.

10.4 Figure 10.42 shows a series RL low pass filter circuit. The cut-off frequency of this filter is 3 kHz. Find the resistance, the magnitude of the transfer function and the phase at 50 kHz frequency.

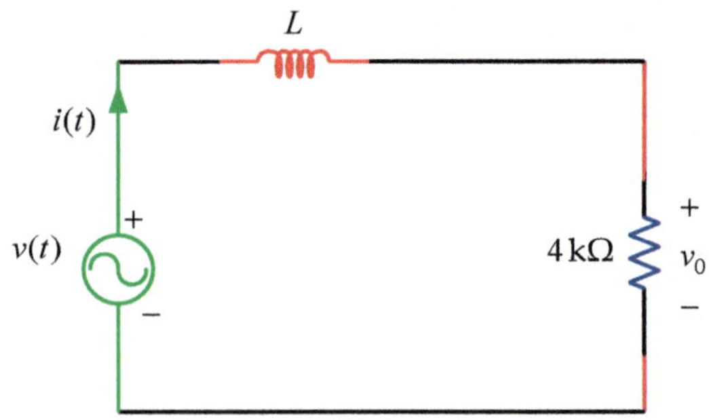

Fig. 10.41 Circuit for Problem 10.3

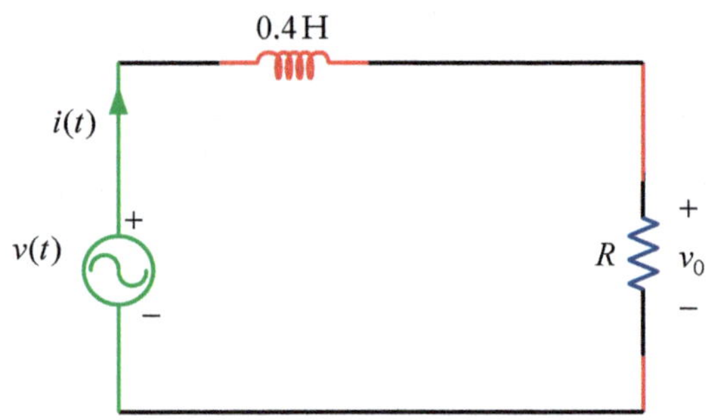

Fig. 10.42 Circuit for Problem 10.4

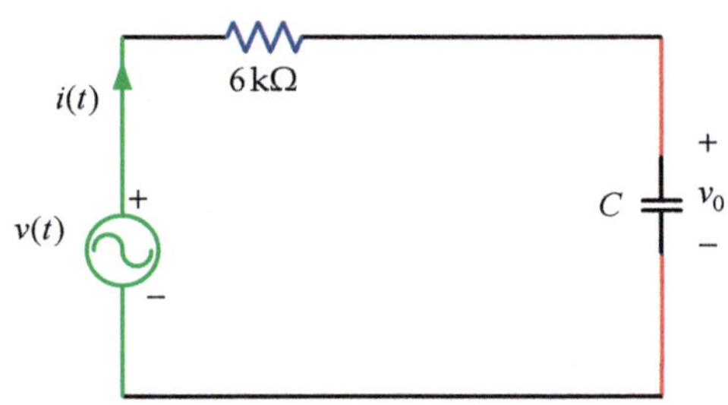

Fig. 10.43 Circuit for Problem 10.5

10.5 A series RC low pass filter circuit is shown in Fig. 10.43. The cut-off frequency of this filter is 5 kHz. Calculate the capacitance.

10.6 Figure 10.44 shows a series RC low pass filter circuit. The cut-off frequency of this filter is 3 kHz. Determine the resistance, magnitude and phase of the transfer function at 50 kHz.

10.7 Figure 10.45 shows a low pass filter. Find the cut-off frequency and the magnitude of the output voltage if the frequency of the input voltage is 2 kHz.

10.8 A series RC low pass filter circuit is shown in Fig. 10.46. Find the cut-off frequency, and the input frequency if the output voltage is 5 V.

Fig. 10.44 Circuit for Problem 10.6

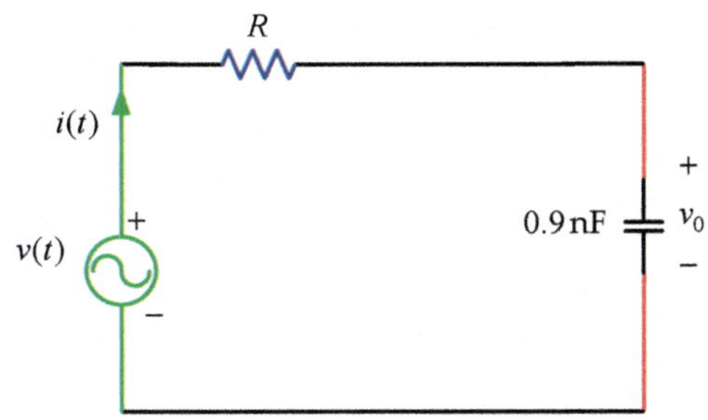

Fig. 10.45 Circuit for Problem 10.7

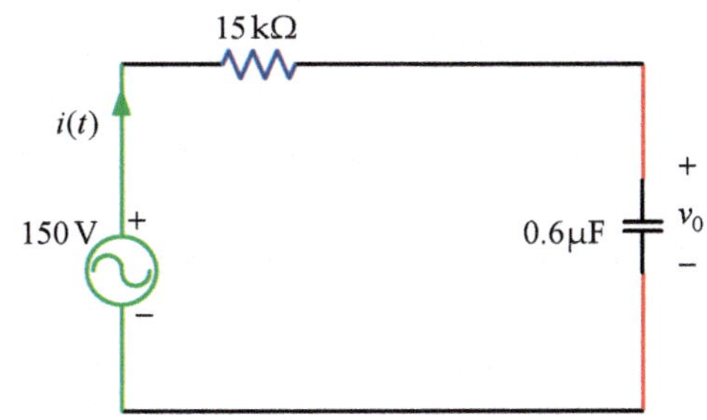

Fig. 10.46 Circuit for Problem 10.8

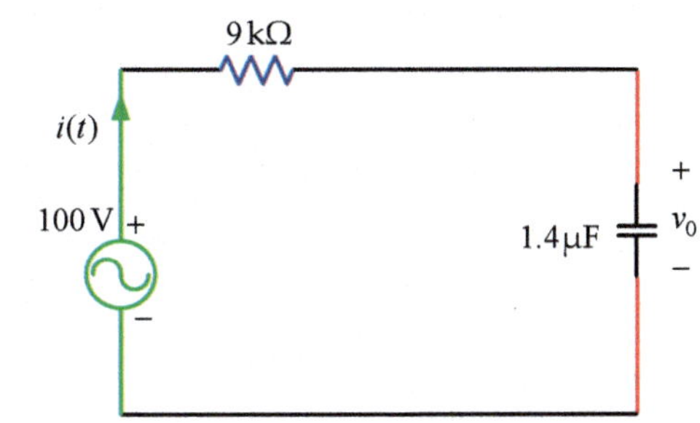

10.9 A series RL high pass filter circuit is shown in Fig. 10.47. Derive the transfer function, cut-off frequency and the value of the resistor if the cut-off frequency is 9 kHz.

10.10 A series RL high pass filter circuit is connected to a load resistance as shown in Fig. 10.48. Derive the transfer function, cut-off frequency and the value of the cut-off frequency if $R = 10\,\Omega$, $R_L = 5\,\Omega$ and $L = 0.5\,\text{H}$.

10.11 A series *RLC* circuit is shown in Fig. 10.20. Find the resonant frequency, quality factor, and bandwidth (Fig. 10.49).

10.12 Figure 10.50 shows a series *RLC* circuit. Calculate the resonant frequency, and quality factor.

Fig. 10.47 Circuit for Problem 10.9

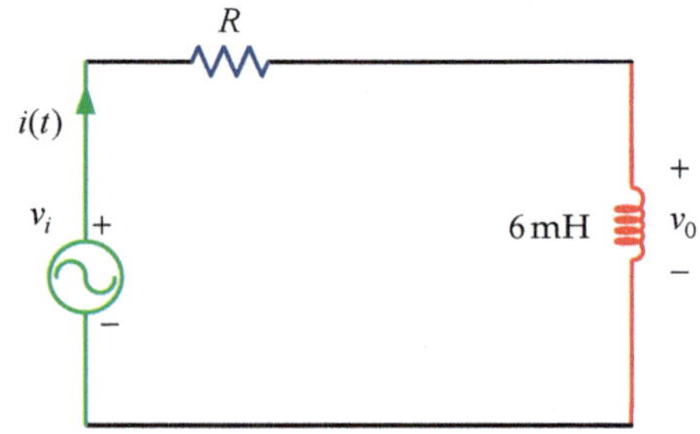

Fig. 10.48 Circuit for Problem 10.9

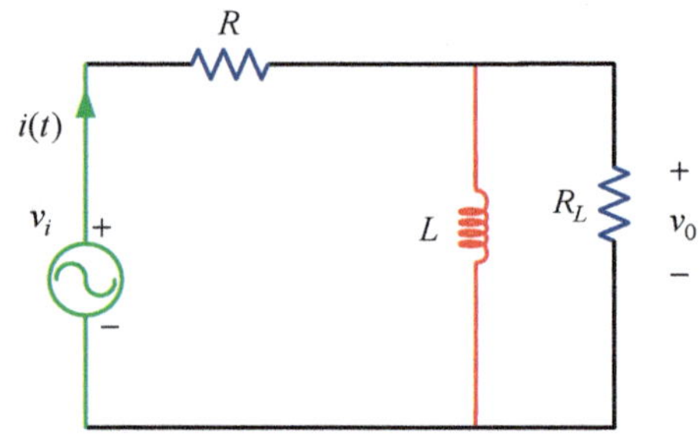

Fig. 10.49 Circuit for Problem 10.11

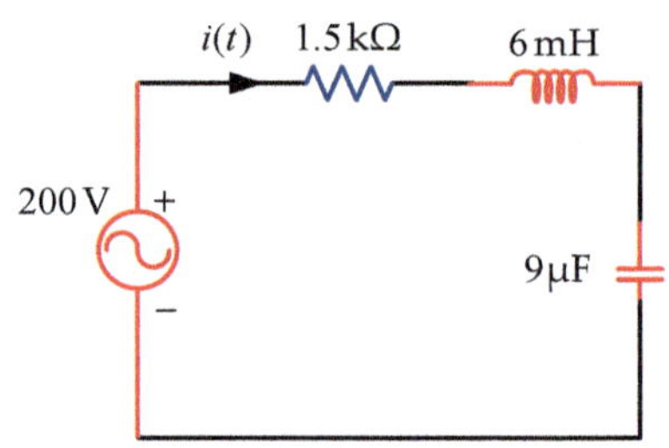

Fig. 10.50 Circuit for Problem 10.12

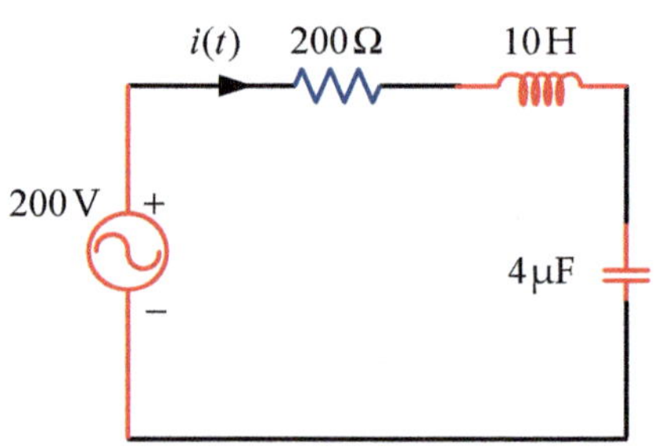

10.13 The quality factor a series RLC circuit is shown in Fig. 10.51. Determine the resistance and capacitance at a resonant frequency of 9 kHz and a quality factor of 10.

10.14 A series RLC circuit is shown in Fig. 10.52. Calculate the resonant frequency, quality factor and cut-off frequencies.

10.15 A series RLC circuit is shown in Fig. 10.53. Find the resistance and inductance at a resonant frequency of 2 kHz and a quality factor of 15.

10.16 For the series RLC circuit shown in Fig. 10.54, the impedance at resonance is 40Ω. Design the circuit by considering the quality factor of 20 and the resonant frequency of 800 Hz.

10.17 A parallel RLC circuit is shown in Fig. 10.55. Calculate the resonant frequency and the resonant impedance.

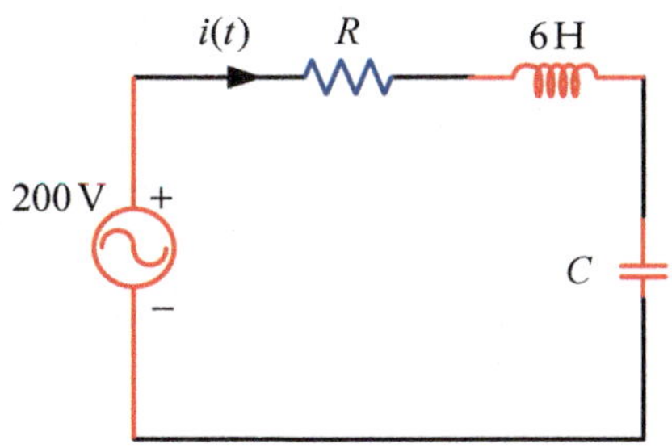

Fig. 10.51 Circuit for Problem 10.13

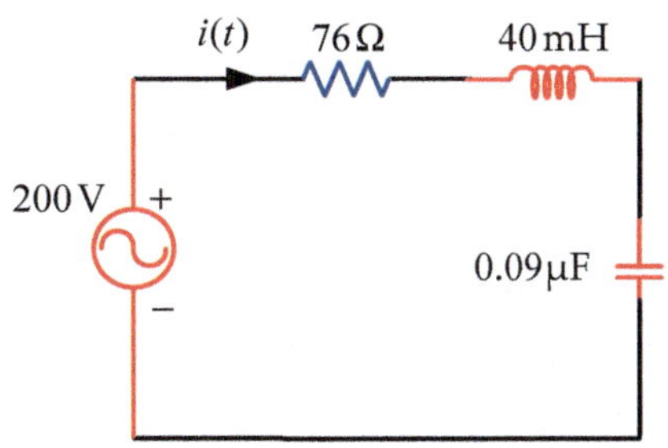

Fig. 10.52 Circuit for Problem 10.14

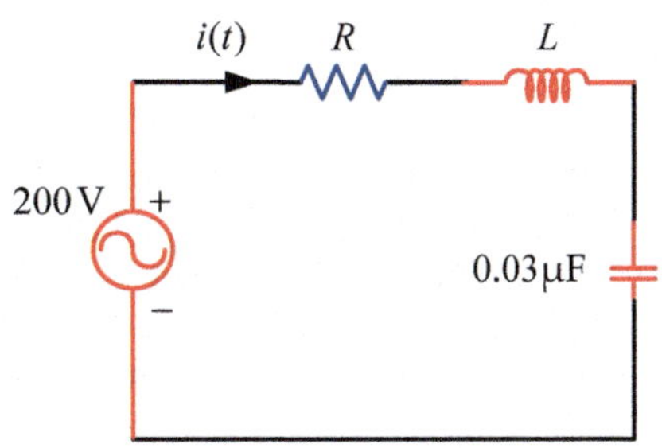

Fig. 10.53 Circuit for Problem 10.15

Fig. 10.54 Circuit for Problem 10.16

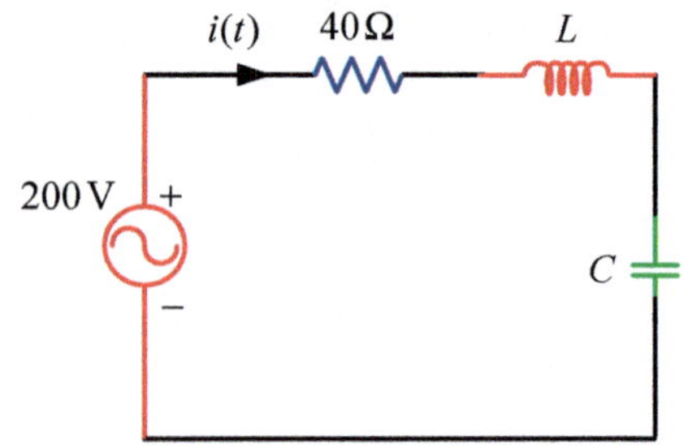

Fig. 10.55 Circuit for Problem 10.17

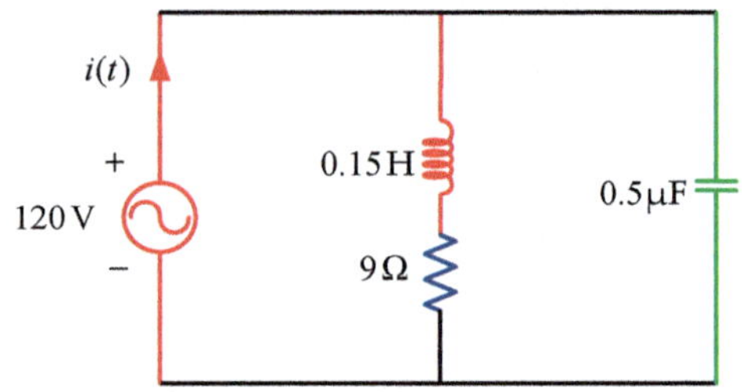

Fig. 10.56 Circuit for Problem 10.18

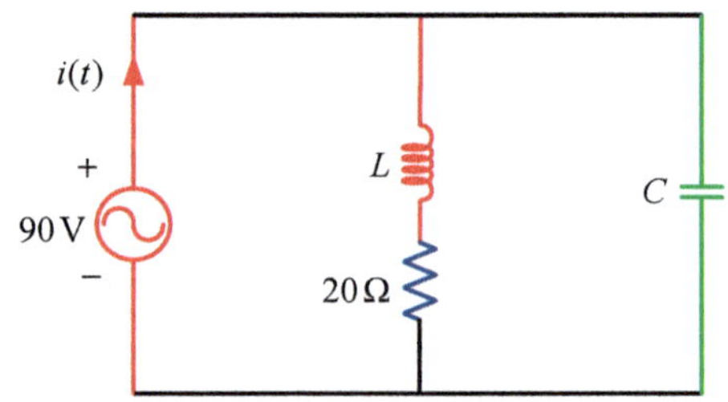

Fig. 10.57 Circuit for Problem 10.19

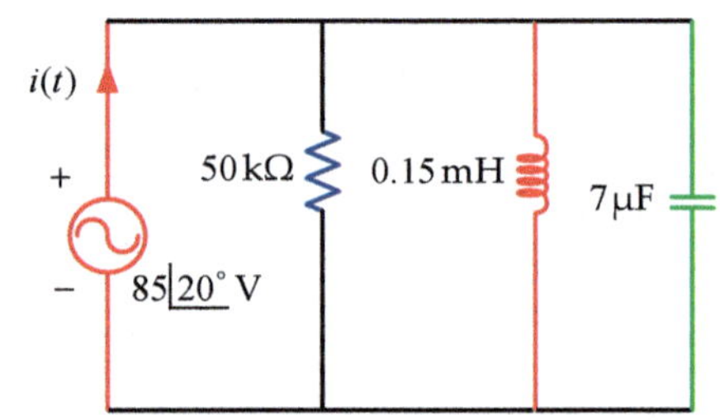

10.18 Figure 10.56 shows a parallel *RLC* circuit. Find the inductance and capacitance for resonant impedance of 40 kΩ and resonant frequency of 800 Hz.

10.19 A parallel *RLC* circuit is shown in Fig. 10.57. Calculate the resonant frequency, bandwidth, quality factor, cut-off frequencies and the average power dissipated at the resonant frequency.

10.20 Draw the Bode plots for the function $H(s) = \frac{s}{s+6}$.

10.21 Draw the Bode magnitude and phase plots for the function $H(s) = \frac{15(s+2)}{(s+7)(s+10)}$.

10.24 Draw the Bode magnitude and phase plots for the function $H(s) = \frac{26s}{(s+9)(s+12)}$.

References

1. J. David Irwin, R. Mark Nelms *Basic Engineering Circuit Analysis*, 11th edn. (Wiley, USA, 2015
2. R.L. Boylestad, *Introductory Circuit Analysis*, 13th ed. (Pearson, 2016)
3. C.K. Alexander, M.N.O. Sadiku, *Fundamentals of Electric Circuits*, 6th edn. (McGraw-Hill Higher Education, January 2016)
4. H.W. Jackson, D. Temple, B.E. Kelly, *Introduction to Electric Circuits*, 9th edn. (Oxford University Press, July 2015)
5. J.W. Nilsson, S.A. Riedel, *Electric Circuits*, 10th edn. (Prentice Hall International Edition, 2015)
6. D. Bell, *Fundamentals of Electric Circuits*, 7th edn. (Oxford University Press, March 2007)
7. G. Rizzoni, J. Kearns, *Principles and Applications of Electrical Engineering*, 6th edn. (McGraw-Hill Education, October 2014)
8. W. Hayt, J. Kemmerly, S. Durbin, *Engineering Circuit Analysis*, 8th edn. (McGraw-Hill Education, 2012)
9. M.A. Salam, *Basic Electrical Circuits*, 2nd edn. (Shroff Publishers & Distributors Pvt. Ltd, India, 2007)

Appendix A
Important Mathematical Formulae

A.1 Introduction

Some important mathematics knowledge along with trigonometric formulae are necessary for analysis of electrical circuit parameters. In this appendix, basic trigonometric formulae, logarithm, differentiation and integration have been discussed.

A.2 Trigonometric Formulae

A right angle triangle *abc* is shown in Fig. A.1. Here, θ is an acute angle. In Fig. A.1, adjacent is abbreviated 'adj', opposite is abbreviated as 'opp' and hypotenuse is abbreviated as 'hyp'.

The following trigonometric functions can be expressed:

$$\sin\theta = \frac{\text{opp}}{\text{hyp}} \qquad \cos\theta = \frac{\text{adj}}{\text{hyp}} \qquad \tan\theta = \frac{\text{opp}}{\text{adj}}$$

$$\text{cosec}\,\theta = \frac{\text{hyp}}{\text{opp}} \qquad \sec\theta = \frac{\text{hyp}}{\text{adj}} \qquad \cot\theta = \frac{\text{adj}}{\text{opp}}$$

In addition, the following formulae can be written:

$$\sin^2\theta + \cos^2\theta = 1 \qquad \sec^2\theta = 1 + \tan^2\theta \qquad \sin 2\theta = 2\sin\theta\cos\theta$$

$$1 + \cos 2\theta = 2\cos^2\theta \qquad 1 - \cos 2\theta = 2\sin^2\theta$$

$$\cos(-\theta) = \cos\theta \qquad \sin(-\theta) = -\sin\theta$$

A.3 Trigonometric Formulae

A well-known sentence 'all students take calculus' before entering engineering faculty is used to formulate as shown in Fig. A.2, where the trigonometric functions have positive sign in each of the four quadrants. The sign of the trigonometric

Md. A. Salam and Q. Mehbubar Rahman, *Fundamentals of Electrical Circuit Analysis*,
https://doi.org/10.1007/978-981-10-8624-3

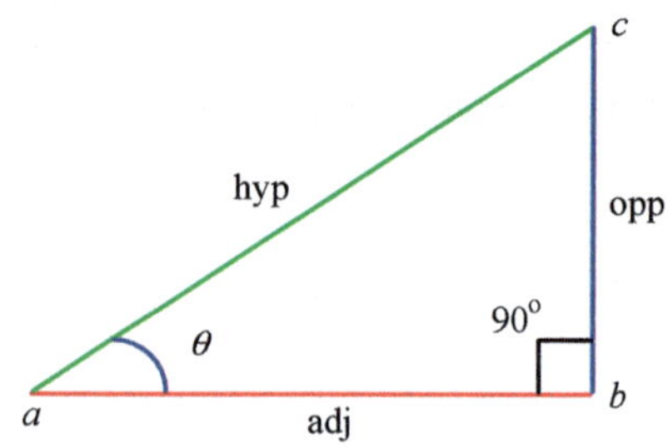

Fig. A.1 Right angle triangle

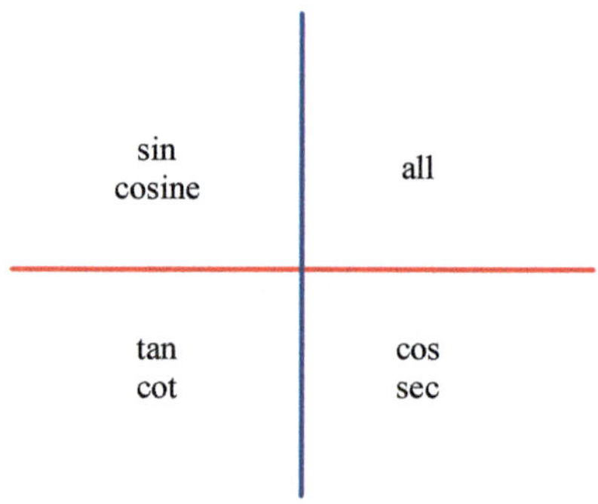

Fig. A.2 Signs of the trigonometric functions

function can be determined using Fig. A.3. If the sign of the angle of the trigonometric function is positive, then it will start from the *Y*-axis and rotate in the clockwise direction, whereas for negative angle, it will start from the *X*-axis and rotate in the anticlockwise direction. The following trigonometric functions can be written:

Fig. A.3 Sign identification for trigonometric function

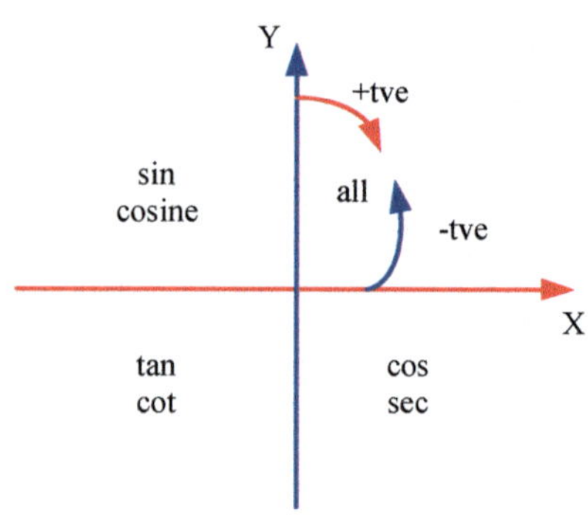

$\sin(90° \pm \theta) = \cos\theta$ $\quad$ $\cos(90° \pm \theta) = \mp \sin\theta$

$\sin(180° \pm \theta) = \mp \sin\theta$ $\quad$ $\cos(180° \pm \theta) = -\cos\theta$

$\sin(\alpha \pm \beta) = \sin\alpha \cos\beta \pm \cos\alpha \sin\beta$ $\quad$ $\cos(\alpha \pm \beta) = \cos\alpha \cos\beta \mp \sin\alpha \sin\beta$

$\sin\alpha \cos\beta = \frac{1}{2}[\sin(\alpha - \beta) + \sin(\alpha + \beta)]$ $\quad$ $\sin\alpha \sin\beta = \frac{1}{2}[\cos(\alpha - \beta) - \cos(\alpha + \beta)]$

$\cos\alpha \cos\beta = \frac{1}{2}[\cos(\alpha - \beta) + \cos(\alpha + \beta)]$

A.4 Exponential and Logarithmic Formulae

Exponential and logarithmic formulae are sometimes used for power analysis in an electrical circuit. These are as follows:

$$e^{\alpha}.\, e^{\beta} = e^{(\alpha+\beta)} \qquad e^0 = 1$$

$$\frac{e^{\alpha}}{e^{\beta}} = e^{(\alpha-\beta)} \qquad (e^{\alpha})^{\beta} = e^{\alpha\beta}$$

$$\log_e(xy) = \log_e x + \log_e y \qquad \log_e\left(\frac{x}{y}\right) = \log_e x - \log_e y$$

$$\log_e(x^n) = n\log_e x \qquad \log_e(1) = 0$$

A.5 Derivative Integral Formulae

In the first few chapters, some worked out examples are dealt with differentiation and integration. Basic formulae given as,

$$\frac{d}{dx}x^n = nx^{n-1} \qquad \frac{d}{dx}e^{xt} = te^{xt}$$

$$\frac{d}{dx}\cos\omega x = -\omega \sin\omega x \qquad \frac{d}{dx}\sin\omega x = \omega \cos\omega x$$

$$\frac{d}{dx}\left(\frac{u}{v}\right) = \frac{v\frac{du}{dx} - u\frac{dv}{dx}}{v^2} \qquad \frac{d}{dx}(1) = 0$$

$$\int \frac{dx}{x} = \ln x + C \qquad \int e^x dx = e^x + C \qquad \int a^x dx = \frac{a^x}{\ln a} + C$$

$$\int \sin x dx = -\cos x + C \qquad \int \cos x dx = \sin x + C \qquad \int \sec^2 x dx = \tan x + C$$

Appendix B
Answers to Practice and Exercise Problems

Chapter 1
Practice problems

1.1 $I = -4e^{-2} + 6t + 1\,A$
1.2 $q = 30\,C$
1.3 $d = 0.06\,cm$
1.4 $q = 0.8\,C$
1.5 $t = 1.6\,s$
1.6 $W = 2.09\,J$
1.7 $N = 4.7\,kWh$
1.8 $Tariff = \$195$
1.9 $P_{in} = 110\,W$, $\eta = 90.91\%$

Exercise problems

1.1 $I = -2e^{-2} + 12t + 3\,A$
1.2 $t = 0.2\,s$
1.3 $I = 4\,A$
1.4 $i = 6t$, 12, $(36 - 6t)$, $q = 48\,C$
1.5 $i = 8t$, $(33 - 5.5t)$, $q = 60\,C$
1.6 $R = 3.3\,\Omega$
1.7 $l = 55.23\,m$
1.8 $d = 0.11\,cm$
1.9 $V = 0.94\,V$
1.10 $t = 4\,s$
1.11 0.14 J
1.12 $N = 2.74\,kWh$, \$0.27

Chapter 2
Practice problems

2.1 2.74 A, 80.54 Ω
2.2 7 mA
2.3 7 A, 4 A
2.4 2 A, 24 V

Md. A. Salam and Q. Mehbubar Rahman, *Fundamentals of Electrical Circuit Analysis*,
https://doi.org/10.1007/978-981-10-8624-3

2.5 5 A, 40 V
2.6 42 V
2.7 7.56 A, 1.64 A
2.8 11.63 Ω, 5.16 A
2.9 1 A, 3 V
2.10 12.44 Ω, 3.62 A
2.11 18.84 Ω, 3.45 A
2.12 10.44 Ω, 3.83 A
2.13 0.86 A
2.14 11.38 A
2.15 0.85 A

Exercise problems

2.1 3.41 A
2.2 3.84 kW
2.3 9 A, 7 A, 10 A
2.4 9 A, 7 A, 6 A
2.5 30 V, −28 V, −38 V
2.6 20 V, 15 V
2.7 4 V, 12 V
2.8 6.25 A, 1.25 A
2.9 3 A, 12 V
2.10 800 W
2.11 8.43 Ω, 4.74 A
2.12 9 Ω, 4.44 A
2.13 8 Ω, 5 A
2.14 8.08 Ω, 55.65 W
2.15 8.04 A
2.16 12.85 A, 22.84 V
2.17 7.48 A, 25.02 V
2.18 1.53 Ω, 29.41 A
2.19 2.83 Ω, 14.13 A
2.20 8.3 Ω, 5.42 A
2.21 7.14 Ω, 6.3 A
2.22 8.09 Ω, 7.42 A
2.23 7.17 Ω, 5.58 A
2.24 3.56 Ω, 14.04 A
2.25 4.66 Ω, 4.29 A
2.26 5.37 Ω, 8.06 A
2.27 1.06 Ω, 18.87 A
2.28 11.08 Ω, 5.42 A, 9.33 V
2.29 6.67 Ω, 6 A
2.30 7.62 W
2.31 1.67 A
2.32 17.78 V

2.33 4.8 A
2.34 8 V
2.35 3.48 W
2.36 12 V
2.37 16.16 V
2.38 25.92 V
2.39 11.44 V
2.40 28.38 V
2.41 20.32 V
2.42 41.49 V
2.43 9 V
2.44 −2.12 V
2.45 23.79 V
2.46 21.7 V
2.47 18.36 V
2.48 6.56 V
2.49 29.51 V
2.50 10.67 V

Chapter 3
Practice problems

3.1 0 A, 2.3 A, 0.48 A
3.2 3.61 A, 2.82 A
3.3 0.44 A, −1.32 A
3.4 5.1 A, 1.1 A, 1.5 A, 38.88 W
3.5 5.5 A, −0.5 A, 181.5 W
3.6 30 V, 13.01 V, 1.15 V, 28.21 W
3.7 30 V, 19.02 V, 12.06 V
3.8 20 V, 5.71 V, 17.71 V, 163.02 W
3.9 16 V, 1.52 V, 0.91 V, 0.18 A

Exercise problems

3.1 3.56 V, −0.42 V, −0.05 V
3.2 1.47 A
3.3 2.57 A
3.4 −1.57 A
3.5 2.26 A
3.6 2.53 A
3.7 6 A, 2.42 A, −0.58 A, 1 W
3.8 4.62 A
3.9 4.83 A
3.10 1.82 A, 0.91 A, 0.91 A
3.11 1.67 A, 0.02 A, 1.65 A
3.12 2.32 A, 1.45 A, 1.45 A
3.13 1.94 A, 1.56 A, 0.6 A

3.14 −6.53 A
3.15 1.5 A, −0.75 A
3.16 5.47 A, 2.9 A, 3.66 A, 0.76 A
3.17 0.34 A, 3.57 A, −0.09 A, 3.23 A
3.18 5.82 V, −2.65 V, 0.97 A
3.19 30 V, 2.4 V, 13.2 V, 1.8 A
3.20 3.07 A
3.21 9.87 A
3.22 2.14 A
3.23 2.26 A
3.24 2.51 A
3.25 3.37 A
3.26 2 A
3.27 2.91 A
3.28 13.34 A

Chapter 4

Practice problems

4.1 9 V
4.2 4 A, 8 A
4.3 −0.4 A
4.4 4.28 A
4.5 0.63 A
4.6 3.27 A
4.7 0.72 A
4.8 3.78 A
4.9 1.71 Ω, 237.17 W

Exercise problems

4.1 3 A
4.2 2 A
4.3 3 V
4.4 3.92 A
4.5 3.08 A
4.6 4.75 A
4.7 3.83 A
4.8 6.76 A
4.9 5.46 A
4.10 4 A
4.11 4.69 A
4.12 2.7 A
4.13 1.96 A
4.14 1.89 A
4.15 5.68 A
4.16 2.34 A

4.17 3.5 A
4.18 1.13 A
4.19 1.15 A
4.20 1.67 A
4.21 1.78 A
4.22 3.79 A
4.23 2.88 A
4.24 100.56 V
4.25 22.85 V
4.26 −3.6 V
4.27 2.79 A
4.28 1.73 A
4.29 5 V, 0.91 A
4.30 6.93 A
4.31 1.86 A
4.32 0.05 A
4.33 2.12 A
4.34 1.57 A
4.35 1.57 A
4.36 1.87 A
4.37 3.51 A
4.38 1.02 A
4.39 5.01 A
4.40 50.27 W
4.41 16.62 V
4.42 2.73 A
4.43 0.78 A
4.44 2.69 A
4.45 6.31 A
4.46 58.17 W
4.47 6.71 A
4.48 2.86 A
4.49 4.12 A
4.50 1.31 A
4.51 4.08 A
4.52 1.86 A
4.53 0.07 A
4.54 1.5 A
4.55 0.43 A
4.56 1.27 A
4.57 6.01 A
4.58 1.22 A
4.59 0.19 A
4.60 0.76 A
4.61 7.05 A

4.62 0.06 A
4.63 0.75 A
4.64 1.3 A
4.65 0.79 A
4.66 3 A
4.67 2.18 A
4.68 1.81 A
4.69 0.54 A
4.70 1.16 A
4.71 1.27 A
4.72 0.55 A
4.73 1.36 A
4.74 0.19 A
4.75 3.12 A
4.76 3.75 A
4.77 1.04 A
4.78 2.56 A
4.79 2.49 A
4.80 2.7 A
4.81 4.85 Ω, 2.74 W
4.82 7.43 Ω, 25.55 W
4.83 1.12 Ω, 86.19 W
4.84 1.05 Ω, 78.95 W
4.85 3.64 Ω, 12.18 W

Chapter 5
Practice problems

5.1 2.24×10^{-10} F
5.2 $-4.66e^{-1000t}$ kV
5.3 33.32 μA
5.4 6.6 mC, 0.72 J
5.5 0.22 mJ, 1.56 mJ, 0.33 mJ
5.6 1.66 μF
5.7 17.14 V, 20 V
5.8 1.98 mm
5.9 1.68×10^{-17} F, 1.34×10^{-15} C, 5.18×10^{-7} C/m^2, 19.51 kV/E, 11.7 kV/m, 39.02 V, 40.95 V
5.10 0.75 m
5.11 0 A
5.12 0.94 m
5.13 $4 + 0.67e^{-0.02t}$ A, $0.33e^{-0.02t} - 3$ A, $0.013e^{-0.02t}$ V
5.14 8.57 A, 65.21 J, 20.45 J
5.15 $12.16(1 - e^{-0.12t})$ V
5.16 $18e^{-\frac{1}{30}t}$ V

5.17 0.33 H

5.18 $18.33e^{-\frac{1}{21}t}$ V

5.19 $6.81 + 0.11e^{-1.94t}$ A

Exercise problems

5.1 3.71×10^{-11} F, 1.48 μC

5.2 $i = 5.65\cos(377t - 20^\circ)$ A

5.3 $-2.4e^{-1500t}$ V

5.4 0 A

5.5 7.5 A

5.6 0.34 J

5.7 3.37 mJ

5.8 0.1 mJ

5.9 3.07 μC

5.10 2.77 mm

5.11 14.47 μH

5.12 0.04 m

5.13 $2.5 - e^{-1.5t}$ A, $3.5 - 2.5e^{-1.5t}$ A

5.14 975.37 J, 25 J

5.15 $26.18e^{-0.018t}$ V

5.16 $3.08.28e^{-0.79t}$ A

Chapter 6
Practice problems

6.1 15 A, 0.98 s

6.2 12.37 A, 306 W

6.3 7.5 A

6.4 1.45 V

6.5 $V = 2\angle 25^\circ$ V, $I = 2\angle 85^\circ$ A

6.6 $1.61\angle 6.5^\circ$

6.7 $V = 25\angle -35^\circ$ V

6.8 $0.1\angle -110^\circ$ A

6.9 $V_c = 6.36\angle -64^\circ$ kV

6.10 $X_c = 0.06\ \Omega$, $1\angle 35.71^\circ$ A, $i(t) = \sin(20t + 35.71^\circ)$ A

6.11 $Y = 0.21\angle 3.55^\circ$ S, $48.3\angle 23.55^\circ$ A, $i(t) = \sin(\omega t + 23.55^\circ)$ A

6.12 $4.08\angle 16.28^\circ\ \Omega$

6.13 $13.95\angle 13.5^\circ\ \Omega$, $15.77\angle 11.5^\circ$ A

Exercise problems

6.1 50 Hz, 0.02 s, 0.05 A,

6.2 25 V, 60 Hz, 0.016 s

6.3 5.83 V, 11.33 W

6.4 5.43 A

6.5 1.59 V, 2.5 V, 1.57, 2, 1.25 W

6.6 5.65 V, 1.11, 1.41, 3.54 W
6.7 1 V, 3.65 V
6.8 2.5 V, 7.9 V
6.9 6 V, 5.03 V
6.10 −1.33 V, 32.33 V, 261.33 W
6.11 3 V, 3.87 V, 3 W
6.12 −6.67 A, 7.05 A, 398.22 W
6.13 1.8 A, 2.45 A
6.14 4 A, 4.24 A
6.15 4.5 A, 3.96 A
6.16 $M = 3.25\angle 40.60^\circ$, $N = 213.47\angle -6.72^\circ$
6.17 $v(t) = 48\sin(\omega t + 25^\circ)$ V, $V = 48\angle 25^\circ$ V
6.18 (i) $i(t) = 3.33\sin 500t$ A, (ii) $i(t) = 8\sin(\omega t + 15^\circ)$ A, (iii) $i(t) = 3.33\cos(314t + 65^\circ)$ A
6.19 $i(t) = -8.49\cos(314t + 12^\circ)$ A, $V = 8\angle 12^\circ$ V, $I = -8.49\angle 12^\circ$ A
6.20 $i(t) = 0.04\sin(377t + 31^\circ)$ A, $I = 0.04\angle 31^\circ$ A
6.21 $i(t) = 9\sin(33t + 124^\circ)$ mA, $I = 9\angle 134^\circ$ mA
6.22 $i(t) = 7.16\sin(377t + 67^\circ)$ mA, $I = 7.16\angle 67^\circ$ mA
6.23 (i) 10.44 Ω, (ii) $15.57\angle 90^\circ$ A, (iii) $i(t) = 15.57\sin(314t + 90^\circ)$ A, $v(t) = 230\sin 314t$ V
6.24 6.28 Ω, $I = 31.61\angle -39.46^\circ$ A
6.25 159.15 Hz, $I = 2.23\angle -60.94^\circ$ A, $V_R = 11.15\angle -60.94^\circ$ V, $V_L = 20.07\angle 29.06^\circ$ V
6.26 (i) 60 Hz, (ii) $15.62\angle 24^\circ$ Ω, (iii) 14.26 Ω, 6.35 Ω
6.27 (i) 11.58 Ω, (ii) $12.61\angle -66.64^\circ$ Ω, (iii) $17.44\angle 91.64^\circ$ A
6.28 $5\angle 53.13^\circ$ Ω, $44\angle -28.13^\circ$ A
6.29 (i) $6.71\angle 63.43^\circ$ Ω, (ii) $3.72\angle -15.43^\circ$ A
(iii) $V_R = 18.6\angle -15.43^\circ$ V, $V_L = 33.48\angle 74.57^\circ$ V, $V_C = 11.16\angle -105.43^\circ$ V
6.30 $17.12\angle 38.22^\circ$ A, $21.86\angle 92.04^\circ$ A, $18.14\angle -38.34^\circ$ A
6.31 $2.78\angle -77.81^\circ$ Ω
6.32 $6.12\angle -39.28^\circ$ Ω
6.33 $6.08\angle -28.25^\circ$ Ω, $0.16\angle 28.25^\circ$ S, $36.18\angle 53.25^\circ$ A
6.34 $7.45\angle 1.73^\circ$ Ω, $29.53\angle 23.27^\circ$ A
6.35 $17.36\angle -0.37^\circ$ Ω, $12.67\angle 25.37^\circ$ A

Chapter 7
Practice problems

7.1 $148.21\angle -12.17^\circ$ V, $123.18\angle -5.24^\circ$ V, $65.28\angle -63.23^\circ$ V
7.2 $244.25\angle 9.69^\circ$ V, $232.46\angle 9.17^\circ$ V, $207.91\angle 35.73^\circ$ V
7.3 $63.32\angle 18.05^\circ$ V, $85.71\angle 159.7^\circ$ V
7.4 $25\angle 10^\circ$ A, $12.13\angle 61.52^\circ$ A, $9.96\angle 63.1^\circ$ A
7.5 $1.48\angle -18.92^\circ$ A, $0.95\angle 10^\circ$ A, $1.17\angle -39.3^\circ$ A, $3.53\angle -39.3^\circ$ V

7.6 $8.06\angle 85.87^\circ$ V
7.7 $3.07\angle 128.22^\circ$ A
7.8 $2.10\angle 61.75^\circ$ A, $2.04\angle 59.74^\circ$ A
7.9 $1.19\angle 108.9^\circ$ A
7.10 $3.55\angle 99.04^\circ$ A

Exercise problems

7.1 $30\angle 10^\circ$ V, $13.54\angle -36.65^\circ$ V
$33.66\angle 157.86^\circ$ V, $8.41\angle 157.86^\circ$ A
7.2 $95.66\angle 44.41^\circ$ V, $84.28\angle 112.69^\circ$ V, $15.12\angle -27.15^\circ$ A
7.3 $147.23\angle 100.22^\circ$ V
7.4 $16.4\angle -18.53^\circ$ V
7.5 $5.23\angle 33.35^\circ$ A
7.6 $16\angle 6.68^\circ$ V
7.7 $0.03\angle -145.62^\circ$ V
7.8 $3.76\angle -29.8^\circ$ A
7.9 $8.86\angle 5.04^\circ$ A
7.10 $15.57\angle -60.28^\circ$ A, $17.6\angle -133.62^\circ$ A
7.11 $17.59\angle 66.36^\circ$ A
7.12 $3.82\angle -59.58^\circ$ A
7.13 $5.31\angle -115.11^\circ$ A
7.14 $10.16\angle -44.32^\circ$ V
7.15 $7.41\angle -54.53^\circ$ A
7.16 $4.27\angle -23.51^\circ$ A
7.17 $2.23\angle -37.86^\circ$ A
7.18 $4.29\angle 45.46^\circ$ A
7.19 $3.53\angle 14.45^\circ$ A
7.20 $2.35\angle -43.15^\circ$ A
7.21 $1.27\angle 69.24^\circ$ A
7.22 $13.38\angle -2.23^\circ$ A
7.23 $34.89\angle 7.14^\circ$ A
7.24 $15.3\angle 14.19^\circ$ A
7.25 $3.6\angle 120.36^\circ$ A
7.26 $10.49\angle -0.69^\circ$ A
7.27 $6.07\angle -14.57^\circ$ A
7.28 $1.68\angle -15.47^\circ$ A
7.29 $2.52\angle -32.06^\circ$ A
7.30 $3.38\angle -13.45^\circ$ A
7.31 $2.4\angle -37.01^\circ$ A
7.32 $1.22\angle 2.71^\circ$ A
7.33 $2.4\angle -37.01^\circ$ A
7.34 $6.32\angle 6.22^\circ$ A
7.35 $3.23\angle -7.08^\circ$ A
7.36 $4.07\angle -27.73^\circ$ A

Chapter 8

Practice problems

8.1 $20\sin(\omega t + 35^\circ)$ V, $93.97 - 100\cos(2\omega t + 50^\circ)$ W

8.2 16.24 W, 6.92 W, 6.48 W, 2.69 W

8.3 600 VA, 1.18 mF

8.4 62.5 W, 39.05 W, 48.81 Var

8.5 12 kW + j3.91kvar, $4.18\angle 18.19^\circ\ \Omega$

8.6 $16.64\angle -63.43^\circ$ VA, 16.64 VA, $12.24\angle 80.53^\circ$ VA

8.7 29.25 kW

8.8 $201.61\angle 52.52^\circ$ V

8.9 $11.35\angle -0.27^\circ\ \Omega$, 3.23 W

Exercise problems

8.1 $6.3 - 6.4\cos(20t - 10^\circ)$ W

8.2 $39.39 - 40\cos(200t - 30^\circ)$ W

8.3 107.17 W, 48.9 W, 9 W, 49.14 W

8.4 109.93 W, 70.56 W, 4.09 W, 25.83 W, 5.77 W

8.5 9.72 W, 9.67 W, 10.21 W, 2.73 W

8.6 1340.62 W, 406.82 W, 494.84 W, 235.22 W, 206.45 W

8.7 276.12 W, 207.09 W, 142.8 W

8.8 3.45 kVA, 13.59 Ω, 0.5 mF

8.9 45.96 W − j38.56 Var, 45.96 W, 38.56 Var

8.10 185.88 W − j49.8 Var, 185.88 W, 49.8 Var

8.11 220.22 W + j440.45 Var, 107.33 W + j36.26 Var, 123.52 W − j112.08 Var, −35.71 W − j197.3 Var, 415.36 W + j167.32 Var

8.12 213.37 W + j426.64 Var, 0.29 W − j76.3 Var, 10.52 W − j43.15 Var, 10.74 W + j19.44 Var 32.63 W + j0.06 Var, 281.19 W + j353.63 Var

8.13 0.78 mF

8.14 167.42 A, $680.94\angle 79.56^\circ$ V

8.15 213.58 A, $854.36\angle 89.71^\circ$ V

8.16 $6.06\angle 32.78^\circ\ \Omega$, 18.08 W

8.17 $5.05\angle -25.88^\circ\ \Omega$, 9.27 W

Chapter 9

Practice problems

9.1 $200\angle 130^\circ$ V, $200\angle -110^\circ$ V

9.2 $115.47\angle -30^\circ$ V, $115.47\angle -150^\circ$ V $115.47\angle 90^\circ$ V

9.3 $33.33\angle -42^\circ$ A, $33.33\angle -162^\circ$ A, $33.33\angle 78^\circ$ A

9.4 $9.07\angle 15^\circ$ A, $10.88\angle -45^\circ$ A, $6.92\angle 115^\circ$ A $12.32\angle -18.56^\circ$ A, $10.09\angle -96.07^\circ$ A, $17.54\angle 127.24^\circ$ A

9.5 $18.6\angle 22.4^\circ$ A, $18.6\angle -97.6^\circ$ A, $18.6\angle 142.4^\circ$ A

9.6 $16.36\angle 35^\circ$ A, $16.36\angle -85^\circ$ A, $16.36\angle 155^\circ$ A
$28.33\angle 5^\circ$ A, $28.33\angle -115^\circ$ A, $28.33\angle 125^\circ$ A

9.7 $32.38\angle 10.86^\circ$ A, $34.62\angle 13.53^\circ$ A, $27.88\angle 11.8^\circ$ A, $12.63\angle -69.09^\circ$ A, $6.8\angle 20.63^\circ$ A, $8.44\angle 126.91^\circ$ A

9.8 $49.08\angle 35^\circ$ A, $49.08\angle -85^\circ$ A, $49.08\angle 155^\circ$ A, 6246.86 W, 18716.73 W, 24020.05 + j11200.73 VA

9.9 112.32 kW

9.10 760.6 W, 459.47 W, 1166.4 W, 2386.47 W

Exercise problems

9.1 $230.94\angle -30^\circ$ V, $230.94\angle -150^\circ$ V, $230.94\angle 90^\circ$ V

9.2 $100\angle -10^\circ$ V, $100\angle -130^\circ$ V, $100\angle 110^\circ$ V

9.3 $33.33\angle -30^\circ$ A, $33.33\angle -150^\circ$ A, $33.33\angle 90^\circ$ A

9.4 $11.82\angle -116.68^\circ$ A, $11.82\angle -236.68^\circ$ A, $11.82\angle 3.32^\circ$ A

9.5 $303.3\angle 55.54^\circ$ V, $303.3\angle -64.46^\circ$ V, $303.3\angle 155.54^\circ$ V

9.6 $852.2\angle 67.92^\circ$ V, $852.2\angle -52.08^\circ$ V, $852.2\angle 187.92^\circ$ V

9.7 $18.09\angle -83.22^\circ\ \Omega$

9.8 $4.42\angle -75.64^\circ\ \Omega$

9.9 $2\angle -20^\circ\ \Omega$

9.10 299.91 W

9.11 $47.63\angle 57^\circ$ A, $47.63\angle -63^\circ$ A, $47.63\angle 177^\circ$ A, $82.49\angle 27^\circ$ A, $82.49\angle 27^\circ$ A, $82.49\angle 147^\circ$ A

9.12 $7.5\angle -63.54^\circ$ A, $7.5\angle -183.54^\circ$ A, $7.5\angle 56.46^\circ$ A, $45\angle -53.54^\circ$ V

9.13 $18.72\angle 18.72^\circ$ A, $18.72\angle -101.28^\circ$ A, $18.72\angle 128.72^\circ$ A, $10.8\angle 48.72^\circ$ A, $10.8\angle -71.28^\circ$ A, $10.8\angle 168.72^\circ$ A

9.14 $50\angle -5^\circ$ A, $50\angle -125^\circ$ A, $50\angle 115^\circ$ A, $86.6\angle -35^\circ$ A, $86.6\angle -155^\circ$ A, $86.6\angle 85^\circ$ A

9.15 $13.86\angle -89.03^\circ$ A, $9.8\angle 134.03^\circ$ A, $6.39\angle 161.56^\circ$ A

9.16 $20\angle 45^\circ$ A, $10\angle 97^\circ$ A, $5.11\angle -155^\circ$ A, $24.86\angle 40.97^\circ$ A, $15.93\angle -164.64^\circ$ A, $12.55\angle -105.76^\circ$ A

9.17 52.41 kW

9.18 3.9 kW

9.19 14.75 W, 4351.76 W

9.20 $8.08\angle -83.87^\circ$ A, $8.08\angle -203.87^\circ$ A, $8.08\angle 36.13^\circ$ A, 756.72 W − j202.74 Var

9.21 78.73 A, 3719.04 W

9.22 12.2 A

9.23 21.34 A, $5471.57\angle 87.13^\circ$ VA
$9152.74\angle 30.03^\circ$ VA, 247.62Vrms, 0.86 lagging

9.24 $10537.61\angle 53.96^\circ$ VA, 0.58 lagging, 48.5 A

9.25 0.65 lagging, $93.59\angle 49.1^\circ$ A

9.26 2000.35 W, 2207.29 W, 6564.72 W, 10772.37 W

9.27 166.66 W, 750.55 Var, $16.8\angle 77.48^\circ\ \Omega$

Chapter 10
Practice problems

10.1 $\frac{I_0(\omega)}{I_i(\omega)} = \frac{j\omega 0.5(3 + j\omega 4)}{j\omega 1.5 + (j\omega)^2 2 + 1}$

$z_1 = 0$, $z_2 = -0.75$, $p_1 = -0.37 + j0.59$, $p_2 = -0.37 - j0.59$

10.2 531.91 kHz, 14.96 kΩ

10.3 102.88 V, 46.69°

10.4 6.49 kHz, 24.49, 0.26 kHz, 6.36 kHz, 6.63 kHz

10.5 187.56 Hz, 53, 22.23 Hz, 2.25 kW

10.6

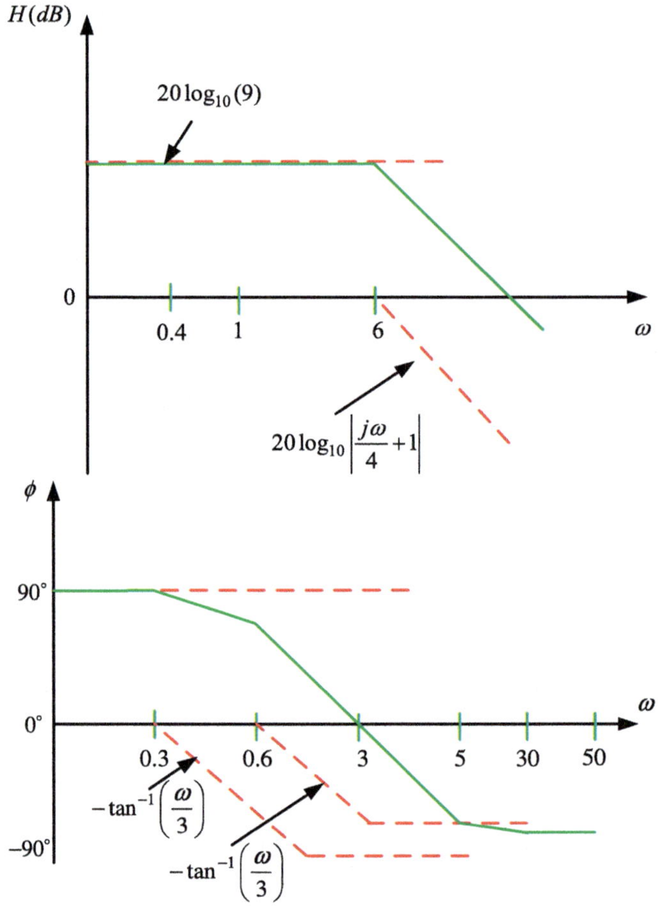

Exercise problems

10.1 $\frac{I_0(\omega)}{I_i(\omega)} = \frac{j\omega 0.95(4 + j\omega 6)}{j\omega 6.65 + (j\omega)^2 5.7 + 1}$

$z_1 = 0$, $z_2 = -0.66$, $p_1 = -0.17$, $p_2 = -0.98$

10.2 $\dfrac{I_0(\omega)}{I_i(\omega)} = \dfrac{(1 + j\omega 3.2)}{j\omega 4.8 + (j\omega)^2 2.4 + 1}$

$z_1 = -0.31$, $p_1 = -0.24$, $p_2 = -1.76$

10.3 0.63 H, 0.01

10.4 7.54 kΩ, 0.06, −86.56°

10.5 5.3 nF

10.6 58.94 kΩ, 0.06, −86.56°

10.7 17.68 Hz, 1.32 V

10.8 12.63 Hz, 252.28 Hz

10.9 $H(\omega) = \dfrac{j\omega}{j\omega + \frac{R}{L}}$, $\omega_c = \frac{R}{L}$, 339.29 Ω

10.10 $H(\omega) = \dfrac{j\omega\left[\frac{R_L L}{R_L + j\omega L}\right]}{j\omega\left[\frac{R_L L}{R_L + j\omega L}\right] + R} = \dfrac{j\omega\left[\frac{R_L}{R_L + j\omega L}\right]}{j\omega\left[\frac{R_L}{R_L + j\omega L}\right] + \frac{R}{L}}$

6.66 Hz

10.11 684.89 Hz, 0.017, 250 kHz

10.12 25.16 Hz, 7.9

10.13 52.12 pF, 33.92 kΩ

10.14 2652.58 Hz, 8.77, 2501.39 Hz, 2803.77 Hz

10.15 0.21 H, 176.38 Ω

10.16 0.24 μF, 0.16 H

10.17 581.07 Hz, 33.33 kΩ

10.18 0.22 μF, 0.17 H

10.19 4.91 kHz, 34.15, 903.38 Hz

Index

Md. A. Salam and Q. Mehbubar Rahman, *Fundamentals of Electrical Circuit Analysis*,
https://doi.org/10.1007/978-981-10-8624-3

Printed by Printforce, the Netherlands